Principles
of
Robot
Motion

Intelligent Robotics and Autonomous Agents
Ronald C. Arkin, editor

Principles
of
Robot Motion

Theory, Algorithms,
and Implementation

Howie Choset, Kevin Lynch, Seth Hutchinson,

George Kantor, Wolfram Burgard, Lydia Kavraki,

and Sebastian Thrun

A Bradford Book
The MIT Press
Cambridge, Massachusetts
London, England

MIT Press books may be purchased at special quantity discounts for business or sales promotional use. For information, please email special_sales@mitpress.mit.edu or write to Special Sales Department, The MIT Press, 55 Hayward Street, Cambridge, MA 02142.

This book was set in LATEX2e by Interactive Composition Corporation and was printed and bound in the United States of America.

Library of Congress Cataloging-in-Publication Data

Principles of robot motion : theory, algorithms, and implementation/Howie Choset [et al.].
    p. cm.   (Intelligent robotics and autonomous agents)
  "A Bradford book."
  Includes bibliographical references and index.
  ISBN-13   978-0-262-03327-5 (alk. paper)
  1. Robots—Motion. I. Choset, Howie M. II. Series.
TJ211.4.P75 2004
  629.8′92—dc22   2004044906

10 9 8 7 6

To our families

Contents

Foreword

THIS IMPRESSIVE book is the result of a serious undertaking of distinguished motion planning researchers led by Howie Choset. Over the years, motion planning has become a major research theme in robotics. The goal is to enable robots to automatically compute their motions from high-level descriptions of tasks and models acquired through sensing. This goal has recently become even more crucial. On the one hand, robotics has expanded from a largely dominant focus on industrial manufacturing into areas where tasks are less repetitive and environments less structured, for instance, medical surgery, ocean and space exploration, assistance to the elderly, and search-and-rescue. In these areas, it is impractical to explicitly program the robots for each new goal. On the other hand, the need for automatic motion-planning capabilities has expanded outside the realm of robotics, into domains such as computer animation (e.g., to generate motions of avatars), computer-aided design (e.g., to test that a product can be assembled or serviced), verification of building codes (e.g., to check access of key facilities to the disabled), exploration of virtual environments (to help the user navigate in geometric models made of tens of millions of triangles), or even computational biology (to help analyze important molecular motions, like folding and binding). Today, progress in motion planning is increasingly motivated by these new applications.

By confronting novel and difficult problems, researchers have made considerable progress in recent years. Not only have faster and more robust algorithms been developed and tested, but the range of motion-planning problems has continuously expanded. In the '80s and part of the '90s, finding collision-free paths was the main or only goal. Today, while obstacle avoidance remains a key issue, other important constraints are considered as well, for instance, visibility, coverage, kinodynamic, optimality, equilibrium, and uncertainty constraints. These constraints make problems more interesting and lead to more useful algorithms. In addition, while research in motion planning used to be neatly divided between theory and practice, this distinction has now largely disappeared. Most recent contributions to the field combine effective algorithms tested on significant problems, along with some formal guarantees of performance.

Although journal and conference papers in motion planning have proliferated, there has not been any comprehensive reference text in more than a decade. This book fills this gap in outstanding fashion. It covers both the early foundations of the field and the recent theoretical and practical progress that has been made. It beautifully demonstrates how the enduring contributions of early researchers in the field, like Lozano-Perez (configuration space) and Reif (complexity of motion planning), have led to a rich and vibrant research area, with ramifications that were unsuspected only a decade ago.

I am usually suspicious of books in which chapters have been written by different authors. But, to my good surprise, this book is more than a standard textbook. The fact that seven authors collaborated on this book attests to the diversity of the research going on in motion planning and the excitement associated with each research topic. Simultaneously, the authors have done excellent work in providing a unified presentation of the core concepts and methodologies, and thus the book can be used as a textbook. This book will serve well the growing community of students, researchers, and engineers interested in motion planning.

Jean-Claude Latombe
Stanford, California

Preface

PEOPLE HAVE always dreamed of building intelligent machines to perform tasks. Today, these machines are called robots, derived from the Czech word *robota* meaning servitude or drudgery. Robots have inspired the imagination of many, appearing in mythology, literature, and popular movies. Some popular robotic characters include Robby the Robot, R2D2 and C3P0, Golem, Pushpack, Wanky and Fanny, Gundam and Lt. Cmdr. Data. Just like their literary counterparts, robots can take on many forms and constructing them involves addressing many challenges in engineering, computer science, cognitive science, language, and so on. Regardless of the form of the robot or the task it must perform, robots must maneuver through our world. This book is about automatic planning of robot motions. However, the approaches developed in this book are not limited to robots: recently, they have been used for "designing" pharmaceutical drugs, planning routes on circuit boards, and directing digital actors in the graphics world.

The robot motion field and its applications have become incredibly broad, and this is why the book has seven co-authors. This type of book requires a broad spectrum of expertise. However, it should be stressed that this is indeed a textbook and not a collection of independent chapters put together in an edited volume. Each author participated in writing each of the chapters and all of the chapters are integrated with each other.

This book is aimed at the advanced undergraduate or new graduate student interested in robot motion, and it may be read in a variety of ways. Our goal in writing in this book is threefold: to create an updated textbook and reference for robot motion, to make the fundamental mathematics behind robot motion accessible to the novice, and to stress implementation relating low-level details to high-level algorithmic concepts.

Since the robot motion field is indeed broad, this book cannot cover all the topics, nor do we believe that any book can contain exhaustive coverage on robot motion. We do, however, point the reader to Jean-Claude Latombe's *Robot Motion Planning* [262].

Latombe's book was one of the first text and reference books aimed at the motion-planning community and it certainly was a guide for us when writing this book. In the decade since Latombe's book was published, there have been great advances in the motion-planning field, particularly in probabilistic methods, mechanical systems, and sensor-based planning, so we intended to create a text with these new advances. However, there are many topics not included in our text that are included in his, including assembly planning, geometric methods in dealing with uncertainty, multiple moving obstacles, approximate cell decompositions, and obstacle representations.

We also believe that concepts from control theory and statistical reasoning have gained greater relevance to robot motion. Therefore, we have included an appendix briefly reviewing linear control systems which serves as background for our presentation on Kalman filtering. Our description of Kalman filtering differs from others in that it relies on a rich geometric foundation. We present a comprehensive description of Bayesian-based approaches. Concepts from mechanics and dynamics have also had great impact on robot motion. We have included a chapter on dynamics which serves as a basis for our description of trajectory planning and planning for underactuated robots.

This book can be read from cover to cover. In doing so, there are four logical components to the book: geometric motion planning approaches (chapters 2 through 6), probabilistic methods (chapters 7, 8, and 9), mechanical control systems (chapters 10, 11, and 12), and the appendices. Covering the entire book could require a full year course. However, not all of the topics in this book need be covered for a course on robot motion. For semester-long courses, the following themes are suggested:

Theme	Chapter and Appendix Sequence
Path Planning	3, 4, G, 5, 7, and 6
Mobile Robotics	2, H, 3, 4, 5, D, and 6
Mechanical Control Systems	3, 10, 11, and 12
Position Estimation	I, J, 8, and 9

The algorithms and approaches presented in this book are based on geometry and thus rest on a solid mathematical basis. Beyond anything superficial, in order to understand the many motion-planning algorithms, one must understand these mathematical underpinnings. One of the goals of this book is to make mathematical concepts more accessible to students of computer science and engineering. In this book, we introduce the intuition behind new mathematical concepts on an "as needed" basis to understand both how and why certain motion planning algorithms work. Some salient concepts are formally defined in each chapter and the appendices contain overviews of some basic topics in more detail. The idea here is that the reader can develop an understanding of motion planning algorithms without getting bogged down by mathematical details, but can turn to them in the appendices when necessary. It is our hope that the reader will gain enough new knowledge in algebra, graph theory, geometry, topology, probability, filtering, and so on, to be able to read the state of the art literature in robot motion.

We discuss implementation issues and it is important to note that such issues are not mere details, but pose deep theoretical problems as well. In chapters 2, 4, 5, and 6, we discuss specific issues on how to integrate range sensor information into a planner. The Kalman Filtering (chapter 8) and Bayesian-based (chapter 9) approaches have been widely used in the robot motion field to deal with positioning and sensor uncertainty. Finally, we discuss in chapters 11 and 12 issues involving kinematic and dynamic contraints that real robots experience.

We have also included pseudocode for many of the algorithms presented throughout the book. In appendix H, we have included a discussion of graph search with detailed examples to enable the novice to implement some standard graph search approaches, with applicability well beyond robot motion. Finally, at the end of each chapter, we present problems that stress implementation.

Acknowledgments

WE FIRST and foremost want to thank our students, who were incredibly supportive of us when writing this book. We would like to thank the members of the Biorobotics/ Sensor Based Planning Lab at Carnegie Mellon, especially Ji Yeong Lee; the Laboratory for Intelligent Mechanical Systems at Northwestern; the robotics group in the Beckman Institute at the University of Illinois, Urbana Champaign; the Physical and Biological Computing Group at Rice, especially Andrew Ladd and Erion Plaku; the Lab for Autonomous Intelligent Systems at the University of Freiburg, especially Dirk Hähnel and Cyrill Stachniss; and the Stanford and Carnegie Mellon Learning Labs, for their contributions and efforts for this book.

We thank Alfred Anthony Rizzi and Matt Mason for their inspiration and support, and Ken Goldberg and Jean-Claude Latombe for their input and advice. For input in the form of figures or feedback on drafts, we thank Ercan Acar, Srinivas Akella, Nancy Amato, Serkan Apaydin, Prasad Atkar, Denise Bafman, Devin Balkcom, Francesco Bullo, Joel Burdick, Prasun Choudhury, Cynthia Cobb, Dave Conner, Al Costa, Frank Dellaert, Bruce Donald, Dieter Fox, Bob Grabowski, Aaron Greenfield, David Hsu, Pekka Isto, James Kuffner, Christian Laugier, Jean-Paul Laumond, Steve LaValle, Brad Lisien, Julie Nord, Jim Ostrowski, Nancy Pollard, Cedric Pradalier, Ionnis Rekleitis, Elie Shammas, Thierry Simeon, Sarjun Skaff, and M. Dick Tsuyuki.

We are also indebted to the many students who helped debug this text for us. Portions of this text were used in Carnegie Mellon's Sensor Based Motion Planning Course, Carnegie Mellon's Mechanical Control Systems reading group, Northwestern's ME 450 Geometry in Robotics, University of Illinois' ECE 450 Advanced Robotic Planning, and University of Freiburg's Autonomous Systems course.

1 *Introduction*

SOME OF the most significant challenges confronting autonomous robotics lie in the area of automatic motion planning. The goal is to be able to specify a task in a high-level language and have the robot automatically compile this specification into a set of low-level motion primitives, or feedback controllers, to accomplish the task. The prototypical task is to find a path for a robot, whether it is a robot arm, a mobile robot, or a magically free-flying piano, from one configuration to another while avoiding obstacles. From this early *piano mover's problem,* motion planning has evolved to address a huge number of variations on the problem, allowing applications in areas such as animation of digital characters, surgical planning, automatic verification of factory layouts, mapping of unexplored environments, navigation of changing environments, assembly sequencing, and drug design. New applications bring new considerations that must be addressed in the design of motion planning algorithms.

Since actions in the physical world are subject to physical laws, uncertainty, and geometric constraints, the design and analysis of motion planning algorithms raises a unique combination of questions in mechanics, control theory, computational and differential geometry, and computer science. The impact of automatic motion planning, therefore, goes beyond its obvious utility in applications. The possibility of building computer-controlled mechanical systems that can sense, plan their own motions, and execute them has contributed to the development of our math and science base by asking fundamental theoretical questions that otherwise might never have been posed.

This book addresses the theory and practice of robot motion planning, with an eye toward applications. To focus the discussion, and to point out some of the important concepts in motion planning, let's first look at a few motivating examples.

Piano Mover's Problem

The classic path planning problem is the piano mover's problem [373]. Given a three-dimensional rigid body, for example a polyhedron, and a known set of obstacles, the problem is to find a collision-free *path* for the omnidirectional free-flying body from a start configuration to a goal configuration. The obstacles are assumed to be stationary and perfectly known, and execution of the planned path is exact. This is called *offline* planning, because planning is finished in advance of execution. Variations on this problem are the *sofa mover's problem,* where the body moves in a plane among planar obstacles, and the *generalized mover's problem,* where the robot may consist of a set of rigid bodies linked at joints, e.g., a robot arm.

The key problem is to make sure no point on the robot hits an obstacle, so we need a way to represent the location of all the points on the robot. This representation is the *configuration* of the robot, and the *configuration space* is the space of all configurations the robot can achieve. An example of a configuration is the set of joint angles for a robot arm or the one orientation and two position variables for a sofa in the plane. The configuration space is generally *non-Euclidean,* meaning that it does not look like an n-dimensional Euclidean space $\mathbb{R}^n$. The dimension of the configuration space is equal to the number of independent variables in the representation of the configuration, also known as the *degrees of freedom* (DOF). The piano has six degrees of freedom: three to represent the position (x-y-z) and three to represent the orientation (roll-pitch-yaw). The problem is to find a curve in the configuration space that connects the start and goal points and avoids all *configuration space obstacles* that arise due to obstacles in the space.

The Mini AERCam

NASA's Johnson Space Center is developing the Mini AERCam, or Autonomous Extravehicular Robotic Camera, for visual inspection tasks in space (figure 1.1). It is a free-flying robot equipped with twelve cold gas thrusters, allowing it to generate a force and torque in any direction. When operating in autonomous mode, it must be able to navigate in a potentially complex three-dimensional environment. In this respect the problem is similar to the piano mover's problem. Since we have to apply thrusts to cause motion, however, we need to plan not only the path the robot is to follow, but also the speed along the path. This is called a *trajectory,* and the thruster inputs are determined by the *dynamics* of the robot. In the piano mover's problem, we only worried about geometric or *kinematic* issues.

Figure 1.1   NASA's Mini AERCam free-flying video inspection robot.

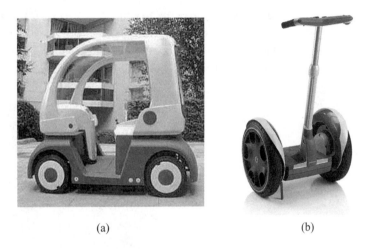

(a)   (b)

Figure 1.2   (a) The CyCab. (b) The Segway Human Transporter.

Personal Transport Vehicles

Small personal transport vehicles may become a primary means of transportation in pedestrian-filled urban environments where the size, speed, noise, and pollution of automobiles is undesirable. One concept is the CyCab [355], a small vehicle designed by a consortium of institutions in France to transport up to two people at speeds up to 30 km/h (figure 1.2a). Another concept is the Segway HT, designed to carry a single rider at speeds up to 20 km/h (figure 1.2b).

To simplify control of vehicles in crowded environments, one capability under study is automatic parallel parking. The rider would initiate the parallel-parking procedure, and the onboard computer would take over from there. Such systems will soon be commercially available in automobiles. On the surface, this problem sounds like the sofa mover's problem, since both involve a body moving in the plane among obstacles. The difference is that cars and the vehicles in figure 1.2 cannot instantaneously slide sideways like the sofa. The velocity constraint preventing instantaneous sideways motion is called a *nonholonomic* constraint, and the motion planner must take this constraint into account. Systems without velocity constraints, such as the sofa, are omnidirectional in the configuration space.

Museum Tour Guides

In 1997, a mobile robot named RHINO served as a fully autonomous tour-guide at the Deutsches Museum Bonn (figure 1.3). RHINO was able to lead museum visitors from one exhibit to the next by calculating a path using a stored map of the museum. Because the perfect execution model of the piano mover's problem is unrealistic in this setting, RHINO had to be able to *localize* itself by comparing its sensor readings to its stored

Figure 1.3   RHINO, the interactive mobile tour-guide robot.

Figure 1.4   The Mars rover Sojourner. http://mars.jpl.nasa.gov/MPF/rover/sojourner.html.

map. To deal with uncertainty and changes in the environment, RHINO employed a *sensor-based* planning approach, interleaving sensing, planning, and action.

Planetary Exploration

One of the most exciting successes in robot deployment was a mobile robot, called Sojourner (figure 1.4), which landed on Mars on July 4, 1997. Sojourner provided up-close images of Martian terrain surrounding the lander. Sojourner did not move very far from the lander and was able to rely on motion plans generated offline on Earth and uploaded. Sojourner was followed by its fantastically successful cousins, Spirit and Opportunity, rovers that landed on Mars in January 2004 and have provided a treasure trove of scientific data. In the future, robots will explore larger areas and thus will require significantly more autonomy. Beyond navigation capability, such robots will have to be able to generate a map of the environment using sensor information. *Mapping* an unknown space with a robot that experiences positioning error is an especially challenging "chicken and egg" problem—without a map the robot cannot determine its own position, and without knowledge about its own position the robot cannot compute the map. This problem is often called *simultaneous localization and mapping* or simply *SLAM*.

Demining

Mine fields stifle economic development and result in tragic injuries and deaths each year. As recently as 1994, 2.5 million mines were placed worldwide while only 100,000 were removed.

Robots can play a key role in quickly and safely demining an area. The crucial first step is finding the mines. In demining, a robot must pass a mine-detecting sensor over all points in the region that might conceal a mine. To do this, the robot must traverse a carefully planned path through the target region. The robot requires a *coverage* path planner to find a motion that passes the sensor over every point in the field. If the planner is guaranteed to find a path that covers every point in the field when such a path exists, then we call the planner *complete*. Completeness is obviously a crucial requirement for this task.

Coverage has other applications including floor cleaning [116], lawn mowing [198], unexploded ordnance hunting [260], and harvesting [341]. In all of these applications, the robot must simultaneously localize itself to ensure complete coverage.

Fixed-base Robot Arms in Industry

In highly structured spaces, fixed-base robot arms perform a variety of tasks, including assembly, welding, and painting. In painting, for example, the robot must deposit a uniform coating over all points on a target surface (figure 1.5). This coverage problem presents new challenges because (1) ensuring equal paint deposition is a more severe requirement than mere coverage, (2) the surface is not usually flat, and (3) the robot must properly coordinate its internal degrees of freedom to drive the paint atomizer over the surface.

Industrial robot installations are clearly driven by economic factors, so there is a high priority on minimizing task execution time. This motivates motion planners that return *time-optimal* motion plans. Other kinds of tasks may benefit from other kinds of optimality, such as energy- or fuel-optimality for mobile robots.

Figure 1.5   ABB painting robot named Tobe.

Figure 1.6   The Hirose active cord.

Figure 1.7   The Carnegie Mellon snake robot Holt mounted on a mobile base.

Snake Robots for Urban Search and Rescue

When a robot has more degrees of freedom than required to complete its task, the robot is called *redundant*. When a robot has many extra degrees of freedom, then it is called *hyper-redundant*. These robots have multidimensional non-Euclidean configuration spaces. Hyper-redundant serial mechanisms look like elephant trunks or snakes (figures 1.6 and 1.7), and they can use their extra degrees of freedom to thread through tightly packed volumes to reach locations inaccessible to humans and conventional machines. These robots may be particularly well-suited to urban search and rescue, where it is of paramount importance to locate survivors in densely packed rubble as quickly and safely as possible.

Robots in Surgery

Robots are increasingly used in surgery applications. In noninvasive stereotactic radiosurgery, high-energy radiation beams are cross-fired at brain tumors. In certain cases,

Figure 1.8   The motions of a digital actor are computed automatically. (Courtesy of J.C. Latombe)

these beams are delivered with high accuracy, using six degrees of freedom robotic arms (e.g., the CyberKnife system [1]). Robots are also used in invasive procedures. They often enhance the surgeon's ability to perform technically precise maneuvers. For example, the da Vinci Surgical System [2] can assist in advanced surgical techniques such as cutting and suturing. The ZEUS System [3] can assist in the control of blunt retractors, graspers, and stabilizers. Clearly, as robotics advances, more and more of these systems will be developed to improve our healthcare.

Digital Actors

Algorithms developed for motion planning or sensor interpretation are not just for robots anymore. In the entertainment industry, motion planning has found a wide variety of applications in the generation of motion for digital actors, opening the way to exciting scenarios in video games, animation, and virtual environments (figure 1.8).

Drug Design

An important problem in drug design and the study of disease is understanding how a protein folds to its native or most stable configuration. By considering the protein as an articulated linkage (figure 1.9), researchers are using motion planning to identify likely folding pathways from a long straight chain to a tightly folded configuration. In pharmaceutical drug design, proteins are combined with smaller molecules to form complexes that are vital for the prevention and cure of disease. Motion planning

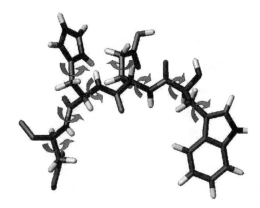

Figure 1.9 A molecule represented as an articulated linkage.

methods are used to analyze molecular binding motions, allowing the automated testing of drugs before they are synthesized in the laboratory.

1.1 Overview of Concepts in Motion Planning

The previous examples touched on a number of ways to characterize the motion planning problem and the algorithm used to address it. Here we summarize some of the important concepts. Our characterization of a motion planner is according to the task it addresses, properties of the robot solving the task, and properties of the algorithm.[1] We focus on topics that are covered in this book (table 1.1).

Task

The most important characterization of a motion planner is according to the problem it solves. This book considers four tasks: *navigation, coverage, localization,* and *mapping. Navigation* is the problem of finding a collision-free motion for the robot system from one configuration (or state) to another. The robot could be a robot arm, a mobile robot, or something else. *Coverage* is the problem of passing a sensor or tool over all points in a space, such as in demining or painting. *Localization* is the problem of using a map to interpret sensor data to determine the configuration of the robot. *Mapping* is the problem of exploring and sensing an unknown environment

---

1. This classification into three categories is somewhat arbitrary but will be convenient for introduction.

Task	Robot	Algorithm
Navigate	Configuration space, degree of freedom	Optimal/nonoptimal motions
Map	Kinematic/dynamic	Computational complexity
Cover	Omnidirectional or	Completeness
Localize	motion constraints	(resolution, probabilistic)
		Online/offline
		Sensor-based/world model

Table 1.1   Some of the concepts covered in this book.

to construct a representation that is useful for navigation, coverage, or localization. Localization and mapping can be combined, as in SLAM.

There are a number of interesting motion planning tasks not covered in detail in this book, such as navigation among moving obstacles, manipulation and grasp planning, assembly planning, and coordination of multiple robots. Nonetheless, algorithms in this book can be adapted to those problems.

Properties of the Robot

The form of an effective motion planner depends heavily on properties of the robot solving the task. For example, the robot and the environment determine the number of *degrees of freedom* of the system and the shape of the *configuration space*. Once we understand the robot's configuration space, we can ask if the robot is free to move instantaneously in any direction in its configuration space (in the absence of obstacles). If so, we call the robot omnidirectional. If the robot is subject to velocity constraints, such as a car that cannot translate sideways, both the constraint and the robot are called *nonholonomic*. Finally, the robot could be modeled using *kinematic* equations, with velocities as controls, or using *dynamic* equations of motion, with forces as controls.

Properties of the Algorithm

Once the task and the robot system is defined, we can choose between algorithms based on how they solve the problem. For example, does the planner find motions that are *optimal* in some way, such as in length, execution time, or energy consumption? Or does it simply find a solution satisfying the constraints? In addition to the quality of the output of the planner, we can ask questions about the *computational complexity* of the planner. Are the memory requirements and running time of the algorithm constant, polynomial, or exponential in the "size" of the problem description? The size of the problem description could be the number of degrees of freedom of the robot system, the amount of memory needed to describe the robot and the obstacles in the environment, etc., and the complexity can be defined in terms of the worst case or the average case. If we expect to scale up the size of the inputs, a planner is often only considered practical if it runs in time polynomial or better in the inputs. When a polynomial-time algorithm has been found for a problem that previously could only be solved in exponential time, some key insight into the problem has typically been gained.

Some planners are *complete,* meaning that they will always find a solution to the motion planning problem when one exists or indicate failure in finite time. This is a very powerful and desirable property. For the motion planning problem, as the number of degrees of freedom increases, complete solutions may become computationally intractable. Therefore, we can seek weaker forms of completeness. One such form is *resolution completeness.* It means that if a solution exists at a given resolution of discretization, the planner will find a solution. Another weaker form of completeness is *probabilistic completeness.* It means that the probability of finding a solution (if one exists) converges to 1 as time goes to infinity.

Optimality, completeness, and computational complexity naturally trade off with each other. We must be willing to accept increased computational complexity if we demand optimal motion plans or completeness from our planner.

We say a planner is *offline* if it constructs the plan in advance, based on a known model of the environment, and then hands the plan off to an executor. The planner is *online* if it incrementally constructs the plan while the robot is executing. In this case, the planner can be *sensor-based,* meaning that it interleaves sensing, computation, and action. The distinction between offline algorithms and online sensor-based algorithms can be somewhat murky; if an offline planner runs quickly enough, for example, then it can be used in a feedback loop to continually replan when new sensor data updates the environment model. The primary distinction is computation time, and practically speaking, algorithms are often designed and discussed with this distinction in mind. A similar issue arises in control theory when attempting to distinguish between feedforward control (commands based on a reference trajectory and dynamic model) and feedback control (commands based on error from the desired trajectory), as techniques like *model predictive control* essentially use fast feedforward control generation in a closed loop. In this book we will not discuss the low-level feedback controllers needed to actually implement robot motions, but we will assume they are available.

1.2   Overview of the Book

Chapter 2 dives right into a class of simple and intuitive "Bug" algorithms requiring a minimum of mathematical background to implement and analyze. The task is to navigate a point mobile robot to a known goal location in a plane filled with unknown static obstacles. The Bug algorithms are sensor-based—the robot uses a contact sensor or a range sensor to determine when it is touching or approaching an obstacle, as well as odometry or other sensing to know its current position in the plane. It has two basic motion primitives, moving in a straight line and following a boundary, and it switches between these based on sensor data. These simple algorithms guarantee that the robot will arrive at the goal if it is reachable.

To move beyond simple point robots, in chapter 3 we study the configuration space of more general robot systems, including rigid bodies and robot arms. The mathematical foundations in this chapter allow us to view general path planning problems as finding paths through configuration space. We study the dimension (degrees of freedom), topology, and parameterizations of non-Euclidean configuration spaces, as well as representations of these configuration spaces as surfaces embedded in higher-dimensional Euclidean spaces. The forward kinematic map is introduced to relate one choice of configuration variables to another. The differential of this map, often called the Jacobian, is used to relate the velocites in the two coordinate systems. Material in this chapter is referenced throughout the remainder of the book.

Chapter 4 describes a class of navigation algorithms based on *artificial potential functions*. In this approach we set up a virtual potential field in the configuration space to make obstacles repulsive and the goal configuration attractive to the robot. The robot then simply follows the downhill gradient of the artificial potential. For some navigation problems, it is possible to design the potential field to ensure that following the gradient will always take the robot to the goal. If calculating such a potential field is difficult or impossible, we can instead use one that is easy to caclulate but may have the undesirable property of local minima, locations where the robot gets "stuck." In this case, we can simply use the potential field to guide a search-based planner. Potential fields can be generated offline, using a model of the environment, or calculated in real-time using current sensor readings. Purely reactive gradient-following potential field approaches always run the risk of getting stuck in local minima, however.

In chapter 5, we introduce more concise representations of the robot's free space that a planner can use to plan paths between two configurations. These structures are called *roadmaps*. A planner can also use a roadmap to explore an unknown space. By using sensors to incrementally construct the roadmap, the robot can then use the roadmap for future navigation problems. This chapter describes several roadmaps including the visibility graph, the generalized Voronoi diagram, and Canny's original roadmap. Chapter 6 describes an alternative representation of the free space called a *cell decomposition* which consists of a set of cells of the free space and a graph of cells with connections between adjacent cells. A cell decomposition is useful for coverage tasks, and it can be computed offline or incrementally using sensor data.

Constructing complete and exact roadmaps of an environment is generally quite computationally complex. Therefore, chapter 7 develops sampling-based algorithms that trade completeness guarantees for a reduction of the running time of the planner. This chapter highlights recent work in probabilistic roadmaps, expansive-spaces trees, and rapidly-exploring random trees and the broad range of motion planning problems to which they are applicable.

Probabilistic reasoning can also address the problems of sensor uncertainty and positioning error that plague mobile robot deployment. We can model these uncertainties and errors as probability distributions and use *Kalman filtering* (chapter 8) and *Bayesian estimation* (chapter 9) to address localization, mapping, and SLAM problems.

Just as the description of configuration space in chapter 3 provides many of the kinematic and geometric tools used in path planning, the description of second-order *robot dynamics* in chapter 10 is necessary for feasible trajectory planning, i.e., finding motions parameterized by time. We can then pose time- and energy-optimal *trajectory planning* problems for dynamic systems subject to actuator limits, as described in chapter 11.

Chapter 11 assumes that the robot has an actuator for each degree of freedom. In chapter 12 we remove that assumption and consider robot systems subject to *nonholonomic* (velocity) constraints and/or acceleration constraints due to missing actuators, or *underactuation*. We study the reachable states for such systems, i.e., controllability, using tools from differential geometry. The chapter ends by describing planners that find motions for systems such as cars, cars pulling trailers, and spacecraft or robot arms with missing actuators.

1.3   Mathematical Style

Our goal is to present topics in an intuitive manner while helping the reader appreciate the deeper mathematical concepts. Often we suppress mathematical rigor, however, when intuition is sufficient. In many places proofs of theorems are omitted, and the reader is referred to the original papers. For the most part, mathematical concepts are introduced as they are needed. Supplementary mathematical material is deferred to the appendices to allow the reader to focus on the main concepts of the chapter.

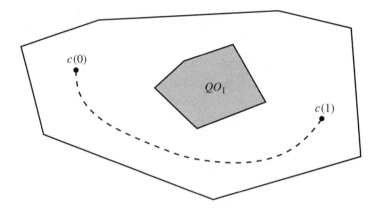

Figure 1.10   A path is a curve in the free configuration space $\mathcal{Q}_{\text{free}}$ connecting $c(0)$ to $c(1)$.

Throughout this book, robots are assumed to operate in a planar ($\mathbb{R}^2$) or three-dimensional ($\mathbb{R}^3$) ambient space, sometimes called the *workspace* $\mathcal{W}$. This workspace will often contain obstacles; let $\mathcal{WO}_i$ be the ith obstacle. The *free workspace* is the set of points $\mathcal{W}_{\text{free}} = \mathcal{W} \backslash \bigcup_i \mathcal{WO}_i$ where the $\backslash$ is a subtraction operator.

Motion planning, however, does not usually occur in the workspace. Instead, it occurs in the configuration space $\mathcal{Q}$ (also called C-space), the set of all robot configurations. We will use the notation $R(q)$ to denote the set of points of the ambient space occupied by the robot at configuration q. An obstacle in the configuration space corresponds to configurations of the robot that intersect an obstacle in the workspace, i.e., $\mathcal{QO}_i = \{q \mid R(q) \bigcap \mathcal{WO}_i \neq \emptyset\}$. Now we can define the *free configuration space* as $\mathcal{Q}_{\text{free}} = \mathcal{Q} \backslash \bigcup_i \mathcal{QO}_i$. We sometimes simply refer to "free space" when the meaning is unambiguous.

In this book we make a distinction between *path planning* and *motion planning*. A *path* is a continuous curve on the configuration space. It is represented by a continuous function that maps some path parameter, usually taken to be in the unit interval [0, 1], to a curve in $\mathcal{Q}_{\text{free}}$ (figure 1.10). The choice of unit interval is arbitrary; any parameterization would suffice. The solution to the *path planning* problem is a continuous function $c \in C^0$ (see appendix C for a definition of continuous functions) such that

(1.1) $c : [0, 1] \rightarrow \mathcal{Q}$ where $c(0) = q_{\text{start}}$, $c(1) = q_{\text{goal}}$ and $c(s) \in \mathcal{Q}_{\text{free}}$ $\forall s \in [0, 1]$.

When the path is parameterized by time t, then $c(t)$ is a trajectory, and velocities and accelerations can be computed by taking the first and second derivatives with respect to time. This means that c should be at least twice-differentiable, i.e., in the class C^2. Finding a feasible trajectory is called trajectory planning or *motion planning*.

In this book, configuration, velocity, and force vectors will be written as column vectors when they are involved in any matrix algebra. For example, a configuration $q \in \mathbb{R}^n$ will be written in coordinates as $q = [q_1, q_2, \ldots, q_n]^T$. When the vector will not be used in any computation, we may simply refer to it as a tuple of coordinates, e.g., $q = (q_1, q_2, \ldots, q_n)$, without bothering to make it a column vector.

2 *Bug Algorithms*

EVEN A simple planner can present interesting and difficult issues. The Bug1 and Bug2 algorithms [301] are among the earliest and simplest sensor-based planners with provable guarantees. These algorithms assume the robot is a point operating in the plane with a contact sensor or a zero range sensor to detect obstacles. When the robot has a finite range (nonzero range) sensor, then the Tangent Bug algorithm [217] is a Bug derivative that can use that sensor information to find shorter paths to the goal. The Bug and Bug-like algorithms are straightforward to implement; moreover, a simple analysis shows that their success is guaranteed, when possible. These algorithms require two behaviors: move on a straight line and follow a boundary. To handle boundary-following, we introduce a curve-tracing technique based on the implicit function theorem at the end of this chapter. This technique is general to following any path, but we focus on following a boundary at a fixed distance.

2.1 Bug1 and Bug2

Perhaps the most straight forward path planning approach is to move toward the goal, unless an obstacle is encountered, in which case, circumnavigate the obstacle until motion toward the goal is once again allowable. Essentially, the Bug1 algorithm formalizes the "common sense" idea of moving toward the goal and going around obstacles. The robot is assumed to be a point with perfect positioning (no positioning error) with a contact sensor that can detect an obstacle boundary if the point robot "touches" it. The robot can also measure the distance $d(x, y)$ between any two points x

and y. Finally, assume that the workspace is *bounded*. Let $B_r(x)$ denote a ball of radius r centered on x, i.e., $B_r(x) = \{y \in \mathbb{R}^2 \mid d(x, y) < r\}$. The fact that the workspace is bounded implies that for all $x \in \mathcal{W}$, there exists an $r < \infty$ such that $\mathcal{W} \subset B_r(x)$.

The start and goal are labeled q_{start} and q_{goal}, respectively. Let $q_0^L = q_{\text{start}}$ and the m-line be the line segment that connects q_i^L to q_{goal}. Initially, $i = 0$. The Bug1 algorithm exhibits two behaviors: motion-to-goal and boundary-following. During motion-to-goal, the robot moves along the m-line toward q_{goal} until it either encounters the goal or an obstacle. If the robot encounters an obstacle, let q_1^H be the point where the robot first encounters an obstacle and call this point a *hit point*. The robot then circumnavigates the obstacle until it returns to q_1^H. Then, the robot determines the closest point to the goal on the perimeter of the obstacle and traverses to this point. This point is called a *leave point* and is labeled q_1^L. From q_1^L, the robot heads straight toward the goal again, i.e., it reinvokes the motion-to-goal behavior. If the line that connects q_1^L and the goal intersects the current obstacle, then there is no path to the goal; note that this intersection would occur immediately "after" leaving q_1^L. Otherwise, the index i is incremented and this procedure is then repeated for q_i^L and q_i^H until the goal is reached or the planner determines that the robot cannot reach the goal (figures 2.1, 2.2). Finally, if the line to the goal "grazes" an obstacle, the robot need not invoke a boundary following behavior, but rather continues onward toward the goal. See algorithm 1 for a description of the Bug1 approach.

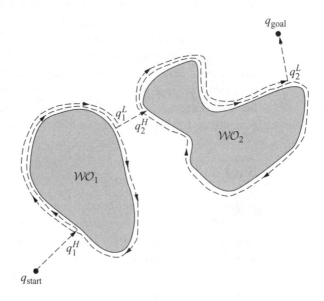

Figure 2.1   The Bug1 algorithm successfully finds the goal.

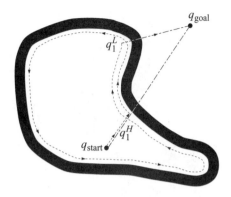

Figure 2.2   The Bug1 algorithm reports the goal is unreachable.

Algorithm 1 Bug1 Algorithm

Input: A point robot with a tactile sensor
Output: A path to the q_{goal} or a conclusion no such path exists

---

1: **while** Forever **do**
2:    **repeat**
3:        From q_{i-1}^L, move toward q_{goal}.
4:    **until** q_{goal} is reached **or** an obstacle is encountered at q_i^H.
5:    **if** Goal is reached **then**
6:        Exit.
7:    **end if**
8:    **repeat**
9:        Follow the obstacle boundary.
10:    **until** q_{goal} is reached **or** q_i^H is re-encountered.
11:    Determine the point q_i^L on the perimeter that has the shortest distance to the goal.
12:    Go to q_i^L.
13:    **if** the robot were to move toward the goal **then**
14:        Conclude q_{goal} is not reachable and exit.
15:    **end if**
16: **end while**

Like its Bug1 sibling, the Bug2 algorithm exhibits two behaviors: motion-to-goal and boundary-following. During motion-to-goal, the robot moves toward the goal on the *m*-line; however, in Bug2 the *m*-line connects q_{start} and q_{goal}, and thus remains fixed. The boundary-following behavior is invoked if the robot encounters an obstacle,

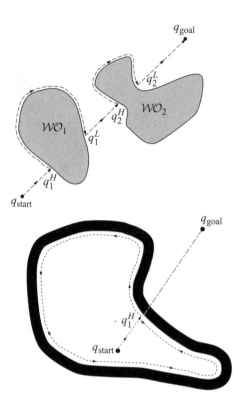

Figure 2.3   (Top) The Bug2 algorithm finds a path to the goal. (Bottom) The Bug2 algorithm reports failure.

but this behavior is different from that of Bug1. For Bug2, the robot circumnavigates the obstacle until it reaches a new point on the m-line closer to the goal than the initial point of contact with the obstacle. At this time, the robot proceeds toward the goal, repeating this process if it encounters an object. If the robot re-encounters the original departure point from the m-line, then there is no path to the goal (figures 2.3, 2.4). Let $x \in \mathcal{W}_{\text{free}} \subset \mathbb{R}^2$ be the current position of the robot, $i = 1$, and q_0^L be the start location. See algorithm 2 for a description of the Bug2 approach.

At first glance, it seems that Bug2 is a more effective algorithm than Bug1 because the robot does not have to entirely circumnavigate the obstacles; however, this is not always the case. This can be seen by comparing the lengths of the paths found by the two algorithms. For Bug1, when the ith obstacle is encountered, the robot completely circumnavigates the boundary, and then returns to the leave point. In the worst case, the robot must traverse half the perimeter, p_i, of the obstacle to reach this leave point.

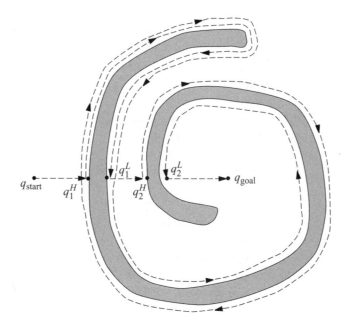

Figure 2.4   Bug2 Algorithm.

Moreover, in the worst case, the robot encounters all n obstacles. If there are no obstacles, the robot must traverse a distance of length $d(q_{\text{start}}, q_{\text{goal}})$. Thus, we obtain

$$(2.1) \qquad L_{\text{Bug1}} \leq d(q_{\text{start}}, q_{\text{goal}}) + 1.5 \sum_{i=1}^{n} p_i.$$

For Bug2, the path length is a bit more complicated. Suppose that the line through q_{start} and q_{goal} intersects the ith obstacle n_i times. Then, there are at most n_i leave points for this obstacle, since the robot may only leave the obstacle when it returns to a point on this line. It is easy to see that half of these intersection points are not valid leave points because they lie on the "wrong side" of the obstacle, i.e., moving toward the goal would cause a collision. In the worst case, the robot will traverse nearly the entire perimeter of the obstacle for each leave point. Thus, we obtain

$$(2.2) \qquad L_{\text{Bug2}} \leq d(q_{\text{start}}, q_{\text{goal}}) + \frac{1}{2} \sum_{i=1}^{n} n_i\, p_i.$$

Naturally, (2.2) is an upper-bound because the summation is over all of the obstacles as opposed to over the set of obstacles that are encountered by the robot.

Algorithm 2 Bug2 Algorithm

Input: A point robot with a tactile sensor

Output: A path to q_{goal} or a conclusion no such path exists

---

1: **while** True **do**
2:    **repeat**
3:       From q_{i-1}^L, move toward q_{goal} along m-line.
4:    **until**
      q_{goal} is reached **or**
      an obstacle is encountered at *hit point q_i^H*.
5:    Turn left (or right).
6:    **repeat**
7:       Follow boundary
8:    **until**
9:       q_{goal} is reached **or**
10:       q_i^H is re-encountered **or**
11:       m-line is re-encountered at a point m such that
12:       $m \neq q_i^H$ (robot did not reach the hit point),
13:       $d(m, q_{\text{goal}}) < d(m, q_i^H)$ (robot is closer), and
14:       If robot moves toward goal, it would not hit the obstacle
15:    Let $q_{i+1}^L = m$
16:    Increment i
17: **end while**

A casual examination of (2.1) and (2.2) shows that L_{Bug2} can be arbitrarily longer than L_{Bug1}. This can be achieved by constructing an obstacle whose boundary has many intersections with the m-line. Thus, as the "complexity" of the obstacle increases, it becomes increasingly likely that Bug1 could outperform Bug2 (figure 2.4).

In fact, Bug1 and Bug2 illustrate two basic approaches to search problems. For each obstacle that it encounters, Bug1 performs an *exhaustive search* to find the optimal leave point. This requires that Bug1 traverse the entire perimeter of the obstacle, but having done so, it is certain to have found the optimal leave point. In contrast, Bug2 uses an *opportunistic* approach. When Bug2 finds a leave point that is better than any it has seen before, it commits to that leave point. Such an algorithm is also called *greedy,* since it opts for the first promising option that is found. When the obstacles are simple, the greedy approach of Bug2 gives a quick payoff, but when the obstacles are complex, the more conservative approach of Bug1 often yields better performance.

2.2   Tangent Bug

Tangent Bug [216] serves as an improvement to the Bug2 algorithm in that it determines a shorter path to the goal using a range sensor with a 360 degree infinite
orientation resolution. Sometimes orientation is called *azimuth*. We model this range
sensor with the *raw distance function* $\rho : \mathbb{R}^2 \times S^1 \to \mathbb{R}$. Consider a point robot
situated at $x \in \mathbb{R}^2$ with rays radially emanating from it. For each $\theta \in S^1$, the value
$\rho(x, \theta)$ is the distance to the closest obstacle along the ray from x at an angle θ. More
formally,

$$
\begin{aligned}
\rho(x, \theta) = &\min_{\lambda \in [0, \infty]} d(x, x + \lambda[\cos\theta, \sin\theta]^T), \\
&\text{such that } x + \lambda[\cos\theta, \sin\theta]^T \in \bigcup_i \mathcal{WO}_i.
\end{aligned}
$$

(2.3)

Note that there are infinitely many $\theta \in S^1$ and hence the infinite resolution. This
assumption is approximated with a finite number of range sensors situated along the
circumference of a circular mobile robot which we have modeled as a point.

Since real sensors have limited range, we define the *saturated raw distance function,*
denoted $\rho_R : \mathbb{R}^2 \times S^1 \to \mathbb{R}$, which takes on the same values as ρ when the obstacle
is within sensing range, and has a value of infinity when the ray lengths are greater
than the sensing range, R, meaning that the obstacles are outside the sensing range.
More formally,

$$
\rho_R(x, \theta) = \begin{cases} \rho(x, \theta), & \text{if } \rho(x, \theta) < R \\ \infty, & \text{otherwise.} \end{cases}
$$

The Tangent Bug planner assumes that the robot can detect discontinuities in ρ_R as
depicted in figure 2.5. For a fixed $x \in \mathbb{R}^2$, an *interval of continuity* is defined to be a
connected set of points $x + \rho(x, \theta)[\cos\theta, \sin\theta]^T$ on the boundary of the free space
where $\rho_R(x, \theta)$ is finite and continuous with respect to θ.

The endpoints of these intervals occur where $\rho_R(x, \theta)$ loses continuity, either as
a result of one obstacle blocking another or the sensor reaching its range limit. The
endpoints are denoted O_i. Figure 2.6 contains an example where ρ_R loses continuity. The points $O_1, O_2, O_3, O_5, O_6, O_7$, and O_8 correspond to losses of continuity
associated with obstacles blocking other portions of $\mathcal{W}_{\text{free}}$; note the rays are tangent
to the obstacles here. The point O_4 is a discontinuity because the obstacle boundary
falls out of range of the sensor. The sets of points on the boundary of the free space
between O_1 and O_2, O_3 and O_4, O_5 and O_6, O_7 and O_8 are the intervals of continuity.

Just like the other Bugs, Tangent Bug iterates between two behaviors: motion-
to-goal and boundary-following. However, these behaviors are different than in the
Bug1 and Bug2 approaches. Although motion-to-goal directs the robot to the goal,

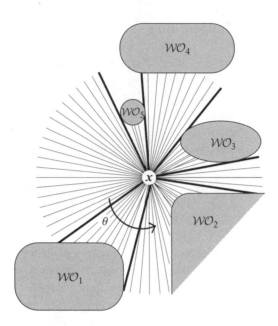

Figure 2.5 The thin lines are values of the raw distance function, $\rho_R(x, \theta)$, for a fixed $x \in \mathbb{R}^2$, and the thick lines indicate discontinuities, which arise either because an obstacle occludes another or the sensing range is reached. Note that the segments terminating in free space represent infinitely long rays.

this behavior may have a phase where the robot follows the boundary. Likewise, the boundary-following behavior may have a phase where the robot does not follow the boundary.

The robot initially invokes the motion-to-goal behavior, which itself has two parts. First, the robot moves in a straight line toward the goal until it senses an obstacle R units away and directly between it and the goal. This means that a line segment connecting the robot and goal must intersect an interval of continuity. For example, in figure 2.7, $\mathcal{WO}_2$ is within sensing range, but does not block the goal, but $\mathcal{WO}_1$ does. When the robot initially senses an obstacle, the circle of radius R becomes tangent to the obstacle. Immediately after, this tangent point splits into two O_i's, which are the endpoints of the interval. If the obstacle is in front of the robot, then this interval intersects the segment connecting the robot and the goal.

The robot then moves toward the O_i that maximally decreases a heuristic distance to the goal. An example of a heuristic distance is the sum $d(x, O_i) + d(O_i, q_{\text{goal}})$. (The heuristic distance can be more complicated when factoring in available information

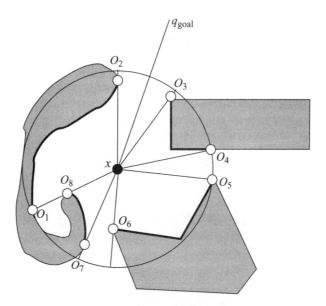

Figure 2.6 The points of discontinuity of $\rho_R(x, \theta)$ correspond to points O_i on the obstacles. The thick solid curves represent connected components of the range of $\rho_R(x, \theta)$, i.e., the intervals of continuity. In this example, the robot, to the best of its sensing range, believes there is a straight-line path to the goal.

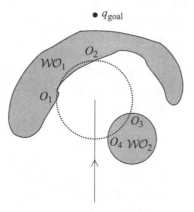

Figure 2.7 The vertical represents the path of the robot and the dotted circle its sensing range. Currently, the robot is located at the "top" of the line segment. The points O_i represent the points of discontinuity of the saturated raw distance function. Note that the robot passes by $\mathcal{WO}_2$.

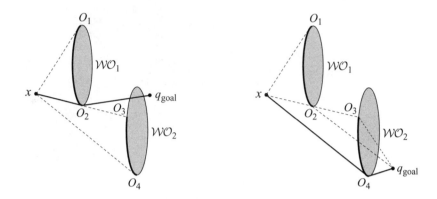

Figure 2.8   (Left) The planner selects O_2 as a subgoal for the robot. (Right) The planner selects O_4 as a subgoal for the robot. Note the line segment between O_4 and q_{goal} cuts through the obstacle.

with regard to the obstacles.) In figure 2.8 (left), the robot sees $\mathcal{WO}_1$ and drives to O_2 because $i = 2$ minimizes $d(x, O_i) + d(O_i, q_{goal})$. When the robot is located at x, it cannot know that $\mathcal{WO}_2$ blocks the path from O_2 to the goal. In figure 2.8 (right), when the robot is located at x but the goal is different, it has enough sensor information to conclude that $\mathcal{WO}_2$ indeed blocks a path from O_2 to the goal, and therefore drives toward O_4. So, even though driving toward O_2 may initially minimize $d(x, O_i) + d(O_i, q_{goal})$ more than driving toward O_4, the planner effectively assigns an infinite cost to $d(O_2, q_{goal})$ because it has enough information to conclude that any path through O_2 will be suboptimal.

The set $\{O_i\}$ is continuously updated as the robot moves toward a particular O_i, which can be seen in figure 2.9. When $t = 1$, the robot has not sensed the obstacle, hence the robot moves toward the goal. When $t = 2$, the robot initially senses the obstacle, depicted by a thick solid curve. The robot continues to move toward the goal, but off to the side of the obstacle heading toward the discontinuity in ρ. For $t = 3$ and $t = 4$, the robot senses more of the obstacle and continues to decrease distance toward the goal while hugging the boundary.

The robot undergoes the motion-to-goal behavior until it can no longer decrease the heuristic distance to the goal. Put differently, it finds a point that is like a local minimum of $d(\cdot, O_i) + d(O_i, q_{goal})$ restricted to the path that motion-to-goal dictates.

When the robot switches to boundary-following, it finds the point M on the sensed portion of the obstacle which has the shortest distance on the obstacle to the goal. Note that if the sensor range is zero, then M is the same as the hit point from the Bug1 and Bug2 algorithms. This sensed obstacle is also called the *followed obstacle*. We make a distinction between the followed obstacle and the *blocking obstacle*. Let x be

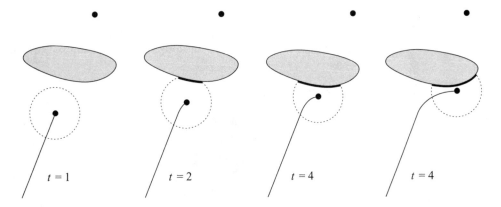

Figure 2.9 Demonstration of motion-to-goal behavior for a robot with a finite sensor range moving toward a goal which is "above" the light gray obstacle.

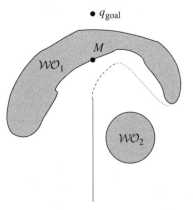

Figure 2.10 The workspace is the same as in figure 2.7. The solid and dashed segments represent the path generated by motion-to-goal and the dotted path represents the boundary-following path. Note that *M* is the "local minimum" point.

the current position of the robot. The blocking obstacle is the closest obstacle within sensor range that intersects the segment $(1 - \lambda)x + \lambda q_{goal}$ $\forall \lambda \in [0, 1]$. Initially, the blocking obstacle and the followed obstacle are the same.

Now the robot moves in the same direction as if it were in the motion-to-goal behavior. It continuously moves toward the O_i on the followed obstacle in the chosen direction (figure 2.10). While undergoing this motion, the planner also updates two values: $d_{followed}$ and d_{reach}. The value $d_{followed}$ is the shortest distance between the boundary which had been sensed and the goal. Let Λ be all of the points within line of sight of x with range R that are on the followed obstacle $\mathcal{WO}_f$, i.e., $\Lambda = \{y \in \partial \mathcal{WO}_f : \lambda x + (1 - \lambda)y \in \mathcal{Q}_{\text{free}} \; \forall \lambda \in [0, 1]\}$. The value d_{reach} is the distance between the goal and the closest point on the followed obstacle that is within line of sight of the robot, i.e.,

$$d_{\text{reach}} = \min_{c \in \Lambda} d(q_{\text{goal}}, c).$$

When $d_{\text{reach}} < d_{\text{followed}}$, the robot terminates the boundary-following behavior.

Let T be the point where a circle centered at x of radius R intersects the segment that connects x and q_{goal}. This is the point on the periphery of the sensing range that is closest to the goal when the robot is located at x. Starting with $x = q_{\text{start}}$ and $d_{\text{leave}} = d(q_{\text{start}}, q_{\text{goal}})$, see algorithm 3.

Algorithm 3 Tangent Bug Algorithm

Input: A point robot with a range sensor
Output: A path to the q_{goal} or a conclusion no such path exists

---

1: **while** True **do**
2:    **repeat**
3:      Continuously move toward the point $n \in \{T, O_i\}$ which minimizes $d(x, n) + d(n, q_{\text{goal}})$
4:    **until**

- the goal is encountered **or**

- The direction that minimizes $d(x, n) + d(n, q_{\text{goal}})$ begins to increase $d(x, q_{\text{goal}})$, i.e., the robot detects a "local minimum" of $d(\cdot, q_{\text{goal}})$.

5:    Chose a boundary following direction which continues in the same direction as the most recent motion-to-goal direction.
6:    **repeat**
7:      Continuously update d_{reach}, d_{followed}, and $\{O_i\}$.
8:      Continuously moves toward $n \in \{O_i\}$ that is in the chosen boundary direction.
9:    **until**

- The goal is reached.

- The robot completes a cycle around the obstacle in which case the goal cannot be achieved.

- $d_{\text{reach}} < d_{\text{followed}}$

10: **end while**

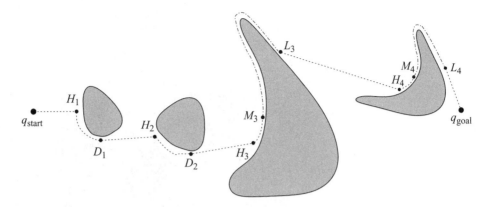

Figure 2.11   The path generated by Tangent Bug with zero sensor range. The dashed lines correspond to the motion-to-goal behavior and the dotted lines correspond to boundary-following.

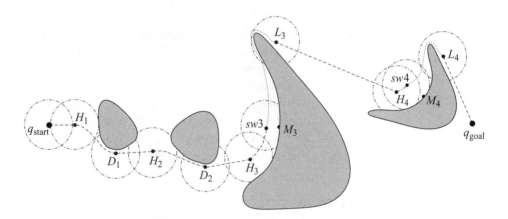

Figure 2.12   Path generated by Tangent Bug with finite sensor range. The dashed lines correspond to the motion-to-goal behavior and the dotted lines correspond to boundary-following. The dashed-dotted circles correspond to the sensor range of the robot.

See figures 2.11, 2.12 for example runs. Figure 2.11 contains a path for a robot with zero sensor range. Here the robot invokes a motion-to-goal behavior until it encounters the first obstacle at hit point H_1. Unlike Bug1 and Bug2, encountering a hit point does not change the behavior mode for the robot. The robot continues with the motion-to-goal behavior by turning right and following the boundary of the first obstacle. The robot turned right because that direction minimized its heuristic distance to the goal. The robot departs this boundary at a *depart point* D_1. The robot

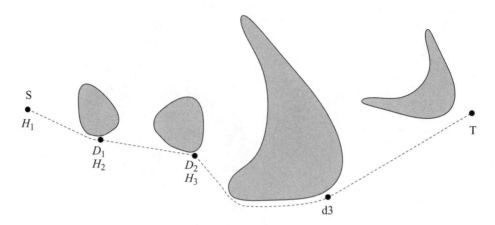

Figure 2.13 Path generated by Tangent Bug with infinite sensor range. The dashed-lines correspond to the motion-to-goal behavior and there is no boundary-following.

continues with the motion-to-goal behavior, maneuvering around a second obstacle, until it encounters the third obstacle at H_3. The robot turns left and continues to invoke the motion-to-goal behavior until it reaches M_3, a minimum point. Now, the planner invokes the boundary-following behavior until the robot reaches L_3. Note that since we have zero sensing range, d_{reach} is the distance between the robot and the goal. The procedure continues until the robot reaches the goal. Only at M_i and L_i does the robot switch between behaviors. Figures 2.12, 2.13 contain examples where the robot has a finite and infininte sensing ranges, respectively.

2.3   Implementation

Essentially, the bug algorithms have two behaviors: drive toward a point and follow an obstacle. The first behavior is simply a form of gradient descent of $d(\cdot, n)$ where n is either q_{goal} or an O_i. The second behavior, boundary-following, presents a challenge because the obstacle boundary is not known *a priori*. Therefore, the robot planner must rely on sensor information to determine the path. However, we must concede that the full path to the goal will not be determined from one sensor reading: the sensing range of the robot may be limited and the robot may not be able to "see" the entire world from one vantage point. So, the robot planner has to be incremental. We must determine first what information the robot requires and then where the robot should move to acquire more information. This is indeed the challenge of sensor-based planning. Ideally, we would like this approach to be reactive with sensory information feeding into a simple algorithm that outputs translational and rotational velocity for the robot.

There are three questions: What information does the robot require to circumnavigate the obstacle? How does the robot infer this information from its sensor data? How does the robot use this information to determine (locally) a path?

2.3.1   What Information: The Tangent Line

If the obstacle were flat, such as a long wall in a corridor, then following the obstacle is trivial: simply move parallel to the obstacle. This is readily implemented using a sensing system that can determine the obstacle's surface normal $n(x)$, and hence a direction parallel to its surface. However, the world is not necessarily populated with flat obstacles; many have nonzero curvature. The robot can follow a path that is consistently orthogonal to the surface normal; this direction can be written as $n(x)^{\perp}$ and the resulting path satisfies $\dot{c}(t) = v$ where v is a basis vector in $(n(c(t)))^{\perp}$. The sign of v is based on the "previous" direction of $\dot{c}$.

Consistently determining the surface normal can be quite challenging and therefore for implementation, we can assume that obstacles are "locally flat." This means the sensing system determines the surface normal, the robot moves orthogonal to this normal for a short distance, and then the process repeats. In a sense, the robot determines the sequence of short straight-line segments to follow, based on sensor information.

This flat line, loosely speaking, is the tangent (figure 2.14). It is a linear approximation of the curve at the point where the tangent intersects the curve. The tangent

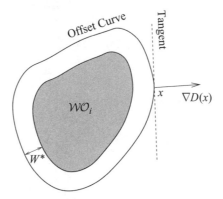

Figure 2.14   The solid curve is the offset curve. The dashed line represents the tangent to the offset curve at x.

can also be viewed as a first-order approximation to the function that describes the curve. Let $c : [0, 1] \rightarrow \mathcal{W}_{\text{free}}$ be the function that defines a path. Let $x = c(s_0)$ for a $s_0 \in [0, 1]$. The tangent at x is $\frac{dc}{ds}\big|_{s=s_0}$. The tangent space can be viewed as a line whose basis vector is $\frac{dc}{ds}\big|_{s=s_0}$, i.e., $\left\{ \alpha \frac{dc}{ds}\big|_{s=s_0} \,\middle|\, \alpha \in \mathbb{R} \right\}$.

2.3.2 How to Infer Information with Sensors: Distance and Gradient

The next step is to infer the tangent from sensor data. Instead of thinking of the robot as a point in the plane, let's think of it as a circular base which has a fine array of tactile sensors radially distributed along its circumference (figure 2.15). When the robot contacts an obstacle, the direction from the contacted sensor to the robot's center approximates the surface normal. With this information, the robot can determine a sequence of tangents to follow the obstacle.

Unfortunately, using a tactile sensor to prescribe a path requires the robot to collide with obstacles, which endangers the obstacles and the robot. Instead, the robot should follow a path at a safe distance $\mathcal{W}^* \in \mathbb{R}$ from the nearest obstacle. Such a path is called

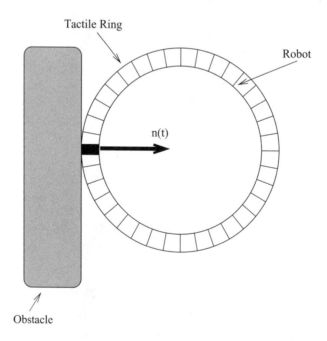

Figure 2.15 A fine-resolution tactile sensor.

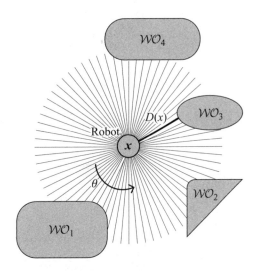

Figure 2.16   The global minimum of the rays determines the distance to the closest obstacle; the gradient points in a direction away from the obstacle along the ray.

an *offset curve* [381]. Let $D(x)$ be the distance from x to the closest obstacle, i.e.,

(2.4) $D(x) = \min_{c \in \bigcup_i WO_i} d(x, c).$

To measure this distance with a mobile robot equipped with an onboard range sensing ring, we use the raw distance function again. However, instead of looking for discontinuities, we look for the global minimum. In other words, $D(x) = \min_s \rho(x, s)$ (figure 2.16).

We will need to use the gradient of distance. In general, the gradient is a vector that points in the direction that maximally increases the value of a function. See appendix C.5 for more details. Typically, the ith component of the gradient vector is the partial derivative of the function with respect to its ith coordinate. In the plane, $\nabla D(x) = [\frac{\partial D(x)}{\partial x_1} \quad \frac{\partial D(x)}{\partial x_2}]^T$ which points in the direction that increases distance the most. Finally, the gradient is the unit direction associated with the smallest value of the raw distance function. Since the raw distance function seemingly approximates a sensing system with individual range sensing elements radially distributed around the perimeter of the robot, an algorithm defined in terms of D can often be implemented using realistic sensors.

There are many choices for range sensors; here, we investigate the use of ultrasonic sensors (figure 2.17), which are commonly found on mobile robots. Conventional ultrasonic sensors measure distance using time of flight. When the speed of sound

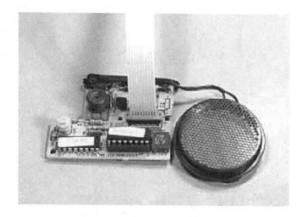

Figure 2.17 The disk on the right is the standard Polaroid ultrasonic transducer found on many mobile robots; the circuitry on the left drives the transducer.

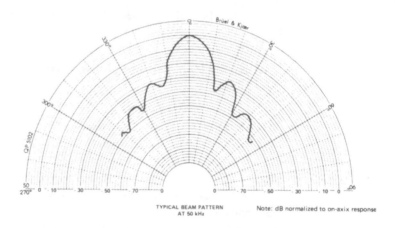

Figure 2.18 Beam pattern for the Polaroid transducer.

in air is constant, the time that the ultrasound requires to leave the transducer, strike an obstacle, and return is proportional to the distance to the point of reflection on the obstacle [113]. This obstacle, however, can be located anywhere along the angular spread of the sonar sensor's beam pattern (figure 2.18). Therefore, the distance information that sonars provide is fairly accurate in depth, but not in azimuth. The beam pattern can be approximated with a cone (figure 2.19). For the commonly used Polaroid transducer, the arcbase is 22.5 degrees. When the reading of the sensor is d, the point of reflection can be anywhere along the arc base of length $\frac{2\pi d 22.5}{360}$.

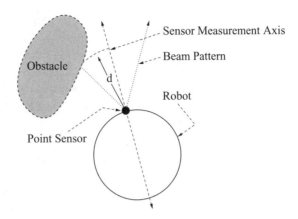

Figure 2.19   Centerline model.

Initially, assume that the echo originates from the center of the sonar cone. We acknowledge that this is a naive model, and we term this the *centerline model* (figure 2.19). The ultrasonic sensor with the smallest reading approximates the global minimum of the raw distance function, and hence $D(x)$. The direction that this sensor is facing approximates the negated gradient $-\nabla D(x)$ because this sensor faces the closest obstacle. The tangent is then the line orthogonal to the direction associated with the smallest sensor reading.

2.3.3   How to Process Sensor Information: Continuation Methods

The tangent to the offset curve is $(\nabla D(x))^{\perp}$, the line orthogonal to $\nabla D(x)$ (figure 2.14). The vector $\nabla D(x)$ points in the direction that maximally increases distance; likewise, the vector $-\nabla D(x)$ points in the direction that maximally decreases distance; they both point along the same line, but in opposite directions. Therefore, the vector $(\nabla D(x))^{\perp}$ points in the direction that locally maintains distance; it is perpendicular to both $\nabla D(x)$ and $-\nabla D(x)$. This would be the tangent of the offset curve which maintains distance to the nearby obstacle.

Another way to see why $(\nabla D(x))^{\perp}$ is the tangent is to look at the definition of the offset curve. For a safety distance $\mathcal{W}^*$, we can define the offset curve implicitly as the set of points where $G(x) = D(x) - \mathcal{W}^*$ maps to zero. The set of nonzero points (or vectors) that map to zero is called the *null space* of a map. For a curve implicitly defined by G, the tangent space at a point x is the null space of $DG(x)$, the Jacobian of G [410]. In general, the i, jth component of the Jacobian matrix is the partial derivative of the ith component function with respect to the jth coordinate and thus the Jacobian is a mapping between tangent spaces. Since in this case, G is a real-valued function ($i = 1$), the Jacobian is just a row vector $DD(x)$. Here, we are reusing the symbol D. The reader is forced to use context to determine if D means distance or differential.

In Euclidean spaces, the ith component of a single-row Jacobian equals the ith component of the gradient and thus $\nabla D(x) = (DD(x))^T$. Therefore, since the tangent space is the null space of $DD(x)$, the tangent for boundary-following in the plane is the line orthogonal to $\nabla D(x)$, i.e., $(\nabla D(x))^\perp$, and can be derived from sensor information.

Using distance information, the robot can determine the tangent direction to the offset curve. If the obstacles are flat, then the offset curve is also flat, and simply following the tangent is sufficient to follow the boundary of an unknown obstacle. Consider, instead, an obstacle with curvature. We can, however, assume that the obstacle is locally flat. The robot can then move along the tangent for a short distance, but since the obstacle has curvature, the robot will not follow the offset curve, i.e., it will "fall off" of the offset curve. To reaccess the offset curve, the robot moves either toward or away from the obstacle until it reaches the safety distance W^*. In doing so, the robot is moving along a line defined by $\nabla D(x)$, which can be derived from sensor information.

Essentially, the robot is performing a numerical procedure of prediction and correction. The robot uses the tangent to locally predict the shape of the offset curve and then invokes a correction procedure once the tangent approximation is not valid. Note that the robot does not explicitly trace the path but instead "hovers" around it, resulting in a sampling of the path, not the path itself (figure 2.20).

A numerical tracing procedure can be posed as one which traces the roots of the expression $G(x) = 0$, where in this case $G(x) = D(x) - W^*$. Numerical curve-tracing techniques rest on the *implicit function theorem* [9, 232, 307] which locally defines a curve that is implicitly defined by a map $G : Y \times \mathbb{R} \to Y$. Specifically, the roots of G locally define a curve parameterized by $\lambda \in \mathbb{R}$. See appendix D for a formal definition.

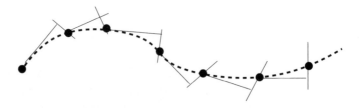

Figure 2.20 The dashed line is the actual path, but the robot follows the thin black lines, predicting and correcting along the path. The black circles are samples along the path.

For boundary following at a safety distance W^*, the function $G(y, \lambda) = D(y, \lambda) - W^*$ implicitly defines the offset curve. Note that the λ-coordinate corresponds to a tangent direction and the y-coordinates to the line or hyperplane orthogonal to the tangent. Let Y denote this hyperplane and $D_Y G$ be the matrix formed by taking the derivative of $G(x) = D(x) - W^* = 0$ with respect to the y-coordinates. It takes the form $D_Y G(x) = D_Y D(x)$ where D_Y denotes the differential with respect to the y-coordinates. If $D_Y G(y, \lambda)$ is surjective at $x = (\lambda, y)^T$, then the implicit function theorem states that the roots of $G(y, \lambda)$ locally define a curve that follows the boundary at a distance W^* as λ is varied, i.e., $y(\lambda)$.

By numerically tracing the roots of G, we can locally construct a path. While there are a number of curve-tracing techniques [232], let us consider an adaptation of a common predictor-corrector scheme. Assume that the robot is located at a point x which is a fixed distance W^* away from the boundary. The robot takes a "small" step, $\Delta\lambda$, in the λ-direction (i.e., the tangent to the local path). In general, this *prediction step* takes the robot off the offset path. Next, a *correction method* is used to bring the robot back onto the offset path. If $\Delta\lambda$ is small, then the local path will intersect a *correcting plane,* which is a plane orthogonal to the λ-direction at a distance $\Delta\lambda$ away from the origin.

The correction step finds the location where the offset path intersects the correcting plane and is an application of the Newton convergence theorem [232]. See appendix D.2 for a more formal definition of this theorem. The Newton convergence theorem also requires that $D_Y G(y, \lambda)$ be full rank at every (y, λ) in a neighborhood of the offset path. This is true because for $G(x) = D(x) - W^*$, $[0 \ \ D_Y G(y, \lambda)]^T = DG(y, \lambda)$. Since $DG(y, \lambda)$ is full rank, so must be $D_Y G(y, \lambda)$ on the offset curve. Since the set of nonsingular matrices is an open set, we know there is a neighborhood around each (y, λ) in the offset path where $DG(y, \lambda)$ is full rank and hence we can use the iterative Newton method to implement the corrector step. If y^h and λ^h are the hth estimates of y and λ, the $h + 1$st iteration is defined as

(2.5) $\quad y^{h+1} = y^h - (D_Y G)^{-1} G(y^h, \lambda^h),$

where $D_Y G$ is evaluated at (y^h, λ^h). Note that since we are working in a Euclidean space, we can determine $D_Y G$ solely from distance gradient, and hence, sensor information.

Problems

1. Prove that $D(x)$ is the global minimum of $\rho(x, s)$ with respect to s.

2. What are the tradeoffs between the Bug1 and Bug2 algorithms?

3. Extend the Bug1 and Bug2 algorithms to a two-link manipulator.

4. What is the difference between the Tangent Bug algorithm with zero range detector and Bug2? Draw examples.

5. What are the differences between the path in figure 2.11 and the paths that Bug1 and Bug2 would have generated?

6. The Bug algorithms also assume the planner knows the location of the goal and the robot has perfect positioning. Redesign one of the Bug algorithms to relax the assumption of perfect positioning. Feel free to introduce a new type of "reasonable" sensor (not a high-resolution Global Positioning System).

7. In the Bug1 algorithm, prove or disprove that the robot does not encounter any obstacle that does not intersect the disk of radius $d(q_{start}, q_{goal})$ centered at q_{goal}.

8. What assumptions do the Bug1, Bug2, and Tangent Bug algorithms make on robot localization, both in position and orientation?

9. Prove the completeness of the Tangent Bug algorithm.

10. Adapt the Tangent Bug algorithm so that it has a limited field of view sensor, i.e., it does not have a 360 degree field of view range sensor.

11. Write out $D_Y G$ for boundary following in the planar case.

12. Let $G_1(x) = D(x) + 1$ and let $G_2(x) = D(x) + 2$. Why are their Jacobians the same?

13. Let $G(x, y) = y^3 + y - x^2$. Write out a y as a function of x in an interval about the origin for the curve defined by $G(x, y) = 0$.

3 *Configuration Space*

To CREATE motion plans for robots, we must be able to specify the position of the robot. More specifically, we must be able to give a specification of the location of every point on the robot, since we need to ensure that no point on the robot collides with an obstacle. This raises some fundamental questions: How much information is required to completely specify the position of every point on the robot? How should this information be represented? What are the mathematical properties of these representations? How can obstacles in the robot's world be taken into consideration while planning the path of a robot?

In this chapter, we begin to address these questions. We first discuss exactly what is meant by a configuration of a robot and introduce the concept of the configuration space, one of the most important concepts in robot motion planning. We then briefly discuss how obstacles in the robot's environment restrict the set of admissible paths. We then begin a more rigorous investigation of the properties of the configuration space, including its dimension, how it sometimes can be represented by a differentiable manifold, and how manifolds can be represented by embeddings and parameterizations. We conclude the chapter by discussing mappings between different representations of the configuration, and the *Jacobian* of these mappings, which relates velocities in the different representations.

3.1   Specifying a Robot's Configuration

To make our discussion more precise, we introduce the following definitions. The *configuration* of a robot system is a complete specification of the position of every point of that system. The *configuration space,* or *C-space,* of the robot system is the space of all possible configurations of the system. Thus a configuration is simply a point in this abstract configuration space. Throughout the text, we use q to denote a configuration and $\mathcal{Q}$ to denote the configuration space.[1] The number of *degrees of freedom* of a robot system is the dimension of the configuration space, or the minimum number of parameters needed to specify the configuration.

To illustrate these definitions, consider a circular mobile robot that can translate without rotating in the plane. A simple way to represent the robot's configuration is to specify the location of its center, (x, y), relative to some fixed coordinate frame. If we know the radius r of the robot, we can easily determine from the configuration $q = (x, y)$ the set of points occupied by the robot. We will use the notation $R(q)$ to denote this set. When we define the configuration as $q = (x, y)$, we have

$$R(x, y) = \{(x', y') \mid (x - x')^2 + (y - y')^2 \leq r^2\},$$

and we see that these two parameters, x and y, are sufficient to completely determine the configuration of the circular robot. Therefore, for the circular mobile robot, we can represent the configuration space by $\mathbb{R}^2$ once we have chosen a coordinate frame in the plane.

Robots move in a two- or three-dimensional Euclidean ambient space, represented by $\mathbb{R}^2$ or $\mathbb{R}^3$, respectively. We sometimes refer to this ambient space as the *workspace.* Other times we have a more specific meaning for "workspace." For example, for a robot arm, often we call the workspace the set of points of the ambient space reachable by a point on the hand or *end effector* (see figure 3.3). For the translating mobile robot described above, the workspace and the configuration space are both two-dimensional Euclidean spaces, but it is important to keep in mind that these are different spaces. This becomes clear when we consider even slightly more complicated robots, as we see next.

Consider a two-joint planar robot arm, as shown in figure 3.1. A point on the first link of the arm is pinned, so that the only possible motion of the first link is a rotation about this joint. Likewise, the base of the second link is pinned to a point at the end of the first link, and the only possible motion of the second link is a rotation about this joint. Therefore, if we specify the parameters θ_1 and θ_2, as shown in figure 3.1, we

---

1. While q is used almost universally to denote a configuration, the configuration space is sometimes denoted by $\mathcal{C}$, particularly in the path-planning community.

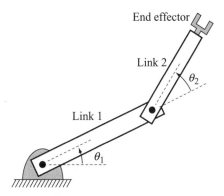

Figure 3.1   The angles θ_1 and θ_2 specify the configuration of the two-joint robot.

have specified the configuration of the arm. For now we will assume no joint limits, so the two links can move over each other.

Each joint angle θ_i corresponds to a point on the unit circle S^1, and the configuration space is $S^1 \times S^1 = T^2$, the two-dimensional torus. It is common to picture a torus as the surface of a doughnut because a torus has a natural embedding in $\mathbb{R}^3$, just as a circle S^1 has a natural embedding in $\mathbb{R}^2$. By cutting the torus along the $\theta_1 = 0$ and $\theta_2 = 0$ curves, we can flatten the torus onto the plane, as shown in figure 3.2. With this planar representation, we are identifying points on S^1 by points in the interval $[0, 2\pi) \subset \mathbb{R}$. While this representation covers all points in S^1, the interval $[0, 2\pi)$, being a subset of the real line, does not naturally wrap around like S^1, so there is a discontinuity in the representation. As we discuss in section 3.4, this is because S^1 is topologically different from any interval of $\mathbb{R}$.

We define the workspace of the two-joint manipulator to be the reachable points by the end effector. The workspace for our two-joint manipulator is an annulus (figure 3.3), which is a subset of $\mathbb{R}^2$. All points in the interior of the annulus are reachable in two ways, with the arm in a *right-arm* and a *left-arm* configuration, sometimes called *elbow-up* and *elbow-down*. Therefore, the position of the end effector is *not* a valid configuration (not a complete description of the location of all points of the robot), so the annulus is not a configuration space for this robot.

So we have seen that the configuration spaces of both the translating mobile robot and the two-joint manipulator are two-dimensional, but they are quite different. The torus T^2 is doughnut-shaped with finite area, while $\mathbb{R}^2$ is flat with infinite area. We delve further into these sorts of differences when we discuss topology in section 3.4.

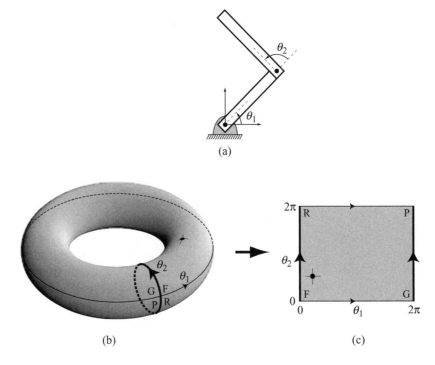

(a)

(b)   (c)

Figure 3.2 (a) A two-joint manipulator. (b) The configuration of the robot is represented as a point on the toral configuration space. (c) The torus can be cut and flattened onto the plane. This planar representation has "wraparound" features where the edge FR is connected to GP, etc.

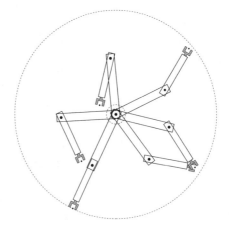

Figure 3.3 The workspace for this two-joint manipulator is an annulus, a disk with a smaller disk removed from it. Note that all points in the interior of the annulus are reachable with a right-arm configuration and a left-arm configuration.

3.2   Obstacles and the Configuration Space

Equipped with our understanding of configurations and of configuration spaces, we can define the path-planning problem to be that of determining a continuous mapping, $c : [0, 1] \rightarrow Q$, such that no configuration in the path causes a collision between the robot and an obstacle. It is useful to define explicitly the set of configurations for which such a collision occurs. We define a *configuration space obstacle* QO_i to be the set of configurations at which the robot intersects an obstacle WO_i in the workspace, i.e.,

$$QO_i = \{q \in Q \mid R(q) \bigcap WO_i \neq \emptyset\}.$$

The *free space* or *free configuration space* Q_{free} is the set of configurations at which the robot does not intersect any obstacle, i.e.,

$$Q_{\text{free}} = Q \backslash (\bigcup_i QO_i).$$

With this notation, we define a *free path* to be a continuous mapping $c : [0, 1] \rightarrow Q_{\text{free}}$, and a *semifree path* to be a continuous mapping $c : [0, 1] \rightarrow \text{cl}(Q_{\text{free}})$, in which $\text{cl}(Q_{\text{free}})$ denotes the closure of Q_{free}. A free path does not allow contact between the robot and obstacles, while a semifree path allows the robot to contact the boundary of an obstacle. We assume that Q_{free} is open unless otherwise noted.

We now examine how obstacles in the workspace can be mapped into the configuration space for the robots that we discussed above.

3.2.1   Circular Mobile Robot

Consider the circular mobile robot in an environment with a single polygonal obstacle in the workspace, as shown in figure 3.4. In figure 3.4(b), we slide the robot around the obstacle to find the constraints the obstacle places on the configuration of the robot, i.e., the possible locations of the robot's reference point. We have chosen to use the center of the robot, but could easily choose another point. Figure 3.4(c) shows the resulting obstacle in the configuration space. Motion planning for the circular robot in figure 3.4(a) is now equivalent to motion planning for a point in the configuration space, as shown in figure 3.4(c).

Figure 3.5 shows three mobile robots of different radii in the same environment. In each case, the robot is trying to find a path from one configuration to another. To transform the workspace obstacles into configuration obstacles, we "grow" the polygon outward and the walls inward. The problem is now to find a path for the point robot in the configuration space. We see that the growing process has disconnected the free configuration space Q_{free} for the largest robot, showing that there is no solution for this robot.

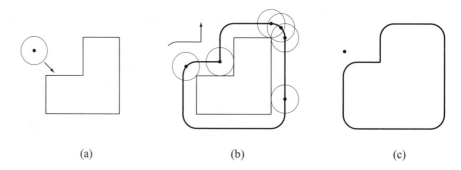

(a)   (b)   (c)

Figure 3.4   (a) The circular mobile robot approaches the workspace obstacle. (b) By sliding the mobile robot around the obstacle and keeping track of the curve traced out by the reference point, we construct the configuration space obstacle. (c) Motion planning for the robot in the workspace representation in (a) has been transformed into motion planning for a point robot in the configuration space.

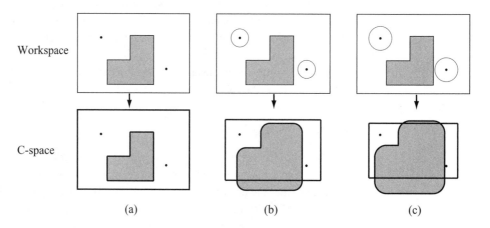

Figure 3.5   The top row shows the workspace and the bottom row shows the configuration space for (a) a point mobile robot, (b) a circular mobile robot, and (c) a larger circular mobile robot.

Although the example in figure 3.5 is quite simple, the main point is that it is easier to think about points moving around than bodies with volume. Keep in mind that although both the workspace and the configuration space for this system can be represented by $\mathbb{R}^2$, and the obstacles appear to simply "grow" in this example, the configuration space and workspace are different spaces, and the transformation from workspace obstacles to configuration space obstacles is not always so simple.

For example, appendix F discusses how to generate configuration space obstacles for a polygon that translates and rotates among polygonal obstacles in the plane. The two-joint arm example is examined next.

3.2.2   Two-Joint Planar Arm

For the case of the circular mobile robot in a world populated with polygonal obstacles, it is easy to compute configuration space obstacles. When the robot is even slightly more complex, it becomes much more difficult to do so. For this reason, grid-based representations of the configuration space are sometimes used. Consider the case of the two-joint planar arm, for which $\mathcal{Q} = T^2$. We can define a grid on the surface of the torus, and for each point on this grid we can perform a fairly simple test to see if the corresponding configuration causes a collision between the arm and any obstacle in the workspace. If we let each grid point be represented by a pixel, we can visualize the configuration space obstacle by "coloring" pixels appropriately. This method was used to obtain figures 3.6, 3.7, and 3.8.[2] In each of the figures, the image on the left depicts a two-joint arm in a planar workspace, while the image on the right depicts the configuration space. In each case, the arm on the left is depicted in several configurations, and these are indicated in the configuration spaces on the right.

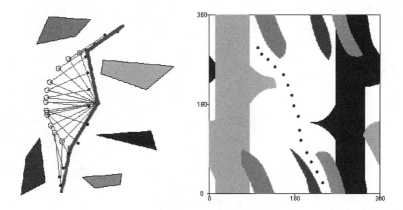

Figure 3.6   (Left) A path for the two-joint manipulator through its workspace, where the start and goal configurations are darkened. (Right) The path in the configuration space.

2. These figures were obtained using the applet at http://ford.ieor.berkeley.edu/cspace/.

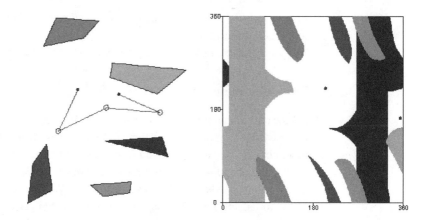

Figure 3.7   (Left) The workspace for a two-joint manipulator where the start and goal configurations are shown. (Right) The configuration space shows there is no free path between the start and goal configurations.

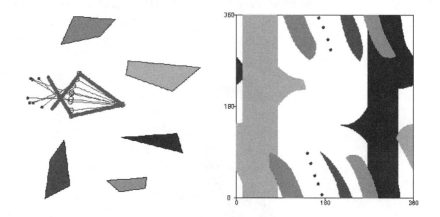

Figure 3.8   (Left) The workspace for a two-joint manipulator where the start and goal configurations are darkened. (Right) The path shows the wraparound of the planar representation of the configuration space.

While pictures such as those in figures 3.6, 3.7, and 3.8 are useful for visualizing configuration space obstacles, they are *not* sufficient for planning collision-free motions. The reason for this is that the grid only encodes collision information for the discrete set of points lying on the grid. A path includes not only grid points, but also the points on the curves that connect the grid points. One possible remedy for this problem is to "thicken" the robot when we test it at a grid point, so that if the thickened robot is collision-free, then paths to adjacent grid points are also collision-free.[3] We could also choose to ignore this problem by choosing a grid resolution that is "high enough."

3.3 The Dimension of the Configuration Space

In our introduction to configuration space above, we restricted our attention to two-dimensional configuration spaces that are easy to visualize. For each example, it was fairly straightforward to conclude that there were only two degrees of freedom: for a translating robot, the configuration was specified by a point in the familiar Euclidean plane, while for the two-joint arm the two joint angles gave a complete specification of the arm's position. In this section we determine the number of degrees of freedom of more complex systems by considering constraints on the motions of individual points of the systems.

As a first example, suppose the robot is a point that can move in the plane. The configuration can be given by two coordinates, typically $q = (x, y) \in Q = \mathbb{R}^2$, once we have chosen a reference coordinate frame fixed somewhere in space. Thus the robot has two degrees of freedom; the configuration space is two-dimensional.

Now consider a system consisting of three point robots, A, B, and C, that are free to move in the plane. Since the points can move independently, in order to specify the configuration of the system we need to specify the configuration of each of A, B, and C. By simply generalizing the case for a single point, we see that a configuration for this system can be given by the six coordinates x_A, y_A, x_B, y_B, x_C, and y_C (assuming that the points can overlap). The system has six degrees of freedom, and in this case we have $Q = \mathbb{R}^6$.

Real robots are typically modeled as a set of rigid bodies connected by joints (or a single rigid body for the case of most mobile robots), not a set of points that are free to move independently. So, suppose now that the robot is a planar rigid body that can both translate and rotate in the plane. Define A, B, and C to be three distinct points that are fixed to the body. To place the body in the plane, we are first free to choose the position of A by choosing its coordinates (x_A, y_A). Now we wish to choose the coordinates of B, (x_B, y_B), but the rigidity of the body requires that this point maintain a constant distance $d(A, B)$ from A:

$$d(A, B) = \sqrt{(x_A - x_B)^2 + (y_A - y_B)^2}.$$

---

3. This approach is called *conservative,* as a motion planner using this approach will never find an incorrect solution, but it might miss solutions when they exist. As a result, the planner can only be resolution complete, not complete.

This equation constrains B to lie somewhere on a circle of radius $d(A, B)$ centered at (x_A, y_A), and our only freedom in placing B is the angle θ from A to B.

Now when we try to choose coordinates (x_C, y_C) for C, we see that our choice is subject to two constraints:

$$d(A, C) = \sqrt{(x_A - x_C)^2 + (y_A - y_C)^2}$$
$$d(B, C) = \sqrt{(x_B - x_C)^2 + (y_B - y_C)^2}$$

In other words, C has already been placed for us. In fact, every point on the body has been placed once we have chosen (x_A, y_A, θ), making this a good representation of the configuration. The body has three degrees of freedom, and its configuration space is $\mathbb{R}^2 \times S^1$.

Each of the distance constraints above is an example of a *holonomic constraint*. A holonomic constraint is one that can be expressed purely as a function g of the configuration variables (and possibly time), i.e., of the form

$$g(q, t) = 0.$$

Each linearly independent holonomic constraint on a system reduces the dimension of the system's configuration space by one. Thus a system described by n coordinates subject to m independent holonomic constraints has an $(n - m)$-dimensional configuration space. In this case, we normally attempt to represent the configuration space by a smaller set of $n - m$ coordinates subject to no constraints, e.g., the coordinates (x_A, y_A, θ) for the planar body above.

Nonholonomic constraints are velocity constraints of the form

$$g(q, \dot{q}, t) = 0$$

which do not reduce the dimension of the configuration space. Nonholonomic constraints are left to chapter 12.

We can apply the counting method above to determine the number of degrees of freedom of a three-dimensional rigid body. Choose three noncollinear points on the body, A, B, C. The coordinates (x_A, y_A, z_A) of point A are first chosen arbitrarily. After fixing A, the distance constraint from A forces B to lie on the two-dimensional surface of a sphere centered at A. After both A and B are fixed, the distance constraints from A and B force C to lie on a one-dimensional circle about the axis formed by A and B. Once this point is chosen, all other points on the body are fixed. Thus the configuration of a rigid body in space can be described by nine coordinates subject to three constraints, yielding a six-dimensional configuration space.

We have already seen that a rigid body moving in a plane has three degrees of freedom, but we can arrive at this same conclusion if we imagine a spatial rigid body with six degrees of freedom with a set of constraints that restricts it to a plane. Choose this book as an example, using three corners of the back cover as points A, B and C. The book can be confined to a plane (e.g., the plane of a tabletop) using the three holonomic constraints

$$z_A = z_B = z_C = 0.$$

The two-joint planar arm can be shown to have two degrees of freedom by this (somewhat indirect) counting method. Each of the two links can be thought of as a rigid spatial body with six degrees of freedom. Six constraints restrict the bodies to a plane (three for each link), two constraints restrict a point on the first link (the first joint) to be at a fixed location in the plane, and once the angle of the first link is chosen, two constraints restrict a point on the second link (the second joint) to be at a fixed location. Therefore we have (12 coordinates) − (10 constraints) = 2 degrees of freedom.

Of course we usually count the degrees of freedom of an open-chain jointed robot, also known as a *serial mechanism,* by adding the degrees of freedom at each joint. Common joints with one degree of freedom are *revolute* (R) joints, joints which rotate about an axis, and *prismatic* (P) joints, joints which allow translational motion along an axis. Our two-joint robot is sometimes called an RR or 2R robot, indicating that both joints are revolute. An RP robot, on the other hand, has a revolute joint followed by a prismatic joint. Another common joint is a *spherical* (ball-and-socket) joint, which has three degrees of freedom.

A *closed-chain* robot, also known as a *parallel mechanism,* is one where the links form one or more closed loops. If the mechanism has k links, then one is designated as a stationary "ground" link, and $k − 1$ links are movable. To determine the number of degrees of freedom, note that each movable link has N degrees of freedom, where $N = 6$ for a spatial mechanism and $N = 3$ for a planar mechanism. Therefore the system has $N(k − 1)$ degrees of freedom before the joints are taken into account. Now each of the n joints between the links places $N − f_i$ constraints on the feasible motions of the links, where f_i is the number of degrees of freedom at joint i (e.g., $f_i = 1$ for a revolute joint, and $f_i = 3$ for a spherical joint). Therefore, the *mobility* M of the mechanism, or its number of degrees of freedom, is given by

$$M = N(k − 1) − \sum_{i=1}^{n}(N − f_i)$$

(3.1)
$$= N(k − n − 1) + \sum_{i=1}^{n} f_i.$$

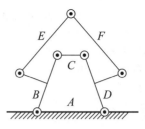

Figure 3.9   A planar mechanism with six links (*A* through *F*), seven revolute joints, and one degree of freedom.

This is known as *Grübler's formula* for closed chains, and it is only valid if the constraints due to the joints are independent. In the planar mechanism of figure 3.9, there are seven joints, each with one degree of freedom, and six links, yielding a mobility of $3(6 - 7 - 1) + 7 = 1$.

As an application of the ideas in this section, determine the number of degrees of freedom of your arm by adding the degrees of freedom at your shoulder, elbow, and wrist. To test your answer, place your palm flat down on a table with your elbow bent. Without moving your torso or your palm, you should find that it is still possible to move your arm. This internal freedom means that your arm is *redundant* with respect to the task of positioning your hand (or a rigid body grasped by your hand) in space; an infinity of arm configurations puts your hand in the same place.[4] How many internal degrees of freedom do you have? How many holonomic constraints were placed on your arm's configuration when you fixed your hand's position and orientation? The sum of the number of constraints and internal degrees of freedom is the number of degrees of freedom of your (unconstrained) arm, and you should find that your arm has more than six degrees of freedom. A robot is said to be *hyper-redundant* with respect to a task if it has many more degrees of freedom than required for the task. (There is no strict definition of "many" here.)

3.4   The Topology of the Configuration Space

Now that we understand how to determine the dimension of a configuration space, we can begin to explore its topology and geometry, each of which plays a vital role in developing and analyzing motion-planning algorithms. Some basic concepts from topology are discussed in appendixes B and C.

---

4. Provided your arm is away from its joint limits.

Figure 3.10 The surfaces of the coffee mug and the torus are topologically equivalent.

Topology is a branch of mathematics that considers properties of objects that do not change when the objects are subjected to arbitrary continuous transformations, such as stretching or bending. For this reason, topology is sometimes referred to as "rubber sheet geometry." Imagine a polygon drawn on a rubber sheet. As the sheet is stretched in various directions, the polygon's shape changes; however, certain properties of the polygon do not change. For example, points that are inside the polygon do not move to the outside of the polygon simply because the sheet is stretched.

Two spaces are topologically different if cutting or pasting is required to turn one into the other, as cutting and pasting are not continuous transformations. For example, the configuration spaces of the circular mobile robot ($\mathbb{R}^2$) and the two-joint planar arm (T^2) are topologically different. If we imagine T^2 as the surface of a rubber doughnut, we see that no matter how we stretch or deform the doughnut (without tearing it), the doughnut will always have a hole in it. Also, if we imagine $\mathbb{R}^2$ as an infinite rubber sheet, there is no way to stretch it (without tearing it) such that a hole will appear in the sheet. To a topologist, all rubber doughnuts are the same, regardless of how they are stretched or deformed (figure 3.10). Likewise, all rubber sheet versions of $\mathbb{R}^2$ are the same.

One reason that we care about the topology of configuration space is that it will affect our representation of the space. Another reason is that if we can derive a path-planning algorithm for one kind of topological space, then that algorithm may carry over to other spaces that are topologically equivalent (see, e.g., chapter 4, section 4.6).

Since topology is concerned with properties that are preserved under continuous transformations, we begin our study of the topology of configuration spaces by describing two types of continuous transformations: homeomorphisms and diffeomorphisms. Appendix C provides an introduction to differentiable transformations.

3.4.1 Homeomorphisms and Diffeomorphisms

A mapping $\phi : S \rightarrow T$ is a rule that places elements of S into correspondence with elements of T. We respectively define the *image* of S under ϕ and the *preimage*

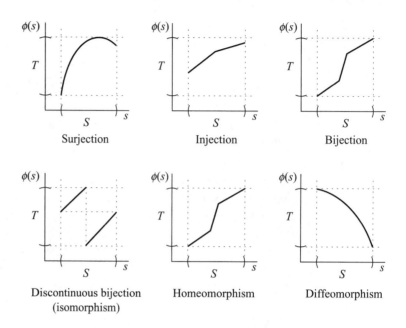

Figure 3.11   Representative ways of looking at surjective, injective, and bijective mappings. Bijections may become homeomorphisms or diffeomorphisms if they are sufficiently differentiable.

of T by

$$\phi(S) = \{\phi(s) \mid s \in S\} \quad \text{and} \quad \phi^{-1}(T) = \{s \mid \phi(s) \in T\}.$$

If $\phi(S) = T$ (i.e., every element of T is covered by the mapping) then we say that ϕ is *surjective* or *onto*. If ϕ puts each element of T into correspondence with at most one element of S, i.e., for any $t \in T$, $\phi^{-1}(t)$ consists of at most one element in S, then we say that ϕ is *injective (one-to-one)*. If ϕ is injective, then when $s_1 \neq s_2$ we have $\phi(s_1) \neq \phi(s_2)$ for $s_1, s_2 \in S$. Maps that are both surjective and injective are said to be *bijective*. Figure 3.11 illustrates these definitions. As another example, the map $\sin : (-\frac{\pi}{2}, \frac{\pi}{2}) \to (-1, 1)$ is bijective, whereas $\sin : \mathbb{R} \to [-1, 1]$ is only surjective. Bijective maps have the property that their inverse exists at all points in the range T, and thus they allow us to move easily back and forth between the two spaces S and T. In our case, we will use bijective maps to move back and forth between configuration spaces (whose geometry can be quite complicated) and Euclidean spaces.

We will consider two important classes of bijective mappings.

DEFINITION 3.4.1   *If $\phi : S \to T$ is a bijection, and both ϕ and ϕ^{-1} are continuous, then ϕ is a* homeomorphism. *When such a ϕ exists, S and T are said to be* homeomorphic.

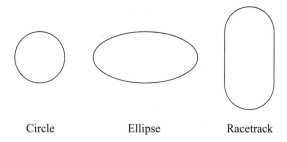

Circle   Ellipse   Racetrack

Figure 3.12   A circle, an ellipse, and a racetrack.

A mapping $\phi : U \to V$ is said to be *smooth* if all partial derivatives of ϕ, of all orders, are well defined (i.e., ϕ is of class C^∞). With the notion of smoothness, we define a second type of bijection.

DEFINITION 3.4.2   *A smooth map $\phi : U \to V$ is a* diffeomorphism *if ϕ is bijective and ϕ^{-1} is smooth. When such a ϕ exists, U and V are said to be* diffeomorphic.

The condition for diffeomorphisms (smoothness) is stronger that the condition for homeomorphisms (continuity), and thus all diffeomorphisms are homeomorphisms.

To illustrate these ideas, consider three one-dimensional surfaces: a circle, denoted by $\mathcal{M}_c$; an ellipse, denoted by $\mathcal{M}_e$; and a "racetrack," denoted by $\mathcal{M}_r$. The racetrack consists of two half-circles connected by straight lines (figure 3.12). We define these shapes mathematically as

(3.2) $\mathcal{M}_c = \{(x, y) \mid f_c(x, y) = x^2 + y^2 - 1 = 0\}$

(3.3) $\mathcal{M}_e = \{(x, y) \mid f_e(x, y) = \dfrac{x^2}{4} + y^2 - 1 = 0\}$

(3.4) $\mathcal{M}_r = \{(x, y) \mid f_r(x, y) = 0\}$

with

(3.5) $f_r(x, y) = \begin{cases} x - 1 & : -1 \le y \le 1, \ x > 0 \\ (y + 1)^2 + x^2 - 1 : & y < -1 \\ (y - 1)^2 + x^2 - 1 : & y > 1 \\ x + 1 & : -1 \le y \le 1, \ x < 0 \end{cases}.$

Note that these surfaces are implicitly defined as being the set of points that satisfy some equation $f(x, y) = 0$.

In some ways, these three surfaces are similar. For example, they are all simple, closed curves in the plane; all of $f_c(x, y)$, $f_e(x, y)$, and $f_r(x, y)$ are continuous. In other ways, they seem quite different. For example, both $f_c(x, y)$ and $f_e(x, y)$ are continuously differentiable, while $f_r(x, y)$ is not. We can more precisely state the similarities and differences between these surfaces using the concepts of homeomorphism and diffeomorphism. In particular, it can be shown that $\mathcal{M}_c$, $\mathcal{M}_e$, and $\mathcal{M}_r$ are all homeomorphic to each other. For example, the map $\phi : \mathcal{M}_e \rightarrow \mathcal{M}_c$ given by

$$\phi(x, y) = \left[\frac{x}{\sqrt{x^2 + y^2}}, \frac{y}{\sqrt{x^2 + y^2}} \right]^T$$

is a homeomorphism.

For this choice of ϕ, both ϕ and ϕ^{-1} are smooth, and therefore, $\mathcal{M}_c$ is diffeomorphic to $\mathcal{M}_e$. Neither $\mathcal{M}_c$ nor $\mathcal{M}_e$ is diffeomorphic to $\mathcal{M}_r$, however. This is because $f_r(x, y)$ is not continuously differentiable, while both $f_c(x, y)$ and $f_e(x, y)$ are. This can be seen by examining the curvatures of the circle, ellipse, and racetrack. For the circle, the curvature is constant (and thus continuous), and for the ellipse, curvature is continuous. For the racetrack, there are discontinuities in curvature (at the points $(-1, 1), (-1, -1), (1, 1), (1, -1)$), and therefore there is no smooth mapping from either the circle or the ellipse to the racetrack.

We are often concerned only with the local properties of configuration spaces. Local properties are defined on *neighborhoods*. For metric spaces[5], neighborhoods are most easily defined in terms of open balls. For a point p of some manifold $\mathcal{M}$, we define an open ball of radius ϵ by

$$B_\epsilon(p) = \{p' \in \mathcal{M} \mid d(p, p') < \epsilon\},$$

where d is a metric on $\mathcal{M}$.[6] A neighborhood of a point $p \in \mathcal{M}$ is any subset $\mathcal{U} \subseteq \mathcal{M}$ with $p \in \mathcal{U}$ such that for every $p' \in \mathcal{U}$, there exists an open ball $B_\epsilon(p') \subset \mathcal{U}$. Any open ball is itself a neighborhood. The open disk in the plane is an example. For the point (x_0, y_0) in the plane, an open ball defined by the Euclidean metric is

$$B_\epsilon(x_0, y_0) = \{(x, y) \mid (x - x_0)^2 + (y - y_0)^2 < \epsilon^2\}.$$

We say that S is *locally diffeomorphic* (resp. *locally homeomorphic*) to T if for each $p \in S$ there exists a diffeomorphism (resp. homeomorphism) f from S to T on some neighborhood $\mathcal{U}$ with $p \in \mathcal{U}$.

The sphere presents a familiar example of these concepts. At any point on the sphere, there exists a neighborhood of that point that is diffeomorphic to the plane. It is not surprising that people once believed the world was flat — they were only looking at their neighborhoods!

---

5. A metric space is a space equipped with a distance metric. See appendix C.
6. One can define all topological properties, including neighborhoods, without resorting to the use of metrics, but for our purposes, it will be easier to assume a metric on the configuration space and exploit the metric properties.

Let us now reflect on the two examples from the beginning of this chapter. For the circular mobile robot, the workspace and the configuration space are diffeomorphic. This is easy to see, since both are copies of $\mathbb{R}^2$. In this case, the identity map $\phi(x) = x$ is a perfectly fine global diffeomorphism between the workspace and configuration space. In contrast, the two-joint manipulator has a configuration space that is T^2, the torus. The torus T^2 is not diffeomorphic to $\mathbb{R}^2$, but it is locally diffeomorphic. If the revolute joints in the two-joint manipulator have lower and upper limits, $\theta_i^\ell < \theta_i < \theta_i^u$, so that they cannot perform a complete revolution, however, then the configuration space of the two-joint manipulator becomes an open subset of the torus, which *is* diffeomorphic to $\mathbb{R}^2$ (globally). This follows from the fact that each joint angle lies in an *open* interval of $\mathbb{R}^1$, and we can "stretch" that open interval to cover the line. An example of such a stretching function is $\tan : (-\frac{\pi}{2}, \frac{\pi}{2}) \to \mathbb{R}$.

3.4.2 Differentiable Manifolds

For all of the configuration spaces that we have seen so far, we have been able to uniquely specify a configuration by n parameters, where n is the dimension of the configuration space (two for the planar two-joint arm, three for a polygon in the plane, etc.). The reason that we could do so was that these configuration spaces were all "locally like" n-dimensional Euclidean spaces. Such spaces, called *manifolds,* are a central topic of topology.

DEFINITION 3.4.3 (Manifold) *A set S is a k-dimensional manifold if it is locally homeomorphic to $\mathbb{R}^k$, meaning that each point in S possesses a neighborhood that is homeomorphic to an open set in $\mathbb{R}^k$.*

While a general k-dimensional manifold is locally homeomorphic to $\mathbb{R}^k$, the configuration spaces that we will consider are locally diffeomorphic to $\mathbb{R}^k$, an even stronger relationship. In fact, when we parameterized configurations in section 3.1, we were merely constructing diffeomorphisms from configuration spaces to $\mathbb{R}^2$. If we construct enough of these diffeomorphisms (so that every configuration in Q is in the domain of at least one of them), and if these diffeomorphisms are compatible with one another (an idea that we will formalize shortly), then this set of diffeomorphisms together with the configuration space define a *differentiable manifold*. We now make these ideas more precise.

DEFINITION 3.4.4 (Chart) *A pair (U, ϕ), such that U is an open set in a k-dimensional manifold and ϕ is a diffeomorphism from U to some open set in $\mathbb{R}^k$, is called a* chart.

The use of the term *chart* is analogous to its use in cartography, since the subset U is "charted" onto $\mathbb{R}^k$ in much the same way that cartographers chart subsets of the globe onto a plane when creating maps. Charts are sometimes referred to as *coordinate systems* because each point in the set U is assigned a set of coordinates in a Euclidean space [?]. The inverse diffeomorphism, $\phi^{-1} : \mathbb{R}^k \to U$, is referred to as a *parameterization* of the manifold.

As an example, consider the one-dimensional manifold $S^1 = \{x = (x_1, x_2) \in \mathbb{R}^2 \mid x_1^2 + x_2^2 = 1\}$. For any point $x \in S^1$ we can define a neighborhood that is diffeomorphic to $\mathbb{R}$. For example, consider the upper portion of the circle, $U_1 = \{x \in S^1 \mid x_2 > 0\}$. The chart $\phi_1 : U_1 \to (-1, 1)$ is given by $\phi_1(x) = x_1$, and thus x_1 can be used to define a local coordinate system for the upper semicircle. In the other direction, the upper portion of the circle can be parameterized by $z \in (-1, 1) \subset \mathbb{R}$, with $\phi_1^{-1}(z) = (z, (1 - z^2)^{\frac{1}{2}})$, which maps the open unit interval to the upper semicircle. But S^1 is not globally diffeomorphic to $\mathbb{R}^1$; we cannot find a single chart whose domain includes all of S^1.

We have already used this terminology in section 3.1, when we referred to θ_1, θ_2 as parameters that represent a configuration of the two-joint arm. Recall that $(\theta_1, \theta_2) \in \mathbb{R}^2$, and that when considered as a representation of the configuration, they define a point in T^2, the configuration space, which is a manifold. We now see that in section 3.1, when we represented a configuration of the planar arm by the pair (θ_1, θ_2), we were in fact creating a chart from a subset of the configuration space to a subset of $\mathbb{R}^2$.

A single mapping from T^2 to $\mathbb{R}^2$ shown in figure 3.2 encounters continuity problems at $\theta_i = \{0, 2\pi\}$. For many interesting configuration spaces, it will be the case that we cannot construct a single chart whose domain contains the entire configuration space. In these cases, we construct a collection of charts that cover the configuration space. We are not free to choose these charts arbitrarily; any two charts in this collection must be compatible for parts of the manifold on which their domains overlap. Two charts with such compatibility are said to be C^∞-related (figure 3.13).

DEFINITION 3.4.5 (C^∞-related)   *Let (U, ϕ) and (V, ψ) be two charts on a k-dimensional manifold. Let X be the image of $U \cap V$ under ϕ, and Y be the image of $U \cap V$ under ψ, i.e.,*

$$X = \{\phi(x) \in \mathbb{R}^k \mid x \in U \cap V\}$$
$$Y = \{\psi(y) \in \mathbb{R}^k \mid y \in U \cap V\}.$$

These two charts are said to be C^∞-related if both of the composite functions

$$\psi \circ \phi^{-1} : X \to Y,$$
$$\phi \circ \psi^{-1} : Y \to X$$

are C^∞.

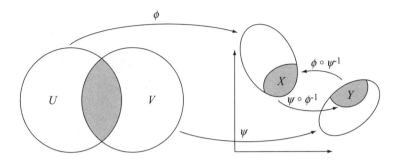

Figure 3.13   The charts (U, ϕ) and (V, ψ) map open sets on the k-dimensional manifold to open sets in $\mathbb{R}^k$.

If two charts are C^∞-related, we can switch back and forth between them in a smooth way when their domains overlap. This idea will be made more concrete in the example of S^1 below.

A set of charts that are C^∞-related, and whose domains cover the entire configuration space $\mathcal{Q}$, form an *atlas* for $\mathcal{Q}$. An atlas is sometimes referred to as the differentiable structure for $\mathcal{Q}$. Together, the atlas and $\mathcal{Q}$ comprise a *differentiable manifold*. There are other ways to define differentiable manifolds, as we will see in section 3.5.

As an example, consider again the one-dimensional manifold S^1. Above, we defined a single chart, (U_1, ϕ_1). If we define three more charts, we can construct an atlas for S^1. These four charts are given by

$$U_1 = \{x \in S^1 \mid x_2 > 0\}, \quad \phi_1(x) = x_1$$
$$U_2 = \{x \in S^1 \mid x_2 < 0\}, \quad \phi_2(x) = x_1$$
$$U_3 = \{x \in S^1 \mid x_1 > 0\}, \quad \phi_3(x) = x_2$$
$$U_4 = \{x \in S^1 \mid x_1 < 0\}, \quad \phi_4(x) = x_2.$$

The corresponding parameterizations are given by $\phi_i^{-1} : (-1, 1) \to U_i$, with

$$\phi_1^{-1}(z) = (z, (1 - z^2)^{\frac{1}{2}})$$
$$\phi_2^{-1}(z) = (z, (z^2 - 1)^{\frac{1}{2}})$$
$$\phi_3^{-1}(z) = ((1 - z^2)^{\frac{1}{2}}, z)$$
$$\phi_4^{-1}(z) = ((z^2 - 1)^{\frac{1}{2}}, z).$$

It is clear that the U_i cover S^1, so to verify that these charts form an atlas it is only necessary to show that they are C^∞-related (figure 3.14). Note that

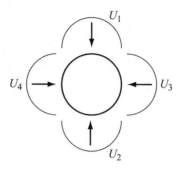

Figure 3.14   Four charts covering the circle S^1.

$U_1 \cap U_2 = U_3 \cap U_4 = \emptyset$, so we need only check the four pairs of composite maps:

$$\phi_1 \circ \phi_3^{-1} : (0, 1) \rightarrow (0, 1), \qquad \phi_3 \circ \phi_1^{-1} : (0, 1) \rightarrow (0, 1)$$
$$\phi_1 \circ \phi_4^{-1} : (0, 1) \rightarrow (-1, 0), \qquad \phi_4 \circ \phi_1^{-1} : (-1, 0) \rightarrow (0, 1)$$
$$\phi_2 \circ \phi_3^{-1} : (-1, 0) \rightarrow (0, 1), \qquad \phi_3 \circ \phi_2^{-1} : (0, 1) \rightarrow (-1, 0)$$
$$\phi_2 \circ \phi_4^{-1} : (-1, 0) \rightarrow (-1, 0), \qquad \phi_4 \circ \phi_2^{-1} : (-1, 0) \rightarrow (-1, 0).$$

In each case, $\phi_i \circ \phi_j^{-1}(z)$ is smooth on each of the open unit intervals that define the domains for the composite mappings given above. For example, $\phi_1 \circ \phi_3^{-1}(z) = (1 - z^2)^{\frac{1}{2}}$.

This collection of four charts is not minimal; it is straightforward to find two charts to cover S^1 (see problem 9).

3.4.3   Connectedness and Compactness

We say that a manifold is *path-connected,* or just *connected,* if there exists a path between any two points of the manifold.[7] All of the obstacle-free configuration spaces that we will consider in this text, e.g., $\mathbb{R}^n$, S^n, and T^n, are connected. The presence of obstacles, however, can disconnect the free configuration space $\mathcal{Q}_{\text{free}}$. In this case, the free configuration space is broken into a set of *connected components,* the maximal connected subspaces. In figure 3.5(c), for example, obstacles break the mobile robot's free configuration space into two connected components. There can be no solution to a motion-planning problem if q_{start} and q_{goal} do not lie in the same connected component of $\mathcal{Q}_{\text{free}}$.

---

7. For more general spaces, the concepts of path-connectedness and connectedness are not equivalent, but for a manifold they are the same. More generally, a space is connected if there is no continuous mapping from the space to a discrete set of more than one element.

A space is *compact*[8] if it resembles a closed, bounded subset of $\mathbb{R}^n$. A space is closed if it includes all of its limit points. As examples, the half-open interval $[0, 1) \subset \mathbb{R}$ is bounded but not compact, while the closed interval $[0, 1]$ is bounded and compact. The space $\mathbb{R}^n$ is not bounded and therefore not compact. The spaces S^n and T^n are both compact, as they can be expressed as closed and bounded subsets of Euclidean spaces. The unit circle S^1, e.g., can be expressed as a closed and bounded subset of $\mathbb{R}^2$.

In configuration spaces with obstacles or joint limits, the modeling of the obstacles may affect whether the space is compact or not. For a revolute joint subject to joint limits, the set of joint configurations is compact if the joint is allowed to hit the limits, but not compact if the joint can only approach the limits arbitrarily closely.

The product of compact configuration spaces is also compact. For a noncompact space $\mathcal{M}_1 \times \mathcal{M}_2$, if $\mathcal{M}_1$ is compact, then it is called the *compact factor* of the space. Compact and noncompact spaces cannot be diffeomorphic.

3.4.4 Not All Configuration Spaces Are Manifolds

We are focusing on configuration spaces that are manifolds, and more specifically differentiable manifolds, but it is important to keep in mind that not all configuration spaces are manifolds. As a simple example, the closed unit square $[0, 1] \times [0, 1] \subset \mathbb{R}^2$ is not a manifold, but a *manifold with boundary* obtained by pasting the one-dimensional boundary on the two-dimensional open set $(0, 1) \times (0, 1)$. Also, some parallel mechanisms with one degree of freedom have configurations from which there are two distinct possible motion directions (i.e., the configuration space is a self-intersecting figure eight). It is beyond the scope of this chapter to discuss other types of configuration spaces, but be aware: if you cannot show it to be a manifold, it may not be!

3.5 Embeddings of Manifolds in $\mathbb{R}^n$

Although a k-dimensional manifold can be represented using as few as k parameters, we have seen above that doing so may require multiple charts. An alternative is to use a representation with "extra" numbers, subject to constraints, to achieve a single global representation. As an example, S^1 is a one-dimensional manifold that we can

---

8. In topology, a space is defined to be compact if every open cover of the space admits a finite subcover, but we will not use these concepts here.

represent as $S^1 = \{(x, y) \mid x^2 + y^2 = 1\}$; we "embed" S^1 in $\mathbb{R}^2$. The fact that we cannot find a single chart for all of S^1 tells us that we cannot embed S^1 in $\mathbb{R}^1$. Likewise, although the torus T^2 is a two-dimensional manifold, it is not possible to embed the torus in $\mathbb{R}^2$, which is why we typically illustrate the torus as a doughnut shape in $\mathbb{R}^3$.

The manifolds S^1 and T^2 can be viewed as *submanifolds* of $\mathbb{R}^2$ and $\mathbb{R}^3$, respectively. Submanifolds are smooth subsets of an ambient space that inherit the differentiability properties of the ambient space. Submanifolds are often created by a smooth set of equality constraints on $\mathbb{R}^n$, as we see in the example of the circle S^1 above. Any differentiable manifold can be viewed as an embedded submanifold of $\mathbb{R}^n$ for large enough n.

When we are confronted with a configuration space that does not permit a single global coordinate chart, we are faced with a choice. We can either use a single set of parameters and suffer the consequences of singularities and discontinuities in the representation, use multiple charts to construct an atlas, or use a single global representation by embedding the configuration space in a higher-dimensional space. One advantage of the last approach is that the representation can facilitate other operations. Important examples are representations of orientation using complex numbers and quaternions (see appendix E) and matrix representations for the configuration of a rigid body in the plane or in space, as discussed next.

3.5.1   Matrix Representations of Rigid-Body Configuration

It is often convenient to represent the position and orientation of a rigid body using an $m \times m$ matrix of real numbers. The m^2 entries of this matrix must satisfy a number of smooth equality constraints, making the manifold of such matrices a submanifold of $\mathbb{R}^{m^2}$. One advantage of such a representation is that these matrices can be multiplied to get another matrix in the manifold. More precisely, these matrices form a *group* with the group operation of matrix multiplication.[9] Simple matrix multiplication can be used to change the reference frame of a representation or to rotate and translate a configuration.

We describe the orientation of a rigid body in n-dimensional space ($n = 2$ or 3) by the matrix groups $SO(n)$, and the position and orientation by the matrix groups $SE(n)$. After describing these representations abstractly, we look in detail at examples

---

9. In fact, the matrix representations in this section are *Lie groups,* as (1) they are differentiable manifolds which are also groups, (2) the group operation is C^∞, and (3) the mapping from an element of the group to its inverse is C^∞.

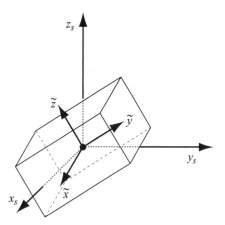

Figure 3.15   The rotation matrix for a body is obtained by expressing the unit vectors $\tilde{x}$, $\tilde{y}$, and $\tilde{z}$ of the body x-y-z frame in a stationary frame x_s-y_s-z_s.

of the use of $SE(n)$ for representing the configuration of a body, for changing the reference frame of the representation, and for translating and rotating a configuration.

Orientation: $SO(2)$ and $SO(3)$

Figure 3.15 shows a rigid body with a frame x-y-z attached to it. Our representation of the orientation of the body will be as a 3×3 matrix

$$R = \begin{bmatrix} \tilde{x}_1 & \tilde{y}_1 & \tilde{z}_1 \\ \tilde{x}_2 & \tilde{y}_2 & \tilde{z}_2 \\ \tilde{x}_3 & \tilde{y}_3 & \tilde{z}_3 \end{bmatrix} = \begin{bmatrix} R_{11} & R_{12} & R_{13} \\ R_{21} & R_{22} & R_{23} \\ R_{31} & R_{32} & R_{33} \end{bmatrix} \in SO(3),$$

where $\tilde{x} = [\tilde{x}_1, \tilde{x}_2, \tilde{x}_3]^T$ is the unit vector in the body x-direction expressed in a stationary coordinate frame x_s-y_s-z_s. The vectors $\tilde{y}$ and $\tilde{z}$ are defined similarly (see figure 3.15).

The matrix R is often called the *rotation matrix* representation of the orientation. This representation uses nine numbers to represent the three angular degrees of freedom, so there are six independent constraints on the matrix entries: each column (and row) is a unit vector,

$$\|\tilde{x}\| = \|\tilde{y}\| = \|\tilde{z}\| = 1,$$

yielding three constraints, and the columns (and rows) are orthogonal to each other,

$$\tilde{x}^T \tilde{y} = \tilde{y}^T \tilde{z} = \tilde{z}^T \tilde{x} = 0,$$

yielding three more constraints. Because we are assuming right-handed frames,[10] the determinant of R is $+1$. Matrices satisfying these conditions belong to the *special orthogonal group* of 3×3 matrices $SO(3)$. "Special" refers to the fact that the determinant is $+1$, not -1.

In the planar case, R is the 2×2 matrix

$$R = \begin{bmatrix} \tilde{x}_1 & \tilde{y}_1 \\ \tilde{x}_2 & \tilde{y}_2 \end{bmatrix} = \begin{bmatrix} \cos\theta & -\sin\theta \\ \sin\theta & \cos\theta \end{bmatrix} \in SO(2),$$

where θ is the orientation of the x-y frame relative to the x_s-y_s frame.

Generalizing, orientations in n-dimensional space can be written

$$SO(n) = \{R \in \mathbb{R}^{n \times n} \mid RR^T = \mathcal{I}, \ \det(R) = 1\},$$

where $\mathcal{I}$ is the identity matrix. This implies the relation

$$R^T = R^{-1}.$$

Position and Orientation: $SE(2)$ and $SE(3)$

Figure 3.16 shows a rigid body with an attached x-y-z coordinate frame relative to a stationary frame x_s-y_s-z_s. Let $p = [p_1, p_2, p_3]^T \in \mathbb{R}^3$ be the vector from the origin of the stationary frame to the body frame, as measured in the stationary frame, and let $R \in SO(3)$ be the rotation matrix as described above, as if the body frame were translated back to the stationary frame. Then we represent the position and orientation of the body frame relative to the stationary frame as the 4×4 transform matrix

$$T = \begin{bmatrix} R & p \\ 0 & 1 \end{bmatrix} \in SE(3),$$

where the bottom row consists of three zeros and a one. (These "extra" numbers will be needed to allow us to perform matrix multiplications, as we will see shortly.) Since R and p both have three degrees of freedom, the configuration of a rigid body in three-space has six degrees of freedom, as we discovered earlier in the chapter.

Generalizing, the position and orientation of a rigid body in n-dimensional space can be written as an element of the *special Euclidean group $SE(n)$*:

$$SE(n) \equiv \begin{bmatrix} SO(n) & \mathbb{R}^n \\ 0 & 1 \end{bmatrix},$$

where the bottom row consists of n zeros and a one.

---

10. To make a right-handed frame, point straight ahead with your right index finger, point your middle finger 90 degrees to the left, and stick your thumb straight up. Your index finger is pointing in the $+x$ direction, your middle finger is pointing in the $+y$ direction, and your thumb is pointing in the $+z$ direction.

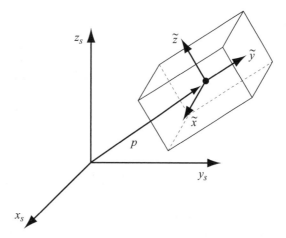

Figure 3.16 The body frame x-y-z relative to a stationary world frame x_s-y_s-z_s.

Uses of the Matrix Representations

The matrix groups $SO(n)$ and $SE(n)$ can be used to

1. represent rigid-body configurations,

2. change the reference frame for the representation of a configuration or a point, and

3. displace (move) a configuration or a point.

When the matrix is used for representing a configuration, we often call it a frame. When it is used for displacement or coordinate change, we often call it a transform. The various uses are best demonstrated by example.

Figure 3.17 shows three coordinate frames on a regular grid of unit spacing. These frames are confined to a plane with their z-axes pointing out of the page. Let T_{AB} be the configuration of frame B relative to frame A, and let T_{BC} be the configuration of frame C relative to frame B. It is clear from the figure that

$$T_{AB} = \begin{bmatrix} R_{AB} & p_{AB} \\ 0 & 1 \end{bmatrix} = \begin{bmatrix} -1 & 0 & 0 & -2 \\ 0 & -1 & 0 & 0 \\ 0 & 0 & 1 & 0 \\ 0 & 0 & 0 & 1 \end{bmatrix}, \quad T_{BC} = \begin{bmatrix} 0 & 1 & 0 & -4 \\ -1 & 0 & 0 & -1 \\ 0 & 0 & 1 & 0 \\ 0 & 0 & 0 & 1 \end{bmatrix}.$$

From these, we can find T_{AC}, the frame C relative to the frame A, by performing a change of reference frame on T_{BC}. This involves premutliplying by T_{AB}, based on

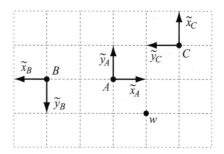

Figure 3.17   Three frames in a plane with their z-axes pointing out of the page.

the rule for coordinate transformations that the second subscript of the matrix on the left cancels with the first subscript of the matrix on the right, if they are the same subscript. In other words,

$$T_{AB}T_{BC} = T_{A\cancel{B}}T_{\cancel{B}C} = T_{AC}.$$

We find that

$$T_{AC} = \begin{bmatrix} -1 & 0 & 0 & -2 \\ 0 & -1 & 0 & 0 \\ 0 & 0 & 1 & 0 \\ 0 & 0 & 0 & 1 \end{bmatrix} \begin{bmatrix} 0 & 1 & 0 & -4 \\ -1 & 0 & 0 & -1 \\ 0 & 0 & 1 & 0 \\ 0 & 0 & 0 & 1 \end{bmatrix} = \begin{bmatrix} 0 & -1 & 0 & 2 \\ 1 & 0 & 0 & 1 \\ 0 & 0 & 1 & 0 \\ 0 & 0 & 0 & 1 \end{bmatrix},$$

which we can verify by inspection.

The representation of the point w in the coordinates of frame C is written w_C. From figure 3.17, we can see that the coordinates of w in C are $[-2, 1, 0]^T$. To facilitate matrix multiplications, however, we will express points in *homogeneous coordinates* by appending a 1 to the end of the vector, i.e.,

$$w_C = [-2, 1, 0, 1]^T.$$

To find the representation of the point w in other frames, we use a modification of the subscript canceling rule to get

$$T_{BC}w_C = T_{B\cancel{C}}w_{\cancel{C}} = w_B = [-3, 1, 0, 1]^T$$

and

$$T_{AB}T_{BC}w_C = T_{A\cancel{C}}w_{\cancel{C}} = w_A = [1, -1, 0, 1]^T,$$

which can be verified by inspection.

Elements of $SE(n)$ can also be used to displace a point. For example, $T_{AB}w_A$ does not satisfy the subscript canceling rule, and the result is not simply a representation

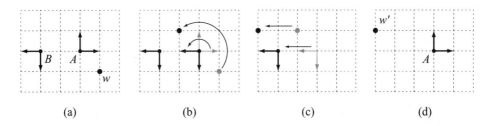

(a)   (b)   (c)   (d)

Figure 3.18 Displacing a point by the transformation T_{AB}. (a) The frames A and B and the point w. (b) Rotating frame A to the orientation of frame B, carrying the point w along with it. (c) Translating frame A to the location of frame B, carrying w along with it. (d) The final point $w'_A = T_{AB}w_A$.

of the point w in a new frame. Instead, the point w is rotated about the origin of the frame A by R_{AB} (expressed in the A frame), and then translated by p_{AB} in the A frame. This is the same motion required to take frame A to frame B. The result is

$$w'_A = T_{AB}w_A = [-3, 1, 0, 1]^T,$$

the location of the transformed point in the frame A. This transformation is shown graphically in figure 3.18.

Finally, we can use elements of $SE(3)$ to displace frames, not just points. For example, given a frame B represented by T_{AB} relative to frame A, and a transform $T_1 \in SE(3)$, then

$$T_{AB'} = T_{AB}T_1 = \begin{bmatrix} R_{AB}R_1 & R_{AB}p_1 + p_{AB} \\ 0 & 1 \end{bmatrix}$$

is the representation of the transformed frame B' relative to A after rotating B about its origin by R_1 (expressed in the B frame) and then translating by p_1 in the original B frame (before it was rotated). On the other hand,

$$T_{AB''} = T_1T_{AB} = \begin{bmatrix} R_1R_{AB} & R_1p_{AB} + p_1 \\ 0 & 1 \end{bmatrix}$$

is the representation of the transformed frame B'' relative to A after rotating B about the origin of A by R_1 (expressed in the A frame) and then translating by p_1 in the A frame. Note that $T_{AB'}$ and $T_{AB''}$ are generally different, as matrix multiplication is not commutative.

If we consider frame B to be attached to a moving body, we call T_1 a *body-frame* transformation if it is multiplied on the right, as the rotation and translation are expressed relative to the body frame B. If A is a stationary world frame, we call T_1 a *world-frame* transformation if it is multiplied on the left, as the rotation and translation

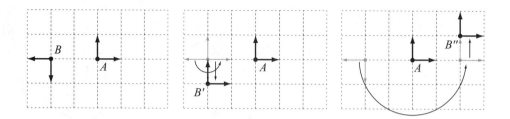

Figure 3.19 (a) The initial frame B relative to A. (b) B' is obtained by rotating about B and then translating in the original y_B-direction. (c) B'' is obtained by rotating about A and then translating in the y_A-direction.

are expressed relative to the fixed A frame. An example is shown in figure 3.19 for

$$T_{AB} = \begin{bmatrix} -1 & 0 & 0 & -2 \\ 0 & -1 & 0 & 0 \\ 0 & 0 & 1 & 0 \\ 0 & 0 & 0 & 1 \end{bmatrix} \quad T_1 = \begin{bmatrix} -1 & 0 & 0 & 0 \\ 0 & -1 & 0 & 1 \\ 0 & 0 & 1 & 0 \\ 0 & 0 & 0 & 1 \end{bmatrix},$$

giving

$$T_{AB'} = T_{AB}T_1 = \begin{bmatrix} 1 & 0 & 0 & -2 \\ 0 & 1 & 0 & -1 \\ 0 & 0 & 1 & 0 \\ 0 & 0 & 0 & 1 \end{bmatrix} \quad T_{AB''} = T_1 T_{AB} = \begin{bmatrix} 1 & 0 & 0 & 2 \\ 0 & 1 & 0 & 1 \\ 0 & 0 & 1 & 0 \\ 0 & 0 & 0 & 1 \end{bmatrix}.$$

Applying n world-frame transformations yields $T_{AB''} = T_n \ldots T_2 T_1 T_{AB}$, while n body-frame transformations yields $T_{AB'} = T_{AB}T_1 T_2 \ldots T_n$.

3.6   Parameterizations of $SO(3)$

We have seen that the nine elements R_{ij} of a rotation matrix $R \in SO(3)$ are subject to six constraints, leaving three rotational degrees of freedom. Thus, we expect that $SO(3)$ can be locally parameterized using three variables. Euler angles are a common parameterization. However, just as we see we cannot find a global parameterization for a circle with a single variable, we cannot build a global parameterization of $SO(3)$ with Euler angles.

Given two coordinate frames $\mathcal{F}_0$ and $\mathcal{F}_1$, we can specify the orientation of frame $\mathcal{F}_1$ relative to frame $\mathcal{F}_0$ by three angles (ϕ, θ, ψ), known as Z-Y-Z Euler angles. These Euler angles are defined by three successive rotations as follows. Initially, the two

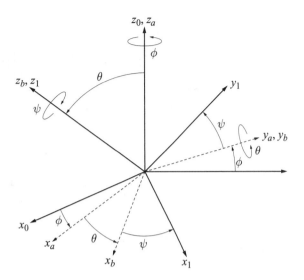

Figure 3.20 Euler angle representation.

frames are coincident. Rotate $\mathcal{F}_0$ about the z-axis by the angle ϕ to obtain frame $\mathcal{F}_a$. Next, rotate frame $\mathcal{F}_a$ about its y-axis by the angle θ to obtain frame $\mathcal{F}_b$. Finally, rotate frame $\mathcal{F}_b$ about its z-axis by the angle ψ to obtain frame $\mathcal{F}_1$. This is illustrated in figure 3.20.

The corresponding rotation matrix R can thus be generated by successive multiplication of rotation matrices that define rotations about coordinate axes,

$$(3.6) \quad R = R_{z,\phi} R_{y,\theta} R_{z,\psi}$$

$$(3.7) \quad = \begin{bmatrix} c_\phi & -s_\phi & 0 \\ s_\phi & c_\phi & 0 \\ 0 & 0 & 1 \end{bmatrix} \begin{bmatrix} c_\theta & 0 & s_\theta \\ 0 & 1 & 0 \\ -s_\theta & 0 & c_\theta \end{bmatrix} \begin{bmatrix} c_\psi & -s_\psi & 0 \\ s_\psi & c_\psi & 0 \\ 0 & 0 & 1 \end{bmatrix}$$

$$(3.8) \quad = \begin{bmatrix} c_\phi c_\theta c_\psi - s_\phi s_\psi & -c_\phi c_\theta s_\psi - s_\phi c_\psi & c_\phi s_\theta \\ s_\phi c_\theta c_\psi + c_\phi s_\psi & -s_\phi c_\theta s_\psi + c_\phi c_\psi & s_\phi s_\theta \\ -s_\theta c_\psi & s_\theta s_\psi & c_\theta \end{bmatrix}.$$

Note that successive rotation matrices are multiplied on the right, as successive rotations are defined about axes in the changing "body" frame.

Parameterization of $SO(3)$ using Euler angles, along with some other representations of $SO(3)$, are described in detail in appendix E.

3.7   Example Configuration Spaces

In most cases, we can model robots as rigid bodies, articulated chains, or combinations of these two. Some common robots and representations of their configuration spaces are given in table 3.1.

When designing a motion planner, it is often important to understand the underlying structure of the robot's configuration space. In particular, we note the following.

- $S^1 \times S^1 \times \cdots \times S^1$ (n times) $= T^n$, the n-dimensional torus

- $S^1 \times S^1 \times \cdots \times S^1$ (n times) $\neq S^n$, the n-dimensional sphere in $\mathbb{R}^{n+1}$

- $S^1 \times S^1 \times S^1 \neq SO(3)$

- $SE(2) \neq \mathbb{R}^3$

- $SE(3) \neq \mathbb{R}^6$

It is sometimes important to know whether a manifold is compact. The manifolds S^n, T^n, and $SO(n)$ are all compact, as are all of their direct products. The manifolds $\mathbb{R}^n$ and $SE(n)$ are not compact, and therefore $\mathbb{R}^n \times \mathcal{M}$ is not compact, regardless of whether or not the manifold $\mathcal{M}$ is compact.

Despite their differences, all of these configuration spaces have an important similarity. When equipped with an atlas, each is a differentiable manifold. In particular,

- $\mathbb{R}^1$ and $SO(2)$ are one-dimensional manifolds;

- $\mathbb{R}^2$, S^2 and T^2 are two-dimensional manifolds;

Type of robot	Representation of $\mathcal{Q}$
Mobile robot translating in the plane	$\mathbb{R}^2$
Mobile robot translating and rotating in the plane	$SE(2)$ or $\mathbb{R}^2 \times S^1$
Rigid body translating in the three-space	$\mathbb{R}^3$
A spacecraft	$SE(3)$ or $\mathbb{R}^3 \times SO(3)$
An n-joint revolute arm	T^n
A planar mobile robot with an attached n-joint arm	$SE(2) \times T^n$

Table 3.1   Some common robots and their configuration spaces.

- $\mathbb{R}^3$, $SE(2)$ and $SO(3)$ are three-dimensional manifolds;

- $\mathbb{R}^6$, T^6 and $SE(3)$ are six-dimensional manifolds.

Thus, for example, all of $\mathbb{R}^3$, $SE(2)$, and $SO(3)$ can be represented locally by a set of three coordinates.

3.8   Transforming Configuration and Velocity Representations

We often need to transform from one representation of the configuration of a robot $q \in \mathcal{Q}$ to some other representation $x \in \mathcal{M}$. A common example occurs when q represents the joint angles of a robot arm and x represents the configuration of the end effector as a rigid body in the ambient space. The representation x is more convenient when planning manipulation tasks in the world, but control of the robot arm is more easily expressed in q variables, so we need an easy way of switching back and forth. It is often the case that $\mathcal{Q}$ and $\mathcal{M}$ are not homeomorphic; the dimension of the two spaces may not even be equal.

Using the robot arm as inspiration, we define the *forward kinematics map* $\phi : \mathcal{Q} \to \mathcal{M}$ and the *inverse kinematics map* $\phi^{-1} : \mathcal{M} \to \mathcal{Q}$. These maps may not be homeomorphisms even if the dimensions of $\mathcal{Q}$ and $\mathcal{M}$ are equal. As the robot system moves, the time derivative $\dot{x} = \frac{dx}{dt}$ is related to the time derivative $\dot{q} = \frac{dq}{dt}$ by

$$\dot{x} = \frac{\partial \phi}{\partial q} \dot{q} = J(q)\dot{q},$$

where J is the *Jacobian* of the map ϕ, also known as the *differential Dϕ* (see appendix C). The Jacobian is also useful for transforming forces expressed in one set of coordinates to another (see chapter 4, section 4.7, and chapter 10).

EXAMPLE 3.8.1   *The 2R robot arm of figure 3.21 has link lengths L_1 and L_2. Its configuration space is $\mathcal{Q} = T^2$, and we represent the configuration by the two joint angles $q = [\theta_1, \theta_2]^T$. The endpoint of the hand in the Cartesian space is $x = [x_1, x_2]^T \in \mathcal{M} \subset \mathbb{R}^2$. In this case, the dimensions of $\mathcal{Q}$ and $\mathcal{M}$ are equal, but they are not homeomorphic. The forward kinematics map $\phi : \mathcal{Q} \to \mathcal{M}$ is*

$$\phi(q) = \begin{bmatrix} \phi_1(q) \\ \phi_2(q) \end{bmatrix} = \begin{bmatrix} L_1 \cos\theta_1 + L_2 \cos(\theta_1 + \theta_2) \\ L_1 \sin\theta_1 + L_2 \sin(\theta_1 + \theta_2) \end{bmatrix}.$$

The inverse kinematics map ϕ^{-1} is one-to-two at most points of $\mathcal{M}$, meaning that the robot can be chosen to be in either the right-arm *or* left-arm *configuration. The inverse kinematics of the 2R arm is most easily found geometrically using the law of cosines and is left for problem 20.*

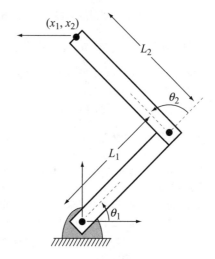

Figure 3.21   The 2R robot arm and the velocity at its endpoint.

The Jacobian of the forward kinematics map is

$$J(q) = \frac{\partial \phi}{\partial q} = \begin{bmatrix} \frac{\partial \phi_1}{\partial \theta_1} & \frac{\partial \phi_1}{\partial \theta_2} \\ \frac{\partial \phi_2}{\partial \theta_1} & \frac{\partial \phi_2}{\partial \theta_2} \end{bmatrix}$$

$$= \begin{bmatrix} -L_1 \sin \theta_1 - L_2 \sin(\theta_1 + \theta_2) & -L_2 \sin(\theta_1 + \theta_2) \\ L_1 \cos \theta_1 + L_2 \cos(\theta_1 + \theta_2) & L_2 \cos(\theta_1 + \theta_2) \end{bmatrix}.$$

Plugging in $L_1 = L_2 = 1$, $\theta_1 = \pi/4$, $\theta_2 = \pi/2$, and $\dot{q} = [1, 0]^T$, as shown in figure 3.21, we find that

$$\dot{x} = J(q)\dot{q} = \begin{bmatrix} -\sqrt{2} & -\sqrt{2}/2 \\ 0 & -\sqrt{2}/2 \end{bmatrix} \begin{bmatrix} 1 \\ 0 \end{bmatrix} = \begin{bmatrix} -\sqrt{2} \\ 0 \end{bmatrix},$$

matching the motion seen in the figure.

When $\sin \theta_2 = 0$, the Jacobian $J(q)$ loses rank, and the robot is said to be in a singular *configuration. In this case, the two-dimensional set of joint velocities $\dot{q}$ maps to a one-dimensional set of endpoint velocities $\dot{x}$ — instantaneous endpoint motion is impossible in one direction.*

EXAMPLE 3.8.2   *A polygon moving in the plane is represented by the configuration $q = [q_1, q_2, q_3]^T \in \mathcal{Q} = \mathbb{R}^2 \times S^1$, where (q_1, q_2) gives the position of a reference frame $\mathcal{F}_p$ attached to the polygon relative to a world frame $\mathcal{F}$, and q_3 gives the*

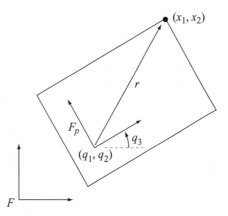

Figure 3.22 The point on the polygon is at r in the polygon frame $\mathcal{F}_p$ and x in the world frame $\mathcal{F}$.

orientation of $\mathcal{F}_p$ relative to $\mathcal{F}$ (see figure 3.22). A point is fixed on the polygon at $r = [r_1, r_2]^T$ in the polygon frame $\mathcal{F}_p$, and let $x = [x_1, x_2]^T \in \mathcal{M} = \mathbb{R}^2$ be the position of this point in the plane. Then the forward kinematics mapping is

$$\begin{bmatrix} x_1 \\ x_2 \end{bmatrix} = \phi(q) = \begin{bmatrix} q_1 \\ q_2 \end{bmatrix} + \begin{bmatrix} \cos q_3 & -\sin q_3 \\ \sin q_3 & \cos q_3 \end{bmatrix} \begin{bmatrix} r_1 \\ r_2 \end{bmatrix},$$

where we recognize the 2×2 rotation matrix. The inverse map ϕ^{-1} in this example is one-to-many, as the dimension of $\mathcal{Q}$ is greater than the dimension of $\mathcal{M}$. The velocities $\dot{x}$ and $\dot{q}$ are related by the Jacobian

$$J(q) = \frac{\partial \phi}{\partial q} = \begin{bmatrix} \frac{\partial \phi_1}{\partial q_1} & \frac{\partial \phi_1}{\partial q_2} & \frac{\partial \phi_1}{\partial q_3} \\ \frac{\partial \phi_2}{\partial q_1} & \frac{\partial \phi_2}{\partial q_2} & \frac{\partial \phi_2}{\partial q_3} \end{bmatrix} = \begin{bmatrix} 1 & 0 & -r_1 \sin q_3 - r_2 \cos q_3 \\ 0 & 1 & r_1 \cos q_3 - r_2 \sin q_3 \end{bmatrix}.$$

Problems

1. Invent your own nontrivial robot system. It could consist of one or more robot arms, mobile platforms, conveyor belts, fixed obstacles, movable objects, etc. Describe the configuration space mathematically. Explain whether or not the configuration space is compact, and if not, describe the compact factors. Describe the connected components of the configuration space. Draw a rough picture of your robot system.

2. Give the dimension of the configuration spaces of the following systems. Explain your answers.

   (a) Two mobile robots rotating and translating in the plane.
   (b) Two translating and rotating planar mobile robots tied together by a rope.
   (c) Two translating and rotating planar mobile robots connected rigidly by a bar.
   (d) The two arms of a single person (with torso stationary) holding on firmly to a car's steering wheel.
   (e) A train on train tracks. What if we include the wheel angles? (The wheels roll without slipping.)
   (f) A spacecraft with a 6R robot arm.
   (g) The end effector of the 6R robot arm of a spacecraft.
   (h) Your legs as you pedal a bicycle (remaining seated with feet fixed to the pedals).
   (i) A sheet of paper.

3. Describe the Bug2 algorithm for a two-joint manipulator. What are the critical differences between the Bug2 algorithm for a mobile base and the two-joint manipulator? What does a straight line mean in the arm's configuration space? Can Bug2 be made to work for the two-joint arm?

4. Prove the configuration space obstacle of a convex mobile robot translating in a plane with a convex obstacle is convex.

5. Prove the union operator propagates from the workspace to the configuration space. That is, the union of two configuration space obstacles is the configuration space obstacle of the union of two workspace obstacles. In other words, assuming Q is a configuration space operator, show that

$$Q(WO_i \bigcup WO_j) = QO_i \bigcup QO_j.$$

6. How many degrees of freedom does a rigid body in n-space have? How many of them are rotational? Prove these two ways: (a) using the method of choosing a number of points on the body and sequentially adding their independent degrees of freedom until each point on the body is fixed, and (b) using the definitions of $SE(n)$ and $SO(n)$.

7. Use cardboard and pushpins to create a closed chain with four links, a four-bar mechanism. One of these bars is considered stationary, or fixed to the ground. Going around the loop, the link lengths between joints are 6.5 (the ground link), 3.0, 1.5, and 3.0 inches (or centimeters) in length. Poke a hole at the midpoint of the 1.5 inch link and trace the path that the hole traces. Describe a good representation of the configuration space of the linkage.

8. Give a homeomorphism from the racetrack to the ellipse in figure 3.12.

9. Find two charts for the unit circle S^1 and prove they form an atlas.

10. Explain why the latitude-longitude chart we often place on the Earth is not a global parameterization. Find two charts for the sphere and prove that they form an atlas.

11. The set of right-arm and left-arm configurations of the 2R manipulator in figure 3.3 each give an annulus of reachable positions by the end effector, neither of which is diffeomorphic to the robot's configuration space. Consider the right-arm and left-arm workspaces as two separate annuluses, and describe how they can be glued together to make a single space that *is* a valid representation of the configuration space. Comment on the topology of this glued space.

12. For the 2R manipulator of figure 3.7, how many connected components of free configuration space are there? Copy the figure, color each of the connected components a different color, and give a drawing of the robot in each of these connected components.

13. Show that compact and noncompact spaces are never diffeomorphic.

14. Find a diffeomorphism from any open interval $(a, b) \in \mathbb{R}$ to the whole real line $\mathbb{R}$.

15. Give an implicit constraint equation $f(x, y, z) = 0$ that embeds a torus in $\mathbb{R}^3$.

16. For $T \in SE(3)$ consisting of the rotation matrix $R \in SO(3)$ and the translation $p \in \mathbb{R}^3$, find the inverse transform T^{-1}, so that $TT^{-1} = T^{-1}T = \mathcal{I}$. Your answer should not contain any matrix inverses.

17. Consider two three-dimensional frames aligned with each other, called A and B. Rotate B 90 degrees about the x-axis of A, then rotate again by 90 degrees about the y-axis of A, then move the origin of B by three units in the z-direction of A. (Make sure you rotate in the right direction! Use the right-hand rule: thumb points along the positive axis, fingers curl in the direction of positive rotation.) Give the matrix $T_{AB} \in SE(3)$ describing the frame B relative to the frame A. Consider the point $x_B = [4, 3, 1, 1]^T$ in homogeneous coordinates in frame B. What is the point x_A (expressed in the frame A)? Consider the point $y_A = [1, 2, 3, 1]^T$ in homogeneous coordinates in frame A, and perform the transformation T_{AB}. Where is the new point y'_A?

18. Write a program to calculate the configuration space for a convex polygonal robot translating in an environment with convex obstacles. The program should read in from a file a counterclockwise list of vertices representing the robot, where $(0, 0)$ is the robot reference point. From a second file, the program should read in a set of obstacles in the workspace. The user enters an orientation for the robot and the program calculates the configuration space obstacles (see, e.g., appendix F). Display the configuration space for different orientations of the robot to show that translating paths between two points may exist for some orientations of the robot, but not for others.

19. Write a program to display the configuration space for a 2R manipulator in a polygonal environment. The program should read in a file containing the location of the base of the robot, the length of the links, and the lists of vertices representing the obstacles. The

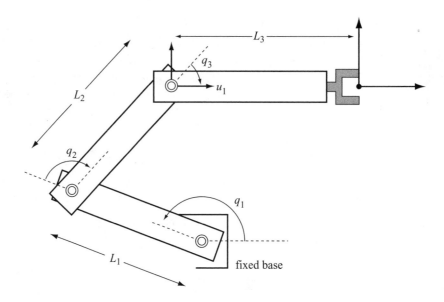

Figure 3.23   A 3R planar robot with a frame attached to the end effector.

program should check for collision at 1 degree increments for each joint and create a plot similar to that shown in figure 3.7.

20. Find the inverse kinematics ϕ^{-1} mapping the end-effector coordinates x to the joint coordinates q for the 2R robot arm in example 3.8.1. Note that for most reachable points x, there will be two solutions, corresponding to the right- and left-arm configurations. Your solution will likely make use of the two-argument arctangent atan2(x_2, x_1), which returns the unique angle in $[-\pi, \pi)$ to the point (x_1, x_2) in the plane, as well as the law of cosines $a^2 = b^2 + c^2 - 2bc \cos A$, where a, b, and c are the lengths of the three edges of a triangle and A is the angle opposite to edge a. Solve for general L_1, L_2 (do not plug in numbers).

21. Give the forward kinematics ϕ for the planar 3R arm shown in figure 3.23, from joint angles q to the position and orientation of the end effector frame in the plane. Find the manipulator Jacobian.

22. For the problem above, show that the forward kinematics mapping is injective, surjective, bijective, or none of these, when viewed as a mapping from T^3 to $\mathbb{R}^2 \times S^1$. Find a "large" set of joint angle ranges $U \subset T^3$ and a set of end-effector configurations $V \subset \mathbb{R}^2 \times S^1$ for which the mapping is a diffeomorphism.

23. The topology of $SE(2)$ is equivalent to $\mathbb{R}^2 \times SO(2)$. Let's represent an element of $\mathbb{R}^2 \times SO(2)$ by (x, R), where $x \in \mathbb{R}^2$, $R \in SO(2)$. As we have seen, we can make $SE(2)$ a group by giving it a group operation, namely, matrix multiplication. We can also make

$\mathbb{R}^2 \times SO(2)$ a group by using the direct product structure to define composition of two elements:

$$(x_1, R_1)(x_2, R_2) = (x_1 + x_2, R_1 R_2) \in \mathbb{R}^2 \times SO(2)$$

We are using vector addition as the group operation on $\mathbb{R}^2$ and matrix multiplication on $SO(2)$. With this group operation, is $\mathbb{R}^2 \times SO(2)$ commutative? Is $SE(2)$ commutative? The spaces $SE(2)$ and $\mathbb{R}^2 \times SO(2)$ are topologically equivalent, but are they equivalent as groups?

4 *Potential Functions*

HAVING SEEN the difficulty of explicitly representing the configuration space, an alternative is to develop search algorithms that incrementally "explore" free space while searching for a path. Already, we have seen that the Bug algorithms maneuver through free space without constructing the configuration space, but the Bug algorithms are limited to simple two-dimensional configuration spaces. Therefore, this chapter introduces navigation planners that apply to a richer class of robots and produce a greater variety of paths than the Bug methods, i.e., they apply to a more general class of configuration spaces, including those that are multidimensional and non-Euclidean.

A *potential function* is a differentiable real-valued function $U : \mathbb{R}^m \to \mathbb{R}$. The value of a potential function can be viewed as energy and hence the gradient of the potential is force. The *gradient* is a vector $\nabla U(q) = DU(q)^T = [\frac{\partial U}{\partial q_1}(q), \ldots, \frac{\partial U}{\partial q_m}(q)]^T$ which points in the direction that locally maximally increases U. See appendix C.5 for a more rigorous definition of the gradient. We use the gradient to define a *vector field,* which assigns a vector to each point on a manifold. A gradient vector field, as its name suggests, assigns the gradient of some function to each point. When U is energy, the gradient vector field has the property that work done along any closed path is zero.

The potential function approach directs a robot as if it were a particle moving in a gradient vector field. Gradients can be intuitively viewed as forces acting on a positively charged particle robot which is attracted to the negatively charged goal. Obstacles also have a positive charge which forms a repulsive force directing the robot away from obstacles. The combination of repulsive and attractive forces hopefully directs the robot from the start location to the goal location while avoiding obstacles (figure 4.1).

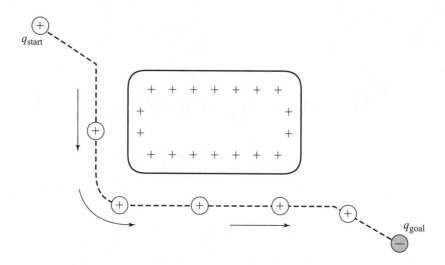

Figure 4.1   The negative charge attracts the robot and the positive charge repels it, resulting in a path, denoted by the dashed line, around the obstacle and to the goal.

Note that in this chapter, we mainly deal with first-order systems (i.e., we ignore dynamics), so we view the gradients as velocity vectors instead of force vectors. Potential functions can be viewed as a landscape where the robots move from a "high-value" state to a "low-value" state. The robot follows a path "downhill" by following the negated gradient of the potential function. Following such a path is called *gradient descent,* i.e.,

$$\dot{c}(t) = -\nabla U(c(t)).$$

The robot terminates motion when it reaches a point where the gradient vanishes, i.e., it has reached a q^* where $\nabla U(q^*) = 0$. Such a point q^* is called a *critical point* of U. The point q^* is either a maximum, minimum, or a saddle point (figure 4.2). One can look at the second derivative to determine the type of critical point. For real-valued functions, this second derivative is the *Hessian* matrix

$$\begin{bmatrix} \frac{\partial^2 U}{\partial q_1^2} & \cdots & \frac{\partial^2 U}{\partial q_1 \partial q_n} \\ \vdots & \ddots & \vdots \\ \frac{\partial^2 U}{\partial q_n \partial q_1} & \cdots & \frac{\partial^2 U}{\partial q_n^2} \end{bmatrix}.$$

When the Hessian is nonsingular at q^*, the critical point at q^* is *non-degenerate,* implying that the critical point is isolated [?]. When the Hessian is positive-definite, the critical point is a local minimum; when the Hessian is negative-definite, then

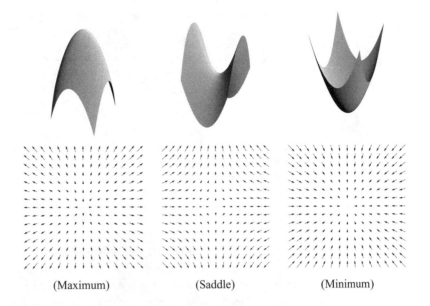

(Maximum)   (Saddle)   (Minimum)

Figure 4.2   Different types of critical points: (Top) Graphs of functions. (Bottom) Gradients of functions.

the critical point is a local maximum. Generally, we consider potential functions whose Hessians are nonsingular, i.e., those which only have isolated critical points. This also means that the potential function is never flat.

For gradient descent methods, we do not have to compute the Hessian because the robot generically terminates its motion at a local minimum, not at a local maximum or a saddle point. Since gradient descent decreases U, the robot cannot arrive at a local maximum, unless of course the robot starts at a maximum. Since we assume that the function is never flat, the set of maxima contains just isolated points, and the likelihood of starting at one is practically zero. However, even if the robot starts at a maximum, any perturbation of the robot position frees the robot, allowing the gradient vector field to induce motion onto the robot. Arriving at a saddle point is also unlikely, because they are unstable as well. Local minima, on the other hand, are stable because after any perturbation from a minimum, gradient descent returns the robot to the minimum. Therefore, the only critical point where the robot can generically terminate is a local minimum. Hopefully this is where the goal is located. See figure 4.3 for an example of a configuration space with its corresponding potential function, along with its energy surface landscape and gradient vector field.

 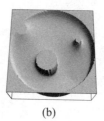  

(a)   (b)   (c)   (d)

Figure 4.3   (a) A configuration space with three circular obstacles bounded by a circle. (b) Potential function energy surface. (c) Contour plot for energy surface. (d) Gradient vectors for potential function.

There are many potential functions other than the attractive/repulsive potential. Many of these potential functions are efficient to compute and can be computed online [234]. Unfortunately, they all suffer one problem—the existence of local minima not corresponding to the goal. This problem means that potential functions may lead the robot to a point which is not the goal; in other words, many potential functions do not lead to complete path planners. Two classes of approaches address this problem: the first class augments the potential field with a search-based planner, and the second defines a potential function with one local minimum, called a *navigation function* [239]. Although complete (or resolution complete), both methods require full knowledge of the configuration space prior to the planning event.

Finally, unless otherwise stated, the algorithms presented in this chapter apply to spaces of arbitrary dimension, even though the figures are drawn in two dimensions. Also, we include some discussion of implementation on a mobile base operating in the plane (i.e., a point in a two-dimensional Euclidean configuration space).

4.1   Additive Attractive/Repulsive Potential

The simplest potential function in $\mathcal{Q}_{\text{free}}$ is the attractive/repulsive potential. The intuition behind the attractive/repulsive potential is straightforward: the goal attracts the robot while the obstacles repel it. The sum of these effects draws the robot to the goal while deflecting it from obstacles. The potential function can be constructed as the sum of attractive and repulsive potentials

$$U(q) = U_{\text{att}}(q) + U_{\text{rep}}(q).$$

The Attractive Potential

There are several criteria that the potential field U_{att} should satisfy. First, U_{att} should be monotonically increasing with distance from q_{goal}. The simplest choice is the *conic potential,* measuring a scaled distance to the goal, i.e., $U(q) = \zeta d(q, q_{goal})$. The ζ is a parameter used to scale the effect of the attractive potential. The attractive gradient is $\nabla U(q) = \frac{\zeta}{d(q,q_{goal})}(q - q_{goal})$. The gradient vector points away from the goal with magnitude ζ at all points of the configuration space except the goal, where it is undefined. Starting from any point other than the goal, by following the negated gradient, a path is traced toward the goal.

When numerically implementing this method, gradient descent may have "chattering" problems since there is a discontinuity in the attractive gradient at the origin. For this reason, we would prefer a potential function that is continuously differentiable, such that the magnitude of the attractive gradient decreases as the robot approaches q_{goal}. The simplest such potential function is one that grows quadratically with the distance to q_{goal}, e.g.,

$$U_{att}(q) = \frac{1}{2}\zeta d^2(q, q_{goal}),$$

with the gradient

$$\nabla U_{att}(q) = \nabla \left(\frac{1}{2}\zeta d^2(q, q_{goal}) \right),$$

$$= \frac{1}{2}\zeta \nabla d^2(q, q_{goal}),$$

(4.1)
$$= \zeta(q - q_{goal}),$$

which is a vector based at q, points away from q_{goal}, and has a magnitude proportional to the distance from q to q_{goal}. The farther away q is from q_{goal}, the bigger the magnitude of the vector. In other words, when the robot is far away from the goal, the robot quickly approaches it; when the robot is close to the goal, the robot slowly approaches it. This feature is useful for mobile robots because it reduces "overshoot" of the goal (resulting from step quantization).

In figure 4.4(a), the goal is in the center and the gradient vectors for various points are drawn. Figure 4.4(b) contains a contour plot for U_{att}; each solid circle corresponds to a set of points q where $U_{att}(q)$ is constant. Finally, figure 4.4(c) plots the graph of the attractive potential.

Note that while the gradient $\nabla U_{att}(q)$ converges linearly to zero as q approaches q_{goal} (which is a desirable property), it grows without bound as q moves away from q_{goal}. If q_{start} is far from q_{goal}, this may produce a desired velocity that is too large. For this reason, we may choose to combine the quadratic and conic potentials so that the

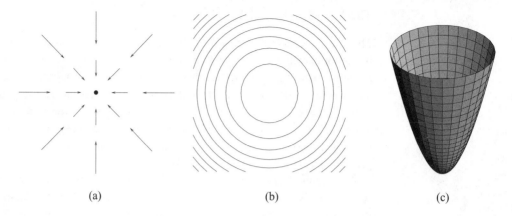

Figure 4.4   (a) Attractive gradient vector field. (b) Attractive potential isocontours. (c) Graph of the attractive potential.

conic potential attracts the robot when it is very distant from q_{goal} and the quadratic potential attracts the robot when it is near q_{goal}. Of course it is necessary that the gradient be defined at the boundary between the conic and quadratic portions. Such a field can be defined by

$$(4.2) \qquad U_{\text{att}}(q) = \begin{cases} \frac{1}{2}\zeta d^2(q, q_{\text{goal}}), & d(q, q_{\text{goal}}) \leq d^*_{\text{goal}}, \\ d^*_{\text{goal}}\zeta d(q, q_{\text{goal}}) - \frac{1}{2}\zeta(d^*_{\text{goal}})^2, & d(q, q_{\text{goal}}) > d^*_{\text{goal}}. \end{cases}$$

and in this case we have

$$(4.3) \qquad \nabla U_{\text{att}}(q) = \begin{cases} \zeta(q - q_{\text{goal}}), & d(q, q_{\text{goal}}) \leq d^*_{\text{goal}}, \\ \frac{d^*_{\text{goal}}\zeta(q-q_{\text{goal}})}{d(q,q_{\text{goal}})}, & d(q, q_{\text{goal}}) > d^*_{\text{goal}}, \end{cases}$$

where d^*_{goal} is the threshold distance from the goal where the planner switches between conic and quadratic potentials. The gradient is well defined at the boundary of the two fields since at the boundary where $d(q, q_{\text{goal}}) = d^*_{\text{goal}}$, the gradient of the quadratic potential is equal to the gradient of the conic potential, $\nabla U_{\text{att}}(q) = \zeta(q - q_{\text{goal}})$.

The Repulsive Potential

A repulsive potential keeps the robot away from an obstacle. The strength of the repulsive force depends upon the robot's proximity to the an obstacle. The closer the robot is to an obstacle, the stronger the repulsive force should be. Therefore, the

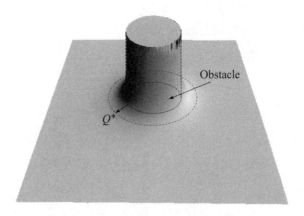

Figure 4.5   The repulsive gradient operates only in a domain near the obstacle.

repulsive potential is usually defined in terms of distance to the closest obstacle $D(q)$, i.e.,

$$(4.4) \qquad U_{\text{rep}}(q) = \begin{cases} \frac{1}{2}\eta \left(\frac{1}{D(q)} - \frac{1}{Q^*} \right)^2, & D(q) \leq Q^*, \\ 0, & D(q) > Q^*, \end{cases}$$

whose gradient is

$$(4.5) \qquad \nabla U_{\text{rep}}(q) = \begin{cases} \eta \left(\frac{1}{Q^*} - \frac{1}{D(q)} \right) \frac{1}{D^2(q)} \nabla D(q), & D(q) \leq Q^*, \\ 0, & D(q) > Q^*, \end{cases}$$

where the $Q^* \in \mathbb{R}$ factor allows the robot to ignore obstacles sufficiently far away from it and the η can be viewed as a gain on the repulsive gradient. These scalars are usually determined by trial and error. (See figure 4.5.)

When numerically implementing this solution, a path may form that oscillates around points that are two-way equidistant from obstacles, i.e., points where D is nonsmooth. To avoid these oscillations, instead of defining the repulsive potential function in terms of distance to the *closest* obstacle, the repulsive potential function is redefined in terms of distances to *individual* obstacles where $d_i(q)$ is the distance to obstacle $\mathcal{QO}_i$, i.e.,

$$(4.6) \qquad d_i(q) = \min_{c \in \mathcal{QO}_i} d(q, c).$$

Note that the min operator returns the smallest $d(q, c)$ for all points c in $\mathcal{QO}_i$.

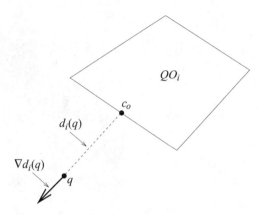

Figure 4.6   The distance between x and $\mathcal{QO}_i$ is the distance to the closest point on $\mathcal{QO}_i$. The gradient is a unit vector pointing away from the nearest point.

It can be shown for convex obstacles $\mathcal{QO}_i$ where c is the closest point to x that the gradient of $d_i(q)$ is

(4.7) $$\nabla d_i(q) = \frac{q - c}{d(q, c)}.$$

The vector $\nabla d_i(q)$ describes the direction that maximally increases the distance to $\mathcal{QO}_i$ from q (figure 4.6).

Now, each obstacle has its own potential function,

$$U_{\text{rep}_i}(q) = \begin{cases} \frac{1}{2} \eta \left(\frac{1}{d_i(q)} - \frac{1}{Q_i^*} \right)^2, & \text{if } d_i(q) \leq Q_i^*, \\ 0, & \text{if } d_i(q) > Q_i^*, \end{cases}$$

where Q_i^* defines the size of the domain of influence for obstacle $\mathcal{QO}_i$. Then $U_{\text{rep}}(q) = \sum_{i=1}^{n} U_{\text{rep}_i}(q)$. Assuming that there are only convex obstacles or nonconvex ones can be decomposed into convex pieces, oscillations do not occur because the planner does not have radical changes in the closest point anymore.

4.2   Gradient Descent

Gradient descent is a well-known approach to optimization problems. The idea is simple. Starting at the initial configuration, take a small step in the direction opposite the gradient. This gives a new configuration, and the process is repeated until the gradient is zero. More formally, we can define a gradient descent algorithm (algorithm 4).

Algorithm 4 Gradient Descent

Input: A means to compute the gradient $\nabla U(q)$ at a point q
Output: A sequence of points $\{q(0), q(1), \ldots, q(i)\}$

1: $q(0) = q_{\text{start}}$
2: $i = 0$
3: **while** $\nabla U(q(i)) \neq 0$ **do**
4:     $q(i+1) = q(i) + \alpha(i) \nabla U(q(i))$
5:     $i = i + 1$
6: **end while**

In algorithm 4, the notation $q(i)$ is used to denote the value of q at the ith iteration and the final path consists of the sequence of iterates $\{q(0), q(1), \ldots, q(i)\}$. The value of the scalar $\alpha(i)$ determines the step size at the i iteration. It is important that $\alpha(i)$ be small enough that the robot is not allowed to "jump into" obstacles, while being large enough that the algorithm does not require excessive computation time. In motion planning problems, the choice for $\alpha(i)$ is often made on an *ad hoc* or empirical basis, perhaps based on the distance to the nearest obstacle or to the goal. A number of systematic methods for choosing $\alpha(i)$ can be found in the optimization literature [45]. Finally, it is unlikely that we will ever exactly satisfy the condition $\nabla U(q(i)) = 0$. For this reason, this condition is often replaced with the more forgiving condition $\|\nabla U(q(i))\| < \epsilon$, in which ϵ is chosen to be sufficiently small, based on the task requirements.

4.3   Computing Distance for Implementation in the Plane

In this section, we discuss some implementation issues in constructing the attractive/repulsive potential function. The attractive potential function is rather straightforward if the robot knows its current location and the goal location. The challenge lies in computing the repulsive function because it requires calculation of distance to obstacles. Therefore, in this section, we discuss two different methods to compute distance, and hence the repulsive potential function. The first method deals with sensor-based implementation on a mobile robot and borrows ideas from chapter 2 in inferring distance information from sensors. The second method assumes the configuration space has been discretized into a grid of pixels and computes distance on the grid.

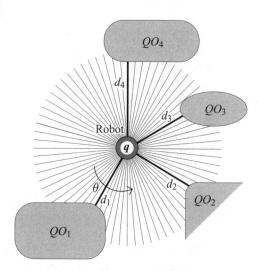

Figure 4.7   Local minima of rays determine the distance to nearby obstacles.

4.3.1   Mobile Robot Implementation

Thus far, the discussion has been general to any configuration space where we can define distance. Now let's consider some issues in implementing these potential functions on a planar mobile robot equipped with range sensors radially distributed around its circumference. These range sensors approximate a value of the raw distance function ρ defined in chapter 2. Whereas $D(q)$ corresponds to the global minimum of the raw distance function ρ, a $d_i(q)$ corresponds to a local minimum with respect to θ of $\rho(q, \theta)$ (figure 4.7). For example, any sensor in the sonar array whose value is less than that of both its left and right neighbors is a local minimum. Such sensors face the closest points on their corresponding obstacles. Therefore, these sensors point in the direction that maximally brings the robot closest to the obstacles, i.e., $-\nabla d_i(q)$. The distance gradient points in the opposite direction. An obstacle distance function may be incorrect if the obstacle is partially occluded by another.

4.3.2   Brushfire Algorithm: A Method to Compute Distance on a Grid

In this subsection, we explain a method for computing distance from a map representation called a *grid,* which is a two-dimensional array of square elements called *pixels.* A pixel has a value of zero if it is completely free of obstacles and one if it is completely or even partially occupied by an obstacle.

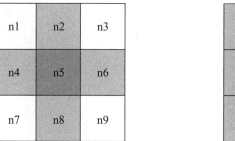
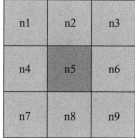

Four-point   Eight-point

Figure 4.8   Four-point vs. eight-point connectivity.

The user or planner has a choice in determining the neighboring relationships of pixels in a grid. When only the north, south, east, and west pixels are considered neighbors, the grid has *four-point connectivity*. When the grid also includes the diagonals as neighbors, then it has *eight-point connectivity* (figure 4.8). Four-point connectivity has the advantage in that it respects the Manhattan distance function (the L^1 metric) because it measures distance as if one were driving in city blocks in midtown Manhattan.

The *brushfire algorithm* uses a grid to approximate distance, and hence the repulsive function. The input to the algorithm is a grid of pixels where the free-space pixels have a value of zero and the obstacles have a value of one. The output is a grid of pixels whose corresponding values each measure distance to the nearest obstacle. These values can then be used to compute a repulsive potential function, as well as its gradient.

In the first step of the brushfire algorithm, all zero-valued pixels neighboring one-valued pixels are labeled with a two. The algorithm can use four-point or eight-point connectivity to determine adjacency. Next, all zero-valued pixels adjacent to two's are labeled with a three. This procedure repeats, i.e., all zero-valued pixels adjacent to an i are labeled with an $i + 1$, as if a brushfire is growing from each of the obstacles until the fire consumes all of the free-space pixels. The procedure terminates when all pixels have an assigned value (figure 4.9).

The brushfire method produces a map of distance values to the nearest obstacle. The gradient of distance at a pixel is determined by finding a neighbor with the lowest pixel value. The gradient is then a vector which points to this neighbor. Note that this vector points in either one of four or one of eight possible directions. If there are multiple neighbors with the same lowest value, simply pick one to define the gradient. Just as the grid is an approximation of the workspace, so is the computed gradient an approximation of the actual distance gradient.

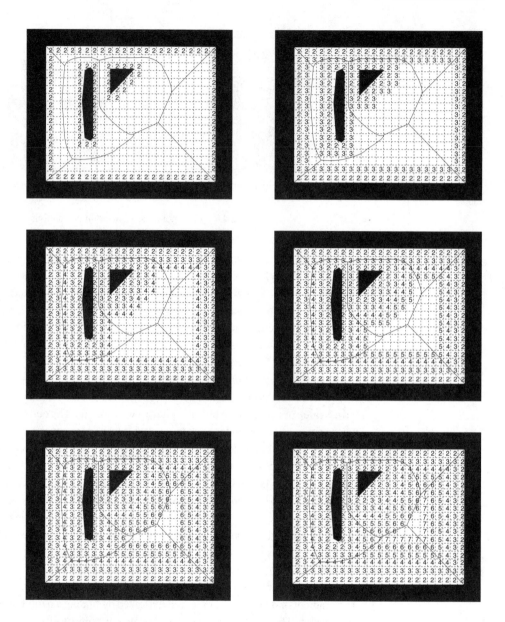

Figure 4.9   Propagation of the brushfire algorithm with eight-point connectivity. The solid lines pass through pixels where fronts collide.

With distance and gradient to the nearest obstacle inhand, a planner can compute the repulsive function. The attractive potential function can be computed analytically and together with the repulsive function, a planner can invoke the additive attractive/ repulsive function described in section 4.1.

It is worth noting that the method described here generalizes into higher dimensions where pixels then become volume elements. For example, in three-dimensions, four-point connectivity generalizes to six-point connectivity and eight-point connectivity generalizes to twenty-six-point connectivity. So, when assigning incremental pixel values to neighboring pixels in higher dimensions, the algorithm choses the appropriate adjacency relationship and then iterates through as described above. Although possible, it would become computationally intractable to compute the brushfire in higher dimensions.

4.4 Local Minima Problem

The problem that plagues all gradient descent algorithms is the possible existence of local minima in the potential field. For appropriate choice of $\alpha(i)$, it can be shown that the gradient descent algorithm is generically guaranteed to converge to a minimum in the field, but there is no guarantee that this minimum will be the global minimum. This means that there is no guarantee that gradient descent will find a path to q_{goal}. In figure 4.10, the robot is initially attracted to the goal as it approaches the horseshoe-shaped obstacle. The goal continues to attract the robot, but the bottom arm of the obstacle deflects the robot upward until the top arm of the horseshoe begins to influence

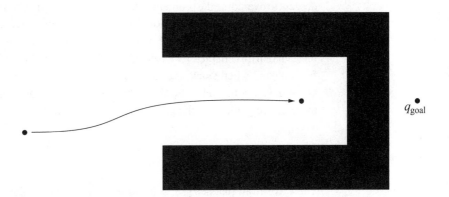

Figure 4.10   Local minimum inside the concavity. The robot moves into the concavity until the repulsive gradient balances out the attractive gradient.

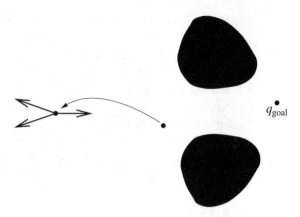

Figure 4.11 Local minimum without concave obstacles. The robot moves away from the two convex obstacles until it reaches a point where the gradient vanishes; at this point, the sum of the attractive gradient and the repulsive gradient is zero.

the robot. At this point, the effect of the top and bottom arms keeps the robot halfway between them and the goal continues to attract the robot. The robot reaches a point where the effect of the obstacle's base counteracts the attraction of the goal. In other words, the robot has reached a q^* where $\nabla U(q^*) = 0$ and q^* is *not* the goal. Note, this problem is not limited to concave obstacles as can be seen in figure 4.11. Local minima present a significant drawback to the attractive/repulsive approach, and thus the attractive/repulsive technique is not complete.

Barraquand and Latombe [37] developed search techniques other than gradient descent to overcome the problem of local minima present when planning with potential functions. Their planner, the Randomized Path Planner (RPP) [37], used a variety of potential functions some of which were simplified expressions of the potentials presented in this chapter. RPP followed the negative gradient of the specified potential function and when stuck at a local minimum, it initiated a series of random walks. Often the random walks allowed RPP to escape the local minimum and in that case, the negative gradient to the goal was followed again.

4.5 Wave-Front Planner

The wave-front planner [38, 208] affords the simplest solution to the local minima problem, but can only be implemented in spaces that are represented as grids. For the sake of discussion, consider a two-dimensional space. Initially, the planner starts with the standard binary grid of zeros corresponding to free space and ones to obstacles. The planner also knows the pixel locations of the start and goal. The goal pixel is labeled with a two. In the first step, all zero-valued pixels neighboring the goal are labeled with a three. Next, all zero-valued pixels adjacent to threes are labeled with four. This procedure essentially grows a wave front from the goal where at each iteration, all pixels on the wave front have the same path length, measured with respect to the grid, to the goal. This procedure terminates when the wave front reaches the pixel that contains the robot start location.

The planner then determines a path via gradient descent on the grid starting from the start. Essentially, the planner determines the path one pixel at a time. Assume that the value of the start pixel is 33. The next pixel in the path is any neighboring pixel whose value is 32. There could be multiple choices; simply pick any one of the choices. The next pixel is then one whose value is 31. Boundedness of the free space (and hence the discretization) and continuity of the distance function ensure that construction of the wave front guarantees that there will always be a neighboring pixel whose value is one less than that of the current pixel and that this procedure forms a path in the grid to the goal, i.e., to the pixel whose value is two.

Figure 4.12 contains six panels that demonstrate various stages of the wave-front propagation using four-point connectivity. Note that all points on the wave front have the same Manhattan distance to the goal. In the lower-left panel, note how the wave-front seemingly collides on itself. We will see later that the point of initial collision corresponds to a saddle point of the function that measures distance to the goal. This point then propagates away from the start as well. The trace of this propagation corresponds to a set of points that have two choices for shortest paths back to the goal, either going around the top of the triangle or below it.

The wave-front planner essentially forms a potential function on the grid which has one local minimum and thus is resolution complete. The planner also determines the shortest path, but at the cost of coming dangerously close to obstacles. The major drawback of this method is that the planner has to search the entire space for a path.

Finally, just like the brushfire method, the wave-front planner generalizes into higher dimensions as well. Consider the three-dimensional case first. Just as a pixel has four edges, a voxel (a three-dimensional pixel) has six faces. Therefore, the analogy to four-point connectivity with pixels is six-point connectivity with voxels. For a voxel with value i, if we assume six-point connectivity, then we assign $i + 1$ to the surrounding six voxels that share a face with the current voxel. Likewise, if we assume twenty-six-point connectivity (analogous to eight-point connectivity), then we assign $i + 1$ to all surrounding voxels that share a face, edge or vertex. It should be noted, however, implementation of the wavefront planner in higher dimensions becomes computationally intractable.

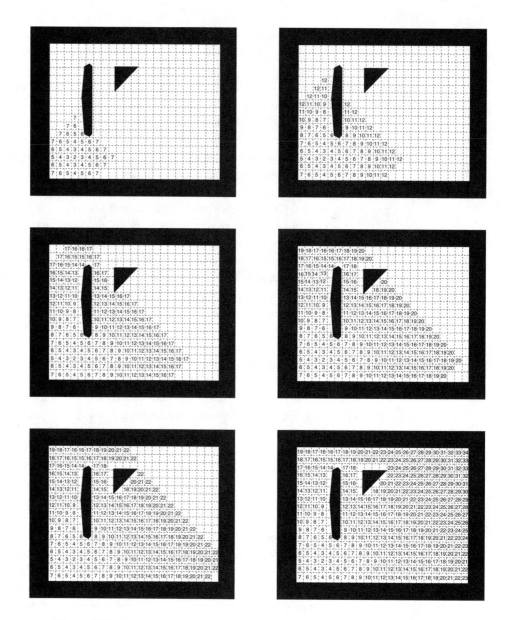

Figure 4.12   Propagation of the wave front using four-point connectivity (assume the start is in the upper-right corner and the goal is the origin of the wave front).

4.6 Navigation Potential Functions

Thus far, we have seen in chapter 2 that the Bug algorithms are complete sensor-based planners that work in unknown spaces, but are limited to planar configuration spaces. Then, at the beginning of this chapter, we have seen that the attractive/repulsive potential function approach applies to a general class of configuration spaces, but suffers from local minima problems, and hence is not complete. The wave-front planner addresses the local minima problem, but requires time and storage exponential in the dimension of the space. In this section, we introduce a new potential that is a function of distance to the obstacles, has only one minimum and applies to a limited class of configuration spaces with dimension two, three, and more. Such potential functions are called *navigation functions,* formally defined in [239, 364].

DEFINITION 4.6.1 *A function* $\varphi : \mathcal{Q}_{\text{free}} \rightarrow [0, 1]$ *is called a* navigation function *if it*

■ *is smooth (or at least C^k for $k \geq 2$),*

■ *has a unique minimum at q_{goal} in the connected component of the free space that contains q_{goal},*

■ *is uniformly maximal on the boundary of the free space, and*

■ *is Morse.*

A *Morse function* is one whose critical points are all non-degenerate. This means that critical points are isolated, and if a Morse function is used for gradient descent, any random perturbation will destabilize saddles and maxima. The navigation function approach represents obstacles as $\mathcal{QO}_i = \{q \mid \beta_i(q) \leq 0\}$; in other words, $\beta_i(q)$ is negative in the interior of $\mathcal{QO}_i$, zero on the boundary of $\mathcal{QO}_i$, and positive in the exterior of $\mathcal{QO}_i$.

4.6.1 Sphere-Space

This approach initially assumes that the configuration space is bounded by a sphere centered at q_0 and has n dim($\mathcal{Q}_{\text{free}}$)-dimensional spherical obstacles centered at $q_1, \ldots q_n$. The obstacle distance functions are easy to define as

$$\beta_0(q) = -d^2(q, q_0) + r_0^2,$$
$$\beta_i(q) = d^2(q, q_i) - r_i^2,$$

where r_i is the radius of the sphere. Note that $\beta_i(q)$ increases continuously as q moves away from the obstacle. Instead of considering the distance to the closest obstacle or the distance to each individual obstacle, we consider

(4.8) $\beta(q) = \displaystyle\prod_{i=0}^{n} \beta_i(q).$

Note that $\beta(q)$ is zero on the boundary of any obstacle, and positive at all points in the interior of the free space. This presumes that the obstacles are disjoint.

This approach uses β to form a repulsive-like function. The attractive portion of the navigation function is a power of distance to the goal, i.e.,

(4.9) $\gamma_\kappa(q) = (d(q, q_{\text{goal}}))^{2\kappa},$

where γ_κ has zero value at the goal and continuously increases as q moves away from the goal. The function $\frac{\gamma_\kappa}{\beta}(q)$ is equal to zero only at the goal, and it goes to infinity as q approaches the boundary of *any* obstacle. More importantly, for a large enough κ, the function $\frac{\gamma_\kappa}{\beta}(q)$ has a unique minimum. This is true because as κ increases, the term $\partial\gamma_\kappa/\partial q$ dominates $\partial\beta/\partial q$, meaning that the gradient of $\frac{\gamma_\kappa}{\beta}$ points toward the goal. Essentially, increasing κ has the effect of making $\frac{\gamma_\kappa}{\beta}$ take the form of a steep bowl centered at the goal. Increasing κ also causes other critical points to gravitate toward the obstacles, as the range of repulsive influence of the obstacles becomes small relative to the overwhelming influence of the attractive field.

Near an obstacle, only that obstacle has a significant effect on the value of $\frac{\gamma_\kappa}{\beta}$. Therefore, the only opportunity for a local minimum to appear is along a radial line between the obstacle and the goal. On this line near the boundary of an obstacle, the Hessian of $\frac{\gamma_\kappa}{\beta}$ cannot be positive definite because $\frac{\gamma_\kappa}{\beta}$ is quickly decreasing in value moving from the obstacle to the goal. Therefore there cannot be any local minimum for large κ, except at the goal [239].

So $\frac{\gamma_\kappa}{\beta}$ has a unique minimum, but unfortunately it can have arbitrarily large values, making it difficult to compute. Therefore, we introduce the analytical switch, which is defined as

(4.10) $\sigma_\lambda(x) = \dfrac{x}{\lambda + x}, \quad \lambda > 0.$

Since $\sigma_\lambda(x)$ is zero at $x = 0$, converges to one as x approaches ∞, and is continuous (figure 4.13), we can use $\sigma_\lambda(x)$ to bound the value of the function $\frac{\gamma_\kappa}{\beta}$, i.e.,

(4.11) $s(q, \lambda) = \left(\sigma_\lambda \circ \dfrac{\gamma_\kappa}{\beta} \right)(q) = \left(\dfrac{\gamma_\kappa}{\lambda\beta + \gamma_\kappa} \right)(q).$

The function $s(q, \lambda)$ has a zero value at the goal, unitary value on the boundary of any obstacle, and varies continuously in the free space. It has a unique minimum for a large enough κ. However, it is still not necessarily a Morse function because it may have degenerate critical points. So, we introduce another function that essentially

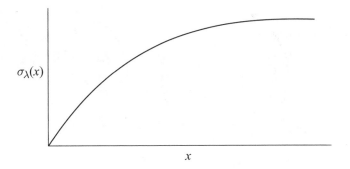

Figure 4.13   Analytic switch function which is used to map the range of a function to the unit interval.

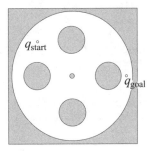

Figure 4.14   Configuration space bounded by a circle with five circle obstacles.

sharpens $s(q, \lambda)$ so its critical points become nondegenerate, i.e., so that $s(q, \lambda)$ can become a Morse function. This sharpening function is

(4.12) $\xi_\kappa(x) = x^{\frac{1}{\kappa}}.$

For $\lambda = 1$, the resulting navigation function on a sphere-world is then

(4.13) $\varphi(q) = \left(\xi_\kappa \circ \sigma_1 \circ \dfrac{\gamma_\kappa}{\beta} \right)(q) = \dfrac{d^2(q, q_{\text{goal}})}{[(d(q, q_{\text{goal}}))^{2\kappa} + \beta(q)]^{1/\kappa}},$

which is guaranteed to have a single minimum at q_{goal} for a sufficiently large κ [239]. Consider the configuration space in figure 4.14. The effect of increasing κ can be seen in figure 4.15, which plots the contour lines for φ as κ increases. For $\kappa = 3$, φ has three local minima, one of which is the global minimum. For $\kappa = 4$ and 6, the local minima become more apparent because it is easier to see the contour lines (actually loops) that encircle the local minima. For $\kappa = 7$ and 8, the "bad" minima are there

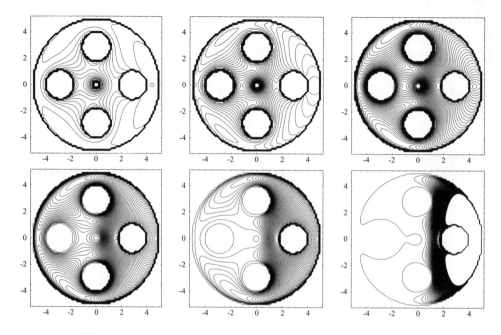

Figure 4.15 Navigation function for a sphere-space with five obstacles for $\kappa = 3$, $\kappa = 4$, $\kappa = 6$, $\kappa = 7$, $\kappa = 8$, and $\kappa = 10$.

but hard to see. Eventually, the "bad" minima morph into saddle points, which are unstable. For $\kappa = 10$, φ has a unique minimum. Therefore, gradient descent will direct the robot to the goal.

We can see the effect of the potential function steepening, critical points gravitating toward the goal, and local minima turning into saddles, in figure 4.16. Unfortunately, this steepening effect has an adverse consequence. The drawback to this particular navigation function is that it is flat near the goal and far away from the goal, but has sharp transitions in between (figure 4.16). This makes implementation of a gradient descent approach quite difficult because of numerical errors.

4.6.2 Star-Space

The result of sphere-spaces is just the first step toward a more general planner. A sphere-space can serve as a "model space" for any configuration space that is diffeomorphic to the sphere-space. Once we have a navigation function for the model space, to find a navigation function for the diffeomorphic configuration space, we need only find the diffeomorphism relating the two spaces.

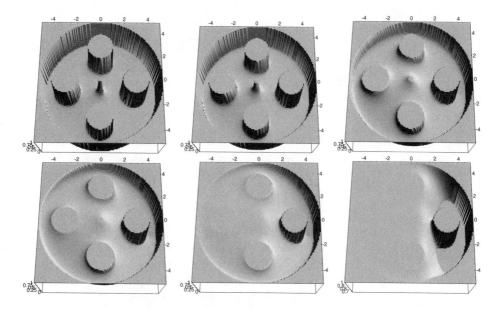

Figure 4.16   Navigation function for a sphere-space with five obstacles for $\kappa = 3$, $\kappa = 4$, $\kappa = 6$, $\kappa = 7$, $\kappa = 8$, and $\kappa = 10$.

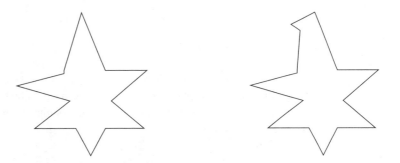

Figure 4.17   (Left) Star-shaped set. (Right) Not a star-shaped set.

In this subsection we consider *star-spaces* consisting of a *star-shaped* configuration space populated by star-shaped obstacles. A star-shaped set S is a set where there exists at least one point that is within line of sight of all other points in the set, i.e.,

$$\exists x \text{ such that } \forall y \in S, \qquad tx + (1-t)y \in S \quad \forall t \in [0, 1].$$

See figure 4.17. All convex sets are star-shaped, but the converse is not true.

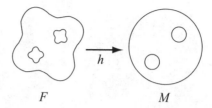

Figure 4.18   The diffeomorphism h maps the star-space F to the sphere-space M.

The approach is to map a configuration space populated by star-shaped obstacles into a space populated by sphere-shaped obstacles. It can be shown [364] that for two free configuration spaces M and F, if $\varphi : M \rightarrow [0, 1]$ is a navigation function on M and there exists a mapping $h : F \rightarrow M$ which is a diffeomorphism, i.e., it is smooth, bijective, and has a smooth inverse, then $\phi = \varphi \circ h$ is a navigation function on F (see figure 4.18). This diffeormorphism ensures that there is a one-to-one correspondence between critical points. We will use this property to define navigation functions in star-spaces using results from sphere-spaces.

The h mapping between the star- and sphere-spaces will be constructed using a translated scaling map

(4.14) $T_i(q) = v_i(q)(q - q_i) + p_i,$

where

(4.15) $v_i(q) = (1 + \beta_i(q))^{1/2} \dfrac{r_i}{d(q, q_i)},$

where q_i is the center of the star-shaped set, and p_i and r_i are, respectively, the center and radius of the spherical obstacle. Here $\beta_i(q)$ defines a star-shaped set such that $\beta_i(q)$ is negative in the interior, zero on the boundary, and positive in the exterior.

Note that if q is in the boundary of the star-shaped obstacle, then $(1 + \beta_i(q)) = 1$, and thus $T_i(q) = r_i \frac{q-q_i}{d(q,q_i)} + p_i$. In other words, $T_i(q)$ maps points on the boundary of the star-shaped set to a sphere.

For the star-shaped obstacle $\mathcal{QO}_i$, we define the analytical switch

(4.16) $s_i(q, \lambda) = \left(\sigma_\lambda \circ \dfrac{\gamma_k \bar{\beta}_i}{\beta_i} \right)(q) = \left(\dfrac{\gamma_k \bar{\beta}_i}{\gamma_k \bar{\beta}_i + \lambda \beta_i} \right)(q),$

where

(4.17) $\bar{\beta}_i = \displaystyle\prod_{j=0, j \neq i}^{n} \beta_j,$

i.e., $\bar{\beta}_i$ is zero on the boundary of the obstacles except the "current" obstacle $\mathcal{QO}_i$. Note that $s_i(q, \lambda)$ is one on the boundary of $\mathcal{QO}_i$, but is zero at the goal and on the boundary of all other obstacles except $\mathcal{QO}_i$.

We define a similar switch for the goal which is one at the goal and zero on the boundary of the free space, i.e.,

$$(4.18) \qquad s_{q_{\text{goal}}}(q, \lambda) = 1 - \sum_{i=0}^{M} s_i .$$

Now, using the above switches and a translated scaling map, we can define a mapping between star-space and sphere-space as

$$(4.19) \qquad h_\lambda(q) = s_{q_{\text{goal}}}(q, \lambda) T_{q_{\text{goal}}}(q) + \sum_{i=0}^{M} s_i(q, \lambda) T_i(q),$$

where $T_{q_{\text{goal}}}(q) = q$ is just the identity map, used for notational consistency.

Note that $h_\lambda(q)$ is exactly $T_i(q)$ on the boundary of the $\mathcal{QO}_i$ because s_i is one on the boundary of $\mathcal{QO}_i$ and for all $j \neq i$, s_j is zero on the boundary of $\mathcal{QO}_i$ (here we include $s_{q_{\text{goal}}}$ as one of the s_j's). In other words, for each i, $h_\lambda(q)$ is T_i on the boundary of obstacle $\mathcal{QO}_i$, which maps the boundary of a star to a sphere. Moreover, $h_\lambda(q)$ is continuous and thus $h_\lambda(q)$ maps the entire star-space to a sphere-space. It can be shown that for a suitable λ, $h_\lambda(q)$ is smooth, bijective, and has a smooth inverse, i.e., is a diffeomorphism [239]. Therefore, since we have a navigation function on a sphere-space, we also have a navigation function on the star-space.

4.7   Potential Functions in Non-Euclidean Spaces

Putting the issue of local minima aside for a moment, another major challenge for implementing potential functions is constructing the configuration space in the first place. This is especially challenging when the configuration space is non-Euclidean and multidimensional. In order to deal with this difficulty, we will define a potential function in the workspace, which is Euclidean, and then lift it to the configuration space. Here, we compute a gradient in the configuration space as a function of gradients in the workspace. To do so, instead of thinking of gradients as velocity vectors, we will now think of them as forces. We then establish a relationship between a workspace force and a configuration space force. Then we apply this relationship to a single rigid-body robot, i.e., we show how to derive a configuration space force using workspace forces acting a rigid-body robot. Finally, we apply this relationship to a multibody robot. We focus the discussion in this section on the attractive/repulsive potentials from section 4.1.

4.7.1   Relationship between Forces in the Workspace and Configuration Space

Since the workspace is a subset of a low-dimensional space (either $\mathbb{R}^2$ or $\mathbb{R}^3$), it is much easier to implement and evaluate potential functions over the workspace than over the configuration space. Now, we will treat the gradient in the workspace as forces. Naturally, workspace potential functions give rise to workspace forces, but ultimately, we will need forces in the configuration space to determine the path for the robot.

Let x and q be coordinate vectors representing a point in the workspace and the configuration of the robot, respectively, where the coordinates x and q are related by the forward kinematics (chapter 3) $x = \phi(q)$. Let f and u denote generalized forces in the workspace and the configuration space, respectively. To represent a force f acting at a point $x = \phi(q)$ in the workspace as a generalized force u acting in the robot's configuration space, we use the principle of virtual work, which essentially says that work (or power) is a coordinate-independent quantity. This means that power measured in workspace coordinates must be equal to power measured in configuration space coordinates. In the workspace, the power done by a force f is the familiar $f^T \dot{x}$. In the configuration space, power is given by $u^T \dot{q}$. From chapter 3, section 3.8, we know that $\dot{x} = J\dot{q}$, where $J = \partial\phi/\partial q$ is the Jacobian of the forward kinematic map. Therefore, the mapping from workspace forces to configuration space forces is given by

$$f^T J\dot{q} = u^T \dot{q}$$
$$f^T J = u^T$$
$$J^T f = u.$$

EXAMPLE 4.7.1 (A Force Acting on a Vertex of a Polygonal Robot)   *Consider the polygonal robot shown in figure 4.19. The vertex a has coordinates $[a_x, a_y]^T$ in the robot's local coordinate frame. Therefore, if the robot's coordinate frame is located at $[x, y]^T$ with orientation θ, the forward kinematic map for vertex a (i.e., the mapping from $q = [x, y, \theta]^T$ to the global coordinates of the vertex a) is given by*

(4.20) $$\phi(q) = \begin{bmatrix} x + a_x \cos\theta - a_y \sin\theta \\ y + a_x \sin\theta + a_y \cos\theta \end{bmatrix}.$$

The corresponding Jacobian matrix is given by

(4.21) $$J(q) = \frac{\partial\phi}{\partial q}(q) = \begin{bmatrix} 1 & 0 & -a_x \sin\theta - a_y \cos\theta \\ 0 & 1 & a_x \cos\theta - a_y \sin\theta \end{bmatrix}.$$

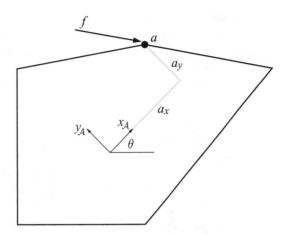

Figure 4.19 The robot $\mathcal{A}$, with coordinate frame oriented at angle θ from the world frame, and vertex a with local coordinates (a_x, a_y).

Therefore, the configuration space force is given by

$$(4.22) \quad \begin{bmatrix} u_x \\ u_y \\ u_\theta \end{bmatrix} = J^T(q) \begin{bmatrix} f_x \\ f_y \end{bmatrix}$$

$$= \begin{bmatrix} 1 & 0 \\ 0 & 1 \\ -a_x \sin\theta - a_y \cos\theta & a_x \cos\theta - a_y \sin\theta \end{bmatrix} \begin{bmatrix} f_x \\ f_y \end{bmatrix}$$

$$(4.23) \quad = \begin{bmatrix} f_x \\ f_y \\ -f_x(a_x \sin\theta + a_y \cos\theta) + f_y(a_x \cos\theta - a_y \sin\theta) \end{bmatrix}$$

and u_θ corresponds to the torque exerted about the origin of the robot frame. Our result for u_θ can be verified by the familiar torque equation $\tau = r \times f$, where r is the vector from the robot's origin to the point of application of f, and $\tau = u_\theta$.

4.7.2 Potential Functions for Rigid-Body Robots

As before, our goal in defining potential functions is to construct a potential function that repels the robot from obstacles, with a global minimum that corresponds to q_{goal}. In the configuration space, this task was conceptually simple because the robot was represented by a single point, which we treated as a point mass under the influence of

a potential field. In the workspace, things are not so simple; the robot has finite area
in the plane and volume in three dimensions. Evaluating the effect of a potential field
on the robot would involve computing an integral over the area/volume defined by
the robot, and this can be quite complex (both mathematically and computationally).
An alternative approach is to select a subset of points on the robot, called control
points, and to define a workspace potential for each of these points. Evaluating the
effect of the potential field on a single point is no different from the evaluations
required in section 4.1. We then use the relationship established in the previous
subsection to convert the individual workspace forces to configuration space forces.
We then add them to get the total configuration space force. As a result, we have
approximately "lifted" the total workspace forces on the robot to a generalized force
in the configuration space.

We need to pick control points $\{r_i\}$ on the robot. The minimum number of control
points depends upon the number of degrees of freedom of the robot. It is the number
of points required to "pin down" the robot. For example, for a rigid-body robot in the
plane, we can fix the position of the robot by fixing the position of two of its points.
For each r_j, the attractive potential is

$$U_{\mathrm{att},j}(q) = \begin{cases} \frac{1}{2}\zeta_i d^2(r_j(q), r_j(q_{\mathrm{goal}})), & d(r_j(q), r_j(q_{\mathrm{goal}})) \le d^*_{\mathrm{goal}} \\ d^*_{\mathrm{goal}}\zeta_j d(r_j(q), r_j(q_{\mathrm{goal}})) - \frac{1}{2}\zeta_j d^*_{\mathrm{goal}}, & d(r_j(q), r_j(q_{\mathrm{goal}})) > d^*_{\mathrm{goal}}. \end{cases}$$

With this potential function, the workspace force for attractive control point r_i is
defined by

$$\nabla U_{\mathrm{att},j}(q) = \begin{cases} \zeta_i(r_j(q) - r_j(q_{\mathrm{goal}})), & d(r_j(q), r_j(q_{\mathrm{goal}})) \le d^*_{\mathrm{goal}}, \\ \frac{d^*_{\mathrm{goal}}\zeta_j(r_j(q)-r_j(q_{\mathrm{goal}}))}{d(r_j(q),r_j(q_{\mathrm{goal}}))}, & d(r_j(q), r_j(q_{\mathrm{goal}})) > d^*_{\mathrm{goal}}. \end{cases}$$

For the workspace repulsive potential fields, we use the same control points $\{r_j\}$,
and define the repulsive potential for r_j as

$$(4.24) \quad U_{\mathrm{repi},j}(q) = \begin{cases} \frac{1}{2}\eta_j \left(\frac{1}{d_i(r_j(q))} - \frac{1}{Q^*_i} \right)^2, & d_i(r_j(q)) \le Q^*_i, \\ 0, & d_i(r_j(q)) > Q^*_i, \end{cases}$$

where $d_i(r_j(q))$ is the shortest distance between the control point r_j and obstacle
$\mathcal{WO}_i$, and Q^*_i is the workspace distance of influence for obstacles. The gradient of
each $U_{\mathrm{repi},j}$ corresponds to a workspace force,

$$(4.25) \quad \nabla U_{\mathrm{repi},j}(q) = \begin{cases} \eta_j \left(\frac{1}{Q^*_i} - \frac{1}{d_i(r_j(q))} \right) \frac{1}{d_i^2(r_j(q))} \nabla d_i(r_j(q)), & d_i(r_j(q)) \le Q^*_i, \\ 0, & d_i(r_j(q)) > Q^*_i. \end{cases}$$

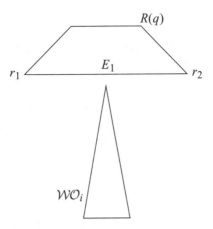

Figure 4.20   The repulsive forces exerted on the robot vertices r_1 and r_2 may not be sufficient
to prevent a collision between edge E_1 and the obstacle.

Often the vertices of the robot are used as the repulsive control points, but it is
important to note that this selection of repulsive control points does not guarantee
that the robot cannot collide with an obstacle. Figure 4.20 shows an example where this
is the case. In this figure, the repulsive control points r_1 and r_2 are very far from the
obstacle $\mathcal{WO}_i$, and therefore the repulsive influence may not be great enough to
prevent the robot edge E_1 from colliding with the obstacle. For this reason, we could
add a *floating* repulsive control point, r_{float}. The floating control point is defined as
that point on the boundary of the robot that is closest to any workspace obstacle.
Obviously the choice of r_{float} depends on the configuration q. For the example shown
in figure 4.20, r_{float} would be located at the center of edge E_1, thus repelling the robot
from the obstacle. The repulsive force acting on r_{float} is defined in the same way as
for the other control points, using (4.25).

The total configuration space force acting on the robot is the sum of the configuration space forces that result from all attractive and repulsive control points, i.e.,

$$U(q) = \sum_{j} U_{\text{att}\,j} + \sum_{i}\sum_{j} U_{\text{rep}\,i\,j}$$

$$(4.26) \qquad = \sum_{j} J_j^T(q)\nabla U_{\text{att}\,i\,j}(q) + \sum_{i}\sum_{j} J_j^T(q)\nabla U_{\text{rep}\,i\,j}$$

in which $J_j(q)$ is the Jacobian matrix for control point r_j. It is essential that the
addition of forces be done in the configuration space and *not* in the workspace.

Path-Planning Algorithm

Having defined a configuration space force, which we will again treat as a velocity, we can use the same gradient descent method for this case as in section 4.1. As before, there are a number of design choices that must be made.

ζ_j controls the relative influence of the attractive potential for control point r_j. It is not necessary that all of the ζ_i be set to the same value. We might choose to weight one of the control points more heavily than the others, producing a "follow the leader" type of motion, in which the leader control point is quickly attracted to its final position, and the robot then reorients itself so that the other attractive control points reach their final positions.

η_j controls the relative influence of the repulsive potential for control point r_j. As with the ζ_i it is not necessary that all of the η_j be set to the same value.

Q_i^* We can define a distinct Q_i^* for each obstacle. In particular, we do not want any obstacle's region of influence to include the goal position of any repulsive control point. We may also wish to assign distinct Q_i^*'s to the obstacles to avoid the possibility of overlapping regions of influence for distinct obstacles.

4.7.3   Path Planning for Articulated Bodies

It is straightforward to extend the methods of the previous subsection to the case of articulated manipulators. Attractive control points are defined on the end effector and repulsive control points are placed on the links. It may be a good idea to use at least one floating control point for each link of the robot, since each link is a rigid body and we would like to prevent the links from colliding with obstacles. For each control point, a Jacobian matrix is computed (see chapter 3, section 3.8). These Jacobians map workspace forces to generalized configuration space forces (joint torques for revolute joints, joint forces for prismatic joints). With these exceptions, the formalism of section 4.7.2 can be directly applied to the path-planning problem for articulated arms (of course the implementation may be a bit more difficult, since the Jacobians may be a bit more difficult to construct, and computing distances to polyhedrons in three dimensions is a bit more involved than computing distances to polygons in the plane). Naturally, this method will be plagued with local minima.

Problems

1. Prove that $d_i(x)$ is a local minimum of $\rho(x, s)$ with respect to s. Show that $D(x)$ can be defined in terms of $d_i(x)$.

2. Does the wave-front planner in a discrete grid yield the shortest distance? (If so, in what metric?)

3. Write a program that determines a path between two points in a planar grid using the wave-front planner. Input from a file a set of obstacles in a known workspace. This file should be a list of vertices and the program should automatically convert the polygonal representation to a binary grid of pixels. Input from the keyboard a start and goal location and write a program to display a meaningful output that a path is indeed determined. Use either four-point or eight-point connectivity.

4. Write a program that determines a path for a planar convex robot that can orient from a start to a final configuration using the wave-front planner. Input from a file a robot and from a separate file a set of obstacles in a known workspace. Input a start and goal configuration from the keyboard. Hand in meaningful output where the robot *must* orient to get from start to goal. Hand in meaningful output where a path is not found.

5. The two-link manipulator in figure 4.21 has no joint limits. Use the wavefront planner to draw the shortest path between the start and goal configurations.

6. Implement, either in simulation or on a mobile robot, a sensor-based attractive/repulsive potential function.

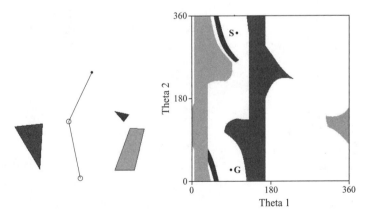

Figure 4.21 (Left) The initial configuration of a two-link manipulator in a polygonal workspace. (Right) The configuration space of the two-link manipulator with a start and goal configuration labeled S and G respectively.

7. Implement the attractive/repulsive potential function for a point robot in a configuration space with the following obstacles

   (a) polygons
   (b) polygons and circles
   (c) polyhedrons
   (d) polyhedrons and spheres
   (e) polyhedrons, spheres, and cylinders.

8. Adapt the attractive/repulsive potential function method to handle moving obstacles.

9. Explain why the paths resulting from the Bug2 algorithm and the navigation potential function look similar.

5 *Roadmaps*

AS DESCRIBED in chapters 2 and 4, a planner plans a path from a particular start configuration to a particular goal configuration. If we knew that many paths were to be planned in the same environment, then it would make sense to construct a data structure once and then use that data structure to plan subsequent paths more quickly. This data structure is often called a *map,* and *mapping* is the task of generating models of robot environments from sensor data. Mapping is important when the robot does not have *a priori* information about its environment and must rely on its sensors to gather information to incrementally construct its map. In the context of indoor systems, three map concepts prevail: topological, geometric, and grids (see figure 5.1).

Topological representations aim at representing environments with graphlike structures, where nodes correspond to "something distinct" and edges represent an adjacency relationship between nodes. For example, places may be locations with specific distinguishing features, such as intersections and T-junctions in an office building, and edges may correspond to specific behaviors or motion commands that enable the robot to move from one location to another, such as wall-following. Recently, it has become popular to augment topological maps with metric information (e.g., relative distance, angle) to help disambiguate places that "look" the same [108, 250, 382, 418] or to use them for navigation [188, 213, 240, 339].

Geometric models use geometric primitives for representing the environment. Mapping then amounts to estimating the parameters of the primitives to best fit the sensor observations. In the past, different representations have been used with great success. Many researchers use line segments [27, 122, 169, 180, 334] to represent parts of the

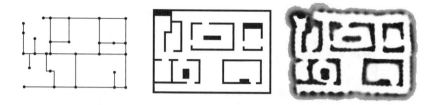

Figure 5.1   Different ways to represent an environment: topologically, geometrically, and using grids.

environment. Popular approaches also represent three-dimensional structures of the environment with triangle meshes [17, 161, 182, 416].

Finally occupancy grids are grid structures, similar as those described in chapter 4, where the value of each pixel corresponds to the likelihood that its corresponding portion of workspace or configuration space is occupied [142]. Occupancy grids were first introduced for mapping unknown spaces with wide-angle ultrasonic sensors; this topic is discussed in chapter 9.

This chapter focuses on a class of topological maps called *roadmaps* [91, 262]. A roadmap is embedded in the free space and hence the nodes and edges of a roadmap also carry physical meaning. For example, a roadmap node corresponds to a specific location and an edge corresponds to a path between neighboring locations. So, in addition to being a graph, a roadmap is a collection of one-dimensional manifolds that captures the salient topology of the free space.

Robots use roadmaps in much the same way people use highway systems. Instead of planning every possible side-street path to a destination, people usually plan their path to a network of highways, then along the highway system, and finally from the highway to their destination. The bulk of the motion occurs on the highway system, which brings the motorist from near the start to near the goal (figure 5.2).

Likewise, using a roadmap, the planner can construct a path between any two points in a connected component of the robot's free space by first finding a collision-free path onto the roadmap, traversing the roadmap to the vicinity of the goal, and then constructing a collision-free path from a point on the roadmap to the goal. The bulk of the motion occurs on the roadmap and thus searching does not occur in a multidimensional space, whether it be the workspace or the configuration space. If the robot knows the roadmap, then it in essence knows the environment. So one way a robot can explore an unknown environment is by relying on sensor data to construct a roadmap and then using that roadmap to plan future excursions into the environment. We now formally define the roadmap.

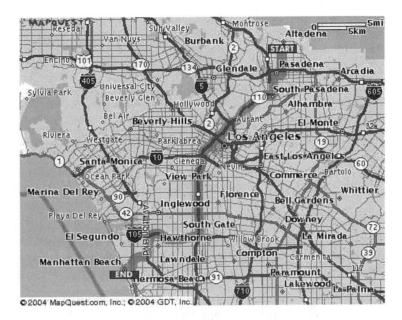

Figure 5.2   Los Angeles freeway system: Planning a path from Pasadena to the Manhattan Beach requires finding a path onto the 110, then to the 105 and 405, and finally from the 405 to the beach. Courtesy of Mapquest.

DEFINITION 5.0.2 (Roadmap)   *A union of one-dimensional curves is a **roadmap** RM if for all q_{start} and q_{goal} in $\mathcal{Q}_{free}$ that can be connected by a path, the following properties hold:*

1. **Accessibility**: *there exists a path from $q_{start} \in \mathcal{Q}_{free}$ to some $q'_{start} \in RM$,*

2. **Departability**: *there exists a path from some $q'_{goal} \in RM$ to $q_{goal} \in \mathcal{Q}_{free}$, and*

3. **Connectivity**: *there exists a path in RM between q'_{start} and q'_{goal}.*

In this chapter, we consider five types of roadmaps: *visibility maps, deformation retracts, retract-like structures, piecewise retracts* and *silhouettes*. All of these roadmaps have a corresponding graph representation. Visibility maps tend to apply to configuration spaces with polygonal obstacles. Nodes of the map are the vertices of the polygons and for visibility maps we can use the terms node and vertex interchangeably. Two nodes of a visibility map share an edge if their corresponding vertices are within line of sight of each other. Deformation retractions are analogous to melting ice or burning grassland. As an arbitrary shaped piece of ice melts, a resulting "stick figure" forms. The ice represents the robot's free space and since the stick figure captures the macroscopic properties of the piece of ice, it can be used for path planning in the robot's free space The representation used for silhouette methods is constructed by repeatedly projecting a shadow of the robot's multidimensional free space onto lower-dimensional spaces until a one-dimensional network is formed.

5.1 Visibility Maps: The Visibility Graph

The defining characteristics of a visibility map are that its nodes share an edge if they are within line of sight of each other, and that all points in the robot's free space are within line of sight of at least one node on the visibility map. This second statement implies that visibility maps, by definition, possess the properties of accessibility and departability. Connectivity must then be explicitly proved for each map for the structure to be a roadmap. In this section, we consider the simplest visibility map, called the *visibility graph* [262, 298].

5.1.1 Visibility Graph Definition

The standard visibility graph is defined in a two-dimensional polygonal configuration space (figure 5.3). The nodes v_i of the visibility graph include the start location, the goal location, and all the vertices of the configuration space obstacles. The graph edges e_{ij} are straight-line segments that connect two line-of-sight nodes v_i and v_j, i.e.,

$$e_{ij} \neq \emptyset \iff s v_i + (1 - s) v_j \in \text{cl}(\mathcal{Q}_{\text{free}}) \ \ \forall s \in [0, 1].$$

Figure 5.3 Polygonal configuration space with a start and goal.

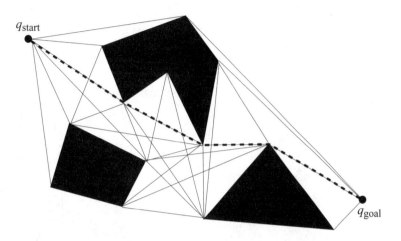

Figure 5.4   The thin solid lines delineate the edges of the visibility graph for the three obstacles represented as filled polygons. The thick dotted line represents the shortest path between the start and goal.

Note that we are embedding the nodes and edges in the free space and that edges of the polygonal obstacles also serve as edges in the visibility graph.

By definition, the visibility graph has the properties of accessibility and departability. We leave it to the reader as an exercise to prove the visibility graph is connected in a connected component of free space. Using the standard two-norm (Euclidean distance), the visibility graph can be searched for the shortest path (figure 5.4) [366]. The visibility graph can be defined for a three dimensional configuration space populated with polyhedral obstacles, but it does not necessarily contain the shortest paths in such a space.

Unfortunately, the visibility graph has many needless edges. The use of *supporting* and *separating* lines can reduce the number of edges. A supporting line is tangent to two obstacles such that both obstacles lie on the same side of the line. For nonsmooth obstacles, such as polygons, a supporting line l can be tangent at a vertex v_i if $\mathcal{B}_\epsilon(v_i) \bigcap l \bigcap \mathcal{QO}_i = v_i$. A separating line is tangent to two obstacles such that the obstacles lie on opposite sides of the line. See figure 5.5 for an example of supporting and separating lines.

The *reduced visibility graph* is soley constructed from supporting and separating lines. In other words, all edges of the original visibility graph that do not lie on a supporting or separating line are removed. Figure 5.6 contains the reduced visibility graph of the example in figure 5.4. The notion of separating and supporting lines can be used to generalize the visibility graph method for curved obstacles [294].

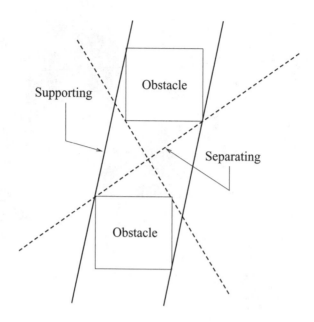

Figure 5.5 Supporting and separating line segments. Note that for polygonal obstacles, we use a nonsmooth notion of tangency.

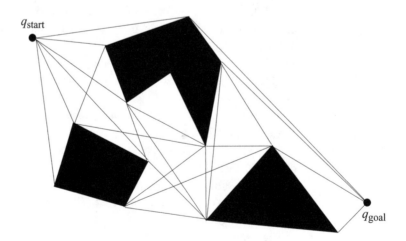

Figure 5.6 Reduced visibility graph.

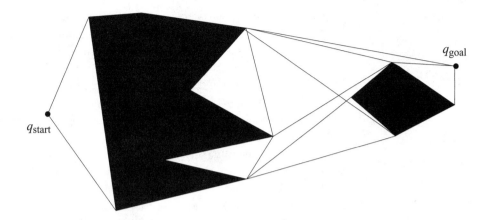

Figure 5.7   Reduced visibility graph with nonconvex obstacles.

At first, the definitions of the supporting and separating lines may seem to only apply to convex obstacles. However, this definition applies to nonconvex shapes as well. Here, we use the notion of local convexity. Recall that convex sets in the plane have the property that for all points on their boundary, there exists a line orthogonal to the surface normal that separates the convex set. This means that the set lies entirely on one side of the line. A set is *locally convex* at a point $\bar{c}$ if the hyperplane tangent to $\bar{c}$ separates the points in a neighborhood of $\bar{c}$ on the boundary of the convex set $\mathcal{QO}_i$. In other words, when N is the surface normal at $\bar{c}$, $\mathcal{QO}_i$ is locally convex at $\bar{c}$ if for all $c \in \left(\mathcal{QO}_i \bigcap \mathrm{nbhd}\,(\bar{c}) \right)$, $(c - \bar{c}) \cdot N \geq 0$ or $(c - \bar{c}) \cdot N \leq 0$. Convex obstacles are locally convex everywhere on the boundary of the set. Figure 5.7 contains a reduced visibility graph for a configuration space with nonconvex obstacles. The reduced visibility graph is beneficial because it has fewer edges making the search for the shortest path more efficient.

5.1.2   Visibility Graph Construction

Let $V = \{v_1, \ldots, v_n\}$ be the set of vertices of the polygons in the configuration space as well as the start and goal configurations. To construct the visibility graph, for each $v \in V$ we must determine which other vertices are visible to v. The most obvious way to make this determination is to test all line segments $\overline{vv_i}$, $v \neq v_i$ to see if they intersect an edge of any polygon. For a particular $\overline{vv_i}$, there are $O(n)$ intersections to check because there are $O(n)$ edges from the obstacles. Now, there are $O(n)$ potential segments emanating from v, so for a particular v, there are $O(n^2)$ tests to determine which vertices are indeed visible from v. This must be done for all $v \in V$ and thus the construction of the visibility graph would have complexity $O(n^3)$.

There is a more efficient way to compute the set of vertices that are visible from v. Imagine a rotating beam of light emanating from a lighthouse beacon. At any moment, the beam illuminates the object that is closest to the lighthouse. Furthermore, as the beam rotates, the obstacle that is illuminated changes only at a finite number of orientations of the beam. If the obstacles in the space are polygons, these orientations occur when the beam is incident on a vertex of some polygon. This insight motivates a class of algorithms known in the computational geometry literature as *plane sweep algorithms*.

A plane sweep algorithm solves a problem by sweeping a line, called the *sweep line*, across the plane, pausing at each of the vertices of the obstacles. At each vertex, the algorithm updates a partial solution to the problem. Plane sweep algorithms are used to efficiently compute the intersections of a set of line segments in the plane, to compute intersections of polygons, and to solve many other computational geometry problems.

For the problem of computing the set of vertices visible from v, we will let the sweep line, l, be a half-line emanating from v, and we will use a rotational sweep, rotating l from 0 to 2π. The key to this algorithm is to incrementally maintain the set of edges that intersect l, sorted in order of increasing distance from v. If a vertex v_i is visible to v, then it should be added to the visibility graph (algorithm 5). It is

---

Algorithm 5 Rotational Plane Sweep Algorithm

Input: A set of vertices $\{v_i\}$ (whose edges do not intersect) and a vertex v
Output: A subset of vertices from $\{v_i\}$ that are within line of sight of v

---

1: For each vertex v_i, calculate α_i, the angle from the horizontal axis to the line segment $\overline{vv_i}$.
2: Create the vertex list $\mathcal{E}$, containing the α_i's sorted in increasing order.
3: Create the active list $\mathcal{S}$, containing the sorted list of edges that intersect the horizontal half-line emanating from v.
4: **for all** α_i **do**
5:    **if** v_i is visible to v **then**
6:       Add the edge (v, v_i) to the visibility graph.
7:    **end if**
8:    **if** v_i is the beginning of an edge, E, not in $\mathcal{S}$ **then**
9:       Insert the E into $\mathcal{S}$.
10:   **end if**
11:   **if** v_i is the end of an edge in $\mathcal{S}$ **then**
12:      Delete the edge from $\mathcal{S}$.
13:   **end if**
14: **end for**

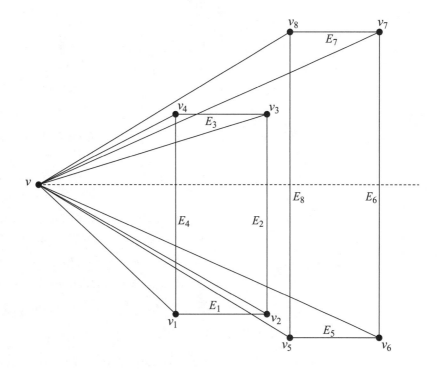

Figure 5.8   An example of the sweep line algorithm at work for an environment containing
two rectangular obstacles.

straightforward to determine if v_i is visible to v. Let $\mathcal{S}$ be the sorted list of edges that
intersects the half-line emanating from v; the set $\mathcal{S}$ is incrementally constructed as
the algorithm runs. If the line segment $\overline{vv_i}$ does not intersect the closest edge in $\mathcal{S}$,
and if l does not lie between the two edges incident on v (the sweep line does not
intersect the interior of the obstacle at v), then v_i is visible from v.

Figure 5.8 shows an example configuration space containing two obstacles with
vertices $v_1, \ldots, v_8$. Table 5.1 shows how the data structures are updated as the algorithm proceeds from initialization to termination. Step 1 of the algorithm determines
the angles, α_i's, at which the line l will pause; such angles correspond to the vertices
of the obstacles. In step 2 of the algorithm, these angles are used to construct the
vertex list, $\mathcal{E}$, and in step 3 the active list $\mathcal{S}$ is initialized. After initialization, $\mathcal{E}$ and $\mathcal{S}$
are the sorted lists:

$\mathcal{E} = \{\alpha_3, \alpha_7, \alpha_4, \alpha_8, \alpha_1, \alpha_5, \alpha_2, \alpha_6, \}$,

$\mathcal{S} = \{E_4, E_2, E_8, E_6\}$.

Vertex	New S	Actions
Initialization	$\{E_4, E_2, E_8, E_6\}$	Sort edges intersecting horizontal half-line
α_3	$\{E_4, E_3, E_8, E_6\}$	Delete E_2 from S. Add E_3 to S.
α_7	$\{E_4, E_3, E_8, E_7\}$	Delete E_6 from S. Add E_7 to S.
α_4	$\{E_8, E_7\}$	Delete E_3 from S. Delete E_4 from S. ADD (v, v_4) to visibility graph
α_8	$\{\}$	Delete E_7 from S. Delete E_8 from S. ADD (v, v_8) to visibility graph
α_1	$\{E_1, E_4\}$	Add E_4 to S. Add E_1 to S. ADD (v, v_1) to visibility graph
α_5	$\{E_4, E_1, E_8, E_5\}$	Add E_8 to S. Add E_5 to S.
α_2	$\{E_4, E_2, E_8, E_5\}$	Delete E_1 from S. Add E_2 to S.
α_6	$\{E_4, E_2, E_8, E_6\}$	Delete E_5 from S. Add E_6 to S.
Termination		

Table 5.1 Table showing the progress of the rotational plane sweep algorithm for the environment of figure 5.8.

At termination, the algorithm has added three new edges to the visibility graph: (v, v_4), (v, v_8), and (v, v_1).

The complexity of algorithm 5 is $O(n^2 \log n)$. The time required by step 1 is $O(n)$, since each vertex must be visited exactly once. For step 2, the required time is $O(n \log n)$, since this is the time required to sort a list of n elements. For step 3, the set of active edges can be computed in $O(n)$ time by merely testing each edge to see if it intersects the horizontal axis. In the worst case, if every edge were to intersect the horizontal axis, this set could be sorted in time $O(n \log n)$. The main loop of the program (step 4) iterates n times (once for each vertex). At each iteration, the algorithm must perform basic bookkeeping operations (insert or delete), but these can be done in time $O(\log n)$ if an appropriate data structure, such as a balanced tree, is used to maintain S Thus, the time required by step 4 is $O(n \log n)$, and therefore the total time complexity of the algorithm is $O(n^2 \log n)$.

Finally, we have not considered here the case when l may simultaneously intersect multiple vertices. In order for this to occur, three vertices must be collinear. When this does occur, the problem can be resolved by slightly perturbing the position of one of the three vertices. When no three vertices are collinear, we say that the polygons are in *general position,* and the general position assumption is common for computational geometry algorithms. It is also possible to modify the visibility test to account for nongeneral configurations, and this is addressed in [124].

5.2   Deformation Retracts: Generalized Voronoi Diagram

The generalized Voronoi diagram (GVD) is the set of points where the distance to the two closest obstacles is the same. Figure 5.9(d) displays an example of the GVD. Path planning is achieved by moving away from the closest point until reaching the GVD, then along the double equidistant GVD to the vicinity of the goal, and then from the GVD to the goal. Since the GVD is defined in terms of distance, one can expect

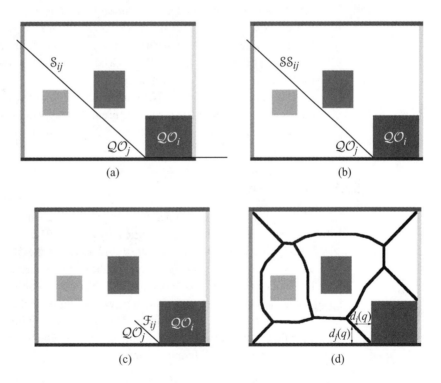

(a)   (b)

(c)   (d)

Figure 5.9   (a) The set $\mathcal{S}_{ij}$ contains points equidistant to two obstacles $\mathcal{QO}_i$ and $\mathcal{QO}_j$. (b) The set $\mathcal{SS}_{ij}$ contains equidistant points with distinct gradients; note that there is no $\mathcal{SS}_{ij}$ structure to the right of the obstacles. We delay discussion of this structure for a moment. (c) The set $\mathcal{F}_{ij}$ has the closest pair of obstacles. (d) The GVD is the union of all such sets.

that a robot equipped with range sensors can incrementally construct the GVD in an unknown space. Once the GVD is constructed, the robot has essentially explored the space because the robot can use the GVD to plan paths in the free space with the GVD.

We show that the GVD is a roadmap because the GVD is a type of *deformation retract*. Deformation retracts are best described by an analogy. Imagine a doughnut-shaped candy: a candy with a hole in the middle of it. As the candy dissolves, eventually a ring remains. This ring captures the topological structure of the candy even though it is significantly smaller than the original. Every point on the ring serves as the center of a corresponding planar disk orthogonal to the ring; each disk is shrunk to a point. In this analogy, the original candy represents the robot's free space and the resulting ring corresponds to a geometric structure called a deformation retract. The function that represents this shrinking process, i.e., the function that maps the filled torus[1] onto a ring, is called a *deformation retraction.*

First in section 5.2.1, we define the GVD and then in section 5.2.2, we show it has the properties of accessibility, connectivity, and departability. In section 5.2.2, we rely on the fact that the GVD is indeed a deformation retract to assure it has the roadmap properties and in section 5.2.3 we describe in more detail as to how the GVD is a deformation retract. Next, in section 5.2.4, we prove that the GVD is indeed one-dimensional. Here, we review the preimage theorem to assert the dimensionality property of the GVD. Finally, in section 5.2.5, we describe three methods to construct the GVD.

5.2.1   GVD Definition

The *Voronoi diagram* is defined for a set of points called *sites* [31]. A *Voronoi region* is the set of points closest to a particular site [31]. The Voronoi diagram is then the set of points equidistant to two sites; it sections off the free space into regions that are closest to a particular site. Points on the Voronoi diagram have two closest sites. In the planar case, the Voronoi diagram is a collection of line segments.

For the purposes of path planning, we can think of the point sites as obstacles, but obstacles are not simple points. Therefore, the definition of a Voronoi region is extended to the *generalized Voronoi region,* $\mathcal{F}_i$, which is the closure of the set of points closest to $\mathcal{QO}_i$. In other words,

(5.1) $\mathcal{F}_i = \{q \in \mathcal{Q}_{\text{free}} \mid d_i(q) \leq d_h(q) \ \ \forall h \neq i\},$

---

1. A torus is two-dimensional structure, and the filled torus is a three-dimensional version, i.e., the convex hull of a torus embedded in $\mathbb{R}^3$.

where $d_i(q)$ is the distance to an obstacle $\mathcal{QO}_i$ from q, i.e., $d_i(q) = \min_{c \in \mathcal{QO}_i} d(q, c)$ (chapter 4, equation (4.6)).

The basic building block of the GVD is the set of points equidistant to two sets $\mathcal{QO}_i$ and $\mathcal{QO}_j$, which we term a *two-equidistant surface* denoted by $\mathcal{S}_{ij} = \{x \in \mathcal{Q} \mid (d_i(q) - d_j(q)) = 0\}$. Note that $d_i(q) - d_j(q) = 0$ is an equivalent way to state $d_i(q) = d_j(q)$ (figure 5.9). A two-equidistant surface pierces obstacles, so we restrict it to the set of points that are both equidistant to $\mathcal{QO}_i$ and $\mathcal{QO}_j$ and have $\mathcal{QO}_i$ and $\mathcal{QO}_j$ as their closest obstacles. This restricted structure is the *two-equidistant face*, which could be denoted by $\mathcal{F}_{ij} = \{q \in \mathcal{S}_{ij} \mid d_i(q) \leq d_h(q) \ \forall h\}^2$. We refine this definition shortly. The union of the two-equidistant faces forms the GVD, i.e.,

$$\text{GVD} = \bigcup_i \bigcup_j \mathcal{F}_{ij}.$$

This definition of the GVD applies to any dimensional spaces. One can see that the GVD partitions the free space into regions $\mathcal{F}_i$ such that points in the interior of one $\mathcal{F}_i$ are closer to $\mathcal{QO}_i$ than to any other obstacle. Points on the GVD have two or more closest obstacles. In the planar case, we term the $\mathcal{F}_{ij}$ as *GVD edges* and they terminate at either *meet points,* the set of points equidistant to three or more obstacles ($\mathcal{F}_{ijk}$), or *boundary points,* the set of points whose distance to the closest obstacle is zero. Boundary points are the endpoints of "spokes" of the GVD.

5.2.2 GVD Roadmap Properties

In $\mathbb{R}^m$, the GVD has the properties of accessibility, connectivity, and departability. In the plane, the GVD is a roadmap because it has these properties and is one-dimensional. We show that the planar GVD is one-dimensional in the next sub-section and the properties of accessibility, connectivity, and departability here. The robot achieves accessibility by moving away from the closest obstacle; it performs gradient ascent of distance D to the closest obstacle, i.e.,

(5.2)
$$\frac{dc(t)}{dt} = \nabla D(c(t)) \quad \text{where} \quad c(0) = q_{\text{start}},$$

until it reaches a point on the GVD.

Equation (5.2) is a first order differential equation implicitly defining the the path $c : [0, 1] \rightarrow \mathcal{Q}_{\text{free}}$. At any point $c(t) \in \mathcal{Q}_{\text{free}}$, the tangent to the path is defined by the gradient of distance to the closest obstacle. The gradient $\nabla D(q)$ points in the direction that maximally increases distance. The tangent of the curve $\frac{dc(t)}{dt}$ is "set"

---

2. Note that we could have written $d_i(q) = d_j(q) \leq d_h(q)$, but the "$= d_j(q)$" is already implied by the $q \in \mathcal{S}_{ij}$.

to the gradient of distance. By constantly following the distance gradient, a path is traced that maximally increases the distance.

LEMMA 5.2.1 (Accessibility of the GVD)   *In an obstacle-bounded environment, gradient ascent of D traces a path from any point in the free space to the GVD.*

Proof   Assume the robot starts at a point q that is not on the GVD. Let $\mathcal{QO}_i$ be the closest obstacle to q. Hence $d_i(q) = D(q)$ and $d_h(q) > d_i(q)$ for all h. The robot traces a path $\frac{dc(t)}{dt} = \nabla d_i(c(t))$ where $c(0) = q$. Since the environment is bounded, continuity of the distance function guarantees that there exists a $\bar{t} \in \mathbb{R}$ and a $\mathcal{QO}_j$, such that $d_i(c(\bar{t})) = d_j(c(\bar{t}))$.   ∎

We use the fact that the GVD is a *deformation retract* to ensure connectivity of the GVD. A deformation retract is the image of a continuous function called a *deformation retraction RM* such that

$RM(q) = q$,   for all q in the GVD,
$RM(q) = q'$,   for any $q \in \mathcal{Q}_{\text{free}}$ and $q' \in \text{GVD}$.

We more formally define the deformation retraction in the next section.

For the GVD, the gradient ascent accessibility procedure *implicitly* defines the deformation retraction without explicitly doing so [340]. In other words, if q is on the GVD, then the image of q is q, i.e., $RM(q) = q$. If q is in the free space but not in the GVD, then the image q is the q' in the GVD that is obtained by moving away from the closest point on the closest obstacle until encountering the GVD, i.e., $RM(q) = q'$.

Connectivity of the GVD is then a consequence of continuity of the RM function. In other words, since RM is continuous, for each connected component of the free space there is a connected component of the GVD. Therefore, there exists a path that connects q_{start} and q_{goal} if and only if there exists a path in the GVD that connects q'_{start} and q'_{goal} where $q'_{\text{start}} = RM(q_{\text{start}})$ and $q'_{\text{goal}} = RM(q_{\text{goal}})$.

Departability is simply accessibility in reverse. However, there are other ways to achieve departability. It can be shown that all points in free space have at least one point on the GVD within line of sight, i.e.,

$$\forall\ q \in \mathcal{Q}_{\text{free}},\ \exists q' \in \text{GVD such that } sq + (1-s)q' \in \mathcal{Q}_{\text{free}}\ \forall s \in [0, 1].$$

This means that if the robot comes within line of sight of the goal, the robot can drive straight toward it. This approach to departability only makes sense if the robot can detect the goal using its on-board sensors.

5.2.3 Deformation Retract Definition

Before defining the deformation retract, we define a weaker structure called a *retract*. For a manifold X, a *retraction* is a continuous function $f : X \to A$ such that $A \subset X$, and $f(a) = a$ for all $a \in A$ [410]. The subset A is the retract. Typically, the dimension of A is less than the dimension of X.

The set of deformation retracts is a subset of the set of retracts and hence the GVD is a retract also. However, the properties of a retract are not sufficient to guarantee that the GVD is a roadmap. It is the fact that that GVD is indeed a deformation retract that makes it a roadmap. Essentially, a deformation retract inherits many topological properties from its ambient space, whereas a retract may not. One important property is that the number of "types" of closed paths in the free space is equal to the number of "types" of closed paths in the deformation retract of the free space.

Let's return to the candy example from the beginning of this section. Although a retract can be a ring, it could also be a single point, a two-dimensional disk orthogonal to the ring, etc. We need to enforce additional properties on the retract so as to guarantee that it captures the topology of its free space and is still one-dimensional. Recall from chapter 3, section 3.4.1 that global diffeomorphisms are mappings that relate spaces that are "topologically similar." Diffeomorphic spaces must have same dimension. Now, we consider spaces that are similar, but of different dimensions.

Let $f : U \to V$ and $g : U \to V$ where U and V are manifolds. A *homotopy* is a continuous function $H : U \times [0, 1] \to V$ such that $H(x, 0) = f(x)$ and $H(x, 1) = g(x)$. An example of H is $H(x, t) = (1 - t)f(x) + tg(x)$. If there exists such a continuous mapping that "deforms" f to g, then f and g are *homotopic,* and the resulting equivalence relation is denoted $f \sim g$. We can also say that two paths f and g are path-homotopic, i.e., $f \sim g$, if they can be continuously deformed into one another. This relation allows for the classification of functions into equivalence classes termed *path-homotopy classes* and are denoted as

$$[c] = \{\bar{c} \in C^0 \mid \bar{c} \sim c\}.$$

where c is a representative element of the class.

Let $A \subset X$ and let $f : X \to A$ be a retraction. A *deformation retraction* is a homotopy $H : X \times [0, 1] \to X$ such that

- $H(x, 0) = x$

- $H(x, 1) \in A$

- $H(a, t) = a$ for $a \in A$ and $t \in [0, 1]$

In other words, H is a homotopy between a retraction and the identity map[3]. Note that all retractions are not necessarily homotopic to the identity map. The retract is now called a *deformation retract*.

We use deformation retractions to smoothly deform, without tearing or pasting X onto a lower, preferably one-dimensional subset A of X. So, as t varies from 0 to 1, a point in X continuously moves through X to a point in A. Moreover, a point y in a neighborhood of x also continuously moves through X to a point in A such that $H(x, t)$ and $H(y, t)$ are close to each other as t varies from 0 to 1. Thefore, the deformation retraction preserves many topological properties of the free space. Thus, while a diffeomorphism preserves the structure of two spaces of the same dimension, a deformation retraction preserves the structure of two spaces of different dimension.

One of the key topological properties of deformation retracts is that they preserve the number of homotopically equivalent closed loops from the ambient space. The number of homotopy equivalence classes of closed loops is called the *first fundamental group,* and is denoted as $\pi_1(X, x_0)$ for loops in X passing through x_0. Since this is a group, it has a group operator ($\star$) that simply concatenates paths. A set X is simply connected if the fundamental group associated with the set, $\pi_1(X, x_0)$, contains only the identity element (e.g., the group only contains one element). If f is a deformation retraction with A as its deformation retract of X, then $\pi_1(X, x_0) = \pi_1(A, f(x_0))$. In other words, the ambient space X and the deformation retract A have the same number of homotopically equivalent closed loops.

Deformation retracts have the properties of connectivity, accessibility, and departability. For each connected component of X, A is a connected set because the image of a connected set under a continuous mapping is a connected set [9]. The deformation retraction determines a path from the start to the deformation retract, as well as a path from the goal to the retract. Let H be the deformation retraction and $H(x, 0) = q_{\text{start}}$. The path to the deformation retract is then defined by $H(x, \cdot) : [0, 1] \rightarrow Q_{\text{free}}$ where $H(x, 1)$ is an element of the deformation retract. Departability is shown in the same manner. Since the deformation retract is connected, there is a path between the retracted start and retracted goal configurations along the deformation retract. Hence, one-dimensional deformation retracts are roadmaps.

The GVD is a retract because the RM (equation (5.2.2)) has been shown to be continuous and maps all points on the GVD to the GVD. Since RM is continuous, the GVD is connected in a connected component of the free space because the image of connected set under a continuous function is a connected set. The GVD is a deformation retract because RM has been shown to be homotopic to the indentity

---

3. Sometimes, a deformation retraction is defined as a retraction that is homotopic to the identity map [207] as opposed to the homotopy.

map. Therefore, RM smoothly deforms the free space onto a one-dimensional subset and defines the accessibility and departability criteria. Finally, since the GVD is a deformation retract, the number of closed-loop path equivalent classes in the GVD equals the number of closed-loop path equivalent classes in the free space because RM preserves the cardinality of the first fundamental group. This makes the GVD a concise representation of the free space.

5.2.4   GVD Dimension: The Preimage Theorem and Critical Points

A key property of a roadmap is that it is one-dimensional. Actually, we show that in the plane, the GVD consists of one-dimensional manifolds. Before we can demonstrate this, we have to take a more careful look at the definition of the GVD. Recall that we are using the distance function d_i to define the GVD, but this function assumes that the obstacles are convex, which is unrealistic in most situations.

At first, it seems to make sense to decompose nonconvex obstacles into convex pieces. This causes problems because there are many ways to construct such a decomposition, thereby resulting in different representations of the free space. Consider the obstacle in figure 5.10. Both decompositions are valid, but unfortunately they give rise to two different definitions of S_{ij}, the set of points equidistant to two obstacles $\mathcal{QO}_i$ and $\mathcal{QO}_j$. There are infinitely many ways to decompose a nonconvex obstacle and hence the possibility for infinitely many representations.

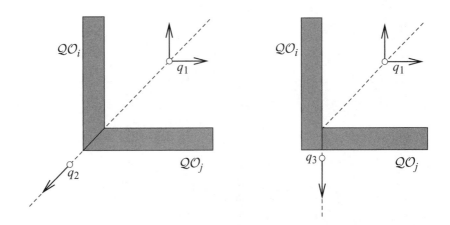

Figure 5.10   A nonconvex obstacle is divided into two pieces, $\mathcal{QO}_i$ and $\mathcal{QO}_j$, but in two different ways. On the left, a diagonal forms the convex obstacles and on the right a vertical cut forms them. Note that in both left and right, the gradient vectors pointing away from the two closest obstacles are distinct at q_1 but they are the same at q_2 and q_3.

It would be nice to have a unique representation of the roadmap, so we refine our definition of the GVD. In figure 5.10, note that there are two portions of $\mathcal{S}_{ij}$: the upper-right portion, which is "between" the two arms of the obstacle and the lower-left portion, which is on the other side of the obstacle. Note that for the portions between the two arms, the gradients to the two closest obstacles are distinct, e.g., $\nabla d_i(q_1) \neq \nabla d_j(q_1)$. However, for the other portions, the gradients line up, e.g., $\nabla d_i(q_2) = \nabla d_j(q_2)$ and $\nabla d_i(q_3) = \nabla d_j(q_3)$. Eliminating the portion of the two-equidistant surface with nondistinct gradient vectors yields a set termed the *two-equidistant surjective surface* denoted as

$$\mathcal{SS}_{ij} = \{ q \in \mathcal{S}_{ij} \mid \nabla d_i(q) \neq \nabla d_j(q) \}.$$

See figure 5.9(b) for an example of a two-equidistant surjective surface defined by a nonconvex obstacle that has been divided into two convex pieces.

This definition of a two-equidistant surjective surface should be salient from a sensor-based perspective. Consider the planar case where distance and gradient vectors can be derived from a laser ranger or a sonar ring which approximates the saturated raw distance function. Recall that the saturated raw distance function corresponds to all of the rays emanating from a single point intersecting as can be seen in figure 5.11.

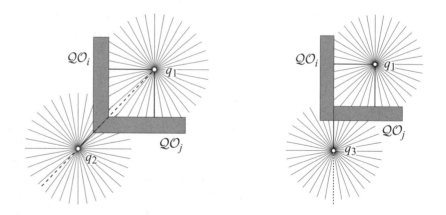

Figure 5.11 A robot is placed at different configurations q_1, q_2 and q_3. It has range sensors radially distributed pointing in a full 360 degrees. The rays emanating from the points correspond to the range readings. Local minima correspond to distance to nearby obstacles. (Left) Use a diagonal cut to break the nonconvex obstacles into convex ones, but a robot cannot see the diagonal cut from q_2 because there is no local minimum. (Right) Use a vertical cut to break the nonconvex obstacles into convex ones, but a robot cannot see the vertical cut from q_3 because there is no local minimum.

On the "inside" of the concavity (at q_1 in figure 5.11), there are two local minima in the raw distance function, whereas on the outside there is one at q_2 and one at q_3. In other words, a robot situated on the "outside" of the obstacle cannot determine from its sensor readings how the obstacle was cut. Another perspective is that on the "inside" of the nonconvex obstacle, the robot "sees" two obstacles and on the "outside," it only "sees" one.

From here, the definition of the two-equidistant face $\mathcal{F}_{ij}$ is modified to be $\mathcal{F}_{ij} = \{q \in \mathcal{SS}_{ij} \mid d_i(q) \le d_h(q) \ \forall h\}$. So the GVD is the set of points equidistant to two obstacles such that the two obstacles are closest and have unique closest points on them.

We are now ready to show that the GVD is indeed one-dimensional. We do this by first rewriting the equidistant relationship $d_i(q) = d_j(q)$ as $d_i(q) - d_j(q) = 0$, which in turn can be written as $(d_i - d_j)(q) = 0$. Intuitively, this one constraint in a two-dimensional space defines a one-dimensional subspace. In other words, equidistance is the preimage of zero under the map $(d_i - d_j) : \mathcal{Q} \to \mathbb{R}$. We use this reformulation to demonstrate that in the plane the GVD comprises one-dimensional manifolds by taking recourse to the preimage theorem [173].

THEOREM 5.2.2 (Preimage Theorem)   *Let M and N be manifolds. Let $G : M \to N \in C^\infty$ and $n \in N$ be a regular value of G. The set $G^{-1}(n) = \{m \in M \mid G(m) = n\}$ is a closed submanifold of M with tangent space given by $T_m G^{-1}(n) = \ker DG(m)$. If N is finitely dimensional, then $\dim(G^{-1}(n)) = \dim(M) - \dim(N)$, i.e., $\dim(G^{-1}(n)) = \dim(M) - \dim(N)$.*

The preimage theorem contains a lot of terminology and notation. A *regular value* is an n where for all $m \in G^{-1}(n)$, the differential $DG(m)$ is surjective (e.g., has full rank). See section C.5.5 for a description of the differential. Next, T_m denotes the *tangent space* at m. So, $T_m M$ is the tangent space at m on the manifold M and $T_p G^{-1}(n)$ is the tangent space at p on the manifold $G^{-1}(n)$, which is a submanifold of M.

A *critical point* is a point where the differential is not surjective and hence loses rank. (For real-valued functions, it is a point where the first derivative vanishes.) Let $\Sigma(G)$ be the set of all critical points of G. For all $q^* \in \Sigma(G)$, $G(q^*)$ are *critical values*. Finally, all points $q \notin \Sigma(G)$ where $DG(q)$ is surjective are termed *regular points* with $G(q)$ as their corresponding *regular values*.

To show that the GVD edges are indeed one-dimensional, we use the preimage theorem to show that they are one-dimensional manifolds. First let's see how the preimage theorem is used to create manifolds. Consider the function $f(x, y) = x^2 + y^2$. The differential $Df(x, y) = [2x, 2y]$. For all $f(x, y) = 91,538$, $Df(x, y) \neq 0$ and thus the preimage of 91,538 under f forms a one-dimensional manifold.

With the GVD, $G = (d_i - d_j)$, and the set of points equidistant to two obstacles is $(d_i - d_j)^{-1}(0)$. However, for all points in the preimage to be regular, DG must be surjective. In other words, $D(d_i - d_j)$ must not be equal to zero. Since in a Euclidean space, $Dd_i(q) = (\nabla d_i(q))^T$, this means $\nabla d_i(q)$ cannot be equal to $\nabla d_j(q)$. However, we are fortunate to have the $\nabla d_i(q) \neq \nabla d_j(q)$ condition in the definition of $\mathcal{SS}_{ij}$. So, in actuality, the $\nabla d_i(q) \neq \nabla d_j(q)$ enforces the surjective condition for the preimage theorem, hence the term *surjective* in the two-equidistant surjective surface. So, by the preimage theorem, $\mathcal{SS}_{ij}$ is one-dimensional in the plane. The set $\mathcal{F}_{ij}$ is a submanifold of $\mathcal{SS}_{ij}$. Therefore, the GVD comprises a set of one-dimensional manifolds (figure 5.9).

5.2.5   Construction of the GVD

We discuss three methods for constructing the planar GVD: the first uses sensor information allowing the robot to construct the GVD in an unknown space; the second assumes the world has polygonal obstacles in which case we can compute complexity information about the GVD; and the final method assumes that the world is a grid allowing for efficient computation.

Sensor-Based Construction of the GVD

Exploring with the GVD is akin to simultaneously generating and exploring a graph that is embedded in the free space. The GVD can be incrementally constructed because it is defined in terms of distance information which is readily provided by range sensors onboard mobile robots. Using such line-of-sight data, the robot initially accesses the GVD and then begins tracing an edge until it encounters a meet point or a boundary point. When the robot encounters a *new* meet point, it marks off the direction from which it came as explored, and then identifies all new GVD edges that emanate from it. From the meet point, the robot explores a new GVD edge until it detects either another meet point or a boundary point. In the case that it detects another *new* meet point, the above branching process recursively repeats. If the robot reaches an *old* meet point, the robot has completed a cycle in the GVD graph and then travels to a meet point with an unexplored edge associated with it. When the robot reaches a boundary node, it simply turns around and returns to a meet point with unexplored GVD edges. When all meet points have no unexplored edges associated with them, exploration is complete.

The robot accesses the GVD by simply moving away from the nearest obstacle until it is equidistant to two obstacles (figure 5.12). Once the robot accesses the GVD, it must incrementally trace the GVD using the same curve tracing technique from

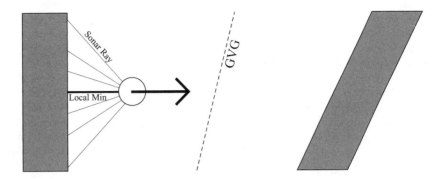

Figure 5.12   The circular disk represents a mobile robot with some of its range sensor readings. The thick ray, labeled local min, corresponds to the smallest sensor reading, and hence the robot will move in the direction indicated by the black arrow to access the GVD, denoted by a dashed line between two nonparallel walls.

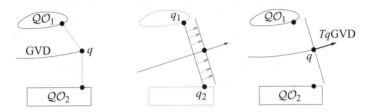

Figure 5.13   The tangent space to the GVD is orthogonal to the line that connects the two closest points on the two closest obstacles.

chapter 2, section 2.3.3, except $G(q) = d_i(q) - d_j(q)$ whose roots are the set of points where $d_i(q) = d_j(q)$. The tangent is the null space of $\nabla G(q)$, which corresponds to a line orthogonal to $\nabla d_i(q) - \nabla d_j(q)$. This is identical to passing a line through the two closest points and taking the vector perpendicular to the line to be the tangent (figure 5.13). A meet point is detected by looking for a sudden change in one of the two closest obstacles.

Polygonal Spaces

In a polygonal environment, obstacles have two features, vertices and edges, thereby making equidistance relationships easy to define. The set of points equidistant to two vertices is a line; the set of points equidistant to two edges is a line; and the set of points equidistant to a vertex and an edge is a parabola. Therefore, by breaking down

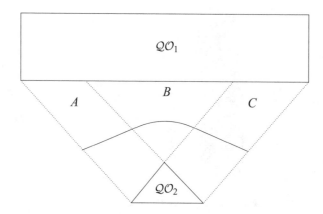

Figure 5.14   GVD edge fragment for two polygonal obstacles.

the free space into regions with the appropriate pair of closest features, one can easily build the GVD. In figure 5.14, regions A and C have a pair of edges as their respective closest features, whereas region B has an edge and vertex as its closest obstacle.

In a polygonal environment with n obstacles and N obstacle vertices, the number of GVD edges falls between $\frac{3(n+1)}{2}$ and $6N + 3n - 3$. The number of nodes on the GVD falls between $\frac{n+5}{2}$ and $4N - n - 2$. See [359] for details.

Grid Configuration Spaces: The Brushfire Method

The method presented in chapter 4, section 4.3.2 can be readily adapted to construct the GVD in a discrete grid. Originally, the input for the brushfire method is a grid of zeros corresponding to free space and ones corresponding to an obstacle. The output of the brushfire method is a discrete map where each pixel in the grid has a value equal to the distance to the closest point on the closest obstacle (the closest pixel with a value of one).

We can view the brushfire method as a wave initially starting at the obstacles and propagating through the free space. As the wave front passes over a pixel, the method assigns a value to the pixel corresponding to how far the wave has traveled. The wave fronts collide at points where the distance to two different obstacles is the same. These are points on the GVD.

The brushfire algorithm can be readily updated to identify the pixels where these collisions occur. Essentially, as the wave propagates, each pixel in the wave front maintains a back pointer to the obstacle pixel from which the wave originated. When the updated brushfire algorithm attempts to assign a "free pixel" with two different back pointers, two wave fronts have collided and the current pixel belongs to the GVD.

5.3   Retract-like Structures: The Generalized Voronoi Graph

Now, we consider the case when $Q = \mathbb{R}^3$. In $\mathbb{R}^3$, the GVD is two-dimensional and therefore reduces the motion planning problem by a single dimension. We use figure 5.9(d) to show this. Imagine extruding the one-dimensional curves in figure 5.9(d) into two-dimensional surfaces in three dimensions; so the one-dimensional curves in figure 5.9(d) become cross sections of two-dimensional sheets. This makes sense because we have a three-dimensional space with one constraint resulting in a two-dimensional subspace. The preimage theorem confirms that the GVD actually comprises two-dimensional manifolds; the dimension of $(d_i - d_j)^{-1}(0)$ is two because $3 - 1 = 2$.

Just as two planes in $\mathbb{R}^3$ generically intersect on a line, two two-equidistant faces intersect and form a one-dimensional manifold. The union of these one-dimensional structures is termed the *generalized Voronoi graph* (GVG) [105, 106]. See figures 5.15 and 5.16 for examples of the GVG in three dimensions.

Figure 5.15   The solid lines represent the GVG for a rectangular enclosure whose ceiling has been removed to expose the interior. Imagine a sphere rolling around touching the removed ceiling, floor, and side wall; the center of this sphere traces a GVG edge.

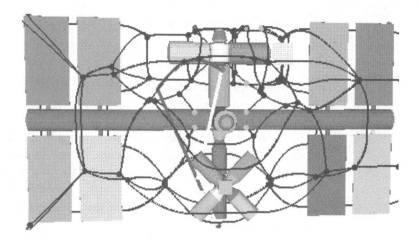

Figure 5.16 The GVG for the International Space Station (in a bounding box). Note that a bounding box was used but is not displayed. Also note how complicated the GVG becomes.

5.3.1 GVG Dimension: Transversality

The GVG edges in $\mathbb{R}^3$ are the set of points equidistant to three obstacles such that the three obstacles are closest and have distinct gradients. Starting with triple equidistance, we define $\mathcal{S}_{ijk} = \{q \mid (d_i - d_j)(q) = 0 \text{ and } (d_i - d_k)(q) = 0\}$. We do not need the additional $(d_j - d_k)(q) = 0$ constraint because $d_i(q) = d_j(q)$ and $d_i(q) = d_k(q)$ imply that $d_j(q) = d_k(q)$. Just like before, we are interested in a subset of $\mathcal{S}_{ijk}$ where the gradients are distinct, and thus

$$\mathcal{SS}_{ijk} = \{q \in \mathcal{S}_{ijk} \mid \nabla d_i(q) \neq \nabla d_j(q), \nabla d_i(q) \neq \nabla d_k(q), \nabla d_j(q) \neq \nabla d_k(q)\}$$
$$= \mathcal{SS}_{ij} \bigcap \mathcal{SS}_{ik} \bigcap \mathcal{SS}_{jk}$$

Note that the transitivity of $d_i(q) = d_j(q)$ and $d_j(q) = d_k(q)$ implies that $d_i(q) = d_k(q)$, but it does not ensure that $\mathcal{SS}_{ijk} = \mathcal{SS}_{ij} \bigcap \mathcal{SS}_{ik}$ because we require *all three gradients* to be distinct. To determine the dimension of the GVG edge, we look at $G : \mathbb{R}^3 \to \mathbb{R}^2$ where

$$G(q) = \begin{bmatrix} (d_i - d_j) \\ (d_i - d_k) \end{bmatrix}(q)$$

whose preimage $G^{-1}(0)$ is the set of points equidistant to three obstacles $\mathcal{QO}_i$, $\mathcal{QO}_j$, and $\mathcal{QO}_k$ when the differential $DG(q)$ is surjective, i.e., does not lose rank. The differential $DG(q)$ can lose rank when either row of $DG(q)$ is zero or the first row is a scalar multiple of the second row in $DG(q)$. We already know by definition that $\nabla d_i(q) \neq$

Figure 5.17 Three ways two lines in the plane can intersect, but only a point-intersection is transversal.

$\nabla d_j(q)$ and $\nabla d_i(q) \neq \nabla d_k(q)$, so all we need to show is that $\nabla(d_i - d_j)(q) \neq \alpha \nabla(d_i - d_k)$ for all $\alpha \in \mathbb{R}$. In other words, we must show that the two rows of $DG(q)$ do not depend upon each other.

We demonstrate this by making a "reasonable" assumption based on *transversality,* a property of how sets intersect. Let's start with a simple example of two intersecting lines in the plane. These lines may intersect in one of three ways: not at all (parallel), at a point (generic), and on a line (overlap) (figure 5.17). The parallel and overlap cases can be viewed as "unstable" because if either line were perturbed a little bit, the intersection would change dimension. The point intersection can be viewed as stable in that if either of the lines were perturbed, a point-type intersection is preserved. We call stable intersections *transversal* and nonstable intersections *nontransversal.* Two lines in three dimensions can never intersect transversally because a generic perturbation can break the intersection to no intersection. In three dimensions, two planes transversally intersect on a line and a plane and a line transversally intersect at a point.

In actuality, transversality is a local property of manifolds. For example, we say that two manifolds may intersect transversally at a point. Since transversality is a local property, we look at the intersection of the tangent spaces, not of the manifolds themselves. If intersection of the tangent spaces is transversal at a point, then the manifolds intersect transversally at that point (figure 5.18). We know from the preimage theorem that the tangent space $T_q G^{-1}(0)$ is given by the set of vectors $\{v \in T_q Q \mid DG(q)v = 0\}$. We assume that surjective equidistant sheets intersect transversally at all points, i.e., $T_q(d_i - d_j)^{-1}(0)$ and $T_q(d_i - d_k)^{-1}(0)$ intersect transversally for all $q \in SS_{ij} \cap SS_{ik}$. If they do not intersect transversally, then after a small perturbation of one of the manifolds, the intersection of the two manifolds will be transversal. In any event, the transversal intersection means that for all $q \in SS_{ij} \cap SS_{ik}$, $\nabla(d_i - d_j)(q) \neq \alpha \nabla(d_i - d_k)(q)$ for all $\alpha \in \mathbb{R}$. Therefore, $DG(q)$ has full rank and we can use the preimage theorem to assure us that SS_{ijk} is indeed a one-dimensional manifold. The GVG in $\mathbb{R}^3$ is then the union of $\mathcal{F}_{ijk} = \{q \in SS_{ijk} \mid d_i(q) \leq d_h(q) \ \forall h\}$, i.e.,

$$(5.3) \qquad \text{GVG} = \bigcup_i \bigcup_j \bigcup_k \mathcal{F}_{ijk}.$$

Structure	Dimension	Codimension	Equidistance	Symbol
GVD	$m - 1$	1	2	$\mathcal{F}_{i_1 i_2}$
GVG	1	$m - 1$	m	$\mathcal{F}_{i_1,\dots,i_m}$

Table 5.2   Comparison of the GVD and the GVG.

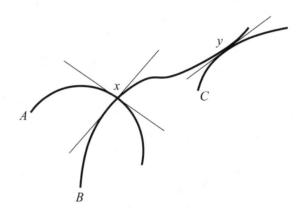

Figure 5.18   The one-dimensional manifolds A and B intersect transversally at x whereas the intersection of B and C at y is not transversal because the tangent spaces at B and C are coincident at y.

In higher dimensions, one can define more equidistant sheets and intersect them to form a GVG. In $\mathbb{R}^m$, the GVG is the set of points equidistant to m obstacles and has dimension one. In contrast, in $\mathbb{R}^m$ the GVD is the set of points equidistant to *two* obstacles and has dimension $m - 1$. Sometimes, an $m - k$-dimensional object lying in an m-dimensional space is said to have *codimension* k; therefore the GVD has codimension one, regardless of the space in which it is defined. When $m = 2$, the GVG and the GVD coincide. For naming convention refer to the GVG as the "one-dimensional" roadmap structure and thus in the plane we will sometimes call it the planar-GVG. See table 5.2.

Now, let's more formally define transversality. Let M_{int} be the intersection of two submanifolds M_1 and M_2 of M. The intersection is said to be *transversal* if $T_x M_1 + T_x M_2 = T_x M$ for all points $x \in M_{\text{int}}$. Therefore, if M_1 and M_2 are finitely dimensional, transversality implies that $\text{codim}(T_x M_1 \bigcap T_x M_2) = \text{codim}(T_x M_1) + \text{codim}(T_x M_2)$ for all $x \in M_{\text{int}}$. For example, two lines in the plane each have codimension one and their intersection has codimension two, which means a zero-dimensional intersection which is a point. Two two-dimensional planes in $\mathbb{R}^4$ have codimension two and intersect at a point, which has codimension four in $\mathbb{R}^4$. The transversality assumption is a generalization of the general position assumption that is commonly assumed in the computational geometry literature.

5.3.2 Retract-like Structure Connectivity

Alas, unlike the case in figure 5.15, the GVG is typically not connected and thus is not a roadmap, as can be seen in the example shown in figure 5.19. Here, there is an outer GVG network of one-dimensional manifolds associated with the rectangular enclosure and there is an inner GVG edge associated with the interior box. We term this latter edge a *GVG cycle* which is a GVG edge that is homeomorphic to S^1. In this section, we first explain why the GVG is not connected and then introduce some techniques that can be used to connect disconnected components of the GVG. For a thorough explanation of these procedures, see [106].

The lack of connectivity of the GVG is not the fault of the GVG definition, but rather a consequence of using deformation retractions: in general, there cannot be a one-dimensional deformation retract of a punctured three-or-more-dimensional space. In other words, whereas in the plane we were able to retract the free space onto the

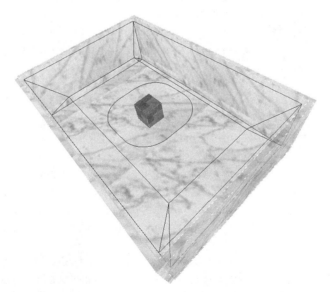

Figure 5.19 A rectangular environment with its ceiling removed to expose a rectangularly shaped box in its interior. The GVG contains two connected components: an outer network similar to the one in figure 5.15 and an inner "halo-like" structure that surrounds the inner box.

GVD with the H mapping (section 5.2.3), in a punctured $\mathbb{R}^3$ there is no continuous function that maps the free space onto a one-dimensional subset that is homotopic to the identity map [63]. The latter condition means that there is no map that "smoothly deforms" the free space onto the one-dimensional structure.

We address the lack of connectivity of the one-dimensional structure by first looking at a connected two-dimensional structure, and then defining one-dimensional structures on the two-dimensional structure to form a roadmap. In $\mathbb{R}^3$ the two-dimensional GVD is connected. In fact, the GVD is a two-dimensional deformation retract of the three-dimensional space. We can exploit this connectivity of the GVD to "patch together" the GVG. Notice that the GVG edges lie on the boundary of the GVD sheets where adjacent GVD sheets intersect. In other words, $\mathcal{F}_{ijk} = \partial\mathcal{F}_{ij} \bigcap \partial\mathcal{F}_{ik} \bigcap \partial\mathcal{F}_{jk}$. Therefore, if, and this is a big if, the boundaries of all two-equidistant sheets were connected, then the resulting GVG would be connected because the GVD is connected. This is the case in figure 5.15 where all two-equidistant faces have connected boundaries. This is not the case in figure 5.19 with the two-equidistant sheet associated with the floor and ceiling; it has a hole in the middle. The boundary of this hole is the GVG edge defined by the floor, ceiling, and interior box. So, our goal now is to connect the boundaries of each of the two-equidistant sheets.

To connect the GVG edges (the boundaries of the two-equidistant faces), we define additional structures called higher-order GVG edges. A second-order GVG edge $\mathcal{F}_{kl}\big|_{\mathcal{F}_{ij}}$ is the set of points where $\mathcal{QO}_i$ and $\mathcal{QO}_j$ are the closest pair of equidistant obstacles and $\mathcal{QO}_k$ and $\mathcal{QO}_l$ are the second-closest, i.e.,

$$\begin{aligned}\mathcal{F}_{kl}\big|_{\mathcal{F}_{ij}} = \{q \mid d_i(q) = d_j(q) \leq d_k(q) = d_l(q) \leq d_h(q) \ \ \forall h \neq i, j, k, l,\\ \text{such that } \nabla d_i(q) \neq \nabla d_j(q) \text{ and } \nabla d_k(q) \neq \nabla d_l(q)\}.\end{aligned}$$

(5.4)

The first line of equation (5.4) establishes the equidistance relationships: a pair of closest obstacles and a pair of second-closest obstacles. The second line of equation (5.4) ensures that the gradients are distinct, a condition necessary for the preimage theorem to assert that $\mathcal{F}_{kl}\big|_{\mathcal{F}_{ij}}$ is a one-dimensional manifold.

The second-order GVG edges are essentially planar-GVG edges but defined on two-equidistant faces. The preimage theorem guarantees that these edges are one-dimensional and terminate (and intersect) at second-order meet points, denoted as $\mathcal{F}_{klp}\big|_{\mathcal{F}_{ij}}$ (figure 5.20).

We call the union of the GVG and second-order GVG the *hierarchical generalized Voronoi graph* (HGVG), which by itself, as can be seen in figure 5.20, is not connected. However, there is a clue in the second-order GVG that directs the planner to "look for" a separate GVG-connected component. Notice in figure 5.20 that there is a network of second-order GVG edges that form a closed-loop path, which we term a *period,* that has a common second-closest obstacle — the box in the middle of the room.

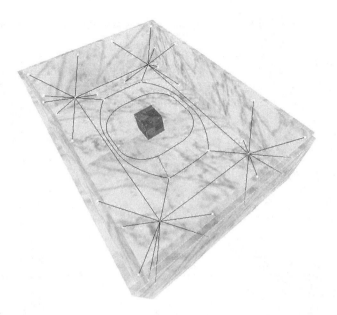

Figure 5.20   The same environment as figure 5.19. The GVG, consisting of the outer network
and the "halo" surrounding the inner box, is drawn. The other lines represent the second order
GVG edges, each drawn on two-equidistant faces. Observe that the second-order GVG edges
form a period that surrounds the GVG cycle in the middle of the free space. Once the period
is determined, a link can be made to the inner GVG cycle.

Once a period is detected, the planner can trace a path that maintains two-way
equidistance between $\mathcal{QO}_i$ and $\mathcal{QO}_j$ while decreasing the distance to $\mathcal{QO}_k$. Such a
path follows, in general, the negative gradient of d_k because we start with $d_k(q) >
d_i(q) = d_j(q)$, and decreasing d_k yields a configuration where $d_k(q) = d_i(q) =
d_j(q)$. However, in order to maintain double equidistance between $\mathcal{QO}_i$ and $\mathcal{QO}_j$,
the negative gradient must be projected onto the two-equidistant face. Hence, the
path is $\dot{c}(t) = -\pi_{T_{c(t)}\mathcal{F}_{ij}} \nabla d_k(c(t))$ where the π operator is projection. Following the
projected negated gradient $-\pi_{T_{c(t)}\mathcal{F}_{ij}} \nabla d_k(c(t))$ traces a path that terminates on a GVG
edge where $d_i(q) = d_j(q) = d_k(q)$ as long as $\pi_{T_{c(t)}\mathcal{F}_{ij}} \nabla d_k(c(t))$ does not vanish. If
$\pi_{T_{c(t)}\mathcal{F}_{ij}} \nabla d_k(c(t))$ goes to zero, then no such GVG edge exists in which case the robot
returns to the second-order period to continue exploration.

   This is just the beginning of what is required for connectivity. Ensuring connectivity
can be quite tedious and challenging. See [106] for details of connectivity of the
HGVG. The HGVG is a type of *retract-like structure* because it is not a retract, but
bears similarities to one.

5.3.3   Lyapunov Control: Sensor-Based Construction of the HGVG

Exploration with the HGVG shares the same key steps as GVD exploration: (1) access the HGVG; (2) explicitly "trace" the HGVG edges; (3) determine the location of nodes; (4) explore the branches emanating from the nodes; and (5) determine when to terminate the tracing procedure. Accessing the GVG (and hence the HGVG) is still gradient ascent, but now it is a sequence of gradient ascent operations. The robot moves away from the closest obstacle until it is two-way equidistant. Then, while maintaining two-way equidistance, the robot increases distance until it is three-way equidistant.

GVG Edge Tracing

Once the robot accesses the GVG, it must incrementally trace GVG edges. Instead of using curve tracing techniques that have discrete steps and discrete corrections, we now derive a control law that smoothly traces the roots of the expression

$$G(q) = \begin{bmatrix} d_1 - d_2 \\ d_1 - d_3 \end{bmatrix} (q) = 0,$$

where d_i is the distance to an object $\mathcal{QO}_i$, and thus if $(d_1 - d_2)(q) = (d_1 - d_3)(q) = 0$, the robot is equidistant to three obstacles and on the GVG. (Likewise, when $\mathcal{Q} = \mathbb{R}^2$, $G(q) = (d_1 - d_2)(q)$, which is zero when the robot is equidistant to two obstacles, a point on the GVD in the plane).

At a point q in the neighborhood of the interior of a GVG edge, the robot steps in the direction

(5.5) $\dot{q} = \alpha v + \beta (DG(q))^\dagger G(q),$

where

- α and β are scalar gains,

- $v \in \text{Null}(DG(q))$, the null space of $DG(q)$,

- $(DG(q))^\dagger$ is the Penrose pseudoinverse of $\nabla G(q)$, i.e.,
  $(DG(q))^\dagger = (DG(q))^T (DG(q)(DG(q))^T)^{-1}$.

Note that when q is on the GVG, $G(q) = 0$ and thus $\dot{q} = \alpha v$ where $v \in \text{Null}(\nabla G(q))$ and is simply the tangent direction of the GVG, as prescribed by the preimage theorem. Since $\nabla G(q)$ is a function of distance gradients, the planner can compute $\nabla G(q)$ solely from range sensor information. This can be done by looking at the n-closest points on the n-closest obstacles, fitting a codimension one plane through these points, and deriving the line orthogonal to this plane. The tangent vector then points along this line.[4]

When q is not on the GVG, then $(\nabla G(q))^{\dagger} G(q) \neq 0$. This term corresponds to the "correction" step which accommodates for curvature in the GVG. Again, this term can easily be determined from sensor data. Whereas the α determines how quickly the robot moves along the GVG, the β represents how aggressively the robot moves back to the GVG, as if α and β were spring constants.

To determine stability of the control law, let $\Gamma = \frac{1}{2} G^T G$ measure the distance a point q is away from the GVG. We look at the first derivative Γ.

$$
\begin{aligned}
\dot{\Gamma}(q) &= G^T(q)\,\dot{G}(q) \\
&= G^T(q)\,DG(q)\,\dot{q} \\
&= G^T(q)\,DG(q)\,(\alpha \mathrm{Null}(DG(q)) + \beta\,(DG^{\dagger}(q))\,G(q)) \\
&= \beta\,G^T(q)\,DG(q)\,DG^{\dagger}(q)\,G(q) \\
&= \beta\,G^T(q)\,DG(q)\,DG^T(q)\,(DG(q)\,DG^T(q))^{-1}\,G(q) \\
&= \beta\,G^T(q)\,G(q).
\end{aligned}
$$

The function Γ is a Lyapunov function [202] for the controller in equation (5.5). Think of a Lyapunov function as an "error function" whose minimal value is zero. Since $(DG(q)\,DG^T(q))$ is invertible in a neighborhood of the GVG [108], if $\beta < 0$, then $\dot{\Gamma}$ is negative. This assures that Γ decreases to zero, meaning that equation (5.5) directs the robot onto the GVG.

Meet Point Homing

While generating the GVG, the robot must precisely locate itself on the meet points. A meet point homing algorithm can be used to stably converge onto the meet point location [109]. The control law for homing onto a meet point is similar to the one for generating GVG edges, except G is now defined as

$$
G(q) = \begin{bmatrix} d_1 - d_2 \\ d_1 - d_3 \\ d_1 - d_4 \end{bmatrix} (q) = 0.
$$

In the planar case, $G(q) = \begin{bmatrix} d_1 - d_2 \\ d_1 - d_3 \end{bmatrix} (q)$. Since it has already been shown to be stable, we use the controller in 5.5 to determine the the path for the robot to home

---

4. Note that there are two choices for this vector, but the planner chooses the direction that directs the robot to continue in the "same" direction.

onto a meet point. Since $\text{Null}(DG(q)) = 0$, the controller is $\dot{q}\beta(DG(q))^{\dagger}G(q)$. Geometrically, this means that when the robot is in the vicinity of the meet point, it draws a sphere through the four closest points on the four closest obstacles (in the planar case, it is a circle through the three closest points). The velocity vector points toward the center of this sphere.

Higher-Order GVG Control Laws

Naturally, by varying the G function, one can trace different structures. A second-order GVG edge has $G(q) = [d_i - d_j, d_k - d_l]^T(q)$. Likewise, a second-order meet point has $G(q) = [d_i - d_j, d_k - d_l, d_k - d_p]^T(q)$.

5.4   Piecewise Retracts: The Rod-Hierarchical Generalized Voronoi Graph

Essentially, the previous roadmaps were defined for a point-robot in a workspace which has a Euclidean configuration space. Now, we turn our attension to defining a roadmap in a non-Euclidean configuration space. Even when full knowledge of the workspace is available prior to the planning event, constructing non-Euclidean configuration spaces can be quite challenging. However, what if the planner has no previous knowledge of the workspace, i.e., it cannot compute the configuration space prior to the planning event? Instead, we define a roadmap for a robot using work-space information. This is of great use because sensors directly provide workspace information.

In this section, we define a roadmap for a line segment operating in the plane. Sometimes we call this line segment a *rod* (figure 5.21). To distinguish among previous roadmaps we have defined, let the point-GVG and the point-HGVG be structures defined for a point robot in a Euclidean configuration space.

Since the configuration space for the rod is $SE(2)$, which is three-dimensional, it makes sense to look at the set of configurations equidistant to three obstacles. However, we measure distance in the *workspace,* not configuration space. To do so, let $R(q) \subset \mathcal{W}$ be the set of points the rod occupies in the workspace when it is at configuration q. At the risk of confusing notation, we re-use the d_i for distance to obstacle $\mathcal{WO}_i$ for the rod robot, i.e.,

$$d_i(q) = \min_{r \in R(q), c \in \mathcal{WO}_i} d(r, c).$$

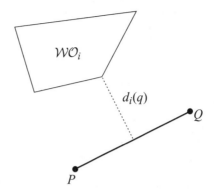

Figure 5.21   The distance from the rod (thick black line) to an obstacle is the distance (dotted line) between the nearest point on the rod to the obstacle and the nearest point on the obstacle to the rod.

Using this notion of distance, a rod-two equidistant face $\mathcal{RF}_{ij}$ is $\{q \in SE(2) \mid d_i(q) - d_j(q) = 0 \text{ and } \nabla d_i(q) \neq \nabla d_j(q)\}$. Then the rod-GVG edge is $\mathcal{RF}_{ijk} = \mathcal{RF}_{ij} \bigcap \mathcal{RF}_{ik} \bigcap \mathcal{RF}_{jk}$ [107].

Just like the point-GVG in $\mathbb{R}^3$, the collection of rod-GVG edges does not necessarily form a connected set. To produce a connected structure we introduce another type of edge, called an *R-edge*. An R-edge is the set of rod configurations that are tangent to the planar point-GVG. The rod-HGVG then comprises rod-GVG edges and R-edges (figure 5.22).

The rod-HGVG is a *piecewise retract* because it is formed by the union of deformation retracts of subsets of the configuration space, the rod-GVG edges, which are then linked together with the *R*-edges to form a connected roadmap. To show that the rod-HGVG is indeed a piecewise retract, we first consider deformation retraction $H_{\text{rod}} : \mathcal{Q}_{\text{free}} \times [0, 1] \rightarrow \text{rod} - \text{GVG}$, which is implicitly defined by a sequence of gradient ascent operations. First the rod is moved away from the closest obstacle, holding its orientation fixed, until it becomes two-way equidistant. Then, while maintaining both two-way equidistance and the fixed orientation, the rod moves until it becomes three-way equidistant, i.e., it arrives at a configuration on the rod-GVG. Note that the orientation of the rod is unchanged by this operation, implying that $\theta(q) = \theta(H_{\text{rod}}(q, s))$ for all $s \in [0, 1]$, where $\theta(q)$ is the orientation of the rod at configuration q.

Naturally, H_{rod} is not continuous over the entire configuration space, but we can restrict ourselves to simple subsets of configuration space that are associated with individual rod-GVG edges. Let $\mathcal{RF}_{ijk}$ be the rod-GVG edge associated with obstacles

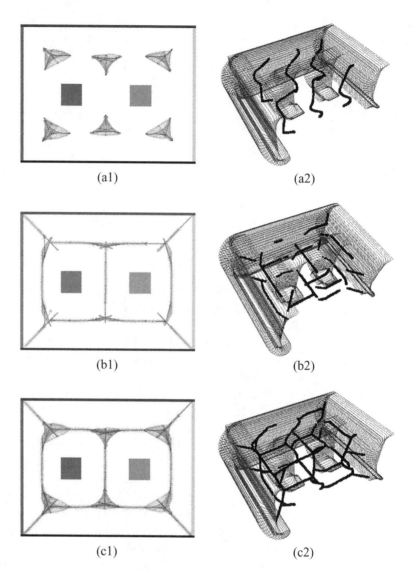

(a1)   (a2)

(b1)   (b2)

(c1)   (c2)

Figure 5.22 Swept volumes (sampled placements) of the rod in a planar workspace and the configurations of the rod in configuration space. (a1) Rod-GVG-edges: each of the clusters represents a set of configurations equidistant to three obstacles. (a2) The configurations of the rod that are equidistant to three obstacles in the *workspace*. (b1) R-edges: the rods are two-way equidistant and tangent to a planar point-GVG edge. (b2) The configurations of the rod that are tangent to the planar point-GVG in the workspace. (c1) Placements of the rod along the rod-HGVG. (c2) The entire rod-HGVG in $SE(2)$.

$\mathcal{WO}_i$, $\mathcal{WO}_j$ and $\mathcal{WO}_k$. It can be shown that H is continuous in the preimage of a connected component of the rod-GVG edge $\mathcal{RF}_{ijk}$ [107]. This preimage, which we denote as a *junction region* J_{ijk}, is contractable and thus has as a retract $\mathcal{RF}_{ijk}$.[5]

Since, for each connected component of a junction region J_{ijk}, there is a connected component of $\mathcal{RF}_{ijk}$, motion planning within one junction region can be accomplished by using $\mathcal{RF}_{ijk}$ as a roadmap. If the union of all the $\mathcal{RF}_{ijk}$'s form a connected set, then planning would be trivial again. However, in general, the $\mathcal{RF}_{ijk}$'s will not form a connected set in $SE(2)$, so we use the R-edges to connect the roadmaps for the junction regions. Intuitively, one can view the rod-GVG edges as being analogous to the nodes of the planar point-GVG, which are connected by the edges of the planar point-GVG. The distinction is that now the nodes are themselves one-dimensional structures, capturing the complete set of rod orientations that are three-way equidistant from a specific set of obstacles (see [107] for a rigorous explanation). Essentially, the R-edges encode the adjacency of the rod-GVG edges by inheriting the correct topological relationships from the plane, allowing us to construct a roadmap of configuration space using the connectivity of the workspace. See [107] for more details.

The rod-HGVG edges can be constructed in a sensor-based fashion using the control laws from section 5.3.3.

5.5   Silhouette Methods

In contrast to looking at equidistance, the silhouette approaches use extrema of a function defined on a codimension one hyperplane called a *slice*[6], which we denote by $\mathcal{Q}_\lambda$. The λ parameterizes the slice; varying the parameter λ has the effect of sweeping the slice through the configuration space. As the slice is swept through the configuration space, for each value of λ, the critical points of a function restricted to the slice are determined. The trace of the critical points as the slice is swept through the configuration space does not necessarily form a connected set. Therefore, the silhouette methods look for another type of critical point and then recursively call the algorithm on a slice passing through these critical points. The resulting network of extremal points forms the roadmap.

---

5. Note that if the rod were "small," $\mathcal{RF}_{ijk}$ would have one connected component which would be homeomorphic to S^1 (figure 5.22(a2)).

6. When the slice is one-dimensional, it can also be called a sweep line (section 5.1).

5.5.1   Canny's Roadmap Algorithm

Roadmap theory in motion planning begins with Canny's work [90]. In addition to developing the roadmap, Canny's work established fundamental complexity bounds using roadmap theory. For an environment populated by obstacles whose boundaries can be represented as p polynomials of maximum degree w for some positive w in configuration space, any navigation path-planning problem can be solved in $p^n (\log p) w^{(O(n^4))}$ time using his roadmap algorithm, where n is the degrees of freedom of the robot (the dimension of the configuration space). The derivation of this result is beyond the scope of this book. See [91, 92] for details.

In this method, the choice of initial sweep direction is arbitrary, but for the sake of discussion, let's choose the q_1-direction. As the slice is swept in the q_1-direction, "extremal points" in the q_2-direction are determined in each slice. The extremal points in the q_2-direction are extrema of the projection function $\pi_2 : \mathbb{R}^{m-1} \to \mathbb{R}$ where $\pi_2(q) = q_2$. The extremal points of π_2 for all of the slices are the *silhouette curves*.

In general, the silhouette curves are not guaranteed to be connected, and hence may not form a roadmap. However, we can look at the slices where the number of silhouette curves changes. These slices are called *critical slices,* and the λ values that parameterize critical slices are *critical values.* The points on the silhouette curves where the silhouette curves are tangent to the critical slices are termed *critical points.*

On the critical slices, the silhouette algorithm is recursively invoked where the new swept slice now has one less dimension than the critical slice, i.e., it has codimension two in the ambient space and codimension one in the critical slice. This slice is swept in the q_2-direction. The new silhouette comprises the trace of extremal points in the q_3-direction. These silhouette curves may not be connected either, so this procedure is recursively invoked on lower-dimensional critical slices until there are no more critical points or the slice has one dimension. In the latter case, the one-dimensional slice is the silhouette; in other words, the roadmap of a one-dimensional set is the set itself. Finally, the union of the resulting silhouette curves forms the roadmap.

Accessibility and departability of the roadmap are achieved by treating the slices that contain q_{start} and q_{goal} as critical $(m-1)$-dimensional slices of the initial sweep. The algorithm simply forms a silhouette network on these slices, possibly reinvoking itself on lower-dimensional slices. Connectivity is proved via an inductive argument [90]. See [189] for details on an example of an implementation of Canny's roadmap.

Figure 5.23 contains an example of a two-dimensional configuration space with a slice being swept through it. The silhouette curves trace the boundary of the environment. Critical points occur when the slice is tangent to the roadmap (and hence the obstacle boundary), as can be seen in figure 5.23. The resulting roadmap is drawn in figure 5.24. A path between a start and goal configuration is determined by first

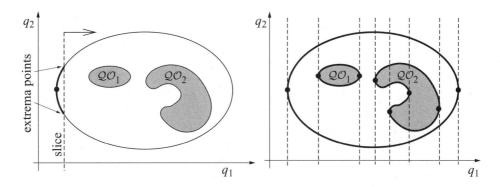

Figure 5.23 Bounded two-dimensional environment with two obstacles $\mathcal{QO}_1$ and $\mathcal{QO}_2$. The left figure contains a single slice, represented by a dashed line, and a partially constructed silhouette. The right figure contains the complete silhouette and slices passing through all critical points.

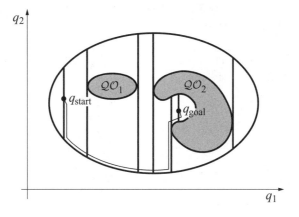

Figure 5.24 Complete silhouette curves traced out with solid lines. A path from start to goal is denoted as a thin curve.

passing a slice through the start and goal, and then including the slice in the roadmap, as can be seen in figure 5.24.

Figure 5.25 contains an example of a two-dimensional surface embedded in $\mathbb{R}^3$. It is an ellipsoid with a hole drilled partially down and then up again. The slice is swept from left to right and extrema are with respect to the in and out of page direction. The silhouette curves comprise an "equator" for the ellipsoid, the perimeter of the holes on the surface and the two curves along the side of the hole. Figure 5.26

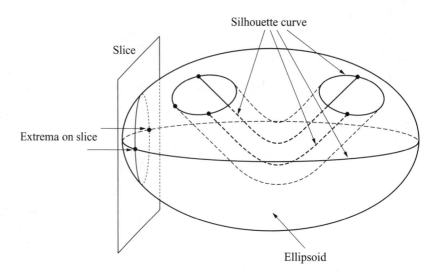

Figure 5.25   Silhouette curves for an ellipsoid with a hole drilled through it that goes down and then bends up.

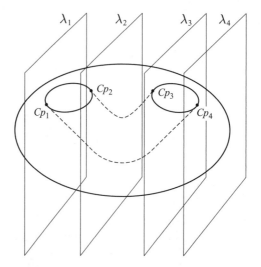

Figure 5.26   Slices passing through the critical points, which are points where the roadmap changes connectivity and is tangent to the slice. Note that the leftmost and rightmost points on the ellipsoid also are critical points but are not displayed.

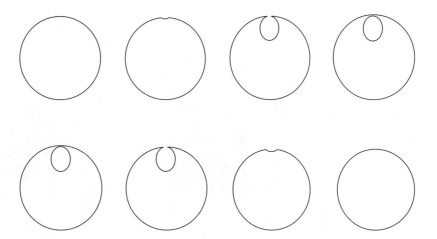

Figure 5.27   The intersection of the slice as it is swept through the ellipsoid with a hole in it displayed in figure 5.26. Starting from the top row on the left, the first two panels (a and b) display the intersection immediately before and after the slice passes through the critical point Cp_1. The next two panels (c and d) display the intersection as the slice passes through critical point Cp_2. The left pair of panels on the bottom row (e and f) correspond to critical point Cp_3 and the right pair (g and h) to Cp_4.

displays the critical slices and critical points for the ellipsoid. Figure 5.27 shows the intersection of the slice and the ellipsoid, immediately before and immediately after, the critical points. Starting from the left in the top row, the first two panels show the intersection just as the slice encounters the first hole. The next two panels show the intersection as the slice finishes passing through the hole. At this critical point, the intersection changes connectivity. Finally, figure 5.28 shows the silhouettes on the two-dimensional slices and the final path for this example between q_{start} and q_{goal}.

Critical Points and Morse Functions

In this section, we define the silhouette curves in terms of critical points of a function. The function has to be Morse [315], as described in chapter 4, section 4.6. The slices themselves are also defined in terms of a function. Originally, Canny suggested that a slice be the preimage of the projection operator π_1. Recall that π_1 projects a point onto its first coordinate, i.e., $\pi_1(q) = q_1$. We denote a slice as $\mathcal{Q}_\lambda = \{x \in \mathcal{Q} \mid \pi_1(q) = \lambda\}$ where $\lambda = q_1 \in \mathbb{R}$. Varying λ has the effect of sweeping the slice through the configuration space and $\bigcup_\lambda \mathcal{Q}_\lambda = \mathcal{Q}$. On each $\mathcal{Q}_\lambda$, we look for extrema of π_2, i.e.,

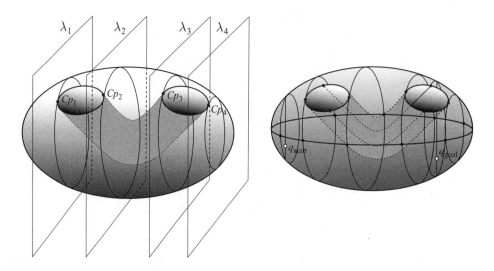

Figure 5.28 (Left) Silhouettes on the two-dimensional slices. (Right) Determining a path with a given start and goal position.

we look for extrema of $\pi_2|_{\mathcal{Q}_\lambda}$, where $\pi_2|_{\mathcal{Q}_\lambda}$ is the projection operator restricted to the slice. Also, recall that $\pi_2(q) = q_2$.

To determine the extrema, we need some machinery to calculate extrema of functions restricted to manifolds. The Lagrange multiplier theorem, stated below, can be used to determine the extrema of real-valued functions restricted to manifolds, which themselves are defined by the preimage of real-valued functions.

LEMMA 5.5.1 (Lagrange Multiplier [410])   *Let S be an n-manifold in $\mathbb{R}^{n+1}$, $S = f^{-1}(c)$ where $f : \mathbb{R}^{n+1} \to \mathbb{R}$ is such that $\nabla f(q) \neq 0$ for all $q \in S$. Suppose $h : \mathbb{R}^{n+1} \to \mathbb{R}$ is a smooth function and $p \in S$ is an extremal point of h on S. Then, there exists a real number μ such that $\nabla h(p) = \mu \nabla f(p)$ (the number μ is called the Lagrange multiplier). In other words, $\nabla f(p)$ is parallel to $\nabla h(p)$ at an extremum p of h on S.*

For example, let's look for extrema of $h = \pi_1$ on a sphere defined by the preimage of any positive scalar under the map $f(q) = q_1^2 + q_2^2 + q_3^2 - 51{,}141$. The gradients are $\nabla h(q) = [1, 0, 0]^T$ and $\nabla f(q) = [2q_1, 2q_2, 2q_3]^T$. These two vectors are parallel when $q_2 = q_3 = 0$ and the only points on $f^{-1}(0)$ that satisfy this condition are on the left-most and right-most points on the sphere (which are in the $q_1 - q_2$ plane).

We could assume that the free space is defined by the preimage of a function f, but there is no guarantee that this function will be real-valued. To determine extrema of a function restricted to such manifolds, Canny generalizes the Lagrange multiplier theorem to handle vector-valued functions.

LEMMA 5.5.2 (Generalized Lagrange Multiplier Theorem [91])   *Let M be the preimage of $f : \mathbb{R}^m \to \mathbb{R}^p$ and $h : \mathbb{R}^m \to \mathbb{R}^n$. The point x is a critical point of $h|_M$ if and only if the following matrix loses its rank [91],*

$$(5.6) \quad D(f,h)_q = \begin{bmatrix} \frac{\partial f_1}{\partial q_1}(q) & \cdots & \frac{\partial f_1}{\partial q_m}(q) \\ \vdots & \ddots & \vdots \\ \frac{\partial f_n}{\partial q_1}(q) & \cdots & \frac{\partial m_n}{\partial q_m}(q) \\ \frac{\partial h_1}{\partial q_1}(q) & \cdots & \frac{\partial h_1}{\partial q_m}(q) \\ \vdots & \ddots & \vdots \\ \frac{\partial h_q}{\partial q_1}(q) & \cdots & \frac{\partial h_p}{\partial q_m}(q) \end{bmatrix}.$$

Clearly, if f and h are real-valued functions (i.e., $n = p = 1$), then lemma 5.5.2 reduces to the Lagrange multiplier theorem because the above two-row matrix only loses rank when one row is a scalar multiple of the other. In other words, the two vectors corresponding to each row are parallel.

Canny introduces one more result, termed the *slice lemma*. The notation can be a bit cumbersome, so let's review it before introducing the slice lemma. Let π_{12} be the projection operator onto the first two coordinates, e.g., $\pi_{12}(q) = (q_1, q_2)$. Then, $\pi_{12}|_S$ is the projection operator restricted to the set S. Finally, $\Sigma(\pi_{12}|_S)$ is the set of critical points of the projection operator restricted to S. The slice lemma then states that the set of critical points of $\pi_{12}|_S$ is the union of the critical points of π_2 on each of the slices, i.e.,

$$\Sigma(\pi_{12}|_S) = \bigcup_\lambda \Sigma\left(\pi_2|_{\pi_1^{-1}(\lambda)}\right).$$

Therefore, the silhouette is the critical set of π_{12}.

With the slice lemma and Canny's generalization in hand, one can produce silhouette curves. Consider again the example of a sphere embedded in $\mathbb{R}^3$. Here, $S = f^{-1}(0)$ where $f(q) = q_1^2 + q_2^2 + q_3^2 - 62{,}370$. We sweep in the q_1-direction and extremize in the q_2-direction, i.e., $h(q) = \pi_{12}(q_1, q_2, q_3)$. Applying lemma 5.5.2,

$$D(f,h) = \begin{bmatrix} 2q_1 & 2q_2 & 2q_3 \\ 1 & 0 & 0 \\ 0 & 1 & 0 \end{bmatrix},$$

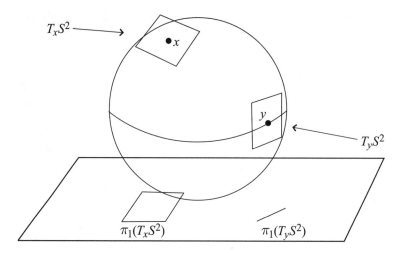

T_xS^2

$\bullet x$

y

T_yS^2

$\pi_1(T_xS^2)$ $\pi_1(T_yS^2)$

Figure 5.29   The tangent spaces of of S^2 at x and y are projected down to a plane. T_xS^2 projects to a two-dimensional space whereas T_yS^2 does not, making y a critical point.

which loses rank on S only when $q_3 = 0$, which corresponds to the unit circle in the $q_1 - q_2$-plane, which is the "equator" of the sphere (figure 5.29). This is the silhouette of the sphere.

Connectivity Changes at Critical Points

Canny's roadmap has two types of critical points: those that define the silhouette itself, as described above, and those that are used to bridge disconnected silhouette curves. We now describe the latter. In particular, we relate the concepts of a "first derivative" vanishing to connectivity changes in the silhouettes.

First, let's consider a planar example. We are now looking for extrema of the slice function $h = \pi_1$, but restricted to the silhouette, which are extrema of $\pi_1|_{\Sigma(\pi_{12})}$. Figure 5.30 depicts two sample critical points, Cp_1 and Cp_2, which are located on the boundaries of the obstacles, $\partial\mathcal{QO}_1$ and $\partial\mathcal{QO}_2$, respectively. Again, in two-dimensions, the silhouettes essentially trace out the boundaries of the free space (and obstacles) and thus $\Sigma(\pi_1|_{\Sigma(\pi_{12})}) = \Sigma(\pi_1|_{\partial\mathcal{Q}_{\text{free}}})$.

We can intuitively show that Cp_1 and Cp_2 are indeed critical points, i.e., $\pi_1|_{\Sigma(\pi_{12})}$ takes its local extrema at Cp_1 and Cp_2. The function $\pi_1(q)$ can be viewed as measuring the distance between a point $q \in \mathcal{Q}$ and the q_2-axis. Therefore, consider a path on $\partial\mathcal{QO}_1$ that passes through Cp_1, as depicted in figure 5.30. Moving along the path toward Cp_1 decreases the value of $\pi_1|_{\Sigma(\pi_{12})}$. After passing through Cp_1, the value

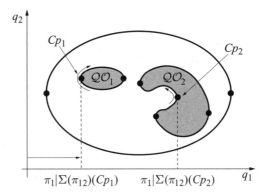

Figure 5.30 The restriction of the slice function $h = \pi_1$ to the silhouette takes a local minimum at Cp_1 and a local maximum at Cp_2. The values $\pi_1|_{\Sigma(\pi_{12})}(Cp_1)$ and $\pi_1|_{\Sigma(\pi_{12})}(Cp_2)$ are plotted on the bottom.

increases. In other words, Cp_1 is a local minimum of $\pi_1|_{\Sigma(\pi_{12})}$. Likewise, Cp_2 is a local maximum of $\pi_1|_{\Sigma(\pi_{12})}$.

Now, let's return the discussion to $\mathbb{R}^m$. Now, we demonstrate that a critical point is indeed a point on the roadmap where the tangent to the roadmap lies in the slice. This is actually a direct result of lemma 5.5.2. Recall that q is a critical point of π_1 restricted to the manifold defined by the preimage of f if $D(f, \pi)(q)$ loses rank. Here, we would like to define the roadmap as the preimage of f, but cannot do so. Instead, we can reason about the differential of f, which is an $m - 1$-by-m matrix. The null space of this matrix is the tangent to the roadmap and the $m - 1$ row vectors of the same matrix form a plane orthogonal to the tangent. Call this plane $T^{\perp}$. Finally, this matrix forms the top $m - 1$ rows of $D(f, h)$. The slice function π_1 has a gradient $[1, 0, \ldots, 0]^T$ and forms the bottom row of $D(f, \pi_1)(q)$. When the tangent lies in the slice plane, the slice plane and $T^{\perp}$ are orthogonal to each other. This means that $\nabla \pi_1(q)$ lies in $T^{\perp}$ which immediately implies that $\nabla \pi_1(q)$ can be written as a linear combination of the first $m - 1$ rows of $D(f, \pi)(q)$. In other words, $D(f, \pi_1)(q)$ loses rank because its bottom row can be written as a linear combination of the top $m - 1$ rows. Therefore q is a critical point.

This can also be seen in three dimensions. Consider the torus in figure 5.31. Here two slices are drawn, one before a critical point and one after. Before the critical point, the intersection of the slice and the torus is diffeomorphic to S^1 and after the intersection it is diffeomorphic to two copies of S^1. In figure 5.32, it can be seen that before the critical point, the roadmap is singly connected and after the intersection it has two connected components.

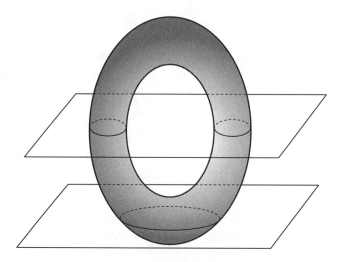

Figure 5.31   Torus with two slices drawn, before and after a critical point.

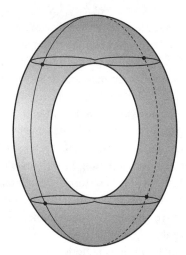

Figure 5.32   Silhouette curves on the torus.

Let's formalize the immediately "before" and "after" statements. Since a real-valued Morse function has a one-dimensional range which can be ordered, the critical values of the Morse function can be ordered as well. Assuming only one critical point per slice, adjacent critical points are those whose critical values are "next" to each other. In other words, let Λ be the set of all critical values. The critical values

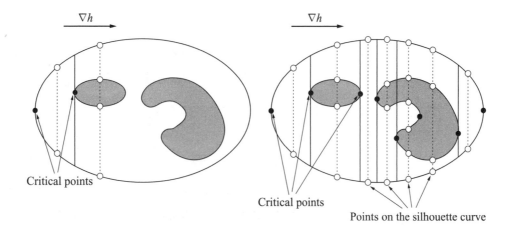

Figure 5.33 The number of silhouette fragments (open circles) changes as the slice passes through critical points (black circles) and remains constant between adjacent critical points.

$\lambda_1, \lambda_2 \in \Lambda$ are *adjacent* if for all critical values in Λ there does not exist a critical value $\bar{\lambda}$ such that $\lambda_1 < \bar{\lambda} < \lambda_2$.

Morse theory asserts that between adjacent critical points of a Morse function, the topology of the manifold on which the Morse function is defined does not change [315]. In the context of the slice function, Morse theory states that there exists a diffeomorphism ϕ such that for all $\lambda_1, \lambda_2 \in (\lambda_*, \lambda^*)$, $\phi(\pi_1|_{\Sigma(\pi_{12}^{-1}(\lambda_1))}) = \phi(\pi_1|_{\Sigma(\pi_{12}^{-1}(\lambda_2))})$, where λ_* and λ^* are adjacent critical values of a real-valued Morse function (figure 5.33).

5.5.2 Opportunistic Path Planner

The *opportunistic path planner* (OPP) generalizes Canny's original roadmap algorithm by tracing the local maxima of any potential function that is Morse on a flat slice as the slice is swept through the configuration space. Canny and Lin [93] suggest that the distance function D evaluated on the slice be used as the potential function. Local maxima on the slice of the distance function are points on the OPP roadmap. The traces of the local maxima as the slice is swept through the workspace or configuration space are termed *freeways*.

The algorithm works as follows: First, a fixed slice direction is chosen. The algorithm initially traces a path from the start to the roadmap by performing gradient ascent on the distance function in the slice that contains the start. Likewise, a path is traced from the goal to the freeway via slice-constrained gradient ascent. These two actions correspond to accessibility and departability.

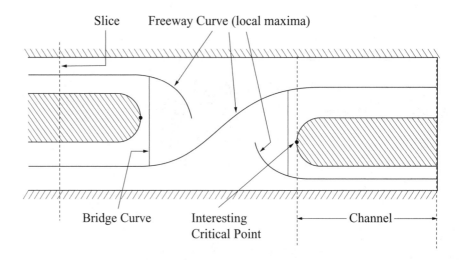

Figure 5.34   Schematic of the OPP planning scheme.

From the point at which the planner accesses the OPP roadmap, the algorithm sweeps a slice through the configuration space tracing local maxima of D constrained to the slice. These local maxima form a freeway. If the start and goal freeways are connected, then the algorithm terminates. In general, the set of freeways will not be connected, and paths between neighboring freeways must be found.

The OPP method uses a slightly different approach from Canny's original roadmap to ensure connectivity of its roadmap. The OPP freeways are connected via *bridge curves*. The bridge curves are constructed in the vicinity of *interesting critical points*. Interesting critical points occur when *channels* (figure 5.34) join or split on slices whose connectivity changes in the free space. Bridge curves are also built when freeways terminate in the free space at bifurcation points (where traces of local maxima and local minima meet). A bridge curve is built leading away from a bifurcation point to another freeway curve.

This procedure is repeated until the start and goal freeway curves are connected, or all interesting critical points and bifurcation points have been explored, in which case there does not exist a path between the start and the goal. The union of bridge and freeway curves, sometimes termed a *skeleton,* forms the one-dimensional roadmap.

Connectivity and Critical Points

Instead of looking for connectivity changes in the roadmap, the OPP method looks for connectivity changes in the slice in the free configuration space. Again, these

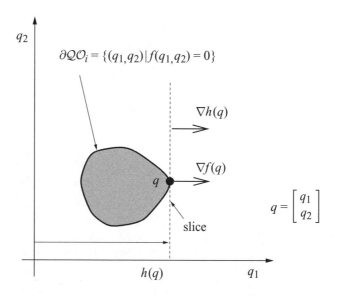

Figure 5.35   At the critical point q, the gradient of the slice function $\nabla h(q)$ is parallel to surface normal of the obstacle $\nabla f(q)$. Also, the slice is tangent to the boundary of the obstacle $\mathcal{QO}_i$ at the critical point q.

connectivity changes correspond to a slice function taking on extremal values. This can be seen in figure 5.30, except now we are looking at the slice function π_1 *restricted to the boundary of the free space,* as opposed to being restricted to the silhouette (both of which coincide in the plane).

Again, $D(f, h)$ loses rank at the critical points. Here, the f function can be used to define the boundaries of the obstacles. In other words, we assume that the boundaries of the obstacles can be represented as the preimage of 0 under the f mapping. Therefore, for $q \in \partial \mathcal{QO}_i$, $\nabla f(q)$ is the surface normal to $\mathcal{QO}_i$ at q. Now, $D(f, h)$ has two rows and loses rank only when $\nabla f(q)$ is parallel to $\nabla h(q)$ which means that the slice gradient is parallel to the surface normal of the obstacle. (Note that we could have used the original Lagrange multiplier theorem here.) See figure 5.35.

Morse theory [315] assures that the topology of the intersection of the boundary and the slice remains constant between critical points, i.e., there exists a diffeomorphism ϕ such that for all $\lambda_1, \lambda_2 \in (\lambda_*, \lambda^*)$, $\phi(h|_{\partial \mathcal{Q}_{\text{free}}}{}^{-1}(\lambda_1)) = \phi(h|_{\partial \mathcal{Q}_{\text{free}}}{}^{-1}(\lambda_2))$, where λ_* and λ^* are adjacent critical values of a real-valued Morse function. Therefore, we are assured that we only need to look for critical points to connect disconnected components of the roadmap.

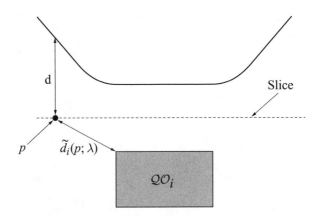

Figure 5.36 The dashed line represents a slice that is hovering above obstacle $\mathcal{QO}_i$. The solid line above the slice is the graph of the distance to the obstacle, but restricted to the slice, i.e., $\tilde{d}_i$.

Nonsmooth Functions

It should be noted that the distance function is nonsmooth. Consider the distance function constrained to a slice $\mathcal{Q}_\lambda = \{q \mid \pi_1(q) = \lambda\}$. Decompose the configuration space coordinates q into "slice coordinates" p and the "sweep coordinate" λ such that $q = [\lambda, p]^T$. The single object distance function *constrained to a slice* is the distance between a point that is in a slice $\mathcal{Q}_\lambda$ and a set $\mathcal{QO}_i$, i.e.,

(5.7) $\quad \tilde{d}_i(p\,;\lambda) = d_i(q) \qquad \text{where } \pi_1(q) = \lambda \quad \text{and} \quad p \in \pi_1^{-1}(\lambda).$

See figure 5.36 for an example of the distance function plotted along a slice. At each slice point, $\tilde{d}_i$ is computed to the closest point of the obstacle.

Typically, a robot's environment is populated with multiple obstacles, and thus we define a distance function for multiple obstacles. The multi-object distance function *constrained to a slice* measures the distance between a point in a slice $\mathcal{Q}_\lambda$ and the *closest* obstacle to that point, i.e.,

(5.8) $\quad \tilde{D}(p\,;\lambda) = \min_i \tilde{d}_i(p\,;\lambda).$

Even when all of the obstacles are smooth and convex, $\tilde{D}$ is not necessarily smooth at the local maxima. For example, in figure 5.37 distance $D(q)$ is plotted along a horizontal slice. On the left-hand side of the slice, since $\mathcal{QO}_1$ is the closest obstacle, $D(q) = d_1(q)$. Likewise, on the right-hand side of the slice, $D(q) = d_2(q)$. When $d_1(q) = d_2(q)$, D is nonsmooth, but for all other points, $D(q)$ is smooth because it inherits the smoothness properties of the single object distance function for convex

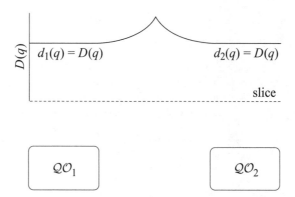

Figure 5.37   Distance function D plotted along a horizontal slice. The slice is represented as a dashed line. The graph of D is overlaid on top of the slice. Note D becomes nonsmooth when $d_1(q) = d_2(q)$, and hence there is not a unique closest obstacle.

sets. Therefore, the gradient vector is either $\nabla d_1(q)$ or $\nabla d_2(q)$ depending upon which obstacle is the unique closest one. However, at the point q^* where D is nonsmooth, the gradient is no longer unique. In fact, it is the set formed by the convex hull of $\nabla d_1(q^*)$ and $\nabla d_2(q^*)$. This gradient is termed a *generalized gradient* [114] and is denoted as

$$\partial D(q^*) = \text{Co}\{\nabla d_i(q^*) \mid i \in Z(q^*)\}$$
$$= \sum_{i \in Z(q^*)} \mu_i \nabla d_i(q^*) \quad \text{where} \quad \sum_{i \in Z(q^*)} \mu_i = 1 \quad \text{and} \quad \mu_i > 0,$$

where Co is the convex hull operator and $Z(q^*)$ is the set of integers that correspond to the indices of the closest obstacles to q^*, i.e., $Z = \{i \mid \text{for all } i \text{ where } d_i(q^*) < d_h(q^*) \text{ for all } h\}$.

With this notion of a generalized gradient, we can establish a calculus for characterizing extrema of a function by looking at the convex hull of the generalized gradient of D [104]. Let 0 be the origin if the tangent space $T_{q^*}\mathbb{R}^m$. If $0 \in \partial D(q^*)$, then q^* is a local maximum. Likewise, if $0 = \partial D(q^*)$, then q^* is a local minimum. It is worth noting that we never had to perform an additional differentiation but were able to characterize the generalized gradient from first-order information.

Problems

1. Prove that the visibility graph is connected.

2. Show an example for which the visibility graph does not produce the shortest path in $\mathbb{R}^3$.

3. How can the visibility graph method be augmented so as to yield the shortest path in $\mathbb{R}^3$?

4. How can the visibility graph in the plane be adapted to handle curved obstacles.

5. Write a program to compute the visibility graph. The program should take as input from a file a list of polygons, which are in turn represented by a list of vertices. The user can input from the keyboard the start and goal configurations. The program then computes the visibility graph and then determines a path from start to goal.

6. Let S be the unit circle defined by the preimage of zero under $f(x, y) = x^2 + y^2 - 1$. Let $g(x, y) = ax^2 + 2bxy + cy^2$ where $a, b, c \in \mathbb{R}$. List the points where g is extremized on S. Draw a picture.

7. Draw the Canny roadmap for the surface configuration space in figure 5.38.

8. Do connectivity changes in the free space in a slice imply connectivity changes in the original Canny roadmap? In the OPP roadmap?

9. What are the benefits of using only the local maxima (and not the other extrema) in the OPP method?

10. The HGVG contains a lot of structure which seemingly can be deleted. Suggest a method to prune this structure.

11. What are the tradeoffs between using roadmaps and pixel-based maps?

12. Prove that for any slice direction, OPP is a subset of the GVG.

13. For the OPP and point-GVG, both in the plane and in $\mathbb{R}^3$, there are useless spokes. If we eliminate them in the planar case, do we still have a topological map? How could we eliminate spokes online?

Figure 5.38   A cylinder with a hole drilled through it.

14. Use the brushfire implementation to compute the planar point-GVG. Beware of jagged edges.

15. The planar point-GVG is defined using a Euclidean distance function and consists of straight line and parabolic segments. One way of thinking of the planar point-GVG is the locus of the centers of circles whose perimeters are tangent to obstacles at two or more points. For the environment below, sketch the GVD using the circle analogy.

16. The definition of the planar point-GVG can be generalized to any convex distance function. Instead of a circle, consider a convex distance function defined by a square (rotated by 45 degrees). For the environment in figure 5.39, sketch the point planar-GVG using both the circle and the square analogy.

17. State at least two advantages and two disadvantages of using potential functions as a sensor-based planner.

18. Consider the real-valued function

$$f(x, y, z) = x^2 + y^2 - z^2.$$

Use the preimage theorem to state the values of c for which $f^{-1}(c)$ is a manifold. For the values of c for which $f^{-1}(c)$ is a manifold, state the dimension of $f^{-1}(c)$. State the values of c for which $f^{-1}(c)$ is connected. Draw pictures of the manifolds for different values of c.

19. Prove that d_i is a convex function when $\mathcal{QO}_i$ is convex.

20. Prove that the generalized Voronoi region is connected in a connected free space.

21. Verify that figure 5.6 contains the reduced visibility graph for figure 5.4.

22. Assume the boundaries of the two-equidistant faces are connected. Prove or disprove that the GVG is connected in $\mathbb{R}^3$.

23. Implement exploration of an unknown workspace using the incremental construction procedures described in this chapter

    (a) Rotate the robot so that the sensor with the smallest sensor reading is pointing "backward." You may use a lookup table here.

    (b) Drive the robot away from the closest obstacle until it is two-way equidistant.

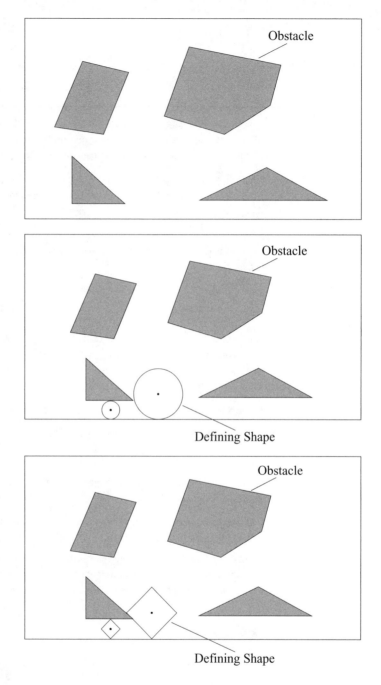

Figure 5.39 Photocopy the above figures to draw planar point-GVG's but with different distance metrics.

(c) Rotate the robot so that it lies in the tangent space of the GVD. You may use a lookup table here.

(d) Drive the robot forward a small distance and test to see if the robot still lies on the GVD (falls into a dead zone that is centered on the GVD).

(e) Rotate the robot by 90 degrees and drive it forward or backward until it is on the GVD and then reorient the robot back into the tangent space.

(f) Trace a GVD-edge until encountering a meet point.

(g) Depart a meet point on a GVD-edge.

(h) Implement the graph data structure for the GVD.

24. Use a local mapping routine to improve upon the exploration procedure described above.

6 *Cell Decompositions*

NEXT, WE consider a different type of representation of the free space called an *exact cell decomposition.* These structures represent the free space by the union of simple regions called *cells.* The shared boundaries of cells often have a physical meaning such as a change in the closest obstacle or a change in line of sight to surrounding obstacles. Two cells are *adjacent* if they share a common boundary. An *adjacency graph,* as its name suggests, encodes the adjacency relationships of the cells, where a node corresponds to a cell and an edge connects nodes of adjacent cells.

Assuming the decomposition is computed, path planning with a cell decomposition is usually done in two steps: first, the planner determines the cells that contain the start and goal, respectively, and then the planner searches for a path within the adjacency graph. Note that the adjacency graph could serve as a roadmap of the free space as well. Therefore, mapping can be achieved by incrementally constructing the adjacency graph.

Cell decompositions, however, distinguish themselves from other methods in that they can be used to achieve coverage. A coverage path planner determines a path that passes an effector (e.g., a robot, a detector, etc.) over all points in a free space. Since each cell has a simple structure, each cell can be covered with simple motions such as back-and-forth farming maneuvers; once the robot visits each cell, coverage is achieved. In other words, coverage can be reduced to finding an exhaustive walk through the adjacency graph. Sensor-based coverage is achieved by simultaneously covering an unknown space and constructing its adjacency graph.

The most popular cell decomposition is the *trapezoidal decomposition* [356]. This decomposition relies heavily on the polygonal representation of the planar configuration space. A more general class of decompositions, which are termed *Morse Decompositions* [12], allow for representations of nonpolygonal and nonplanar spaces. Morse decompositions are based on ideas from Canny's roadmap work. We then consider a broader class of decompositions which includes those based on visibility constraints. One such decomposition serves as a basis for the pursuit/evasion problem which is introduced section 6.3.

6.1   Trapezoidal Decomposition

The trapezoidal decomposition comprises two-dimensional cells that are shaped like trapezoids. Some cells can be shaped like triangles, which can be viewed as degenerate trapezoids where one of the parallel sides has a zero-length edge. Assume a simple (x, y) coordinate system for the planar configuration space, the free space is bounded by a polygon and that all of the obstacles are polygonal. For the sake of explanation, assume that each vertex v_i on all of the polygons has a unique x coordinate, i.e., for all $i \neq j$, $v_{i_x} \neq v_{j_x}$. This assumption is equivalent to saying that the polygons lie in general position (see figure 6.1).

To form the decomposition, at each vertex v draw two segments, one called an upper vertical extension and the other called a lower vertical extension. Here, "up" and "above" correspond to increasing the y coordinate, and likewise "down" and "below" mean decreasing it. The upper and lower vertical extensions start at the vertex and

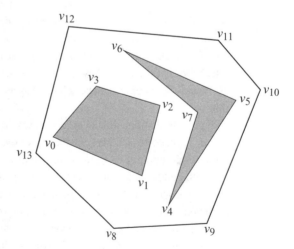

Figure 6.1   Sample polygonal configuration space.

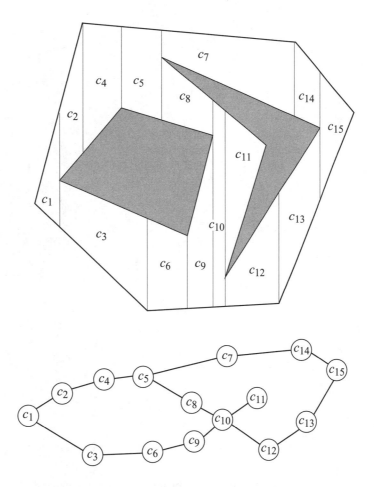

Figure 6.2   Trapezoidal decomposition for the configuration space in figure 6.1.

terminate when they first intersect an edge of the polygon that lies immediately above and below v, respectively. Note that many vertices will have either just an upper or a lower vertical extension. Figure 6.2 contains the trapezoidal decomposition and its adjacency graph for the workspace in figure 6.1. Recall that two cells are adjacent if they share a common boundary, i.e., a common vertical extension.

Once the cells that contain the start and goal are determined, the planner searches the adjacency graph to determine the path. However, the result of the graph search is just a sequence of nodes, not a sequence of points embedded in the free space, and so the next step is to determine the explicit path. Since a trapezoid is a convex set,

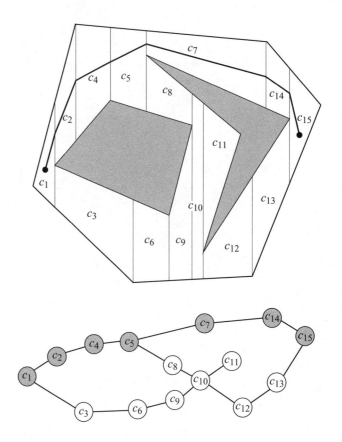

Figure 6.3   The resulting paths in the adjacency graph and free space.

any two points on the boundary of a trapezoidal cell can be connected by a straight-line segment that does not intersect any obstacle. The planner constructs the path, one trapezoid at a time, by connecting the midpoints of the vertical extensions to the centroids of each trapezoid. This yields a connected collision-free path through the free space that is derived from the adjacency graph. To connect the start and goal points, simply draw a straight line to the vertical extensions' midpoints of the appropriate trapezoids (figure 6.3).

The next issue centers on constructing the decomposition itself. The input to the algorithm is a list of polygons, each represented by a list of vertices. The first step is to sort the vertices based on the x-coordinate of each vertex. This takes $O(n \log n)$ time and $O(n)$ storage where n is the number of edges (or vertices) in all of the polygons.

The next step is to determine the vertical extensions. For each vertex v_i, a naive algorithm can intersect a line through v_i with each edge e_j for all j. This will require $O(n)$ time resulting in a $O(n^2)$ time to construct the trapezoidal decomposition. We can do better: the extensions can be determined by sweeping a sweep line (similar to the sweep line in chapter 5, section 5.1 or the slice in Canny's roadmap algorithm in section 5.5.1) through the free space stopping at the vertices, which are sometimes termed *events*. While passing the sweep line, the planner can maintain a list L that contains the "current" edges which the slice intersects.

With the list L, determining the vertical extensions at each event requires $O(n)$ time with a simple search, but if the list is stored in an "efficient" data structure like a balanced tree, then the search requires $O(\log n)$ time. It is easy to determine the y-coordinates of the intersection of the line that passes through v_i and each edge e_i. The trick is to find the appropriate edge or edges for the vertical extensions, i.e., the two edges that v lies between. Let these two edges be called e_{LOWER} and e_{UPPER}.

So as long as the "current" list requires $O(\log n)$ insertions and deletions, as balanced trees do, then keeping track of all the edges that intersect the sweep line, i.e., maintaining L, requires $O(n \, \log n)$ time. Let e_{lower} and e_{upper} be the two edges that contain v (these are *not* e_{LOWER} and e_{UPPER}). The "other" vertex of e_{lower} has a y-coordinate lower than the "other" vertex of e_{upper}. Now, there are four types of events (figure 6.4) that can occur and the type of event determines the appropriate action to take on the list. These events and actions are

- e_{lower} and e_{upper} are both to the left of the sweep line
  - delete e_{lower} and e_{upper} from the list
  - $(\ldots, e_{\text{LOWER}}, e_{\text{lower}}, e_{\text{upper}}, e_{\text{UPPER}}, \ldots) \rightarrow (\ldots, e_{\text{LOWER}}, e_{\text{UPPER}}, \ldots)$

- e_{lower} and e_{upper} are both to the right of the sweep line
  - insert e_{lower} and e_{upper} into the list
  - $(\ldots, e_{\text{LOWER}}, e_{\text{UPPER}}, \ldots) \rightarrow (\ldots, e_{\text{LOWER}}, e_{\text{lower}}, e_{\text{upper}}, e_{\text{UPPER}}, \ldots)$

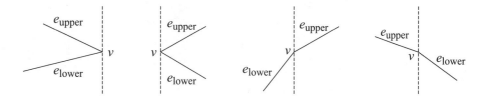

Figure 6.4   The kinds of events.

- e_{lower} is to the left and e_{upper} is to the right of the sweep line
  - delete e_{lower} from the list and insert e_{upper}
  - $(\ldots, e_{\text{LOWER}}, e_{\text{lower}}, e_{\text{UPPER}}, \ldots) \rightarrow (\ldots, e_{\text{LOWER}}, e_{\text{upper}}, e_{\text{UPPER}}, \ldots)$

- e_{lower} is to the right and e_{upper} is to the left of the sweep line
  - delete e_{upper} from the list and insert e_{lower}
  - $(\ldots, e_{\text{LOWER}}, e_{\text{upper}}, e_{\text{UPPER}}, \ldots) \rightarrow (\ldots, e_{\text{LOWER}}, e_{\text{lower}}, e_{\text{UPPER}}, \ldots)$

Figures 6.5 through 6.8 contain examples of a sweep line being swept through a polygonal free space of figure 6.1 with the corresponding list updates at each event.

Finally, we need to determine which cells contain the start and goal. First, we will seemingly construct a finer cell decomposition which will have cells that are subsets of the trapezoid and then infer from there which trapezoids contain the start and goal. Draw a vertical line through all of the events forming "slabs" of the free space. Let w be a point in the free space. Determining which slab contains w requires $O(\log n)$ time. From here, it is easy to determine which edge of the polygonal workspace intersects the slab and thus it requires a second $O(\log n)$ search to determine the ceiling and floor edges that contain w. With the ceiling and floor determined, it is trivial to determine which trapezoid contains w. See [124] for more efficient algorithms.

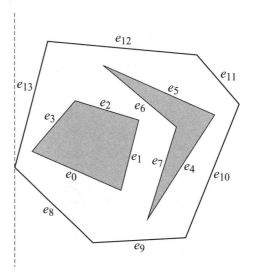

Figure 6.5 $L : \emptyset \rightarrow \{e_8, e_{13}\}$.

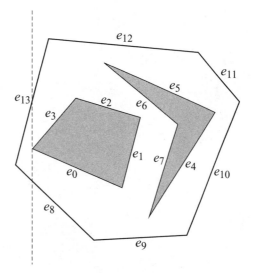

Figure 6.6 $L : \{e_8, e_{13}\} \rightarrow \{e_8, e_0, e_3, e_{13}\}$.

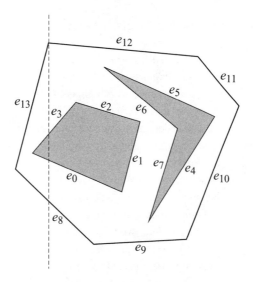

Figure 6.7 $L : \{e_8, e_0, e_3, e_{13}\} \rightarrow \{e_8, e_0, e_3, e_{12}\}$.

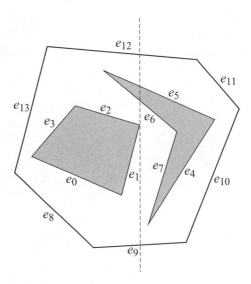

Figure 6.8 $L : \{e_9, e_1, e_2, e_6, e_5, e_{12}\} \rightarrow \{e_9, e_6, e_5, e_{12}\}$.

6.2   Morse Cell Decompositions

Conventional motion planning approaches determine paths between start and goal configurations, such as those described in chapters 2, 4, 5, and elsewhere. However, applications such as robotic demining and floor cleaning require a robot to pass over all points in its free space, i.e., follow a path to *cover* the space. A planner can use an exact cell decomposition to cover an unknown space by simply covering each cell and then using the adjacency graph to ensure each cell is visited and hence covered. This approach requires that each cell can indeed be covered. Naturally, cells with simple structure can easily be covered; e.g., the cells of the trapezoidal decomposition can be covered with simple back-and-forth motions.

Unfortunately, the trapezoidal decomposition may not produce efficient paths for coverage. Here, we measure efficiency in terms of area covered vs. path length traversed. Observe that cells in the trapezoidal decomposition can be "clumped" together to form more efficient coverage paths. Perhaps a bigger drawback to the trapezoidal method is that it fundamentally requires a polygonal workspace, which is not a realistic assumption for many applications.

In this section, we use Morse functions (chapter 5, section 5.5) to define cells that have simple structure and can be defined in nonpolygonal spaces. Recall that a Morse function is one whose critical points are nondegenerate; from a practical perspective,

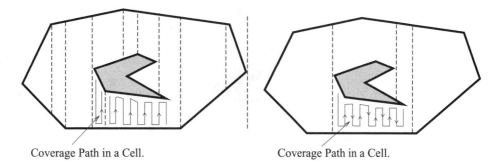

Coverage Path in a Cell.   Coverage Path in a Cell.

Figure 6.9   (left) Trapezoidal decomposition and (right) boustrophedon decomposition for the same space. Each cell, in both decompositions, can be covered with simple back and forth motions. However, note that the coverage path in the boustrophedon decomposition is a little bit shorter. The region below the polygon obstacle requires an extra pass because the planner has to "start over" each time the robot enters a new cell. Since there are fewer cells under the polygonal obstacle in the boustrophedon decomposition, the coverage path is shorter.

this means that critical points are isolated. In this section, we evolve the trapezoidal decomposition to a new decomposition called the *boustrophedon decomposition* and then show that the boustrophedon decomposition is a *Morse decomposition.* Next, we generalize the boustrophedon decomposition to form other Morse decompositions.

6.2.1   Boustrophedon Decomposition

From a coverage perspective, a minor shortcoming of the trapezoidal decomposition is that many small cells are formed that can seemingly be aggregated with neighboring cells. Reorganizing the cells can result in a shorter and more efficient path to cover the same area.

To address this issue, the boustrophedon[1] cell decomposition approach was introduced [110]. The boustrophedon decomposition is formed by considering the vertices at which a vertical line can be extended both up and down in the free space (figure 6.9). We call such vertices *critical points,* and we will show that these correspond to the same critical points in Canny's roadmap described in chapter 5, section 5.5.1.

With the decomposition in hand, the planner determines a coverage path in two steps. First, the planner determines an exhaustive walk through the adjacency graph (figure 6.10). This list can be computed by using a depth-first search algorithm. Once

---

1. The Greek word *boustrophedon* literally means "ox turning" [8]. Typically, when an ox drags a plow in a field, it crosses the full length of the field in a straight-line, turns around, and then traces a new straight line path adjacent to the previous one.

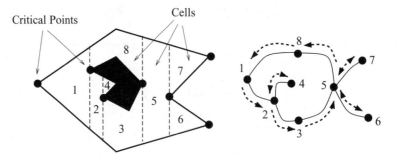

Exhaustive walk 1–2–4–2–3–5–6–5–7–5–8–1

Figure 6.10   The boustrophedon decomposition of a space and its adjacency graph. The nodes represent the cells and the edges indicate the adjacent cells. An exhaustive walk on the graph is generated.

the ordered list of cells is determined, the planner then computes the explicit robot motions within each cell. The path in each cell consists of a repeated sequence of straight-line segments separated by one robot width and short segments connecting the straight line-segments. Typically, these short segments follow the boundary of the environment.

6.2.2   Morse Decomposition Definition

We generalize the boustrophedon decomposition beyond polygons by borrowing ideas from Canny's work [91,93] which first applied a "slicing method" to motion planning, as described in chapter 5. Recall that a *slice* is a is a codimension one manifold denoted by Q_λ. The slices are parameterized by λ (varying λ sweeps a slice through the space). The portion of the slice in the free configuration space, Q_{free}, is denoted by $Q_{\text{free}\lambda}$, i.e., $Q_{\text{free}\lambda} = Q_\lambda \bigcap Q_{\text{free}}$. Recall from chapter 5 that connectivity changes of $Q_{\text{free}\lambda}$ were used to ensure the connectivity of the roadmap. Now, we are going to use the connectivity changes to define cells in a cell decomposition.

Recall from section 5.5 in chapter 5 that the slice can be defined in terms of the preimage of the projection operator $\pi_1 : Q \to \mathbb{R}$. In the plane $\pi_1(x, y) = x$ and the slice $Q_\lambda = \pi_1^{-1}(\lambda)$ corresponds to a vertical slice. Increasing the value of λ sweeps the slice to the right through the plane. As the slice is swept through the target region, obstacles intersect (or stop intersecting) the slice in the free space, severing it into smaller pieces as the slice first encounters an obstacle (or merging smaller pieces into larger pieces as the slice immediately departs an obstacle). The connectivity changes occur at points termed *critical points*. Note that critical points are analogous to vertices which have vertical extensions that can be drawn both up and down.

Slices that contain critical points are termed *critical slices*. It should be emphasized, however, that the slice $\mathcal{Q}_\lambda$ itself does not change connectivity, but rather the slice in the free space $\mathcal{Q}_{\text{free}\lambda}$ changes connectivity at critical points. Naturally, $\mathcal{Q}_{\text{free}\lambda}$ contains one or more connected components, which are termed *slice intervals* and are denoted $\mathcal{Q}_{\text{free}\lambda}^j$ for the jth open connected slice interval. So, $\mathcal{Q}_{\text{free}\lambda} = \bigcup_j \mathcal{Q}_{\text{free}\lambda}^j$. Denote the set of slice intervals that contain a critical point by I^*. Note that a critical point cannot be in the interior of a slice interval; it can only lie at the endpoints of a slice interval. With this, we can define a Morse decomposition

DEFINITION 6.2.1 (Morse Decomposition [12]) *A Morse decomposition is an exact cell decomposition whose cells are the connected components of* $\mathcal{Q}_{\text{free}} \backslash I^*$.

In figure 6.11, the dashed lines are the slice intervals lying in the free space and have end points lying on obstacle boundaries. Each of these slice intervals has at least one critical point on an obstacle boundary as well. When the slice intervals are removed from the free space, the remaining free space is still two-dimensional but is no longer connected. Each connected component is a cell.

One can see that within a cell, the slice interval remains connected and only extends or contracts. Morse theory assures us that between critical slices, "merging" and "severing" of slices do not occur, i.e., the topology of the slice remains constant.

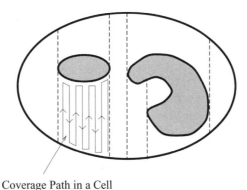

Coverage Path in a Cell

Figure 6.11 The boustrophedon decomposition of a nonpolygonal environment. As we sweep a straight-line slice from left to right, its connectivity in the free space changes first from one to two, then two to one and so forth. At the points where these connectivity changes occur, we locate the cell boundaries in the free space.

This is useful for tasks such as coverage because the robot can trivially perform simple motions between critical points and guarantee complete coverage of a cell (figure 6.11). A coverage path within a cell contains two parts: motion along the slice and motion along the boundary of the obstacles. A bulk of the coverage operation occurs with motions along the slices, sometimes called *laps,* and this motion terminates when the robot encounters an obstacle. Motion along the boundary of the obstacle directs the robot to move "one width[2] over," i.e., increase its slice function value by one robot width while following the cell boundary along an obstacle. We call the distabce between subsequent laps as the *inter-lap distance.*

Cao, Huang, and Hall [96] implicitly use a Morse decomposition to achieve coverage but they assume all obstacles are convex. The Morse decompositions, defined above, assume that the critical points are not degenerate, but Butler et al. [86] present a coverage algorithm that uses decompositions of rectilinear spaces where all critical points are degenerate.

6.2.3   Examples of Morse Decomposition: Variable Slice

The definition of a Morse decomposition is not specific to a particular slice. In the previous section, the slice was defined by the preimage of a real-valued function, which happened to be π_1. We can use this function to define the boustrophedon decomposition which induces back-and-forth coverage pattern. Now we will vary the function that defines the shape of the slice, resulting in different decompositions and hence different patterns by which the free space is covered. Now, we rewrite the definition of the slice as the preimage of a general real-valued function $h : \mathcal{Q} \rightarrow \mathbb{R}$. For the boustrophedon decomposition in the plane, this function is $h(x, y) = x$.

Spiral, Spike, and Squarel Patterns

We can use the function $h(x, y) = \sqrt{x^2 + y^2}$ to produce a pattern of concentric circles in the plane. Critical points occur at points where a circle changes connectivity; this happens when it is tangent to an obstacle. Critical points are then used to form annular or arc-shaped cells and the adjacency graph (figure 6.12, left). As before, a planner determines a coverage path in two steps: first it finds an exhaustive walk through the adjacency graph and then it plans the explicit coverage path in each uncovered cell. The coverage pattern within a cell has three parts: motion along a slice, motion orthogonal to the slice, and motion along the boundary of the cell. The slice here, however, is not a straight-line segment, but rather a circle or subset of a circle. Therefore, in

---

2. Here, width is determined by the size of the detector or end-effector that is being used.

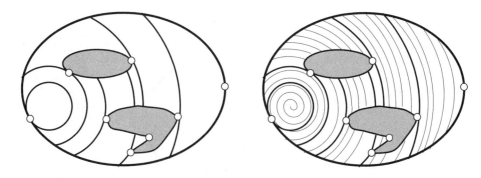

Figure 6.12   (Left) Cell decomposition for $h(x, y) = \sqrt{x^2 + y^2}$ and (right) its associated spiral coverage pattern. The slices are the circles that are the preimages of h. At the critical points, labeled with little open circles (not to be confused with the slices), the circle-shaped slices become tangent to the obstacles. Rather than moving along circular paths and stepping outwardly, the robot follows a spiral pattern.

the plane, a planner initially directs the robot to circumnavigate a circle, move the interlap distance along the radius of the circle, and then circumnavigate a circle of a larger radius. If the robot encounters an obstacle while circumnavigating a circle, the planner simply directs the robot to follow the obstacle boundary until the robot has moved an interlap distance and then follows the circle of a larger radius.

Note that instead of following a circle and stepping outward, the robot can follow a spiral pattern until it encounters critical points (figure 6.12, right). The spiral pattern bypasses the need to step along the radial direction. This yields a path that maximizes the area covered per unit distance traveled in regions sparsely populated with obstacles.

The function $h(x, y) = \tan(\frac{y}{x})$ induces a pattern that is orthogonal to the set of concentric circles (figure 6.13). Using this pattern to perform coverage has the effect of covering the region closest to the center of the pattern more densely. This is useful if the likelihood of finding a desired object is highest at the center of the pattern and the robot's detector experiences false negatives (something is under the detector but the detector does not sense it).

The function $h(x, y) = |x| + |y|$ can be used to produce cells that look like rotated squares or diamonds (figure 6.14). For coverage, instead of driving in concentric squares, we can direct the robot to "spiral" out while looking for critical points, hence the term *squarel*. The resulting pattern is shown in figure 6.14. The squarel pattern serves as an approximation to the spiral pattern that is easier to implement on differential drive robots.

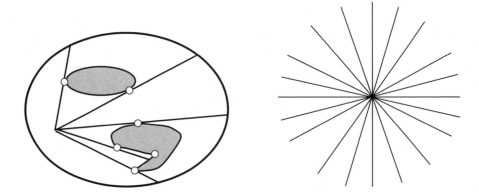

Figure 6.13   Decomposition for $h(x, y) = \tan(\frac{y}{x})$ and a spiked pattern. The free space is sliced like a pie. At the critical points, the slices are tangent to obstacles. The robot can use this pattern to cover more densely the region closest to the center of the pattern.

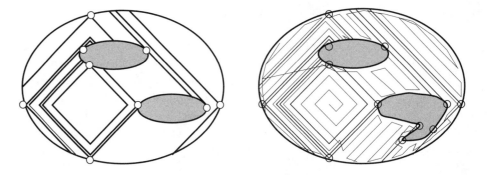

Figure 6.14   Decomposition for $h(x, y) = |x| + |y|$ and a coverage pattern. Squares are the slices and at the critical points the corner of a square touches an obstacle or the side of it becomes tangent to an obstacle. Since it is easier for the robot to move along straight lines rather than circles, this pattern can be used to approximate the spiral pattern.

Note that $h(x, y) = |x| + |y|$ is not smooth so we have to use the formulation of the generalized gradient given in chapter 5, section 5.5.2. The squarel pattern has two parts: a straight line segment and a 90 degree turn, as can be seen in figure 6.14. Note that critical points occur when the flat portions of $\{(x, y) : |x| + |y| = \lambda\}$ become tangent to an obstacle and at some of the 90 degree turn points. At these points, the obstacle surface normal lies in the convex hull of the two flat portions that meet at the 90 degree turn point.

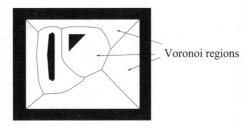

Figure 6.15   GVD of an environment.

Brushfire Decomposition

The brushfire algorithm [262] is a popular technique to construct the GVD (figure 6.15). In chapter 4, section 4.3.2, we described the brushfire algorithm on a grid; here we describe it on a continuous space.

The brushfire algorithm is so named because in implementation, imaginary wave fronts emanate from each obstacle and collide at points on the GVD. By noting the location of the collision points, the algorithm constructs the GVD. The algorithm, however, induces a decomposition that is *not* the Voronoi regions of the GVD. Instead, the decomposition models the topology of the wave fronts as they initially collide with each other and form or destroy new wave fronts. Compare figure 6.15 and figure 6.16.

The distance function D, which measures the distance between the point x and the nearest point c on the closest obstacle $\mathcal{QO}_i$, admits a decomposition termed the *brushfire decomposition*. Each slice of D is a wave front where each point on the front has propagated a distance λ from the closest obstacle. As λ increases, the wave fronts progress. Cells of the brushfire decomposition are formed when these wave fronts initially collide. Figure 6.16 contains a decomposition induced by D where regions of the same color represent a cell. Whereas for the boustrophedon decomposition we are essentially "pushing" a line segment through the cell, here we are "growing" a wave front that originates on the boundary of the environment, which in figure 6.16 has three obstacles: the exterior, the vertical barlike obstacle, and a triangle. These three wave fronts progress until they initially collide with each other, which occurs at critical points. The light gray regions adjacent to the obstacles represent the three newly formed cells. The type of critical points that define the gray regions in figure 6.16 are saddle points. In fact, all of the cells are defined by saddle points of D. Note that since D is nonsmooth, its generalized gradient must be used as well.

To determine the coverage path, the planner again first derives the decomposition, finds an exhaustive walk through the adjacency graph, and then plans the coverage path within each cell. The coverage path within a cell consists of three parts: motion

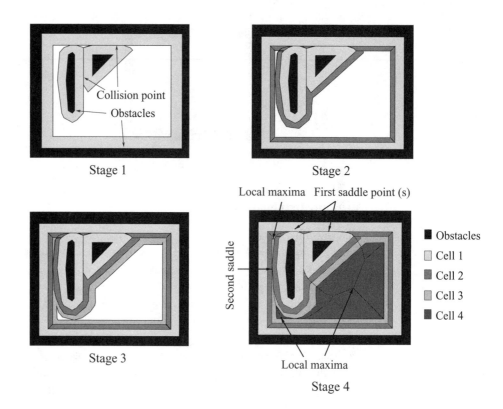

Figure 6.16 Incremental construction of the cells of the brushfire decomposition. The wave fronts collide with each other at the points located on the GVD.

along the slice, motion orthogonal to the slice, and motion along the boundary of the cell (figure 6.17). For motion along the slice, the robot follows a path at a fixed distance from the nearest obstacle. The robot follows an obstacle boundary at a fixed distance until it returns to its starting point or a point where the distance to two obstacles becomes the same. When the robot returns to its starting point, it simply moves away from the closest obstacle by one width and repeats following the obstacle boundary at the new fixed distance. When the robot becomes doubly equidistant, it is on the boundary of the cell, at which point it follows the GVD. The robot follows the GVD until it reaches a point where the distance to the nearest obstacle is an integer multiple of the robot's diameter, at which point the robot then resumes obstacle boundary-following at this fixed distance. Here, the boundary-following motion is much different from before because cell boundaries lie exclusively in the free space with the exception of the first slice $\lambda = 0$.

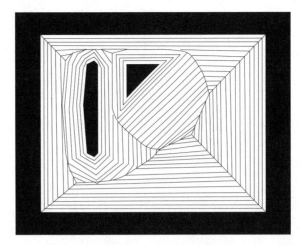

Figure 6.17   Coverage pattern for the brushfire decomposition. To generate this pattern, the robot follows the boundaries of obstacles and thus it has a continuous robust reference for localization. Therefore this pattern is suitable for the robots that are prone to accrue dead-reckoning error. However, the robot relies heavily on long-range sensors.

The pattern induced by the brushfire algorithm is ideally suited for coverage with mobile robots experiencing dead-reckoning error but have a large sensing range. A mobile robot can follow this pattern servoing off of the boundaries of the obstacles by moving forward and maintaining a fixed distance from the boundary (figure 6.17). Since the robot is servoing off of range readings to the obstacle boundary, this method is insensitive to dead-reckoning error. This benefit, however, requires that the robot can indeed measure distance to the obstacle, which could be far away from the robot. This is in contrast to the boustrophedon decomposition approach which requires only very limited sensing range but which is sensitive to dead-reckoning error.

Wave-Front Decomposition

Let $h(x, y)$ be the length of the shortest path between a point (x, y) and a fixed location. The level sets $h^{-1}(\lambda)$ foliate the free space where for a given λ, the set of points in $h^{-1}(\lambda)$ are a distance λ away from the fixed point in the free space. This particular function is sometimes called the wave-front potential, which was described for a discrete space in chapter 4, section 4.5. Imagine a wave front starting at q_{start} and expanding into the free space. The value λ parameterizes each wave front (or level set of h). Once the wave front crosses q_{goal}, the planner can backtrack a path from q_{goal} to q_{start} [208].

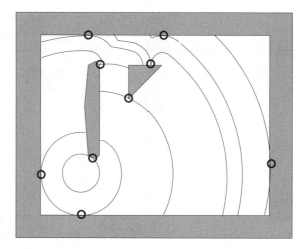

Figure 6.18 Wave-front decomposition defined on a continuous domain. The wave front emanates from a point in the lower-left portion of the figure. Cusp points on the wave fronts originate from the critical points, e.g., the cusp point on the upper boundary of the obstacle located on the left.

The shortest path-length function induces a cell decomposition, as well. Critical points of this function occur both when wave fronts becomes tangent to obstacles and when wave fronts collide (figure 6.18). Note how once the waves collide, they propagate as one wave with a nonsmooth point that originated at the critical point. In fact, this nonsmooth point traces the set of points of equal pathlength to the goal for two classes of paths, one to the right of the obstacle and one to the left. This decomposition is especially useful for coverage by a tethered robot where the robot's tether is incrementally fed and the robot sweeps out curves each at constant tether length.

6.2.4   Sensor-Based Coverage

Now, let's place the robot in an unknown environment, but assume it has the standard range sensor ring as depicted in chapter 2, figures 2.5 and 2.16. The task is to simultaneously cover and explore the unknown space. This can be reduced to concurrently and incrementally covering each cell while constructing the adjacency graph. For sensor-based coverage, however, we incrementally construct a "dual" graph called a *Reeb graph* [154]. This graph is dual in the sense that the nodes of the Reeb graph are the critical points and the edges connect neighboring critical points, i.e., correspond to cells. For the sake of explanation, we limit discussion to Morse decompositions

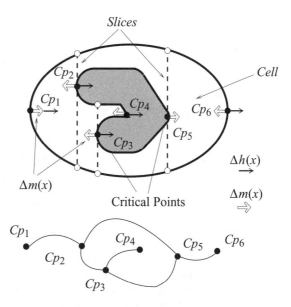

Figure 6.19   A boustrophedon decomposition and its Reeb graph. At the critical points, the surface normals and sweep direction are parallel.

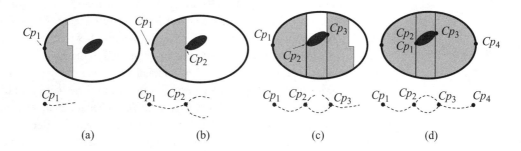

Figure 6.20   Incremental construction of the graph while the robot is covering the space.

defined by $h(x, y) = x$, i.e., the boustrophedon decomposition. See figure 6.19 for an example of the boustrophedon decomposition and its corresponding Reeb graph.

The procedure of concurrently covering the cells and constructing the Reeb graph is depicted in figure 6.20. In figure 6.20(a), the robot starts to cover the space at the critical point Cp_1 and the planner instantiates an edge with only one node. When the robot is done covering the cell between Cp_1 and Cp_2, the planner joins their corresponding nodes with an edge in the graph representation (figure 6.20b). Now the

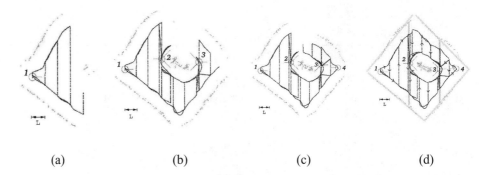

(a)   (b)   (c)   (d)

Figure 6.21 Four stages of coverage in an unknown environment with a robot-size detector on a Nomad Scout named RT Snookums. The coverage path followed by RT Snookums is shown by dotted black lines. We depict the critical points as light gray circles with lines emanating from them. The lines represent the directions of the corresponding adjacent cells. The robot incrementally constructs the graph representation by sensing the critical points 1, 2, 3, 4, 3, 2 (in the order of appearance) while covering the space. In the final stage (d), since all the critical points have explored edges, the robot concludes that it has completely covered the space. For the sake of discussion, we outlined the boundaries of the obstacles and cells in (d). The length scale $L = 0.53$ meters.

robot has two new uncovered cells. Since the space is *a priori* unknown, the planner arbitrarily chooses the lower cell to cover. When the robot reaches Cp_3, nodes of Cp_2 and Cp_3 become connected with an edge and the lower cell is completed (figure 6.20c). At Cp_3, the planner directs the robot to cover the cell to the right of Cp_3. When the robot senses Cp_4, it goes back to Cp_3 and starts to cover the upper cell. When the robot returns to Cp_2, the planner determines that all of the edges of all of the nodes (critical points) have been explored (figure 6.20d). Thus the planner concludes that the robot has completely covered the space. Figure 6.21 shows different stages of this incremental construction in an *a priori* unknown 2.75 meter × 3.65 meter room with a Nomad mobile robot that has a sonar ring.

Two details remain: How does the robot sense a critical point when it encounters one and how does the robot find all of the critical points? Critical point sensing is rather straightforward: the robot looks for points where the surface normals are parallel to the sweep direction. This is a direct consequence of lemma 5.5.2 in chapter 5, section 5.5.1. Here, we are looking for extrema of h on the boundaries of the obstacles. Let m implicitly represent a function whose preimage is the surface boundary. The matrix $D(h, m)(x)$ (chapter 5, section 5.5.1) then loses rank when $\nabla h(x)$ is parallel to $\nabla m(x)$, i.e., the slice normal is parallel to the surface normal (figure 6.19). This can easily be detected by looking at the global minimum of the range sensors in a range sensor ring.

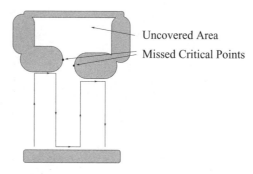

Figure 6.22   Critical points in the ceiling are missed with conventional coverage algorithms.

The final challenge is to ensure that the robot encounters all critical points. Assume the robot starts to cover a cell at one of its defining critical points. While covering the cell, the robot looks for the other critical point that indicates complete coverage of the cell and the next node in the Reeb graph. We term this critical point the *closing critical point*. Since the Reeb graph is connected, the main challenge is to guarantee that the robot finds the closing critical point of each cell.

Most conventional coverage algorithms (e.g., [187, 190, 300]) miss the closing critical point because they perform the bulk of their coverage using a raster scan type of motion: move along a slice or lap to an obstacle, follow the obstacle boundary for a lateral distance equal to interlap spacing, and repeat. This alternates boundary-following between the "ceiling" and "floor" of the cell, as shown in figure 6.22. Unfortunately, this raster scan approach can miss the closing critical point of a cell. In figure 6.22, since the robot did not follow the boundary of the ceiling, it cannot sense the critical points in the ceiling using the critical point sensing method, described above. We may try to solve this problem by making the robot perform boundary-following along the ceiling in the reverse direction so that it will sense the critical points related to the ceiling. We call this motion *reverse boundary following*. However, reverse boundary following motion by itself is still not sufficient.

The robot must undergo additional motion to detect the closing critical point. We present an algorithm called the *cycle algorithm* [11] that ensures that the robot will find the closing critical point while performing coverage. For details, see [11]. Let S_i be the start point of the cycle algorithm; S_i is on (or near) the boundary of free space. From this point the robot looks for critical points via the following phases (figure 6.23):

1. *Forward phase*: The robot follows a slice, i.e., laps, until it encounters an obstacle. Then the robot follows the boundary of the obstacle in the forward sweep direction until either the robot moves laterally one lap width or until the robot encounters a critical point in the floor.

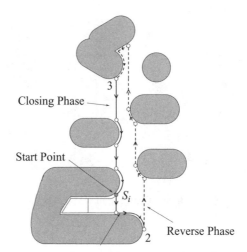

Closing Phase

Start Point

S_i

3

2

Reverse Phase

Forward Phase

Figure 6.23 In this particular arrangement of the obstacles, the robot executes every step of the cycle algorithm. The robot first follows the path between points S_i and 2 during the forward phase. Then it follows the path between points 2 and 3 during the reverse phase. Finally, in the closing phase the robot follows the path between points 3 and S_i.

2. *Reverse phase*: The robot executes one or more laps in the reverse direction, intermixed with reverse boundary-following. Each reverse boundary-following operation terminates when the robot finds a critical point or when the aggregate lateral motion in the reverse direction is one lap width.

3. *Closing phase*: The robot executes one or more laps along the slice, possibly intermixed with boundary-following. Each boundary-following operation terminates when the robot encounters S_i or the slice in which S_i lies.

This algorithm is the most important part of the incremental construction. It guarantees encountering the closing critical point of a cell if it exists between subsequent laps.

6.2.5 Complexity of Coverage

We define complexity of coverage in two ways: first, we establish a relationship among the number of critical points, cells, and obstacles and second we determine an upper-bound on path length given the perimeter the obstacles and the diameter Δ smallest disk that circumscribes the space (figure 6.24). We limit our discussion to coverage with the boustrophedon decomposition. First, we establish a relationship between

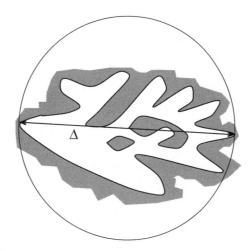

Figure 6.24   To determine the complexity of the algorithm in terms of the environment size, we use the diameter Δ of the "minimal" disk that fully contains the space.

the number of cells, critical points and obstacles. The Reeb graph encodes the cells, critical points, and "obstacles." Note that obstacles (including the outer boundary) are represented with "faces"[3] in the graph. Graph theory uses *Euler's formula* to relate the number of nodes v, edges e and faces f of a planar connected graph [57] by

$$v - e + f = 2.$$

The nodes of the Reeb graph correspond to critical points, its edges represent the cells and its faces depict the obstacles. Moreover, the Reeb graph is connected and planar. Therefore we can use Euler's formula with one modification. Since the outer boundary of the space is, in general, not termed an obstacle, we subtract one from the number of faces to get the number of obstacles. Let N_{cp} be the number of critical points, N_{ce} be the number of cells and N_{ob} be the number of obstacles (figure 6.25). Then

$$N_{ce} = N_{cp} + N_{ob} - 1.$$

This formula tells us that the number of cells increases *linearly* as the robot discovers new critical points.

Next, we calculate an upper bound on the total coverage path length. To simplify the calculation, we analyze lapping, boundary following and backtracking motions

---

3. A plane graph partitions the space into connected regions. Closures of these regions are called faces [57].

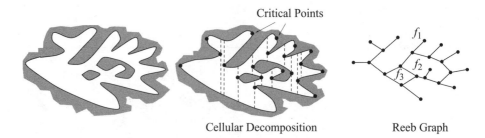

Critical Points

Cellular Decomposition   Reeb Graph

Figure 6.25 In this decomposition example, there are twenty-one critical points (nodes in the graph), $N_{cp} = 21$, and two obstacles (faces f_2, f_3 in the graph; f_1 is the outer boundary), $N_{ob} = 2$. Using the modified Euler's formula $N_{ce} = N_{cp} + N_{ob} - 1$, there must be twenty-two cells (edges in the graph), $N_{ce} = 22$.

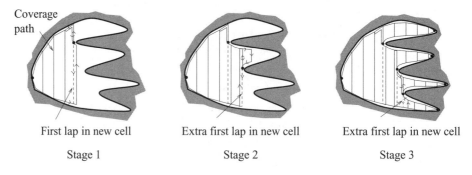

Coverage path

First lap in new cell   Extra first lap in new cell   Extra first lap in new cell

Stage 1   Stage 2   Stage 3

Figure 6.26 When the robot starts to cover a new cell, it performs an "extra" lap starting from a critical point on one of its boundaries.

separately. Since the space is fully contained within a Δ diameter disk, the length of each lapping path can be at most Δ. There must be at least $\lceil \frac{\Delta}{2r} \rceil$ lapping paths where $2r$ is the interlap spacing. However, often there is an additional lap associated with starting the coverage operation within a cell (figure 6.26). Hence, the maximum number of lapping paths is $\lceil \frac{\Delta}{2r} \rceil + N_{ce}$. Since the length of each lapping path is bounded above by Δ, the total path length of the lapping motions is bounded above by $\Delta \lceil \frac{\Delta}{2r} \rceil + \Delta N_{ce}$.

Now we analyze the length of boundary-following paths. Let P_{cell} be the length of the floor and ceiling of a cell. The coverage algorithm guarantees that the robot follows the entire floor and ceiling of a cell along the obstacle boundaries. Therefore, the length of boundary-following paths in a cell is at least P_{cell}. However, the robot, for each cycle, performs an undo-reverse boundary-following motion to get to the

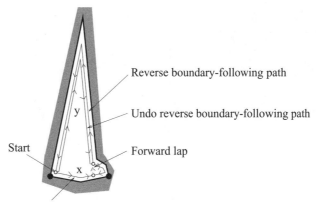

Reverse boundary-following path

Undo reverse boundary-following path

Forward lap

Start

y

x

Forward boundary-following path

Figure 6.27 The total perimeter of the cell is equal to $x + y$ where y is the length of the floor and x is the length of the ceiling and $x \gg y$. The total path length traveled along the boundary of the cell is bounded above by $2x + y$. In the worst case, as x gets much larger than y, this value is equal to $2(x + y)$.

start point. Hence, the lower bound is $1.5 P_{\text{cell}}$. In the worst case, the upper bound becomes $2 P_{\text{cell}}$ (figure 6.27). Then, the total length of the boundary-following paths is less than $2 P_{\text{total}}$ where P_{total} is the length of the perimeter of all of the obstacles and the outer boundary.

After discovering the closing critical point of a cell, the robot backtracks to the closing critical point of a cell with uncovered cells associated with it by boundary-following and (if necessary) lapping (figure 6.28). In the worst case, the length of this backtracking path is $P_{\text{cell}} + \Delta$ (where the robot follows every boundary and the longest slice). When we consider all the backtracking paths, the upper bound becomes $P_{\text{total}} + \Delta N_{ce}$.

The robot starts to cover an uncovered cell from one of its defining critical points. While discovering this critical point by performing the cycle algorithm, the robot covers a small portion of the uncovered cell. The extra boundary-following path followed by the robot to discover the critical point is bounded above by P_{cell}. Hence, the total extra boundary-following path length is bounded above by P_{total}.

When the robot finishes covering a cell, it performs a depth-first search on the Reeb graph to choose an uncovered cell (if any are left) (figure 6.29). The robot reaches the uncovered cell by traversing the covered cells. To traverse a covered cell, the robot performs boundary-following and lapping motions as we explained in section 6.2.1. Within each covered cell, the total path length traveled is bounded above

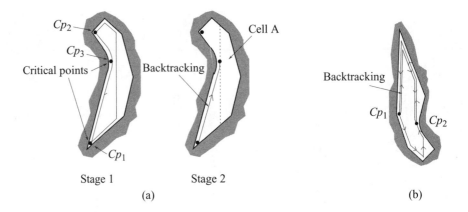

Figure 6.28   (a) The robot starts to cover the space from Cp_1. Along the first cycle path, it discovers the critical points Cp_1, Cp_2, and Cp_3. The robot moves back to Cp_3 by following the boundary of the obstacle to start to cover cell A. (b) The robot travels back to Cp_2 from Cp_1 by boundary-following and lapping. Therefore, the length of the backtracking path is bounded above by $P_{\text{cell}} + D$ for each cell.

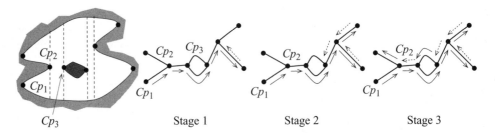

Figure 6.29   The robot starts to cover the space from Cp_1. Whenever the robot finishes covering a cell, a depth-first search is performed on the graph to choose a new cell to cover. On the graph, solid arrows depict the coverage directions and dashed arrows correspond to backtracking directions. The depth-first search on the graph requires a maximum of N_{ce} (number of edges) backtracking.

by $P_{\text{cell}} + 2\Delta$ (figure 6.30). Since we perform a depth-first search on the graph, each cell is traversed at most once [118], and therefore the backtracking path length is bounded by $\sum_{i=1}^{N_{ce}} P_{\text{cell}_i} + 2N_{ce}\Delta$ or $P_{\text{total}} + 2N_{ce}\Delta$.

Combining the above upper bounds, the length of the coverage path is less than

$$\frac{\Delta^2}{2r} + 4\Delta N_{ce} + 5P_{\text{total}},$$

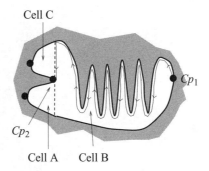

Figure 6.30 After finishing covering cells A and B, the robot needs to travel from Cp_1 to Cp_2 to start to cover cell C. The robot simply follows the boundary of the obstacle either along the ceiling or floor of the cell. In the worst case, the boundary-following path length is bounded above by the length of the perimeter of the obstacles that form the boundary of the cell.

or using the modified Euler's formula,

$$\frac{\Delta^2}{2r} + 4\Delta(N_{cp} + N_{ob}) + 5P_{\text{total}} - 4\Delta.$$

Therefore, the total coverage path length is bounded linearly by the area of the space, the number of critical points, and the length of the perimeter of the obstacles and the outer boundary.

6.3 Visibility-Based Decompositions for Pursuit/Evasion

In the previous section, we used *connectivity changes* (i.e., critical points) to decompose the space into cells. The benefit of using the critical points is that the cells have a structure that is "easy" to cover. In this section, we use changes in line-of-sight related information to define cells. Such cells form *visibility-based decompositions.* Moving from one cell to another corresponds to a change in visibility, e.g., obstacle or target appears or disappears. We can use a visibility-based cell decomposition to address the *pursuit/evasion problem.* This problem, first introduced by Suzuki and Yamashita [403], considers one or more multiple agents called *pursuers* who are searching a bounded free space (usually polygonal) for a single agent called an *evader.* This evader can be a bad guy who is escaping the police or a trapped survivor wandering around a disaster site in need of help in searching rescue for workers. Lavalle, Guibas, and coworkers [172, 273] use a cell decomposition approach to address the pursuit/evasion problem. This decomposition lies in the *workspace,* not the configuration space, and the agents are points in the plane.

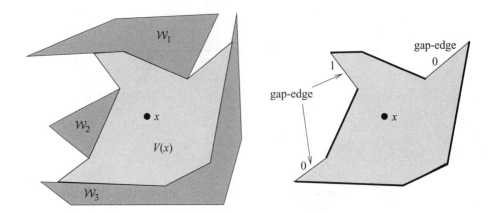

Figure 6.31   The visibility polygon is shaded inside a polygonal world populated by obstacles. The gap-edges and their labels for a generic visibility polygon $V(x)$ are labeled. A clear edge has a 0 label, a contaminated edge has a 1 label. Here, $B(x) = \{010\}$.

In this description, we borrow terminology and notation from [172, 273]. The evader e is caught when any one of the pursuers γ^i becomes within line of sight with it, i.e., there exists an i such that for all $\tau \in [0, 1]$, $\tau e + (1 - \tau)\gamma_i \in \mathcal{W}_{\text{free}}$. Let $e : [0, \infty) \to \mathcal{W}_{\text{free}}$ and $\gamma^i : [0, \infty) \to \mathcal{W}_{\text{free}}$ respectively be the paths that the evader and the ith pursuer follow. An evader is caught at the earliest time t when there exists an i such that for all $\tau \in [0, 1]$, $\tau e(t) + (1-\tau)\gamma_i(t) \in \mathcal{W}_{\text{free}}$. Let us recast the capture condition once more: let $V(x) \subset \mathcal{W}_{\text{free}}$ be the star-shaped set of points that are within line of sight of x (figure 6.31). An evader is caught if there exists an i and t such that $e(t) \in V(\gamma_i(t))$. A *motion strategy* is the collection of the pursuer paths $\gamma = \{\gamma_1, \ldots, \gamma_n\}$ and is termed a *solution strategy* if at least one pursuer catches the evader for all $e(t)$. Finally, let $H(\mathcal{W}_{\text{free}})$ be the minimum number of pursuers required to capture an evader in $\mathcal{W}_{\text{free}}$ in finite time.

We address the pursuit/evasion problem in two steps. Continuing to borrow terminology from [172, 273], we will define qualitatively important subsets of the free space and then use these subsets to define the decomposition. A region of $\mathcal{W}_{\text{free}}$ that *may* contain an evader is termed *contaminated*. If a region is not contaminated, then it is *clear*. However, a region that was contaminated, then cleared, and contaminated again is termed *recontaminated* (figure 6.32).

A visibility polygon now has two types of edges, those that lie on the boundary of obstacles and those that lie in the free space, which we call *gap-edges*. Gap-edges have a zero label if they bound a cleared region and a one if they bound a contaminated region (figure 6.31).

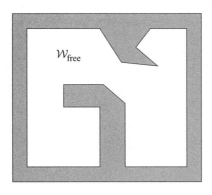 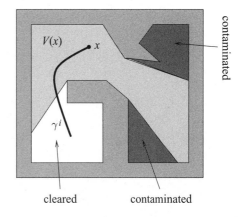

Figure 6.32   A polygonal world with visibility polygon, cleared area, and contaminated area. A cleared area is a region we know the evader is not in. A contaminated region could contain the evader. The ith pursuer's path is also drawn.

Let $B(x)$ denote a binary vector of these gap-edge labels for a particular star-shaped set centered x. The pair $(x, B(x))$ denotes the *information state* and the set of all possible information states is the *information space*. A connected set $v \subset \mathcal{W}_{\text{free}}$ is *conservative* if for all $x \in v$, $B(x)$ remains fixed (figure 6.33). Finally, we can construct an adjacency graph for the conservative regions in a given environment. Hence, the conservative regions form an exact cell decomposition. (figure 6.34)

To construct the conservative regions for a polygonal environment, simply extend rays from each convex vertex of all of the obstacles until they intersect another obstacle. Also, if two vertices are within line of sight of each other, extend two rays, one from each vertex, but in the opposite directions. In other words, for $v_i \in \mathcal{WO}_i$ and $v_j \in \mathcal{WO}_j$, if $\lambda v_i + (1 - \lambda)v_j \in \mathcal{W}_{\text{free}}$ for all $\lambda \in (-\epsilon, 1 + \epsilon)$ for some $\epsilon > 0$, then extend a ray from v_i away from v_j and vice versa until they intersect an obstacle (figure 6.35).

This process forms a cell decomposition of the free space where each cell is a conservative region. This cell decomposition, however, is not sufficient to solve the pursuit/evasion problem. We have to form a cell decomposition in the information space [172,273]. To do this, first identify all of the "transitions" that can occur when an agent passes from one conservative cell to another in the free space decomposition (figure 6.36). If

1. a gap-edge disappears, do nothing;

2. a gap-edge appears, assign it a zero;

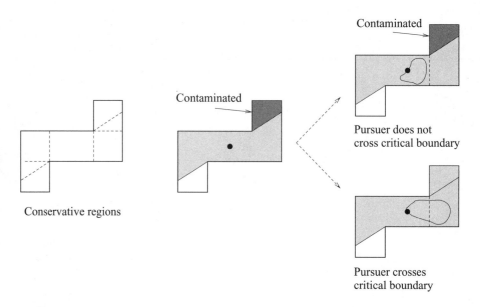

Conservative regions

Contaminated

Pursuer does not
cross critical boundary

Pursuer crosses
critical boundary

Figure 6.33   The leftmost figure denotes all of the conservative regions separated by dashed lines. In the next figure, the pursuer starts off with a visibility polygon represented by a light gray area and a contaminated region by dark gray. If the pursuer crosses the critical boundary, then the contaminated region becomes cleared and the information state changes.

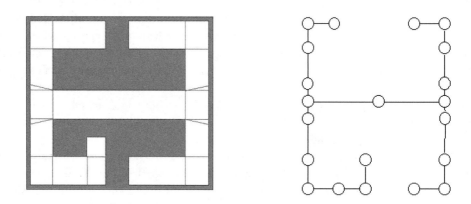

Figure 6.34   Conservative regions and their corresponding adjacency graph.

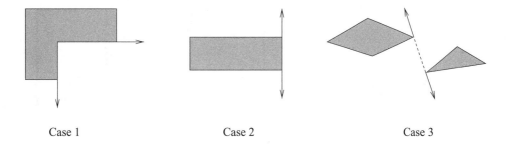

Case 1   Case 2   Case 3

Figure 6.35 The three conservative edge construction cases.

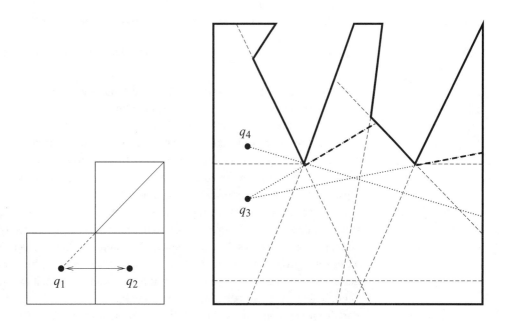

Figure 6.36 (Left) A gap-edge appears or disappears (Right) Multiple gap-edges merge or a gap-edge divides. Moving from q_3 to q_4 causes two edges to merge into one (case 3). From q_4 to q_3, a single gap edge splits into two (case 4).

3. two or more gap-edges merge into one, if any of them had a one label, assign a one to the new edge;

4. a gap-edge divides into multiple gap-edges, assign the new edges the same label as the original;

This transition information serves as a basis for an adjacency graph for a new decomposition. This graph, called the information graph, can be used to solve the

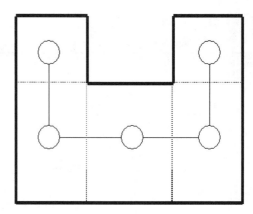

Figure 6.37   A simple space with its corresponding adjacency graph overlaid on top.

pursuit/evasion problem. For each cell of the conservative region decomposition, we generate a sequence of nodes, each corresponding to a possible set of gap-edge labels. Figure 6.37 contains a simple free space with its adjacency graph overlaying on it. Consider the upper right conservative region. For all points in this region, it can only have one gap-edge which could have either a zero or a one label. Note that this gap-edge does not lie in the conservative region, i.e., it is *not* the horizontal line that separates the rightmost conservative regions.

Likewise, the conservative region in the lower-right cell has only one free edge. However, the transition from the upper-right cell to the lower-right cell is limited by the possible transition cases. In other words, if the upper-right cell has a $B(x) = 1$, then the lower-right cell must have $B(x) = 1$, and it cannot be zero. Therefore, the edges of the information graph represent the possible transitions from cell to cell (figure 6.38).

For a single pursuer and single evader in simply-connected spaces, a planner can start from any node in the information graph and then search for a node that has $B(x) = 0$ (a vector of zeros). This determines a path through the conservative regions that is guaranteed to catch the pursuer (figure 6.39).

Now, let's consider the number of pursuers required to find an evader. First, assume that $\mathcal{W}_{\text{free}}$ is a simply-connected polygon. Assume $\mathcal{W}_{\text{free}}$ is partitioned into $\mathcal{W}_{\text{free}1}$ and $\mathcal{W}_{\text{free}2}$ by connecting two vertices of the boundary of $\mathcal{W}_{\text{free}}$. Moreover, if $H(\mathcal{W}_{\text{free}1}) \leq k$, and $H(\mathcal{W}_{\text{free}2}) \leq k$, then $H(\mathcal{W}_{\text{free}}) \leq k + 1$, since $\mathcal{W}_{\text{free}}$ can be cleared by first clearing $\mathcal{W}_{\text{free}1}$ and $\mathcal{W}_{\text{free}2}$ successively using the same k pursuers while keeping one pursuer, called the "static pursuer," at the common boundary between them.

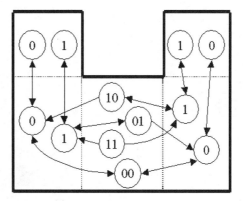

Figure 6.38 The information graph Gi, with all possible routes through the graph shown. Each node in the information graph has all the possible gap-edge labels for the conservative region corresponding to each node.

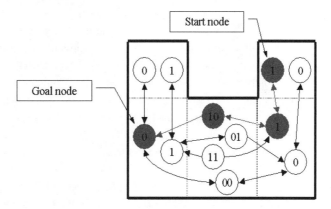

Figure 6.39 The information graph Gi from figure 6.38 with a solution path highlighted. Any node on the information graph with all zeros is a solution. Thus we use a graph search of our choice until we find a node of all zeros.

Since a simply-connected polygon can be partitioned into two pieces such that each component has at least one third of the edges of the original polygon, a simply-connected polygon can be triangulated by recursively connecting two vertices, and the "depth" of such a triangulation is at most $O(\log n)$. Therefore the original polygon can be cleared by clearing each triangle using one pursuer while keeping $O(\log n)$ static pursuers at each "level." Thus, at most $O(\log n)$ pursuers are required to clear a simply-connected polygon (figure 6.40).

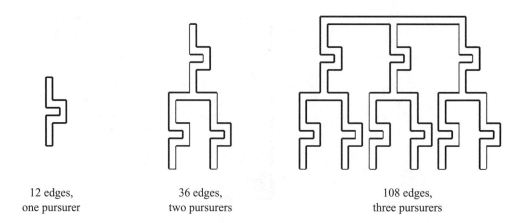

| 12 edges, | 36 edges, | 108 edges, |
| one pursurer | two pursuers | three pursurers |

Figure 6.40 Using a U-shaped space we can show how there are simply-connected free spaces that require order (log n) pursuers to explore. Each U-shape requires a single robot, and we can use another to divide the space into smaller and smaller sections.

Now consider a space with holes. Let h be the number of the holes of a free space $\mathcal{W}_{\text{free}}$, and assume that $\mathcal{W}_{\text{free}}$ is triangulated. Let a *trichromatic triangle* be the triangle that touches three distinct connected components of $\mathcal{W}_{\text{free}}$. If all of the trichromatic triangles were removed, then $\mathcal{W}_{\text{free}}$ would be divided into (disconnected) simply-connected regions. The number of the trichromatic triangles can be determined by forming a graph that has the following properties. The vertices of the graph correspond to the holes of the space, and two of the vertices are connected if there is a trichromatic triangle that touches the boundary of the two holes corresponding to these vertices. Since this graph is planar, it can be shown that the number of edges of this graph and therefore the number of the trichromatic triangles is $O(h)$.

Now consider the dual graph of the triangulation of $\mathcal{W}_{\text{free}}$, but actually consider only the vertices corresponding to the trichromatic triangles. Note that each edge of this graph corresponds to a simply-connected region of $\mathcal{W}_{\text{free}}$, which can be cleared using $O(\log n)$ pursuers using the result above. This graph can be partitioned using $O(\sqrt{h})$ edges into two components so that each component has at least one third of the edges. The $O(\sqrt{h})$ "static" pursuers are placed on the edges that partition the graph. Recursively applying the planar graph separator theorem, and placing static pursuers accordingly (thus total number of static pursuer is $O(\sqrt{h} + \sqrt{2/3h} + \sqrt{4/9h}) = O(\sqrt{h})$, a simply-connected region (i.e., an edge of the dual graph) can be isolated. Since a simply-connected region can be cleared using $O(\log n)$ pursuers, the complete region can be cleared using $O(\sqrt{h} + \log n)$ pursuers (figure 6.41).

4 holes, 111 edges, 4 pursuers

Figure 6.41 An example of a space that requires $O(\sqrt{h} + \log n)$ pursuers, where n is the number of edges and h is the number of holes. This example requires four pursuers. Three pursuers are used to divide the space into simply-connected regions, while the other robot searches.

Problems

1. At the end of section 6.1, we describe a method for locating a single query. Why is this important if it takes less time to locate a single query than to construct the search graph?

2. Write a program that determines a path for a planar convex translation-only robot from a start to final configuration using the trapezoidal decomposition. Input from a file a robot and from a separate file a set of obstacles in a known workspace. Input a start and goal configuration from the keyboard. Use the configuration space generator from chapter 3. Hand in meaningful output.

3. Describe a generalization of the trapezoidal decomposition in three dimensions. What do the cells look like?

4. Consider the trapezoidal decomposition and adjacency graph in figure 6.2. Using the method described in section 6.1, determine a path when q_{start} is in c_5 and q_{goal} is in c_6.

5. Describe a generalization of the boustrophedon decomposition in three dimensions. What do the cells look like?

6. How does the solution to the pursuit/evasion problem change if there are multiple evaders?

7. The solution for the pursuit/evasion problem is described in the workspace, not the configuration space. Why?

8. Figure 6.35 shows the three cases that are used to construct the cell decomposition of conservative cells. Show that the solution to the pursuit/evasion problem, described above, works correctly without case 2. Hint: Show that there are no critical changes in gaps for case 2, except where a gap is "anchored." Note that removing case 2 does, however, lead to concave cells. Show the new decomposition for the workspace in figure 6.34.

9. Adapt the pursuit/evasion problem to the case where the pursuer has a limited field-of-view sensor, then has a limited range sensor, and then a limited range limited field-of-view sensor.

7 *Sampling-Based Algorithms*

DIFFERENT PLANNERS described in chapter 5 build roadmaps in the free (or semi-free) configuration space. Each of these methods relies on an explicit representation of the geometry of Q_{free}. Because of this, as the dimension of the configuration space grows, these planners become impractical. Figure 7.1 shows a path-planning problem that cannot be solved in a reasonable amount of time with the methods presented in chapter 5, but can be solved with the *sampling-based* methods described in this chapter. Sampling-based methods employ a variety of strategies for generating samples (collision-free configurations of the robot) and for connecting the samples with paths to obtain solutions to path-planning problems.

Figures 7.2(a) and (b) show two typical examples from industrial automation that sampling-based planners can routinely solve. Sampling-based planners can also be used to address problems that extend beyond classic path planning. Figure 7.2(c) shows a CAD (computer-aided design) model of an aircraft engine. A planner can be used to determine if a part can be removed from that engine. Such information is extremely important for the correct design of the engine, as certain parts need to be removable for maintainability purposes. In this case, the planner considers the part to be separated as a robot that can move freely in space. Figure 7.2(d) involves an example from computer animation where a planner is used to plan the motion of the human figure. Figures 7.2(e) and (f) provide examples that involve planning with kinematic and dynamic constraints, while figure 7.2(g) displays the folding of a small peptide molecule. This chapter discusses the basics of sampling-based path planning.

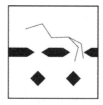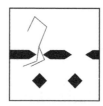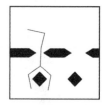

Figure 7.1   Snapshots along a path of a planar manipulator with ten degrees of freedom. The manipulator has a fixed base and its first three links have prismatic joints—they can extend to one and a half times their original length. (From Kavraki [221].)

The Development of Sampling-Based Planners

Sampling-based planners were developed at a time when several complexity results on the path-planning problem were known. The generalized mover's problem, in which the robot consists of a collection of polyhedra freely linked together at various vertices, was proven PSPACE-hard by Reif [361]. Additional study on exact path-planning techniques for the generalized mover's problem led Schwartz and Sharir to an algorithm that was doubly exponential in the degrees of freedom of the robot [373]. This algorithm is based on a cylindrical algebraic decomposition of semi-algebraic descriptions of the configuration space [117]. Recent work in real algebraic geometry renders the algorithm singly exponential [42]. Canny's algorithm [90], which builds a roadmap in the configuration space of the robot, is also singly exponential in the degrees of freedom of the robot. Furthermore, Canny's work showed that the generalized mover's problem was PSPACE-complete [90, 95]. The implementation of the above general algorithms is very difficult and not practical for the planning problems shown in figure 7.2.

The complexity of path-planning algorithms for the generalized mover's problem fueled several thrusts in path-planning research. These included the search for subclasses of the problem for which complete polynomial-time algorithms exist (e.g., [183, 374]), the development of methods that approximated the free configuration space (e.g., [67, 68, 132, 297]), heuristic planners (e.g., [174]), potential-field methods (e.g., [38, 40]), and the early sampling-based planners (e.g., [40, 47, 101, 165, 220, 231, 244]).

The Probabilistic RoadMap planner (PRM) [231] demonstrated the tremendous potential of sampling-based methods. PRM fully exploits the fact that it is cheap to check if a single robot configuration is in $\mathcal{Q}_{\text{free}}$ or not. PRM creates a roadmap in $\mathcal{Q}_{\text{free}}$. It uses rather coarse sampling to obtain the nodes of the roadmap and very fine sampling to obtain the roadmap edges, which are free paths between node configurations. After the roadmap has been generated, planning queries can be answered by connecting the user-defined initial and goal configurations to the roadmap, and by

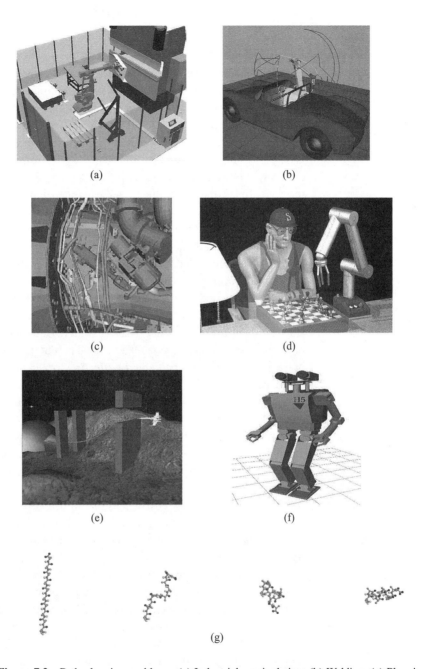

Figure 7.2 Path-planning problems. (a) Industrial manipulation. (b) Welding. (c) Planning removal paths for a part (the "robot") located at the center of the figure. (d) Computer animation. (e) Planning aircraft motion. (f) Humanoid robot. (g) Folding of a small peptide molecule. ((a) From Bohlin and Kavraki [54]; (b) from Hsu and Latombe [196]; (c) courtesy of Latombe; (d) from Koga, Kondo, Kuffner and Latombe [241]; (e) from Kuffner and LaValle [272]; (f) from Kuffner [248]; (g) from Amato [21].)

using the roadmap as in chapter 5 to solve the path-planning problem at hand. Initially, node sampling in PRM was done using a uniform random distribution. This planner is called basic PRM. It was observed that random sampling worked very well for a wide variety of problems [221, 231, 345] and ensured the probabilistic completeness of the planner [221, 229]. However, it was also observed [221] that random sampling is only a baseline sampling for PRM and many other sampling schemes are useful and are bound to be efficient for many planning problems as the analysis of the planner revealed. Today, these sampling schemes range from importance sampling in areas that during the course of calculations are found difficult to explore, to deterministic sampling such as quasirandom sampling and sampling on a grid. This chapter will describe the basic PRM algorithm, several popular node-sampling strategies, as well as their advantages and disadvantages, and popular node-connection strategies.

PRM was conceived as a multiple-query planner. When PRM is used to answer a single query, some modifications are made: the initial and goal configurations are added to the roadmap nodes and the construction of the roadmap is done incrementally and is stopped when the query at hand can be answered. However, PRM may not be the fastest planner to use for single queries. The second part of this chapter describes sampling-based planners that are particularly effective for single-query planning, including the Expansive-Spaces Tree planner (EST) [192, 196] and the Rapidly-exploring Random Tree planner (RRT) [249, 270]. These planners exhibit excellent experimental performance and will be discussed in detail.

Combination of the above methods is also possible and desirable in many cases. The Sampling-Based Roadmap of Trees (SRT) planner [14, 43] constructs a PRM-style roadmap of single-query-planner trees. It has been observed that for very difficult path planning problems, single-query planners need to construct large trees in order to find a solution. In some cases, the cost of constructing a large tree may be higher than the cost of constructing a roadmap of $\mathcal{Q}_{\text{free}}$ with SRT. This illustrates the distinction between multiple-query and single-query planning, and its importance. The SRT planner will be discussed in detail in this chapter.

Despite their simplicity, which is exemplified in the basic PRM planner, sampling-based planners are capable of dealing with robots with many degrees of freedom and with many different constraints. Sampling-based planners can take into account kinematic and dynamic constraints (e.g., [195, 271]), closed-loop kinematics (e.g., [121, 184, 268]), stability constraints (e.g., [64, 247, 248]), reconfigurable robots (e.g., [98, 139, 149]), energy constraints (e.g., [251, 255]), contact constraints (e.g., [210]), visibility constraints (e.g., [123]) and others. Clearly some planners are better at dealing with specific types of constraints than others. For example, as discussed in section 7.5.1, EST and RRT planners are particularly useful for problems that involve kinematic and dynamic constraints. Kinodynamic problems are described in chapters 10, 11, and 12.

PRM, EST, RRT, SRT, and their variants have changed the way path planning is performed for high-dimensional robots. They have also paved the way for the development of planners for problems beyond basic path planning. Because of space limitations, this chapter concentrates on the above planners and some of their variants, and does not include a comprehensive description of all effective sampling-based planning methods.

Characteristics of Sampling-Based Planners

An important characteristic of the planners described in this chapter is that they do not attempt to explicitly construct the boundaries of the configuration space obstacles or represent cells of $\mathcal{Q}_{\text{free}}$. Instead, they rely on a procedure that can decide whether a given configuration of the robot is in collision with the obstacles or not. In some sense, sampling-based planners have very limited access to the configuration space. Efficient collision detection procedures ease the implementation of sampling-based planners and increase the range of their applicability. Furthermore, since collision detection is a separate module, it can be tailored to specific robots and applications. Recent advances in collision detection algorithms have contributed heavily to the success of sampling-based planners. Any future performance improvements in collision checking, which is an active area of research, will also benefit directly the performance of sampling-based planners. Examples of available collision detection packages include GJK [89, 163], SOLID [420, 421], V-Clip [316], I-Collide [115, 290], V-Collide [199], QuickCD [238], PQP [261], RAPID [168], SWIFT [140], SWIFT++ [141], and others [88, 296, 357, 376].

Another important characteristic of sampling-based planners is that they can achieve some form of completeness. Completeness requires that the planner always answers a path-planning query correctly, in asymptotically bounded time. Complete planners cannot be implemented in practice for robots with more than three degrees of freedom due to their high combinatorial complexity. A weaker, but still interesting, form of completeness is the following: if a solution path exists, the planner will eventually find it. If the sampling of the sampling-based planner is random, then this form of completeness is called probabilistic completeness. If the sampling is deterministic, including quasirandom or sampling on a grid, this form of completeness is called resolution completeness with respect to the sampling resolution. Probabilistic completeness was shown for one of the earliest sampling-based planners, called the Randomized Path Planner (RPP) [39, 257], setting a standard for sampling-based methods. PRM was also shown to be probabilistically complete [195, 221–223, 228, 252]. The analysis of the probabilistic completeness for the basic PRM planner [221, 228] is presented in this chapter. The theoretical results relate the probability that PRM fails to find a path to the running time of the planner. Hence there is not only experimental evidence that PRM planners work well; there is also theoretical evidence of why this is the case. The analysis also sheds light on why the basic PRM planner works well on a large class of difficult problems.

Overview of This Chapter

Section 7.1 introduces PRM. In its basic form, PRM constructs a roadmap that represents the connectivity of $\mathcal{Q}_{\text{free}}$. This roadmap can be used for answering multiple queries. Guidelines for the efficient implementation of this planner for a general robot are also given. The guidelines are also relevant for the efficient implementation of the other sampling-based planners described in this chapter. A number of different sampling methods and connection strategies for PRM are then presented. Planners that are optimized for single-queries are described in section 7.2. In general, these planners generate trees in $\mathcal{Q}_{\text{free}}$. Some of the most efficient single-query planners, such as EST and RRT planners, perform a conditional sampling of $\mathcal{Q}_{\text{free}}$: the samples generated depend on the currently constructed tree and the goal configuration. In section 7.2 the EST and RRT planners are described in detail. The combination of the different sampling and connection strategies of sections 7.1 and 7.2 leads to an even more powerful planner, SRT, which is described in section 7.3. An analysis of PRM is given in section 7.4. Various extensions of the generalized mover's problem are then discussed in section 7.5, including applications from computational structural biology.

7.1   Probabilistic Roadmaps

The PRM planner is described in [231]. The planner resulted from the work of independent groups [225, 226, 344, 345, 404] and was further developed in [221, 223, 227, 228]. PRM divides planning into two phases: the learning phase, during which a roadmap in $\mathcal{Q}_{\text{free}}$ is built; and the query phase, during which user-defined query configurations are connected with the precomputed roadmap. The nodes of the roadmap are configurations in $\mathcal{Q}_{\text{free}}$ and the edges of the roadmap correspond to free paths computed by a local planner. The objective of the first phase is to capture the connectivity of $\mathcal{Q}_{\text{free}}$ so that path-planning queries can be answered efficiently.

The basic PRM algorithm presented below can be used to solve high-dimensional problems such as the one in figure 7.1. It has been shown to be probabilistically complete [221, 229, 252]. In this section, the choices for the sampling and connection strategies of PRM are reduced to a bare minimum to facilitate the presentation. The emphasis here is to describe a planner that is easy to implement and works well even with rather high-dimensional problems (5–12 degrees of freedom).

7.1.1 Basic PRM

The basic PRM algorithm first constructs a roadmap in a probabilistic way for a given workspace. The roadmap is represented by an undirected graph $G = (V, E)$. The nodes in V are a set of robot configurations chosen by some method over $\mathcal{Q}_{\text{free}}$. For the moment, assume that the generation of configurations is done randomly from a uniform distribution. The edges in E correspond to paths; an edge (q_1, q_2) corresponds to a collision-free path connecting configurations q_1 and q_2. These paths, which are referred to as local paths, are computed by a local planner. In its simplest form, the local planner connects two configurations by the straight line in $\mathcal{Q}_{\text{free}}$, if such a line exists.

In the query phase, the roadmap is used to solve individual path-planning problems. Given an initial configuration q_{init} and a goal configuration q_{goal}, the method first tries to connect q_{init} and q_{goal} to two nodes q' and q'', respectively, in V. If successful, the planner then searches the graph G for a sequence of edges in E connecting q' to q''. Finally, it transforms this sequence into a feasible path for the robot by recomputing the corresponding local paths and concatenating them. Local paths can be stored in the roadmap but this would increase the storage requirements of the roadmap, a topic which is discussed later in this section.

The roadmap can be reused and further augmented to capture the connectivity of $\mathcal{Q}_{\text{free}}$. Although the learning phase is usually performed before any path-planning query, the two phases can also be interwoven. It is reasonable to spend a considerable amount of time in the learning phase if the roadmap will be used to solve many queries.

Roadmap Construction

To make the presentation more precise, let

- Δ be the local planner that on input $(q, q') \in \mathcal{Q}_{\text{free}} \times \mathcal{Q}_{\text{free}}$ returns either a collision-free path from q to q' or NIL if it cannot find such a path. Assume for the moment that Δ is symmetric and deterministic.

- *dist* be a function $\mathcal{Q} \times \mathcal{Q} \to \mathbb{R}^+ \cup \{0\}$, called the *distance function,* usually a metric on $\mathcal{Q}$.

Algorithm 6 describes the steps of the roadmap construction. For all algorithms described in this chapter, it should be noted that only the main steps are given and that implementation details are missing.

Initially, the graph $G = (V, E)$ is empty. Then, repeatedly, a configuration is sampled from $\mathcal{Q}$. For the moment, assume that the sampling is done according to a uniform random distribution on $\mathcal{Q}$. If the configuration is collision-free, it is added to the roadmap. The process is repeated until n collision-free configurations have been

---

Algorithm 6 Roadmap Construction Algorithm

Input:
 n : number of nodes to put in the roadmap
 k : number of closest neighbors to examine for each configuration
Output:
 A roadmap $G = (V, E)$

1: $V \leftarrow \emptyset$
2: $E \leftarrow \emptyset$
3: **while** $|V| < n$ **do**
4:      **repeat**
5:          $q \leftarrow$ a random configuration in $\mathcal{Q}$
6:      **until** q is collision-free
7:      $V \leftarrow V \cup \{q\}$
8: **end while**
9: **for all** $q \in V$ **do**
10:      $N_q \leftarrow$ the k closest neighbors of q chosen from V according to *dist*
11:      **for all** $q' \in N_q$ **do**
12:          **if** $(q, q') \notin E$ **and** $\Delta(q, q') \neq$ NIL **then**
13:              $E \leftarrow E \cup \{(q, q')\}$
14:          **end if**
15:      **end for**
16: **end for**

---

sampled. For every node $q \in V$, a set N_q of k closest neighbors to the configuration q according to some metric *dist* is chosen from V. The local planner is called to connect q to each node $q' \in N_q$. Whenever Δ succeeds in computing a feasible path between q and q', the edge (q, q') is added to the roadmap. Figure 7.3 shows a roadmap constructed for a point robot in a two-dimensional Euclidean workspace, where Δ is a straight-line planner.

A number of components in algorithm 6 are still unspecified. In particular, it needs to be defined how random configurations are created in line (5), how the closest neighbors are computed in line (10), how the distance function *dist* used in line (10) is chosen, and how local paths are generated in line (12).

Query Phase

During the query phase, paths are found between arbitrary input configurations q_{init} and q_{goal} using the roadmap constructed in the learning phase. Algorithm 7 illustrates this process.

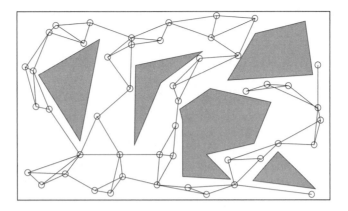

Figure 7.3 An example of a roadmap for a point robot in a two-dimensional Euclidean space. The gray areas are obstacles. The empty circles correspond to the nodes of the roadmap. The straight lines between circles correspond to edges. The number of k closest neighbors for the construction of the roadmap is three. The degree of a node can be greater than three since it may be included in the closest neighbor list of many nodes.

Assume for the moment that Q_{free} is connected and that the roadmap consists of a single connected component. The main question is how to connect q_{init} and q_{goal} to the roadmap. Queries should terminate as quickly as possible, so an inexpensive algorithm is desired here. The strategy used in algorithm 7 to connect q_{init} to the roadmap is to consider the k closest nodes in the roadmap in order of increasing distance from q_{init}, according to the metric *dist,* and try to connect q_{init} to each of them with the local planner until one connection succeeds. The number of closest neighbors considered in algorithm 7 can be different from the one in algorithm 6. The same procedure is used to connect q_{goal} to the roadmap.

If the connection of q_{init} and q_{goal} to the roadmap is successful, the shortest path is found on the roadmap between q_{init} and q_{goal} according to *dist* (e.g., using Dijkstra's algorithm or the A^* algorithm). If one wishes, this path may be improved by running a smoothing postprocessing algorithm. Figure 7.4 shows the solution to a query solved with the roadmap from figure 7.3.

In general, the roadmap may consist of several connected components. This is very likely when Q_{free} is itself not connected, but it may also happen when Q_{free} is connected, and the roadmap has not managed to capture the connectivity of Q_{free}. If the roadmap contains several components, algorithm 7 can be used to connect both q_{init} and q_{goal} to two nodes in the same connected component of the roadmap, e.g., by giving it as input a single connected component of G. All components of G should be considered. If the connection of q_{init} and q_{goal} to the same connected

---

Algorithm 7 Solve Query Algorithm

Input:

  q_{init}: the initial configuration

  q_{goal}: the goal configuration

  k: the number of closest neighbors to examine for each configuration

  $G = (V, E)$: the roadmap computed by algorithm 6

Output:

  A path from q_{init} to q_{goal} or failure

1: $N_{q_{\text{init}}} \leftarrow$ the k closest neighbors of q_{init} from V according to *dist*

2: $N_{q_{\text{goal}}} \leftarrow$ the k closest neighbors of q_{goal} from V according to *dist*

3: $V \leftarrow \{q_{\text{init}}\} \cup \{q_{\text{goal}}\} \cup V$

4: set q' to be the closest neighbor of q_{init} in $N_{q_{\text{init}}}$

5: **repeat**

6:     **if** $\Delta(q_{\text{init}}, q') \neq \texttt{NIL}$ **then**

7:         $E \leftarrow (q_{\text{init}}, q') \cup E$

8:     **else**

9:         set q' to be the next closest neighbor of q_{init} in $N_{q_{\text{init}}}$

10:     **end if**

11: **until** a connection was succesful or the set $N_{q_{\text{init}}}$ is empty

12: set q' to be the closest neighbor of q_{goal} in $N_{q_{\text{goal}}}$

13: **repeat**

14:     **if** $\Delta(q_{\text{goal}}, q') \neq \texttt{NIL}$ **then**

15:         $E \leftarrow (q_{\text{goal}}, q') \cup E$

16:     **else**

17:         set q' to be the next closest neighbor of q_{goal} in $N_{q_{\text{goal}}}$

18:     **end if**

19: **until** a connection was succesful or the set $N_{q_{\text{goal}}}$ is empty

20: $P \leftarrow$ shortest path($q_{\text{init}}, q_{\text{goal}}, G$)

21: **if** P is not empty **then**

22:     **return** P

23: **else**

24:     **return** failure

25: **end if**

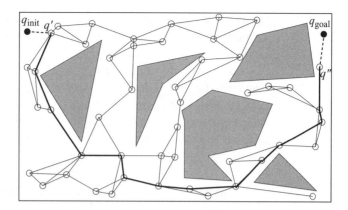

Figure 7.4   An example of how to solve a query with the roadmap from figure 7.3. The configurations q_{init} and q_{goal} are first connected to the roadmap through q' and q''. Then a graph-search algorithm returns the shortest path denoted by the thick black lines.

component of the roadmap succeeds, a path is constructed as in the single-component case. The method returns failure if it cannot connect both q_{init} and q_{goal} to the same roadmap component.

Adding to the Roadmap

If path-planning queries fail frequently, the roadmap may not adequately capture the connectivity of $\mathcal{Q}_{\text{free}}$. When this occurs, the current roadmap can be extended by resuming the construction step algorithm (exclude lines (1) and (2) from algorithm 6 and pass as a parameter the current roadmap). It should be emphasized again that in this section we present a very basic PRM. It has been observed for example, that when trying to connect components biased sampling may be particularly effective [231]. Biased sampling (see Connection Sampling in section 7.1.3) increases the sampling density in areas of $\mathcal{Q}_{\text{free}}$ that have good chances to facilitate component connection.

Directed Roadmaps and Roadmaps That Store Local Paths

So far, it has been assumed that Δ is symmetric and deterministic. It is also possible to use a local planner Δ that is neither symmetric nor deterministic.

In many cases, connecting some configuration q to some configuration q' does not necessarily imply that the opposite can be done. If the local planner takes the robot from q to q' and the robot can also execute the path in reverse to go from q' to q, the roadmap is an undirected graph. Adding the edge (q, q') implies that the edge

(q', q) can also be added. If local paths cannot be reversed, a directed roadmap must be constructed. A separate check must be performed to determine if the edge (q', q) can also be added to the roadmap.

A deterministic local planner will always return the same path between two configurations and the roadmap does not have to store the local path between the two configurations in the corresponding edge. The path can be recomputed if needed to answer a query. On the other hand, if a nondeterministic local planner is used, the roadmap will have to associate with each edge the local path computed by Δ. In general, the use of nondeterministic local planners increases the storage requirements of the roadmap. It permits, however, the use of more powerful local planners, which can be an advantage in certain cases as discussed in section 7.3.

7.1.2   A Practical Implementation of Basic PRM

One of the advantages of the basic PRM algorithm presented in the previous section is that it is easy to implement and performs well for a variety of problems. This section focuses on the details of a successful implementation of basic PRM that scales well for robots with many degrees of freedom. Issues that relate to a practical implementation of a planner, such as smoothing of the final path, are also discussed. These issues pertain to all planners in this chapter. The reader is also referred to [246] for details on implementation details and potential pitfalls.

Sampling Strategy: Uniform Distribution

In basic PRM [231] the nodes of the roadmap constitute a uniform random sampling of $\mathcal{Q}_{\text{free}}$. To obtain a configuration, each translational degree of freedom can be drawn from the interval of allowed values of the corresponding degree of freedom using the uniform probability distribution over this interval. The same principle applies to rotational degrees of freedom but care should be taken not to favor specific orientations because of the representation used (see the example at the end of section 7.1.2 and [246]). The main idea is that the sampling distribution should be symmetry invariant. The sampled configuration is checked for collision. If it is collision-free, the sample is added to the nodes of the roadmap; otherwise, it is discarded. Collision checking can be done using a variety of existing general techniques, as mentioned above.

Sampling from a uniform distribution is the simplest method for generating sample configurations, but other methods could be used, as we describe below. Section 7.4 offers a theoretical explanation of why sampling from a uniform distribution works well for many problems.

Connection Strategy: Selecting Closest Neighbors

Another important choice to be made is that of selecting the set N_q of closest neighbors to a configuration q. Many data structures have been proposed in the field of computational geometry that deal with the problem of efficiently calculating the closest neighbors to a point in a d-dimensional space. A relatively efficient method both in terms of space and time is the kd-tree data structure [124].

A d-dimensional kd-tree uses as input a set S of n points in d dimensions and constructs a binary tree that decomposes space into cells such that no cell contains too many points. A kd-tree is built recursively by splitting S by a plane into two subsets of roughly equal size: S_ℓ, which includes points of S that lie to the left of the plane; and S_r, which includes the remaining points of S. The plane is stored at the root, and the left and right child are recursively constructed with input S_ℓ and S_r, respectively. Figure 7.5 illustrates the construction of a 2-dimensional kd-tree for ten points on a plane.

A kd-tree for a set of n points in d dimensions uses $O(dn)$ storage and can be built in $O(dn \log n)$ time. A rectangular range query takes $O(n^{1-\frac{1}{d}} + m)$ time, where m is the number of reported neighbors. As d grows large, the cost of using kd-trees becomes linear. The rectangular range query time can be reduced considerably by introducing a small approximation error. This modified approach is called Approximate Nearest Neighbor queries (ANN) and is becoming increasingly popular [30].

Distance Functions and Embeddings

Function *dist* is used to resolve the k closest neighbors query. It should be defined so that, for any pair (q', q'') of configurations, $dist(q', q'')$ reflects the likelihood that the

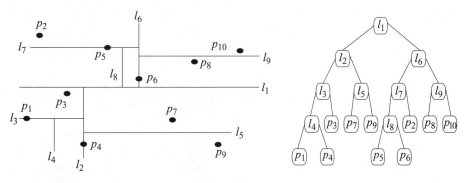

(a) The way the plane is subdivided.   (b) The corresponding binary tree.

Figure 7.5   A kd-tree for ten points on a plane.

local planner will fail to compute a collision-free path between these configurations. One possibility is to define $dist(q', q'')$ as some measure of the workspace region swept by the robot, such as the area or the volume, when it moves in the absence of obstacles along the path $\Delta(q', q'')$. Intuitively, minimizing the swept volume, will minimize the chance of collision with the obstacles. An exact computation of swept areas or volumes is notoriously difficult, which is why heuristic metrics generally attempt to approximate the swept-volume metric (see [19, 246]).

An approximate and inexpensive measure of the swept-region can be constructed as follows. The robot's configurations q' and q'' can be mapped to points in a Euclidean space, emb(q') and emb(q''), respectively, and the Euclidean distance between them can be used, i.e.,

$$dist(q', q'') = \| \, \text{emb}(q') - \text{emb}(q'') \, \| \, .$$

A practical choice for the embedding function is to select $p > 0$ points on the robot, concatenate them, and create a vector whose dimension is p multiplied by the dimension of the workspace of the robot. In order to represent a configuration q in the embedded space, the set of transformations corresponding to this configuration is applied to the p points, and emb(q) is obtained. Distances can be easily defined using the equation above. An example is given at the very end of this section. Note, however, that this choice of embeddings has its shortcomings. In particular, it is not clear what the number p should be. It is also not clear how to choose p points so that the exact shape of the robot is taken into account. Furthermore, as is the case with the swept-volume metric, the embedding does not take into account obstacles. So even when two configurations are close to one another, connecting them may be impossible due to obstacles.

For the case of rigid body motion, an alternative solution is to split *dist* into two components, one that expresses the distance between two configurations due to translation and one due to orientation. For example, if X and R represent the translation and rotation components of the configuration $q = (X, R) \in SE(3)$ respectively, then

$$dist(q', q'') = w_t \|X' - X''\| + w_r f(R', R'')$$

is a weighted metric with the translation component $\|X' - X''\|$ using a standard Euclidean norm, and the positive scalar function $f(R', R'')$ returning typically an approximate measure of the distance between the rotations $R', R'' \in SO(3)$. The rotation distance is scaled relative to the translation distance via the weights w_t and w_r. A reasonable choice of $f(R', R'')$ is the length of the geodesic curve between R' and R''. The selection of an appropriate rotation distance function $f(R', R'')$ depends on the representation for the orientation of the robot, such as Euler angles or quaternions.

One of the difficulties with this method is deciding proper weight values. Furthermore, the extension to articulated bodies is not straightforward. A thorough discussion of metrics for rigid body planning is given in [246].

The choices for the embedding, its dimensionality, and the *dist* can have a great effect on the efficiency of the PRM algorithm. Different problems may require different approaches and there is great interest in the motion-planning community in finding appropriate metrics [19, 246] and embeddings for interesting instances of the generalized mover's problem.

Local Planner

In section 7.1, it was assumed that Δ is symmetric and deterministic. This is a design decision and it is possible to accommodate planners that are nondeterministic, and/or not symmetric.

Another important design decision is related to how fast the local planner should be. There is clearly a tradeoff between the time spent in each individual call of this planner and the number of calls. If a powerful local planner is used, it would often succeed in finding a path when one exists. Hence, relatively few nodes might be required to build a roadmap capturing the connectivity of $\mathcal{Q}_{\text{free}}$ sufficiently well to reliably answer path-planning queries. Such a local planner would probably be rather slow, but this could be somewhat compensated by the small number of calls needed. On the other hand, a very fast planner is likely to be less successful. It will require more configurations to be included in the roadmap and as a result, the local planner is called more times for the connections between nodes. Each call will be cheaper, however. In section 7.3, a roadmap technique that uses a powerful local planner is discussed.

The choice of the local planner also affects the query phase. It is important to be able to connect any given q_{init} and q_{goal} configurations to the roadmap or to detect very quickly that no such connection is possible. This requires that the roadmap be dense enough that it always contains at least some nodes to which it is easy to connect q_{init} and q_{goal}. It thus seems preferable to use a very fast local planner, even if it is not too powerful, and build large roadmaps with configurations widely distributed over $\mathcal{Q}_{\text{free}}$. In addition, if the local planner is very fast, the same planner can be used to connect q_{init} and q_{goal} to the roadmap at query time. Discussions of the use of different local planners can be found in [14, 162, 203, 221].

One popular planner, applicable to all holonomic robots, connects any two given configurations by a straight-line segment in $\mathcal{Q}$ and checks this line segment for collision. Care should be taken to interpolate the translation and rotation components separately (see [246]). There are two commonly-used choices for collision checking, the incremental and the subdivision collision-checking algorithms. In both cases, the line segment, or more generally, any path generated by the local planner between configurations q' and q'', is discretized into a number of configurations $(q_1, \ldots, q_\ell)$, where $q' = q_1$ and $q'' = q_\ell$. The distance between any two consecutive configurations q_i and q_{i+1} is less than some positive constant `step_size`. This value is problem specific and is defined by the user. It is important to note that again sampling is used to determine if a local path is collision-free. But in this case, sampling is done at a much finer level than was done for node generation and this is a very important feature of PRM. In general, the value of `step_size` needs to be very small to guarantee that all collisions are found.

In the case of incremental collision checking, the robot is positioned at q' and moved at each step by `step_size` along the straight line in $\mathcal{Q}$ between q' and q''. A collision check is performed at the end of each step. The algorithm terminates as soon as a collision is detected or when q'' is reached.

In the case of the subdivision collision checking, the middle point q_m of the straight line in $\mathcal{Q}$ between q' and q'' is first checked for collision. Then the algorithm recurses on the straight lines between (q', q_m) and (q_m, q''). The recursion halts when a collision is found or the length of the line segment is less then `step_size`.

In both algorithms, the path is considered collision-free if none of the intermediate configurations yields collision. Neither algorithm has a clear theoretical advantage over the other, but in practice the subdivision collision checking algorithm tends to perform better [162, 367]. The reason is that, in general, shorter paths tend to be collision-free. Subdivision collision checking cuts down the length of the local path as soon as possible. It is also possible to use an adaptive subdivision collision-checking algorithm that dynamically adjusts `step_size`. In [376], `step_size` is determined by relating the distance between the robot and the workspace obstacles to the maximum length of the path traced out by any point on the robot. Furthermore, the method in [376] is exact, i.e., it always finds a collision when a collision exists, whereas the above discretization techniques may miss a collision if `step_size` is too large.

Figure 7.6 illustrates how the incremental and subdivision collision-checking algorithms are sampling the straight line between two configurations q' and q''. In this example, the subdivision algorithm performs a smaller number of collision checks. If the obstacle had been close to q', then the incremental algorithm would have performed a smaller number of collision checks.

Postprocessing Queries

A postprocessing step may be applied to the path connecting q_{init} to q_{goal} to improve its quality according to some criteria. For example, shortness and smoothness might

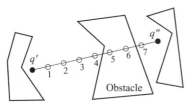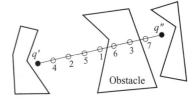

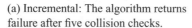

(a) Incremental: The algorithm returns
failure after five collision checks.

(b) Subdivision: The algorithm returns
failure after three collision checks.

Figure 7.6   Sampling along the straight line path between two configurations q' and q''. The numbers correspond to the order in which each strategy checks the samples for collision.

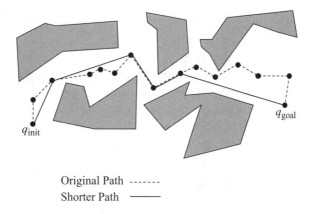

Original Path $------$
Shorter Path $\rule{2em}{0.4pt}$

Figure 7.7   Processing the path returned from PRM to get a shorter path with the greedy approach.

be desirable. Postprocessing is applicable to any path-planning algorithm, but is presented here for completeness of the implementation guidelines of the basic PRM.

From a given path, a shorter path could be obtained by checking whether nonadjacent configurations q_1 and q_2 along the path can be connected with the local planner. This idea has been described often in the literature (e.g., [150, 383]). The points q_1 and q_2 could be chosen randomly. Another alternative would be a greedy approach. Start from q_{init} and try to connect directly to the target q_{goal}. If this step fails, start from the configuration after q_{init} and try again. Repeat until a connection can be made to q_{goal}, say from the point q_0. Now set the target to q_0 and begin again, trying to connect from q_{init} to q_0, and repeat the procedure. This procedure can also be applied toward the opposite direction. Figure 7.7 illustrates the application of the greedy approach in the forward direction to shorten a path in a two-dimensional Euclidean workspace.

There are various reasons why configurations q_1 and q_2 along a path may have not been connected with an edge from the roadmap construction step of PRM. They may not be close according to the distance function *dist,* and the k closest neighbor query may not return them as neighbors. They may, however, be in a relatively uncluttered part of $\mathcal{Q}_{\text{free}}$ and a long edge connecting them may still be possible. These cases will occur more frequently if the Creating Sparse Roadmaps connection strategy has been used (see section 7.1.4).

Instead of shortening the path, a different objective may be to get a path with smooth curvature. A possible approach to this is to use interpolating curves, such as splines, and use the configurations that have been computed by PRM as the interpolation points for the curves. In this case, collision checking is performed along the curves until curves that satisfy both the smoothness properties and the collision avoidance criteria are found.

Postprocessing steps such as path shortening and path smoothing can improve the quality of the path, but can also impose a significant overhead on the time that it takes to report the results of a query. In general, if paths with certain optimality criteria are desired, it is worth trying to build these paths during the roadmap construction phase of PRM. For example, a large dense roadmap will probably yield shorter paths than a smaller and sparser roadmap.

An Example

Figure 7.8(a) shows a motion-planning problem for a robot in a three-dimensional workspace. The robot is a rigid nonconvex polyhedral object; it can freely translate and rotate in the workspace as long as it does not collide with the obstacles. The workspace is made up of a rigid thin wall that has a narrow passage. A bounding box is defined that contains the wall and is small enough so that it does not allow the robot to move from one side of the wall to the other without going through the narrow passage. The goal is to build a roadmap that a planner can use to successively solve motion-planning queries where q_{init} and q_{goal} appear on the two different sides of the wall.

The problem has six degrees of freedom, three translational and three rotational. The configuration $q = (p, r)$ of the robot can be represented by a point p expressing the translational component and a quaternion r (see appendix E) expressing the rotational component. A configuration is generated by picking at random a sample from a uniform distribution from a subset of allowable positions in $\mathbb{R}^3$ and picking a random axis of rotation and a random angle for the quaternion (for details see [246]).

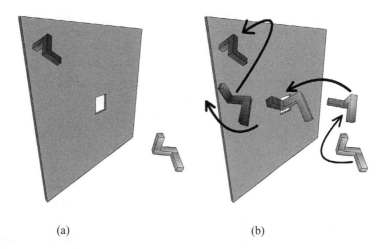

(a)   (b)

Figure 7.8   An example of a motion-planning problem where both the robot and the obstacles are a collection of polyhedral objects in three dimensions. Parts of the robot on the other side of the wall are indicated by the darker color. (a) The initial and goal configuration of the query. (b) A path produced from a PRM with $n = 1000$ and $k = 10$.

In order to find the k closest neighbors of a configuration, configurations are embedded in a space where Euclidean distance is defined. A method that works well in practice is to choose a pair of points on the surface of the robot that have maximum distance and construct a six-dimensional vector emb(q) for the robot's initial configuration. If q' is obtained by applying a translation and rotation transformation to q, then emb(q') is obtained by applying the same transformations to the pair of points in emb(q). The distance metric *dist* is then defined as the Euclidean distance of the two embeddings.

For every configuration and its k closest counterparts, the subdivision collision-checking algorithm is used to check if the straight line in Q is collision-free. Intermediate configurations between $q' = (p', r')$ and $q'' = (p'', r'')$ are obtained by performing linear interpolations on p' and p'' and spherical interpolations on r' and r''. The edge (q', q'') is added to the roadmap when all the intermediate configurations are collision-free.

When the roadmap has been completed, it can be used to solve user-specified queries. The k closest neighbors for the query points are calculated and the local planner attempts to connect q_{init} and q_{goal} to them. As soon as they are connected to the same component, an A^* algorithm is run on the graph to find the path. Figure 7.8(b) shows intermediate configurations of a path returned by the above procedure.

7.1.3   PRM **Sampling Strategies**

Several node-sampling strategies have been developed over the years for PRM. For many path-planning problems, a surprisingly large number of general sampling schemes will provide reasonable results (see e.g., the comparison of sampling schemes given in [162]). The analysis of section 7.4 provides some insight as to why this is the case. Intuitively, many planning problems in the physical world are difficult but not "pathological" (as in the kind of problem one encounters in NP-hardness proofs). Without doubt, however, the choice of the node-sampling strategy can play a significant role in the performance of PRM. This was observed in the original PRM publications which suggested mechanisms to generate samples in a non-uniform way [231]. Increasing the density of sampling in some areas of the free space is referred to as importance sampling and has been repeatedly demonstrated to increase the observed performance of PRM. In this section we describe several node-sampling schemes.

The uniform random sampling used in early work in PRM is the easiest sampling scheme to implement. As a random sampling method, it has the advantage that, in theory, a malicious opponent cannot defeat the planner by constructing carefully crafted inputs. It has the disadvantage, however, that, in difficult planning examples, the running time of PRM might vary across different runs. Nevertheless, random sampling works well in many practical cases involving robots with a large number of degrees of freedom.

There exist cases where uniform random sampling has poor performance. Often, this is the result of the so-called narrow passage problem. If a narrow passage exists in Q_{free} and it is absolutely necessary to go through that passage to solve a query, a sampling-based planner must select a sample from a potentially very small set in order to answer the planning query. A number of different sampling methods have been designed with the narrow passage problem in mind and are described below. The narrow passage problem still remains a challenge for PRM planners and is an active area of research.

The remainder of this section describes sampling strategies that have been developed with the narrow passage problem in mind and then other general sampling strategies. We conclude the section with a brief discussion of how one might select an appropriate sampling scheme for a particular problem.

Sampling Near the Obstacles

Obstacle-based sampling methods sample near the boundary of configuration-space obstacles. The motivation behind this kind of sampling is that narrow passages can be considered as thin corridors in Q_{free} surrounded by obstacles.

OBPRM [18] is one of the first and very successful representatives of obstacle-based sampling methods. Initially, OBPRM generates many configurations at random from a uniform distribution. For each configuration q_{in} found in collision, it generates a random direction v, and the planner finds a free configuration q_{out} in the direction v. Finally, it performs a simple binary search to find the closest free configuration q to the surface of the obstacle. Configuration q is added to the roadmap, while q_{in} and q_{out} are discarded.

The Gaussian sampler [59] addresses the narrow passage problem by sampling from a Gaussian distribution that is biased near the obstacles. The Gaussian distribution is obtained by first generating a configuration q_1 randomly from a uniform distribution. Then a distance step is chosen according to a normal distribution to generate a configuration q_2 at random at distance step from q_1. Both configurations are discarded if both are in collision or if both are collision-free. A sample is added to the roadmap if it is collision-free and the other sample is in collision.

In [194], samples are generated in a dilated $\mathcal{Q}_{free}$ by allowing the robot to penetrate by some small constant distance into the obstacles. The dilation of $\mathcal{Q}_{free}$ widens narrow passages, making it easier for the planner to capture the connectivity of the space. During a second stage, all samples that do not lie in $\mathcal{Q}_{free}$ are pushed into $\mathcal{Q}_{free}$ by performing local resampling operations.

Sampling Inside Narrow Passages

The bridge planner [193] uses a bridge test to sample configurations inside narrow passages. In a bridge test, two configurations q' and q'' are sampled randomly from a uniform distribution in $\mathcal{Q}$. These configurations are considered for addition to the roadmap, but if they are both in collision, then the point q_m halfway between them is added to the roadmap if it is collision free. This is called a bridge test because the line segment between q' and q'' resembles a bridge with q' and q'' inside obstacles acting as piers and the midpoint q_m hovering over $\mathcal{Q}_{free}$. Observe that the geometry of narrow passages makes the construction of short bridges easy, while in open space the construction of short bridges is difficult. This allows the bridge planner to sample points inside narrow passages by favoring the construction of short bridges.

An efficient solution to the narrow passage problem would generate samples that are inside narrow passages but as far away as possible from the obstacles. The Generalized Voronoi Diagrams (GVDs) described in chapter 5 have exactly this property. Although exact computation of the GVD is impractical for high-dimensional configuration spaces, it is possible to find samples on the GVD without computing it explicitly. This can be done by a retraction scheme [427]. The retraction is achieved by a bisection method that moves each sample configuration until it is equidistant from two points on the boundary of $\mathcal{Q}_{\text{free}}$.

A simpler approach is to compute the GVD of the workspace and generate samples that somehow conform to this GVD [155, 171, 191]. For example, the robot can have some predefined handle points (e.g., end-points of the longest diameter of the robot) and sampling can place those handle points as close to the GVD as possible with the hope of aligning the whole robot with narrow passages. The disadvantage of workspace-GVD sampling is that it is in general difficult to generate configurations of the robot close to the GVD (details are given in [155, 171, 191]). The advantage of workspace-GVD sampling is that the GVD captures well narrow passages in the workspace that typically lead to narrow passages in $\mathcal{Q}_{\text{free}}$. Additionally, an approximation of the GVD of the workspace can be computed efficiently using graphics hardware [352] which is one of the reasons why this sampling method is popular for virtual walkthroughs and related simulations.

Visibility-Based Sampling

The goal of the visibility-based PRM [337] is to produce visibility roadmaps with a small number of nodes by structuring the configuration space into visibility domains. The visibility domain of a configuration q includes all configurations that can be connected to q by the local planner. This planner, unlike PRM which accepts all the free configurations generated in the construction stage, adds to the roadmap only those configurations q that satisfy one of two criteria: (1) q cannot be connected to any existing node, i.e., q is a new component, or (2) q connects at least two existing components. In this way, the number of configurations in the roadmap is kept small.

Manipulability-Based Sampling

Manipulability-based sampling [281, 282] is an importance-sampling approach that exploits the manipulability measure associated with the manipulator Jacobian [432]. Intuitively, manipulability characterizes the arm's freedom of motion for a given configuration. The motivation for using manipulability as a bias for sampling is as follows. In regions of the configuration space where manipulability is high, the robot has great dexterity, and therefore relatively fewer samples should be required in these areas. Regions of the configuration space where manipulability is low tend to be near (or to include) singular configurations of the arm. Near singularities, the range of possible motions is reduced, and therefore such regions should be sampled more densely.

Let $J(q)$ denote the manipulator Jacobian matrix (i.e., the matrix that relates velocities of the end effector to joint velocities). For a redundant arm (e.g., an arm with more than six joints for a 3D workspace) the manipulability in configuration q is given by

(7.1) $$\omega(q) = \sqrt{\det J(q) J^T(q)}.$$

To bias sampling, an approximation to the cumulative density function (CDF) for ω is created. Samples are then drawn from a uniform density on the configuration space, and rejected with probability proportional to the associated CDF value of their manipulability value.

Quasirandom Sampling

A number of deterministic (sometimes called quasirandom) alternatives to random sampling have been used [62, 269, 291, 292]. These alternatives were first introduced in the context of Monte Carlo integration and aim to optimize various properties of the distribution of the samples. Before discussing some of these alternatives, we briefly describe two ways to evaluate a set of samples.

Let P be a set of point samples on some space X, and N be the number of points in P. One way to evaluate the quality of the samples in P is to assess how "uniformly" the points in P cover X. This is done with respect to a specific collection of subsets of X, called a *range space,* denoted by $\mathcal{R}$. Let $\mathcal{R}$ be the set of all axis-aligned rectangular subsets of X, and define μ to be the measure (or volume) of a set. Since P contains N points, the difference between the relative volumes of R to X and the fraction of samples contained in $R \in \mathcal{R}$ is given by

$$\left| \frac{\mu(R)}{\mu(X)} - \frac{|P \cap R|}{N} \right|.$$

If we take the supremum of this difference over all $R \in \mathcal{R}$ we obtain the concept of *discrepancy.*

DEFINITION 7.1.1   *The discrepancy of point set P with respect to range space $\mathcal{R}$ over some space X is defined as*

$$D(P, \mathcal{R}) = \sup_{R \in \mathcal{R}} \left| \frac{\mu(R)}{\mu(X)} - \frac{|P \cap R|}{N} \right|.$$

It is not necessary to take $\mathcal{R}$ as the subset of axis-aligned rectangles, but this choice gives an intuitive understanding of discrepancy. Another common choice is to take $\mathcal{R}$ as the set of d-balls, i.e., for each $R \in \mathcal{R}$ we have $R = \{x' \mid \|x - x'\| < \epsilon\}$, for some point x and radius $\epsilon > 0$.

While discrepancy provides a measure of how uniformly points are distributed over the space X, *dispersion* provides a measure of the largest portion of X that contains no points in P. For a given metric ρ, the distance between a point $x \in X$ and a point $p \in P$ is given by $\rho(x, p)$. Thus, $\min_{p \in P} \rho(x, p)$ gives the distance from x to the nearest point in P. If we take ρ to be the Euclidean metric, this gives the largest empty ball centered on x. If we then take the minimization over all points in X, we obtain the size of the largest empty ball in X. This is exactly the concept of dispersion.

DEFINITION 7.1.2   *The* dispersion δ *of point set P with respect to the metric ρ is given by*

$$\delta(P, \rho) = \sup_{x \in X} \min_{p \in P} \rho(x, p).$$

An important result due to Sukharev gives a bound on the number of samples required to achieve a given dispersion. In particular, the *Sukharev sampling criterion* states that when ρ is taken as the L_∞ norm, a set P of N samples on the d-dimensional unit cube will have

$$\delta(P, \rho) \geq \frac{1}{2\lfloor N^{\frac{1}{d}} \rfloor}.$$

So, to achieve a given dispersion value, say δ^*, since N must be an integer, we have

$$\delta^* \geq \frac{1}{2\lfloor N^{\frac{1}{d}} \rfloor} \rightarrow N \geq \left(\frac{1}{2\delta^*}\right)^d,$$

i.e., the number of samples required to achieve a desired dispersion grows exponentially with the dimension of the space. In some sense, this result implies that to minimize dispersion, sampling on a regular grid will yield results that are as good as possible.

Now that we have quantitative measures for the quality of a set of samples, we describe some common ways to generate samples. For the case of $X = [0, 1]$ the *Van der Corput sequence* gives a set of samples that minimizes both dispersion and discrepancy. The nth sample in the sequence is generated as follows. Let $a_i \in \{0, 1\}$ be the coefficients that define the binary representation of n,

$$n = \sum_i a_i 2^i = a_0 + a_1 2 + a_2 2^2 + \cdots.$$

The nth element of the Van der Corput sequence, $\Phi(n)$, is defined as

$$\Phi(n) = \sum_i a_i 2^{-(i+1)} = a_0 2^{-1} + a_1 2^{-2} + \cdots.$$

Figure 7.9(a) shows the first sixteen elements of a Van der Corput sequence.

The Van der Corput sequence can only be used to sample the real line. The *Halton sequence* generalizes the Van der Corput sequence to d dimensions. Let $\{b_i\}$ define

n	n (binary)	$\Phi(n)$ (binary)	$\Phi(n)$	n	$\Phi_2(n)$	$\Phi_1(n)$
0	0	0.0	0	0	0	0
1	1	0.1	1/2	1	1/3	1/2
2	10	0.01	1/4	2	2/3	1/4
3	11	0.11	3/4	3	1/9	3/4
4	100	0.001	1/8	4	4/9	1/8
5	101	0.101	5/8	5	7/9	5/8
6	110	0.011	3/8	6	2/9	3/8
7	111	0.111	7/8	7	5/9	7/8
8	1000	0.0001	1/16	8	8/9	1/16
9	1001	0.1001	9/16	9	1/27	9/16
10	1010	0.0101	5/16	10	10/27	5/16
11	1011	0.1101	13/16	11	19/27	13/16
12	1100	0.0011	3/16	12	4/27	3/16
13	1101	0.1011	11/16	13	13/27	11/16
14	1110	0.0111	7/16	14	22/27	7/16
15	1111	0.1111	15/16	15	7/27	15/16

(a)   (b)

Figure 7.9   (a) Van der Corput sequence, (b) Halton sequence for $d = 2$.

a set of d relatively prime integers, e.g., $b_1 = 2$, $b_2 = 3$, $b_3 = 5$, $b_4 = 7, \ldots$. The integer n has a representation in base b_j given by

$$n = \sum_i a_{ij} b_j^i, \quad a_{ij} \in \{0, 1, \ldots, b_j - 1\}$$

and $\Phi_{b_j}(n)$ is defined as

$$\Phi_{b_j}(n) = \sum a_{ij} b_j^{-(i+1)}.$$

The nth sample is then defined by the coordinates $p_n = (\Phi_{b_1}(n), \Phi_{b_2}(n), \cdots, \Phi_{b_d}(n))$. Figure 7.9(b) shows the first sixteen elements of a Halton sequence for $b_1 = 2$, $b_2 = 3$.

When the range space $\mathcal{R}$ is a set of axis-aligned rectangular subsets of X, the discrepancy for the Halton sequence is bounded by

$$D(P, \mathcal{R}) \leq O \left(\frac{\log^d N}{N} \right).$$

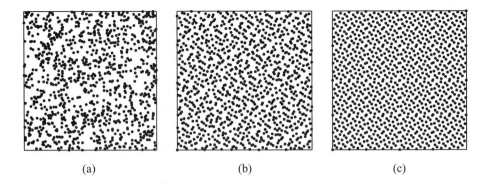

(a)   (b)   (c)

Figure 7.10   These figures shows 1024 samples generated in the plane using (a) a random number generator, (b) a Halton sequence, (c) a Hammersley sequence.

When the range space $\mathcal{R}$ is the set of $d-$balls, the discrepancy is bounded by

$$D(P, \mathcal{R}) \leq O\left(N^{-\frac{(d+1)}{2}}\right).$$

When N is specified, a *Hammersley sequence* (sometimes called a Hammersley point set, since the number of points is known and finite) achieves the best possible asymptotic discrepancy. The nth point in a Hammersley sequence is obtained by using the first $d-1$ coordinates of a point in the Halton sequence, with the ratio n/N as the first coordinate,

$$p_n = (n/N, \Phi_{b_1}(n), \Phi_{b_2}(n), \cdots \Phi_{b_{d-1}}(n)), \qquad n = 0 \ldots N-1.$$

Figure 7.10 shows point sets generated using a random number generator (figure 7.10a), a Halton sequence (figure 7.10b), and a Hammersley sequence (figure 7.10c). Each point set contains 1024 points.

The use of quasirandom sequences has the advantage that the running time is guaranteed to be the same for all the runs due to the deterministic nature of the point generation process. The resulting planner is resolution complete. The analysis of section 7.4 also sheds light as to why quasirandom sequences work well. As with any deterministic sampling method however, it is possible to construct examples where the performance of the planner deteriorates. As a remedy, it has been suggested to perturb the sequence [162]. The perturbation is achieved by choosing a random configuration from a uniform distribution in a small area around the sample point being added to the sequence. The area is gradually reduced as more points are added to the sequence. Certain quasirandom sequences can also be seen as generating points in a multiresolution grid in $\mathcal{Q}$ [269].

Grid-Based Sampling

Grid-based planners have appeared in the early planning literature [244, 274] but did not use some key abstractions of PRM such as the collision checking primitives. The nodes of a grid can be an effective sampling strategy in the PRM setting. Especially when combined with efficient node connection schemes (see section 7.1.4), they can result in powerful planners for problems arising in industrial settings [52]. A natural way of using grid-based search in a PRM is to use a rather coarse resolution for the grid and take advantage of the collision-checking abstraction; moving from one grid node q to a neighboring grid node q' would require collision checking, and hence sampling, at a finer resolution between the nodes. During the query phase, attempts are made to connect q_{init} and q_{goal} to nearby grid points. The resolution of the grid that is used to build the roadmap can be progressively increased either by adding points one at a time or by adding an entire hyperplane of samples chosen to fill the largest gap in the existing grid [52]. Of particular interest for path planning is the use of infinite sequences based on regular structures, which incrementally enhance their resolution. Recent work has demonstrated the use of such sequences for building lattices and other regular structures that have an implicit neighborhood structure, which is very useful for PRMs [269, 291]. A grid-based path-planning algorithm is resolution complete.

Connection Sampling

Connection sampling [221, 231] generates samples that facilitate the connection of the roadmap and can be combined with all previously described sampling methods. Typically, if a small number of configurations is initially generated, there may exist a few disconnected components at the end of the construction step. If the roadmap under construction is disconnected in a place where $\mathcal{Q}_{\text{free}}$ is not, this place may correspond to some difficult area of $\mathcal{Q}_{\text{free}}$, possibly to a narrow passage of $\mathcal{Q}_{\text{free}}$. The idea underlying connection sampling is to select a number of configurations from the roadmap that are likely to lie in such regions and expand them. The expansion of a configuration q involves selecting a new free configuration in the neighborhood of q as described below, adding this configuration to the roadmap, and trying to connect it to other configurations of the roadmap in the same manner as in the construction step. The connection sampling step increases the density of the roadmap in regions of $\mathcal{Q}_{\text{free}}$ that are believed to be difficult. Since the gaps between components of the roadmap are typically located in these regions, the connectivity of the roadmap is likely to increase. Connection sampling thus never creates new components in the roadmap. At worst, it fails to reduce the number of components.

A simple probabilistic scheme can be used for connection sampling. Each configuration q is associated with a heuristic measure of the difficulty of the region around

q expressed by a positive weight $w(q)$. Thus, $w(q)$ is large whenever q is considered to be in a difficult region. Weights are normalized so that their sum for all configurations in the roadmap is one. Then, repeatedly, a configuration q is selected from the roadmap with probability

$$Pr(q \text{ is selected}) = w(q),$$

and then q is expanded. The weights can be computed only once at the beginning of the process and not modified when new configurations are added to the roadmap, or can be modified periodically.

There are several ways to define the heuristic weight $w(q)$ [221, 231]. A function that has been found to work well in practice is the following. Let $\deg(q)$ be the number of configurations to which q is connected. Then,

$$w(q) = \frac{\frac{1}{\deg(q)+1}}{\sum_{q' \in V} \frac{1}{\deg(q')+1}}.$$

The expansion of a configuration q requires the generation of a configuration in the neighborhood of q. Typically, such a configuration can be found easily by selecting values for the degrees of freedom of the robot within a small interval centered at the values of the corresponding degrees of freedom of q. If this fails, a small random-bounce walk may be used to arrive at a new collision-free configuration. For holonomic robots, a random-bounce walk [231] from q consists of repeatedly picking at random a direction of motion and moving in this direction until an obstacle is hit. When a collision occurs, a new random direction is chosen. The above steps are repeated for a number of times. The configuration q' reached by the random-bounce walk and the edge (q, q') are inserted into the roadmap. Moreover, the path computed between q and q' is explicitly stored, since it was generated by a nondeterministic technique. The fact that q' belongs to the same connected component as q is also recorded. Then attempts are made to connect q' to the other connected components of the roadmap in the same way as in the construction step of PRM.

Choosing Among Different Sampling Strategies

Choosing among different sampling strategies is an open issue. Here, we give some very rough guidelines on how to choose a sampling strategy.

The success of PRM should be partly attributed to the fact that for a large range of problems (difficult but not "pathological" problems—see section 7.4) several simple sampling strategies work well. For example, uniform random sampling works well for many problems found in practice involving 3–7 degrees of freedom. If consistency in the running time is an issue, quasirandom sampling and lattice-based sampling provide some advantages. When the dimension grows, and again for problems that do not exhibit pathological behavior, random sampling is the simplest way to go. When problems that have narrow passages are considered, sampling-based strategies that were designed with narrow passages in mind should be used.

Combinations of different sampling methods are possible and in many cases critical for success. If π_A and π_B are two different sampling methods, a weighted hybrid sampling method π can be produced by setting $\pi = (1 - w)\pi_A + w\pi_B$. For example, connection sampling could be used in combination with random sampling [231] or OBPRM sampling. One sampling strategy can also be considered a filter for another. For example, the Gaussian sampler can be used to filter nodes created according to the bridge test [263].

None of the sampling methods described in this chapter provides clearly the best strategy across all planning problems. Sampling should also be considered in relation with the connection strategy used (see section 7.1.4) and the local planner used (see [14, 162, 203, 221] and section 7.3). Finally, it must be emphasized that it is possible to create "pathological" path-planning instances that will be arbitrarily hard for any sampling-based planner.

7.1.4 PRM **Connection Strategies**

An important aspect of PRM is the selection of pairs of configurations that will be tried for connections by a local planner. The objective is to select those configurations for which the local planner is likely to succeed. As has been discussed, one possible choice is to use the local planner to connect every configuration to all of its k closest neighbors. The rationale is that nearby samples lead to short connections that have good chances of being collision free. This section discusses some other approaches, their advantages and disadvantages. Clearly, the function used to select the neighbors and the implemented local planner can drastically affect the performance [19, 246] of any connection strategy described in this section.

Creating Sparse Roadmaps

A method that can speed up the roadmap construction step is to avoid the computation of edges that are part of the same connected component [231, 404]. Since there exists a path between any two configurations in a connected component, the addition of the new edge will not improve the connectivity of the roadmap. Several implementations of this idea have been proposed. The simplest is to connect a configuration with the nearest node in each component that lies close enough. This method avoids many calls to the local planner and consequently speeds up the roadmap construction step.

As the graph is being built, the connected components can be maintained by using a fast disjoint-set data structure [119].

With the above method, no cycles can be created and the resulting graph is a forest, i.e., a collection of trees. Since a query would never succeed due to an edge that is part of a cycle, it is indeed sensible not to consume time and space computing and storing such an edge. In some cases, however, the absence of cycles may lead the query phase to construct unnecessarily long paths. This drawback can be mitigated by applying postprocessing techniques, such as smoothing, on the resulting path. It has been observed however that allowing some redundant edges to be computed during the roadmap construction phase (e.g., two or three per node) can significantly improve the quality of the original path without significant overhead [162]. Recent work shows how to add useful cycles in PRM roadmaps that result in higher quality (shorter) paths [336].

Connecting Connected Components

The roadmap constructed by PRM is aimed at capturing the connectivity of $\mathcal{Q}_{\text{free}}$. In some cases, due to the difficulty of the problem or the inadequate number of samples being generated, the roadmap may consist of several connected components. The quality of the roadmap can be improved by employing strategies aimed at connecting different components of the roadmap. Connection sampling, introduced in section 7.1.3, attempts to connect different components of the roadmap by placing more nodes in difficult regions of $\mathcal{Q}_{\text{free}}$. Section 7.2 describes sampling-based tree planners that can be very effective in connecting different components of the roadmap. This is exploited in the planner described in section 7.3. Random walks and powerful planners such as RPP [40] can also be used to connect components [221]. Other strategies are described in [323].

Lazy Evaluation

The idea behind lazy evaluation is to speed up performance by doing collision checks only when it is absolutely necessary. Lazy evaluation can be applied to almost all the sampling-based planners presented in this chapter [52–54]. In this section, lazy evaluation is described as a node connection scheme. It has also given rise to very effective planners that will be described in the next section.

When lazy evaluation is employed, PRM operates on a roadmap G, whose nodes and paths have not been fully evaluated. It is assumed that all nodes and all edges of a node to its k neighbors are free of collisions. Once PRM is presented with a query, it connects q_{init} and q_{goal} to two close nodes of G. The planner then performs a graph search to find the shortest path between q_{init} and q_{goal}, according to the distance function used. Then the path is checked as follows. First, the nodes of G on the path are checked for collision. If a node is found in collision, it is removed from G together with all the edges originating from it. This procedure is repeated until a path with free nodes is discovered. The edges of that path are then checked. In order to avoid unnecessary collision checks, however, all edges along the path are first checked at a coarse resolution, and then at each iteration the resolution becomes finer and finer until it reaches the desired discretization. If an edge is found in collision, it is removed from G. The process of finding paths, checking their nodes and then checking their edges is repeated until a free path is found or all nodes of G have been visited. Once it is decided that a node of G is in $\mathcal{Q}_{\text{free}}$, this information is recorded to avoid future checks. For the edges, the resolution at which they have been checked for collision is also recorded so that if an edge is part of a future path, collision checks are not replicated. If no path is found and the nodes of G have been exhausted, new nodes and edges can be added to G. The new nodes can be sampled not only randomly but also from the difficult regions of $\mathcal{Q}_{\text{free}}$ [54]. This kind of sampling is similar to the connection sampling strategy of PRM described in section 7.1.3.

A related lazy scheme [335] assigns a probability to each edge of being collision free. This probability is computed by taking into account the resolution at which the edge has been checked. The edge probabilities can be used to search for a path in G that has good chances of being in $\mathcal{Q}_{\text{free}}$.

7.2   Single-Query Sampling-Based Planners

PRM was originally presented as a multiple-query planner: the goal was to create a roadmap that captures the connectivity of $\mathcal{Q}_{\text{free}}$ and then answer multiple user-defined queries very fast. In many planning instances, the answer to a single query is of interest and these instances are best served by single-query planners. Single-query planners attempt to solve a query as fast as possible and do not focus on the exploration of the entire $\mathcal{Q}_{\text{free}}$.

Many efficient single-query sampling-based planners exist. Some of them preceded PRM. One of the first widely used sampling-based planners was RPP [40]. RPP works by constructing potential fields over the workspace that attract control points of the robot to their corresponding positions in the goal configuration while pushing these robot points away from the obstacles (see also chapter 4). The workspace potentials are combined using an arbitration function to generate a configuration space potential. Starting from the initial configuration RPP performs a gradient motion until it reaches a local minimum. If the goal configuration has not been reached, RPP executes a series of random motions to escape the local minimum. In this way, RPP incrementally builds a graph of local minima, where the path joining two local minima is obtained by concatenating a random motion and a gradient descent motion. "Ariadne's clew" is another algorithm that uses samples in the configuration space [47, 48]. The algorithm works by interleaving the exploration of Q with searches for paths to the goal configuration. "Ariadne's clew" builds a tree from the initial configuration. During exploration, new configurations are placed in Q_{free} as far as possible from one another. The selection of configurations can be difficult and is done through genetic optimization. For each new configuration, a local search is performed to determine if the goal configuration is reachable from it. Many other algorithms (e.g., [33, 102, 165, 204]) explored the idea of planning by generating sample points in Q_{free}, but will not be presented in this chapter due to space limitations. The planner in [204] called the 2Z-method bears some similarities with PRM.

PRM itself can also be used as single-query planner. In that case, q_{init} and q_{goal} should be inserted to the roadmap at the beginning. The planner should check periodically if the given query can be solved, that is if q_{init} and q_{goal} belong to the same component of the roadmap. At that point, the construction of the roadmap should be aborted. The sampling and connections strategies described in section 7.1 are all applicable here. In particular, the careful application of lazy evaluation has yielded an effective single-query PRM planner, which is called LazyPRM [52–54]. LazyPRM "creates" a roadmap whose nodes and edges have not been checked for collision. The planner performs a standard search to find a path from the initial to the goal configuration and starts checking the path for collisions as described in section 7.1.4. The planner stops when a collision-free path has been found and it was shown experimentally that this was achieved well before the roadmap was fully checked [53].

This section describes two planners that were designed primarily for single-query planning. The planners are Expansive-Spaces Trees (ESTs) [192, 195, 196, 235] and Rapidly-exploring Random Trees (RRTs) [249, 270–272]. These planners also have the advantage that they are very efficient for kinodynamic planning (see section 7.5 and chapters 10, 11, and 12). For the moment, we concentrate on geometric path-planning.

ESTs and RRTs bias the sampling of configurations by maintaining two trees, T_{init} and T_{goal}, rooted at q_{init} and q_{goal} configurations, respectively, and then growing the trees toward each other until they are merged into one. It is possible to construct only a single tree rooted at q_{init} that grows toward q_{goal}, but, for geometric path-planning, this is usually less efficient than maintaining two trees. In the construction step, new configurations are sampled from Q_{free} near the boundaries of the two trees. A configuration is added to a tree only if it can be connected by the local planner to some existing configuration in the tree. In the merging step, a local planner attempts to connect pairs of configurations selected from both trees. If successful, the two trees become one connected component and a path from q_{init} to q_{goal} is returned.

For answering a single query, it is necessary to cover only the parts of $\mathcal{Q}_{\text{free}}$ relevant to the query. ESTs and RRTs developed sampling strategies that bias the sampling of the configurations toward the unexplored components of $\mathcal{Q}_{\text{free}}$ relevant to the query. The introduced sampling methods are fundamentally conditional: the generation of a new configuration depends on the initial and goal configuration and any previously generated configurations. The planners, however, are faced with the following dilemma: although it is important to search the part of $\mathcal{Q}_{\text{free}}$ that is relevant to the given query, the planners need to demonstrate that their sampling can potentially cover the whole $\mathcal{Q}_{\text{free}}$. This is necessary for ensuring probabilistic completeness. ESTs are a purely forward projection/propagation method. An EST pushes the constructed tree to unexplored parts of $\mathcal{Q}_{\text{free}}$ by sampling points away from densely sampled areas. A rigorous analysis shows that $\mathcal{Q}_{\text{free}}$ will be covered under certain assumptions [192]. RRTs employ a steering strategy that pulls the tree to unexplored parts of $\mathcal{Q}_{\text{free}}$. An RRT attempts to expand toward points in the free configuration space away from the tree. The algorithm has been shown to be probabilistically complete under certain assumptions [271]. Figure 7.11 shows a single tree expanded from q_{init} using a variant of EST [350].

At the end of this section, the SBL [367] planner is described. SBL is a bi-directional EST that uses lazy evaluation for its node connection strategy. This allows the planner to explore the free space very efficiently and at the same time reduce the number of collision checks with further performance improvements over traditional ESTs.

Figure 7.11   Tree generated by a tree-based motion planner for docking a space shuttle at the space station. (From Phillips and Kavraki [350].)

7.2.1   Expansive-Spaces Trees

ESTs were initially developed as an efficient single-query planner that covers the space between q_{init} and q_{goal} rapidly [192, 195, 196, 235]. The developers of the algorithm did not use the acronym EST in their original publications. The acronym was later adopted and was inspired by the notion of "expansive" space used in the theoretical analysis of the algorithm. EST was initially geared toward kinodynamic problems, and for these problems a single tree is typically built (see section 7.5.1). A number of recent planners are based on or use ESTs [14, 350, 367]. The EST algorithm has been shown to be probabilistically complete [192].

Construction of Trees

Let T be one of the trees T_{init} or T_{goal} rooted at q_{init} and q_{goal}, respectively. The planner first selects a configuration q in T from which to grow T and then samples a random configuration, q_{rand}, from a uniform distribution in the neighborhood of q. Configuration q is selected at random with probability $\pi_T(q)$. The local planner Δ (see section 7.1) attempts a connection between q and q_{rand}. If successful, q_{rand} is added to the vertices of T and (q, q_{rand}) is added to the edges of T. The process is repeated until a specified number of configurations has been added to T. The pseudocode is given in algorithms 8 and 9. Figure 7.12 illustrates this method in the simple case of a point robot in a two-dimensional Euclidean workspace.

Recall that in the roadmap construction of PRM, algorithm 6 in section 7.1, a new random configuration in $\mathcal{Q}_{\text{free}}$ is never rejected but it is immediately added to the

---

Algorithm 8 Build EST Algorithm

Input:

   q_0: the configuration where the tree is rooted

   n : the number of attempts to expand the tree

Output:

   A tree $T = (V, E)$ that is rooted at q_0 and has $\leq n$ configurations

1: $V \leftarrow \{q_0\}$

2: $E \leftarrow \emptyset$

3: **for** $i = 1$ to n **do**

4:     $q \leftarrow$ a randomly chosen configuration from T with probability $\pi_T(q)$

5:     extend EST (T, q)

6: **end for**

7: **return** T

Algorithm 9 Extend EST Algorithm

Input:

   $T = (V, E)$: an EST

   q: a configuration from which to grow T

Output:

   A new configuration q_{new} in the neighborhood of q, or NIL in case of failure

---

1: $q_{\text{new}} \leftarrow$ a random collision-free configuration from the neighborhood of q
2: **if** $\Delta(q, q_{\text{new}})$ **then**
3:     $V \leftarrow V \cup \{q_{\text{new}}\}$
4:     $E \leftarrow E \cup \{(q, q_{\text{new}})\}$
5:     **return** q_{new}
6: **end if**
7: **return** NIL

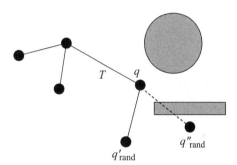

Figure 7.12 Adding a new configuration to an EST. Suppose q is selected and q'_{rand} is created in its neighborhood. The local planner succeeds in connecting q to q'_{rand}. Configuration q'_{rand} and the edge (q, q'_{rand}) are added to the tree T. Had q''_{rand} been created, no nodes or edges would have been added to T, as the local planner would have failed to connect q and q''_{rand}.

roadmap. No attempts are made to connect it to existing configurations in the roadmap. In contrast, in the construction step of EST, a new configuration is added to T only if Δ succeeds in connecting it to an existing configuration in T. It follows then that there is a path from the root of T to every configuration in T.

Guiding the Sampling

The effectiveness of EST relies on the ability to avoid oversampling any region of Q_{free}, especially the neighborhoods of q_{init} and q_{goal}. Hence, careful consideration is given to the choice of the probability density function π_T. Ideally, the function π_T should be chosen such that the sampled configurations constitute a rather uniform covering of the connected components of $\mathcal{Q}_{\text{free}}$ containing q_{init} and q_{goal}. A good choice of π is biased toward configurations of T whose neighborhoods are not dense. There are several ways to measure the density of a neighborhood. One that works well in practice associates with each configuration q of T a weight, $w_T(q)$, that counts the number of configurations within some predefined neighborhood of q. If $\pi_T(q)$ is defined to be inversely proportional to $w_T(q)$, then configurations with sparse neighborhoods are more likely to be picked by the planner and used as input to algorithm 9.

The naive method to compute $\pi_T(q)$ enumerates all the configurations of T and tests if they are close to q. This method takes linear time in the number of configurations, n, in the tree T and works well only for relatively small n. A reasonable approximation to $\pi_T(q)$ can be obtained by imposing a grid on $\mathcal{Q}$. At each iteration, the planner selects the configuration from which to grow the tree by choosing at random a cell and a configuration from this cell. This method was used in [367] and is described in subsection 7.2.3.

Several other π_T functions have been proposed. In [349, 350], $\pi_T(q)$ is defined to be a function of the order in which q is generated, its number of neighbors, its out degree, and an A^* cost function A^*_{cost}. The A^*_{cost} is commonly used in graph search to focus the search toward paths with low cost and is computed as the sum of the total cost from the root of the tree to q and the estimated cost from q to the goal configuration. The above weight function combines in a natural way standard EST heuristics with potential field methods.

Merging of Trees

The merging of the trees is achieved by pushing the exploration of the space from one tree toward the space explored by the other tree. Initially, a configuration in T_{init} is used as described in algorithm 9 to produce a new configuration q. Then the local planner attempts to connect q to its closest k configurations in T_{goal}. If a connection is successful, the two trees are merged. Otherwise, the trees are swapped and the process is repeated for a specified number of times. Figure 7.13 illustrates the merging of two EST trees in a simple case of a two-dimensional Euclidean space.

The merging of the two trees is obtained by connecting some configuration $q_1 \in T_{\text{init}}$ to some configuration $q_2 \in T_{\text{goal}}$ by using the local planner Δ. Thus, the path between q_{init} and q_{goal}, which are the roots of the corresponding trees, is obtained by concatenating the path from q_{init} to q_1 in T_{init} to the path from q_2 to q_{goal} in T_{goal}.

Care should be taken when implementing ESTs. A successful implementation requires a fast update of π_T as new configurations are added to T. The linear cost of the

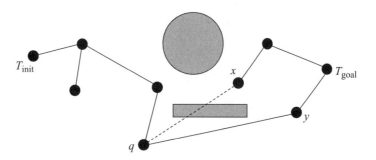

Figure 7.13   Merging two EST trees. Configuration q is just added to the first tree, T_{init}. The local planner attempts to connect q to its closest configurations x and y in the second tree, T_{goal}. The local planner fails to connect q to x, but succeeds in the case of y.

naive method is too high and grid-based approaches or hashing methods (such as those described in section 7.2.3) must be employed for large n and high-dimensional Q.

7.2.2   Rapidly-Exploring Random Trees

RRTs were introduced as a single-query planning algorithm that efficiently covers the space between q_{init} and q_{goal} [249,270–272]. The planner was again initially developed for kinodynamic motion planning, where, as in the case of ESTs, a single tree is built. The applicability of RRTs extends beyond kinodynamic planning problems. The RRT algorithm has been shown to be probabilistically complete [271].

Construction of Trees

Let T be one of the trees T_{init} or T_{goal} rooted at q_{init} and q_{goal}, respectively. Each tree T is incrementally extended. At each iteration, a random configuration, q_{rand}, is sampled uniformly in Q_{free}. The nearest configuration, q_{near}, to q_{rand} in T is found and an attempt is made to make progress from q_{near} toward q_{rand}. Usually this entails moving q_{near} a distance `step_size` in the straight line defined by q_{near} and q_{rand}. This newly generated configuration, q_{new}, if it is collision-free, is then added to the vertices of T, and the edge $(q_{\text{near}}, q_{\text{new}})$ is added to the edges of T. The pseudocode is given in algorithms 10 and 11. Figure 7.14 illustrates the extension step of an RRT for a point robot operating in a two-dimensional Euclidean workspace.

The sampling is done by algorithm 11, which produces a new configuration, q_{new}, as a result of moving some configuration q_{near} by `step_size` toward a configuration q_{rand}. A natural question to consider is how `step_size` is determined.

Algorithm 10 Build RRT Algorithm

Input:

    q_0: the configuration where the tree is rooted

    n : the number of attempts to expand the tree

Output:

    A tree $T = (V, E)$ that is rooted at q_0 and has $\leq n$ configurations

---
1: $V \leftarrow \{q_0\}$

2: $E \leftarrow \emptyset$

3: **for** $i = 1$ to n **do**

4:     $q_{\text{rand}} \leftarrow$ a randomly chosen free configuration

5:     extend RRT (T, q_{rand})

6: **end for**

7: **return** T

Algorithm 11 Extend RRT Algorithm

Input:

    $T = (V, E)$: an RRT

    q: a configuration toward which the tree T is grown

Output:

    A new configuration q_{new} toward q, or NIL in case of failure

---
1: $q_{\text{near}} \leftarrow$ closest neighbor of q in T

2: $q_{\text{new}} \leftarrow$ progress q_{near} by step_size along the straight line in $\mathcal{Q}$ between q_{near} and q_{rand}

3: **if** q_{new} is collision-free **then**

4:     $V \leftarrow V \cup \{q_{\text{new}}\}$

5:     $E \leftarrow E \cup \{(q_{\text{near}}, q_{\text{new}})\}$

6:     **return** q_{new}

7: **end if**

8: **return** NIL

One possible way is to choose step_size dynamically based on the distance between q_{near} and q_{rand} as given by the distance function used. It makes sense to choose a large value for step_size if the two configurations are far from colliding, and small otherwise. RRT is sensitive to the distance function used, since it is this function that determines step_size and guides the sampling. A discussion of the metrics and their effects on RRTs is found in [103]. It is also interesting to consider a greedier alternative that tries to move q_{new} as close to q_{rand} as possible. This method, algorithm 12, calls

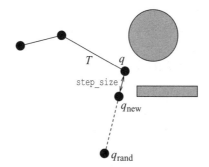

Figure 7.14   Adding a new configuration to an RRT. Configuration q_{rand} is selected randomly from a uniform distribution in $\mathcal{Q}_{free}$. Configuration q is the closest configuration in T to q_{rand} (this configuration is denoted as q_{near} in the algorithm). Configuration q_{new} is obtained by moving q by step_size toward q_{rand}. Only q_{new} and the edge (q, q_{new}) are added to the RRT.

Algorithm 12 Connect RRT Algorithm

Input:

  $T = (V, E)$: an RRT

  q: a configuration toward which the tree T is grown

Output:

  connected if q is connected to T; failure otherwise

1: **repeat**
2:   $q_{new} \leftarrow$ extend RRT (T, q)
3: **until** $(q_{new} = q$ **or** $q_{new} = $ NIL$)$
4: **if** $q_{new} = q$ **then**
5:   **return** connected
6: **else**
7:   **return** failure
8: **end if**

---

algorithm 11 until q_{new} reaches q_{rand} or no progress is possible. If algorithm 12 is called in algorithm 10, line 5, an RRT with greater than n nodes may be created.

It is important to note the tradeoff that exists between the exploration of $\mathcal{Q}$ and the number of samples added to the tree, especially for high-dimensional problems. If step_size is small, then the exploration steps are short and the nodes of the tree are close together. A successful call to algorithm 12 results in many nodes being added to the tree. As the number of nodes becomes large, memory consumption is increased and finding the nearest neighbor becomes expensive, which in turn reduces the performance of the planner. In such cases, it may be better to add only the last sample of the Extend RRT iteration to the tree and no intermediate samples.

Guiding the Sampling

Selecting a node uniformly at random in step 4 of algorithm 10 is a basic mechanism of RRT. It is also of interest to consider other sampling functions that are biased toward the connected components of $\mathcal{Q}$ that contain q_{init} or q_{goal}. Let's consider the case of q_{goal}. At an extreme, a very greedy sampling function can be defined that sets q_{rand} to q_{goal} (if the tree being built is rooted at q_{init}) or to q_{init} (if the tree being built is rooted at q_{goal}). The problem with this approach is that it introduces too much bias, and eventually RRT ends up behaving like a randomized potential field planner that gets stuck in local minima. It seems, therefore, that a suitable choice is a sampling function that alternates, according to some probability distribution, between uniform samples and samples biased toward regions that contain the initial or the goal configuration. Experimental evidence [249, 271] has shown that setting q_{rand} to q_{goal} with probability p, or randomly generating q_{rand} with probability $1 - p$ from a uniform distribution, works well. Even for small values of p, such as 0.05, the tree rooted at q_{init} converges much faster to q_{goal} than when just uniform sampling is used. This simple function can be further improved by sampling in a region around q_{goal} instead of setting q_{rand} to q_{goal}. The region around q_{goal} is defined by the vertices of the RRT closest to q_{goal} at each iteration of the construction step.

Merging of Trees

In the merging step, RRT tries to connect the two trees, T_{init} and T_{goal}, rooted at q_{init} and q_{goal}, respectively. This is achieved by growing the trees toward each other. Initially, a random configuration, q_{rand}, is generated. RRT extends one tree toward q_{rand} and as a result obtains a new configuration, q_{new}. Then the planner attempts to extend the closest node to q_{new} in the other tree toward q_{new}. If successful, the planner terminates, otherwise the two trees are swapped and the process is repeated a certain number of times. Figure 7.15 illustrates a simple case of merging two RRTs in a two-dimensional Euclidean space.

The merge algorithm, as presented in algorithm 13, uses algorithm 11. Recall that algorithm 11 produces a new configuration that is only `step_size` away from the nearest node in the existing RRT. By replacing algorithm 11 in lines (3) and (5) with algorithm 12, new configurations are produced farther away. This greedier approach has been reported to work well [249, 271]. It is also reasonable to replace only one of the algorithm 11 calls, and thus obtain a balance between the two approaches. The

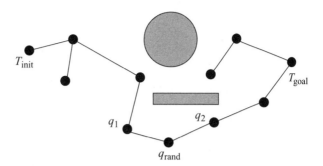

Figure 7.15   Merging two RRTs. Configuration q_{rand} is generated randomly from a uniform distribution in Q_{free}. Configuration q_1 was extended to q_{rand}. q_2 is the closest configuration to q_{rand} in T_{goal}. It was possible to extend q_2 to q_{rand}. As a result, T_{init} and T_{goal} were merged.

Algorithm 13 Merge RRT Algorithm

Input:

   T_1: first RRT

   T_2: second RRT

   ℓ: number of attempts allowed to merge T_1 and T_2

Output:

   `merged` if the two RRTs are connected to each other; `failure` otherwise

---

1: **for** $i = 1$ to ℓ **do**

2:     $q_{rand} \leftarrow$ a randomly chosen free configuration

3:     $q_{new,1} \leftarrow$ extend RRT (T_1, q_{rand})

4:     **if** $q_{new,1} \neq$ NIL **then**

5:        $q_{new,2} \leftarrow$ extend RRT $(T_2, q_{new,1})$

6:        **if** $q_{new,1} = q_{new,2}$ **then**

7:           **return** `merged`

8:        **end if**

9:        SWAP(T_1, T_2)

10:    **end if**

11: **end for**

12: **return** `failure`

---

choice of whether a greedier or more balanced approach is used for the exploration depends on the particular problem being solved. Discussions can be found in [249, 271]. Once the two RRTs are merged together, a path from q_{init} to q_{goal} is obtained in the same way as in the case of ESTs.

The implementation of RRT is easier than that of EST. Unlike EST, RRT does not compute the number of configurations lying inside a predefined neighborhood of a node and it does not maintain a probability distribution for its configurations.

7.2.3 Connection Strategies and the SBL Planner

Lazy evaluation, which was introduced as a connection strategy in section 7.1.4, can also be used in the context of tree-building sampling methods. A combination of lazy evaluation and ESTs has been presented in the context of the Single-query, Bi-directional, Lazy collision-checking (SBL) planner [367].

SBL constructs two EST trees rooted at q_{init} and q_{goal}. SBL creates new samples according to the EST criteria but does not immediately test connections between samples for collisions. A connection between two configurations is checked exactly once and this is done only when the connection is part of the path joining the two trees together (if such a path is found). This results in substantial time-savings as reported in [367].

It is also worth noting that SBL uses a clever way to find which configurations to expand and hence guide the sampling (see section 7.2.1). In the original EST, a configuration is chosen for expansion according to the density of sampling in its neighboorhood. Finding neighbors is in general an expensive operation as the dimension increases. SBL imposes a coarse grid on $\mathcal{Q}$. It then picks randomly a non-empty cell in the grid and a sample from that cell. The probability to pick a certain sample is greater if this sample lies in a cell with few nodes. This simple technique allows a fast implementation of SBL and is applicable to all EST-based planners. Details on how this technique helps in the connection of trees grown from the initial and goal configurations are given in [367].

EST and RRT employ excellent sampling and connection schemes that can be further exploited to obtain even more powerful planners, as discussed in the next section.

7.3 Integration of Planners: Sampling-Based Roadmap of Trees

This section shows how to effectively combine a sampling-based method primarily designed for multiple-query planning (PRM) with sampling-based tree methods primarily designed for single-query planning (EST, RRT, and others). The Sampling-Based Roadmap of Trees (SRT) planner [14, 43, 353] takes advantage of the local sampling schemes of tree planners to populate a PRM-like roadmap. SRT replaces the local planner of PRM with a single-query sampling-based tree planner enabling it to solve problems that other planners cannot.

A question arises as to whether SRT is a multiple-query or single-query planner. SRT can be seen as a multiple-query planner, since once the roadmap is constructed, SRT can use the roadmap to answer multiple queries. SRT can also be seen as a single-query planner because for certain very difficult problems, the cost of constructing a roadmap and solving a query by SRT is less than that of any single-query planner solving the same query. This is why in section 7.1 it was pointed out that the distinction of planners to multiple-query and single-query planners is very useful for describing the planners, but always needs to be placed in perspective given the planning problem at hand.

As in the PRM formulation, SRT constructs a roadmap aiming at capturing the connectivity of Q_{free}. The nodes of the roadmap are not single configurations but trees, as illustrated in figure 7.16. Connections between trees are computed by a bi-directional tree algorithm such as EST or RRT. Recall that a roadmap is an undirected graph $G = (V, E)$ over a finite set of configurations $V \subset Q_{\text{free}}$, and each edge $(q', q'') \in E$ represents a local path from q' to q''. SRT constructs a roadmap of trees. The undirected graph $G_T = (V_T, E_T)$ is an induced subgraph of G which is defined by partitioning G into a set of subgraphs $T_1, \ldots, T_n$, which are trees, and contracting them into the vertices of G_T. In other words, $V_T = \{T_1, \ldots, T_n\}$ and $(T_i, T_j) \in E_T$ if

Figure 7.16 An example of a roadmap for a point robot in a two-dimensional workspace. The dark gray areas are obstacles. Each node of the roadmap is a tree rooted at the black squares. The thin-solid lines indicate connections between configurations of the same tree. The thick-dashed lines indicate connections between configurations of two different trees. The light gray areas delineate the separate trees.

there exist configurations $q_i \in T_i$ and $q_j \in T_j$ such that q_i and q_j have been connected by a local path.

Adding Trees to the Roadmap

In SRT, the trees of the roadmap G_T are computed by sampling their roots uniformly at random in $\mathcal{Q}_{\text{free}}$, and then growing the trees using a sampling-based tree planner, such as algorithms 8 and 10. Note that in principle any of the node-sampling strategies of PRM described in section 7.1.3 can be applied.

Adding Edges to the Roadmap

The roadmap construction is not yet complete since no edges have been computed. An edge between two trees indicates that they are merged into one. For each tree T_i, a set N_{T_i} consisting of closest and random tree neighbors is computed and a connection is attempted between T_i and each tree T_j in N_{T_i}. As in PRM, SRT may choose to avoid the computation of candidate edges that cannot decrease the number of connected components in G_T. In fact, any of the PRM connection strategies of section 7.1.4 can be applied here. In order to determine the closest neighbors, each tree T_i defines a representative configuration q_{T_i} which is computed as an aggregate of the configurations in T_i. The distance between two trees T_i and T_j is defined as $dist(q_{T_i}, q_{T_j})$. It has been observed experimentally in [14,43] that the consideration of random neighbors offsets some of the problems introduced by the distance function used.

Computation of candidate edges is typically carried out by a sampling-based tree planner. First, for each candidate edge (T_i, T_j), a number of close pairs of configurations of T_i and T_j are quickly checked with a fast deterministic local planner. If a local path is found, no further computation takes place. Otherwise, the sampling-based tree planner used to add trees to the roadmap should be employed. During tree connection, additional configurations are typically added to the trees T_i and T_j. If the connection is successful, the edge (T_i, T_j) is added to E_T and the graph components to which T_i and T_j belonged are merged into one. Note that the trees T_i and T_j are connected when some configuration $q_i \in T_i$ is connected to some configuration $q_j \in T_j$.

The pseudocode is given in algorithm 14. In addition to RRT and EST, other sampling-based tree planners, such as [251, 350], can be used with SRT. It is also possible to incorporate lazy evaluation into SRT by using a planner similar to SBL for tree expansion and edge computations.

Algorithm 14 Connect SRT Algorithm

Input:

   V_T: a set of trees

   k: number of closest neighbors to examine for each tree

   r: number of random neighbors to examine for each tree

Output:

   A roadmap $G_T = (V_T, E_T)$ of trees

1: $E_T \leftarrow \emptyset$

2: **for all** $T_i \in V_T$ **do**

3:      $N_{T_i} \leftarrow k$ nearest and r random neighbors of T_i in V_T

4:      **for all** $T_j \in N_{T_i}$ **do**

5:          **if** T_i and T_j are not in the same connected component of G_T **then**

6:              merged $\leftarrow$ FALSE

7:              $S_i \leftarrow$ a set of randomly chosen configurations from T_i

8:              **for all** $q_i \in S_i$ and merged $=$ FALSE **do**

9:                  $q_j \leftarrow$ closest configuration in T_j to q_i

10:                 **if** $\Delta(q_i, q_j)$ **then**

11:                     $E_T \leftarrow E_T \cup \{(T_i, T_j)\}$

12:                     merged $\leftarrow$ TRUE

13:                 **end if**

14:             **end for**

15:             **if** merged $=$ FALSE **and** Merge Trees (T_i, T_j) **then**

16:                 $E_T \leftarrow E_T \cup \{(T_i, T_j)\}$

17:             **end if**

18:         **end if**

19:     **end for**

20: **end for**

Answering Queries

As in PRM, the construction of the roadmap enables SRT to answer multiple queries efficiently if needed. Given q_{init} and q_{goal}, the trees T_{init} and T_{goal} rooted at q_{init} and q_{goal}, respectively, are grown for a small number of iterations and added to the roadmap. Neighbors of T_{init} and T_{goal} are computed as a union of the k closest and r random trees, as described previously. The tree-connection algorithm alternates between attempts to connect T_{init} and T_{goal} to each of their respective neighbor trees. A path is found if at any point T_{init} and T_{goal} lie in the same connected component of the roadmap. In order to determine the sequence of configurations that define a path from q_{init} to q_{goal}, it is necessary to find the sequence of trees that define a path from T_{init} to T_{goal} and then concatenate the local paths between any two consecutive trees. Path smoothing can be applied to the resulting path to improve the quality of the output.

Parameters of SRT

A nice feature of SRT is that it can behave exactly as PRM, RRT, or EST. That is, if the number of configurations in a tree is one, the number of close pairs is one and the number of iterations to run the bi-directional tree planner is zero (denoted by Merge Trees in line (15) of algorithm 14), then SRT behaves as PRM. If the number of trees in the roadmap is zero and the number of close pairs is zero, then SRT behaves as RRT or EST depending on the type of tree. SRT provides a framework where successful sampling schemes can be efficiently combined.

Parallel SRT

SRT is significantly more decoupled than tree planners such as ESTs and RRTs. Unlike ESTs and RRTs, where the generation of one configuration depends on all previously generated configurations, the trees of SRT can be generated independently of one another. This decoupling allows for an efficient parallelization of SRT [14]. By increasing the power of the local planner and by using trees as nodes of the roadmap, SRT distributes its computation evenly among processors, requires little communication, and can be used to solve very high-dimensional problems and problems that exceed the resources available to the sequential implementation [14]. Adding trees to the roadmap can be parallelized efficiently, since there are no dependencies between the different trees. Adding edges to the roadmap is harder to parallelize efficiently. Since trees can change after an edge computation and since computing an edge requires direct knowledge of both trees, the edge computations cannot be efficiently parallelized without some effort [14]. Furthermore, if any computation pruning according to the sparse roadmaps heuristic is done (see section 7.1.4), this will entail control flow dependencies throughout the computation of the edges.

7.4 Analysis of PRM

The planners discussed in this chapter sample points in $\mathcal{Q}_{\text{free}}$ and connect them using a local planner. As opposed to exact motion-planning algorithms, such as [90, 306, 361, 373, 375], it is possible that PRM and other sampling-based motion planners can report falsely that no path exists. It would seem that the correctness of the motion planner has been sacrificed in favor of good experimental performance. This, however, is not exactly the case. Rather than being a purely heuristic technique, a weaker completeness property, called probabilistic completeness, can be proved to hold for PRM as was discussed in the introductory section of this chapter.

This section deals with probabilistic completeness proofs and analyses of the basic PRM planner. In the basic PRM planner, samples are chosen from a uniform random distribution. Although the presented results are for an idealized version of PRM, it is strongly conjectured that probabilistic completeness results can be extended to conditional random sampling and to deterministic sampling, in the latter case, in the form of resolution completeness results.

Suppose that $a, b \in \mathcal{Q}_{\text{free}}$ can be connected by a path in $\mathcal{Q}_{\text{free}}$. PRM is considered to be probabilistically complete, if for any given (a, b)

$$\lim_{n \to \infty} Pr[(a, b) \text{ FAILURE}] = 0,$$

where $Pr[(a, b) \text{ FAILURE}]$ denotes the probability that PRM fails to answer the query (a, b) after a roadmap with n samples has been constructed. The number of samples gives a measure of the work that needs to be done and hence it can be used as a measure of the complexity of the algorithm.

The results presented in this section apply to the basic PRM algorithm. Section 7.4.1 analyzes the operation of PRM in a Euclidean space. Using this analysis, it is possible to gain an estimate on how much work (as measured by the number of generated samples) is needed to produce paths with certain properties. Section 7.4.2 shows how certain goodness properties of the underlying space affect the performance of PRM. It is this analysis that sheds light on why PRM works well with extremely simple sampling strategies such as uniform sampling. Experimental observations indicate that many of the path-planning problems that arise in physical settings have goodness properties, such as the ones described in section 7.4.2, that may not require elaborate sampling schemes. Both analyses prove probabilistic completeness for PRM. Section 7.4.3 shows an equivalence between the probabilistic completeness of PRM and a much simpler planner.

7.4.1   PRM Operating in $\mathbb{R}^d$

This section provides an analysis [222,223] of PRM operating in Euclidean $\mathbb{R}^d$. Assuming that a path between two different configurations a and b exists, it is shown that the probability of PRM failing to connect a and b depends on (1) the length of the known path, (2) the distance of the path from the obstacles, and (3) the number of configurations in the roadmap. Connecting a and b by a long path requires a larger number of intermediate configurations to be present in the roadmap. Paths that are closer to obstacles are harder to obtain because of potential collisions. Similarly, paths that are inside narrow passages are harder to obtain because the probability of placing random configurations inside narrow passages is small. The probabilistic completeness of PRM is proved by tiling the known path with a set of carefully chosen balls and showing that generating a point in each ball ensures that a path between a and b will be found.

Let $\mathcal{Q}_{\text{free}}$ be an open subset of $[0, 1]^d$ and let *dist* be the Euclidean metric on $\mathbb{R}^d$. The local planner of PRM connects points $a, b \in \mathcal{Q}_{\text{free}}$ when the straight-line $\overline{ab}$ lies in $\mathcal{Q}_{\text{free}}$. A path γ in $\mathcal{Q}_{\text{free}}$ from a to b consists of a continuous map $\gamma : [0, 1] \rightarrow \mathcal{Q}_{\text{free}}$, where $\gamma(0) = a$ and $\gamma(1) = b$. The clearance of a path, denoted $\mathtt{clr}(\gamma)$, is the farthest distance away from the path at which a given point can be guaranteed to be in $\mathcal{Q}_{\text{free}}$. If a path γ lies in $\mathcal{Q}_{\text{free}}$, then $\mathtt{clr}(\gamma) > 0$.

The measure μ denotes the volume of a region of space, e.g., $\mu([0, 1]^d) = 1$. For any measurable subset $A \subset \mathbb{R}^d$, $\mu(A)$ is its volume. For example, an open ball of radius ϵ centered at x is denoted by $B_\epsilon(x)$ and its volume is given by $\mu(B_\epsilon(x))$. The uniform distribution is used by PRM to sample points. If $A \subset \mathcal{Q}_{\text{free}}$ is a measurable subset and x is a random point chosen from $\mathcal{Q}_{\text{free}}$ by the point-sampling function of PRM, then

$$Pr(x \in A) = \frac{\mu(A)}{\mu(\mathcal{Q}_{\text{free}})}.$$

THEOREM 7.4.1 *Let* $a, b \in \mathcal{Q}_{\text{free}}$ *such that there exists a path* γ *between a and b lying in* $\mathcal{Q}_{\text{free}}$. *Then the probability that* PRM *correctly answers the query* (a, b) *after generating n configurations is given by*

$$Pr[(a, b)\ \text{SUCCESS}] = 1 - Pr[(a, b)\ \text{FAILURE}] \geq 1 - \left\lceil \frac{2L}{\rho} \right\rceil e^{-\sigma \rho^d n},$$

where L is the length of the path γ, $\rho = \mathtt{clr}(\gamma)$, $B_1(\cdot)$ *is the unit ball in* $\mathbb{R}^d$ *and*

$$\sigma = \frac{\mu(B_1(\cdot))}{2^d \mu(\mathcal{Q}_{\text{free}})}.$$

Proof Let $\rho = \mathtt{clr}(\gamma)$ and note that $\rho > 0$. Let $m = \left\lceil \frac{2L}{\rho} \right\rceil$ and observe that there are m points on the path $a = x_1, \ldots, x_m = b$ such that $dist(x_i, x_{i+1}) < \rho/2$. Let $y_i \in B_{\rho/2}(x_i)$ and $y_{i+1} \in B_{\rho/2}(x_{i+1})$. Then the line segment $\overline{y_i y_{i+1}}$ must lie inside $\mathcal{Q}_{\text{free}}$ since both endpoints lie in the ball $B_\rho(x_i)$. An illustration of this basic fact is given in figure 7.17. Let $V \subset \mathcal{Q}_{\text{free}}$ be a set of n configurations generated uniformly at random by PRM. If there is a subset of configurations $\{y_1, \ldots, y_m\} \subset V$ such that $y_i \in B_{\rho/2}(x_i)$, then a path from a to b will be contained in the roadmap. Let $I_1, \ldots, I_m$ be a set of indicator variables such that each I_i witnesses the event that there is a $y \in V$

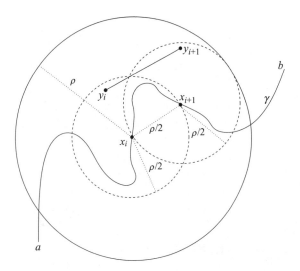

Figure 7.17 Points y_i and y_{i+1} are inside the $\rho/2$ balls and straight-line $\overline{y_i y_{i+1}}$ is in $\mathcal{Q}_{\text{free}}$.

and $y \in B_{\rho/2}(x_i)$. It follows that PRM succeeds in answering the query (a, b) if $I_i = 1$ for all $1 \leq i \leq m$. Therefore,

$$Pr[(a, b) \text{ FAILURE}] \leq Pr \left(\bigvee_{i=1}^{m} I_i = 0 \right) \leq \sum_{i=1}^{m} Pr[I_i = 0],$$

where the last inequality follows from the union bound [119].

The events $I_i = 0$ are independent since the samples are independent. The probability of a given $I_i = 0$ is computed by observing that the probability of a single randomly generated point falling in $B_{\rho/2}(x_i)$ is $\mu(B_{\rho/2}(x_i))/\mu(\mathcal{Q}_{\text{free}})$. It follows that the probability that none of the n uniform, independent samples falls in $B_{\rho/2}(x_i)$ satisfies

$$Pr[I_i = 0] = \left(1 - \frac{\mu(B_{\rho/2}(x_i))}{\mu(\mathcal{Q}_{\text{free}})} \right)^n.$$

Since the sampling is uniform and independent, then

$$Pr[(a, b) \text{ FAILURE}] \leq \left\lceil \frac{2L}{\rho} \right\rceil \left(1 - \frac{\mu(B_{\rho/2}(\cdot))}{\mu(\mathcal{Q}_{\text{free}})} \right)^n.$$

However

$$\frac{\mu(B_{\rho/2}(\cdot))}{\mu(\mathcal{Q}_{\text{free}})} = \frac{\left(\frac{\rho}{2} \right)^d \mu(B_1(\cdot))}{\mu(\mathcal{Q}_{\text{free}})} = \sigma \rho^d,$$

for σ defined as in the statement of this theorem. The bound is obtained by using the relation $(1 - \beta)^n \leq e^{-\beta n}$ for $0 \leq \beta \leq 1$:

$$Pr[(a, b) \text{ FAILURE}] \leq \left\lceil \frac{2L}{\rho} \right\rceil e^{-\sigma \rho^d n}.$$

∎

As shown from the proof above, a better estimate for $Pr[(a, b)$ FAILURE] is available than the exponential bound given in theorem 7.4.1. The exponential bound is a simplification that allows the direct calculation of n when the user wishes to specify an acceptable value for $Pr[(a, b)$ FAILURE]. The proof of theorem 7.4.1 can be extended to take into account that clearance can vary along the path [223]. Theorem 7.4.1 implies that PRM is probabilistically complete. Moreover, the probability of failure converges exponentially quickly to 0.

7.4.2 $(\epsilon, \alpha, \beta)$-Expansiveness

This section argues how PRM roadmaps capture the connectivity of $\mathcal{Q}_{\text{free}}$ based on the analysis of [192, 196, 197, 228, 229]. A principal intuition behind PRM has been that in spaces that are not "pathologically" difficult, that is in spaces where reasonable assumptions about connectivity hold, the planner will do well even with simple sampling schemes such as random sampling.

Observe that, in the general case, $\mathcal{Q}_{\text{free}}$ can be broken into a union of disjoint connected components $\{\mathcal{Q}_{\text{free}1}, \ldots, \mathcal{Q}_{\text{free}i}, \ldots\}$. Let $G = (V, E)$ be the roadmap constructed by PRM with uniform sampling. For each $\mathcal{Q}_{\text{free}i}$, let $V_i = V \cap \mathcal{Q}_{\text{free}i}$ and let G_i be the subgraph of G induced by V_i. In the rest of this section, it is shown how to determine the number of configurations that should be generated to ensure that, with probability exceeding a given constant, each G_i is connected.

Given a subset S of $\mathcal{Q}_{\text{free}}$, the reachable set from S is the set of configurations in $\mathcal{Q}_{\text{free}}$ that are visible from any configuration in S. Figures 7.18(a) and (b) show an example.

DEFINITION 7.4.1   *Let $S \subset \mathcal{Q}_{\text{free}}$. The reachable set of S is defined as*

$$\texttt{reach}(S) = \{x \in \mathcal{Q}_{\text{free}} \mid \exists y \in S \text{ such that } \overline{xy} \subset \mathcal{Q}_{\text{free}}\}.$$

The shorthand $\texttt{reach}(x)$ *is used instead of* $\texttt{reach}(\{x\})$ *when* $x \in \mathcal{Q}_{\text{free}}$.

A space $\mathcal{Q}_{\text{free}}$ is ϵ-good if the volume of $\mathcal{Q}_{\text{free}}$ that each point in $\mathcal{Q}_{\text{free}}$ can reach is at least an ϵ fraction of the total free volume of $\mathcal{Q}_{\text{free}}$.

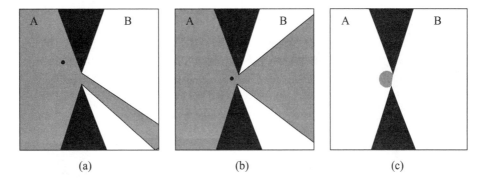

Figure 7.18 The areas in black indicate the obstacles. In (a) and (b), the areas in gray indicate the reachability sets of the two points represented by black circles. In (c), the set A has a small lookout (the gray area), because only a small subset of points in A near the narrow passage can see a large fraction of points in B. (From Hsu [192].)

DEFINITION 7.4.2 *Let ϵ be a constant in $(0, 1]$. A space $\mathcal{Q}_{\text{free}}$ is ϵ-good if for all $x \in \mathcal{Q}_{\text{free}}$,*

$$\mu(\text{reach}(x)) \geq \epsilon \mu(\mathcal{Q}_{\text{free}}).$$

The β-lookout of a subset S of a connected component of $\mathcal{Q}_{\text{free}i}$ is the subset of S for which each configuration in that subset can reach more than a β fraction of $\mathcal{Q}_{\text{free}i} \setminus S$. An example is given in figure 7.18(c).

DEFINITION 7.4.3 *Let β be a constant in $(0, 1]$ and let S be a subset of a connected component $\mathcal{Q}_{\text{free}i}$ of $\mathcal{Q}_{\text{free}}$. The β-lookout set of S is defined as*

$$\text{lookout}_\beta(S) = \{x \in S \mid \mu(\text{reach}(x) \setminus S) \geq \beta\mu(\mathcal{Q}_{\text{free}i} \setminus S)\}.$$

The following definition captures how reachability spreads across the space.

DEFINITION 7.4.4 *Let ϵ, α and β be constants in $(0, 1]$. A space is $(\epsilon, \alpha, \beta)$-expansive if*

1. it is ϵ-good, and

2. for any connected subset of $S \subset \mathcal{Q}_{\text{free}}$, $\mu(\text{lookout}_\beta(S)) \geq \alpha\mu(S)$.

The first condition of definition 7.4.4 ensures that a certain fraction of $\mathcal{Q}_{\text{free}}$ is visible from any configuration in $\mathcal{Q}_{\text{free}}$. The second condition ensures that each subset $S \subseteq \mathcal{Q}_{\text{free}i}$ has a large lookout set. It is reasonable to think of S as the union of the

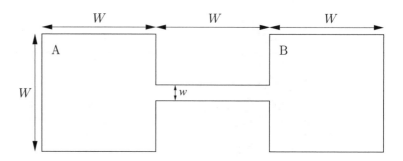

Figure 7.19 An example of an $(\epsilon, \alpha, \beta)$-expansive $\mathcal{Q}_{\text{free}}$ with $\epsilon, \alpha, \beta \approx w/W$. The points with the smallest ϵ are located in the narrow passage between square A and square B. Each such point sees only a subset of $\mathcal{Q}_{\text{free}}$ of volume approximately $3wW$. Hence $\epsilon \approx w/W$. A point near the top right corner of square A sees the entire square; but only a subset of A, of approximate volume wW, contains points that each see a set of volume $2wW$; hence $\alpha \approx w/W$ and $\beta \approx w/W$.   (From Hsu [192].)

reachability sets of a set V of points. Large values of α and β indicate that it is easy to choose random points from S such that adding them to V results in significant expansion of S. This is desirable since it allows for a quick exploration of the entire space. Figure 7.19 gives an example of an expansive space and indicates the values of ϵ, α and β.

We now introduce the concept of a linking sequence, which will be used in the development that follows.

DEFINITION 7.4.5   *A linking sequence of length ℓ for a configuration $x \in \mathcal{Q}_{\text{free}}$ is a set of configurations $x_1 = x, x_2, \ldots, x_\ell$ with an associated sequence of reachable sets $X_1 = \texttt{reach}(x_1), X_2, \ldots, X_\ell \subset \mathcal{Q}_{\text{free}}$, where for all $1 < i \le \ell$,*

$$x_i \in \texttt{lookout}_\beta(X_{i-1}) \quad and \quad X_i = X_{i-1} \cup \texttt{reach}(x_i).$$

The proof of the main result relies on two technical lemmas, whose proofs are given in [192]. Lemma 7.4.6 gives a bound on the probability of sampling a linking sequence for a given configuration x in terms of α, ϵ, and t, the length of the linking sequence.

LEMMA 7.4.6   *Let V be a set of n configurations chosen independently and uniformly at random from $\mathcal{Q}_{\text{free}}$. Let $s = 1/\alpha\epsilon$. Given any configuration $x \in V$, there exists a linking sequence in V of length t for x with probability at least $1 - se^{-(n-t-1)/s}$.*

Lemma 7.4.7 gives a lower bound on the volume of V_t for an arbitrary linking sequence of length t.

LEMMA 7.4.7 *Let $x_1 = x, x_2, \ldots, x_t$ be a length t linking sequence for $x \in \mathcal{Q}_{\text{free}i}$, where $\mathcal{Q}_{\text{free}i}$ is a connected component of $\mathcal{Q}_{\text{free}}$. Let $X_1, X_2, \ldots, X_t$ be the associated reachable sets. If $t \geq \beta^{-1} \ln(4)$, then*

$$\mu(X_t) \geq \frac{3\mu(\mathcal{Q}_{\text{free}i})}{4}.$$

The main result of this section follows. Given a number δ, the theorem finds n such that if $2n + 2$ configurations are sampled, then each subgraph G_i is a connected graph with probability at least $1 - \delta$. This indicates that the connectivity of the roadmap G conforms to the connectivity of $\mathcal{Q}_{\text{free}}$. It means that, with high probability, no two connected components of G lie in the same connected component of $\mathcal{Q}_{\text{free}}$.

THEOREM 7.4.2 *Let δ be a constant in $(0, 1]$. Suppose a set V of $2n + 2$ configurations for*

$$n = \left\lceil \frac{8 \ln \left(\frac{8}{\epsilon \alpha \delta} \right)}{\epsilon \alpha} + \frac{3}{\beta} \right\rceil,$$

is chosen independently and uniformly at random from $\mathcal{Q}_{\text{free}}$. Then, with probability at least $1 - \delta$, each subgraph G_i is a connected graph.

Proof Let x and y be any two configurations in the same connected component $\mathcal{Q}_{\text{free}i}$. Divide the remaining configurations into two sets V' and V'' of n configurations each. By lemma 7.4.6, there is a linking sequence of length t for x in V' with probability at least $1 - se^{-(n-t)/s}$. The same holds true for y and V'. Let $X_t(x)$ and $X_t(y)$ be the reachability sets determined by the linking sequences of length t of x and y. By choosing $t \geq 1.5\beta$, lemma 7.4.7 is applied to ensure that $\mu(X_t(x))$ and $\mu(X_t(y))$ are larger than $3\mu(\mathcal{Q}_{\text{free}i})/4$. It follows that $\mu(X_t(x) \cap X_t(y)) \geq \mu(\mathcal{Q}_{\text{free}i})/2$. It is known that $\mu(\mathcal{Q}_{\text{free}i}) \geq \epsilon$, because $\mathcal{Q}_{\text{free}i}$ is an ϵ-good space; the visibility region of any point in $\mathcal{Q}_{\text{free}i}$ must have volume at least ϵ. Since the configurations in V'' are sampled independently and uniformly at random, it follows that with probability at least $1 - (1 - \epsilon/2)^n \geq 1 - e^{-n\epsilon/2}$, there is a configuration in V'' that lies in the intersection of the reachability sets. This means that there is a path from x to y in G_i.

Let B be the event that x and y fail to connect in G_i. By applying a union bound and by the linking sequence construction, it follows that

$$Pr[B] \leq 2se^{-(n-t)/s} + e^{-n\epsilon/2}.$$

By choosing $n \geq 2t$ and recalling that $s = 1/\alpha\epsilon$,

$$Pr[B] \leq 2se^{-n/2s} + e^{-n\epsilon/2} \leq 2se^{-n/2s} + e^{-n/2\alpha\epsilon} \leq 3se^{-n/2s}.$$

A graph G_i will fail to be connected if some pair $x, y \in V_i$ fails to be connected. There are at most $\binom{n}{2}$ such pairs and the probability of this occurring is at most

$$\binom{n}{2} Pr[B] \leq \binom{n}{2} 3se^{-n/2s} \leq 2n^2 se^{-n/2s} \leq 2se^{(-n-4s\ln n)/2s} \leq 2se^{-n/4s},$$

where the last inequality follows from the observation that $n/2 \geq 4s \ln n$ for $n \geq 8s \ln(8s)$. By requiring that $n \geq 8s \ln(8s/\delta)$, it follows that

$$2se^{-n/4s} \leq 2se^{-2\ln(8s/\delta)} \leq 2s \left(\frac{\delta^2}{8s} \right) \leq \delta.$$

It is sufficient to choose $n \geq 8s \ln(8s/\delta) + 2t$ for this argument to succeed. By substituting $s = 1/\alpha\epsilon$ and $t = 1.5\beta$ into this expression, the stated result is obtained.   ∎

Theorem 7.4.2 implies probabilistic completeness, although some additional argumentation is needed. The main limitation of the above analysis is the reliance on the α, β, and ϵ constants being nonzero. This will be true for any polyhedral space. Since any configuration space can be well approximated with a polyhedral space without changing its connectivity, theorem 7.4.2 holds. A detailed analysis can be found in [37, 192, 196, 197].

7.4.3 Abstract Path Tiling

In this section, theorem 7.4.1 is generalized by reducing the set of assumptions to a bare minimum. The new assumptions are sufficient for defining the planner's sampling scheme and the notion of reachability. In fact, the structural requirements for the configuration space are very simple and are captured by the mathematical abstraction of a probability space: essentially a space over which probability can be defined. In the new framework, the balls used to tile a path in theorem 7.4.1 can be replaced with arbitrary sets of strictly positive measure. These sets are not necessarily connected or open. The analysis is introduced in order to consider PRM operating on motion-planning problems with difficult configuration spaces, and with complex local planners such as those arising from motion planning with dynamics, deformable objects, objects with contact, and others [252]. The framework presented in this section enables a rigorous treatment of asymmetric reachability, nonmanifold configuration spaces, and sampling from arbitrary distributions. Hence, it reveals the applicability of the PRM scheme to problems beyond basic path-planning. A detailed analysis can be found in [252].

As before, the distribution for configuration generation is encoded with the probability measure μ. So if $A \subset \mathcal{Q}_{\text{free}}$, then $\mu(A)$ is the probability that a random sample

---

Algorithm 15 Random Incremental Algorithm

1: $x_0 \leftarrow x$

2: $\ell \leftarrow 0$

3: **loop**

4:     Check if $x_\ell R y$, if so return $x_0, \ldots, x_\ell, y$ as the computed path

5:     Generate $x_{\ell+1}$ at random from distribution μ

6:     Check if $x_\ell R x_{\ell+1}$, if not **return** no path

7:     $\ell \leftarrow \ell + 1$

8: **end loop**

---

from $\mathcal{Q}_{\text{free}}$ lies in A. The local path planner is further generalized away from a straight-line planner and is instead replaced with an arbitrary binary relation, R. Informally, $x R y$ means y can be reached using the local planner from x. Note that $x R y$ need not imply $y R x$. More precisely, the set $R \subset \mathcal{Q}_{\text{free}} \times \mathcal{Q}_{\text{free}}$ is the set of all query configurations that can be connected by the local planner. For example, if $\mathcal{Q}_{\text{free}} \subset [0, 1]^d$ and the local planner is a straight-line planner, then

$$(x, y) \in R \Leftrightarrow \overline{xy} \subset \mathcal{Q}_{\text{free}}.$$

It is required that R is measurable in $\mathcal{Q}_{\text{free}} \times \mathcal{Q}_{\text{free}}$. Membership in R is written interchangeably as $(x, y) \in R$ or $x R y$.

This section develops two distinct ideas from these definitions. A simple motion planner based on random incremental construction of a path is stated in algorithm 15. First, it will be shown that PRM is probabilistically complete if and only if algorithm 15 can answer correctly on every query with nonzero probability. Second, it is proved that probabilistic completeness implies a bound on $Pr[(x, y) \text{ FAILURE}]$ similar to the one stated in theorem 7.4.1.

THEOREM 7.4.3 *Algorithm 15 succeeds with nonzero probability on every query if and only if* PRM *is probabilistically complete. Furthermore, if* PRM *is probabilistically complete, then there exist constants $\ell \geq 0$ and $p > 0$ such that*

$$Pr[(x, y) \text{ FAILURE}] \leq \ell e^{-pn},$$

where n is the number of configurations in the roadmap.

Proof First, the equivalence between PRM and algorithm 15 is proven. Suppose that algorithm 15 succeeds on query (x, y) with probability $P > 0$. The probability of generating each intermediate point along the path from x to y is the same for algorithm 15 and PRM, since they both sample randomly from the same distribution μ. Hence, PRM succeeds on query (x, y) with probability $P > 0$.

For the converse, suppose that after constructing the roadmap, PRM succeeds on query (x, y) with probability $P > 0$. Choose n to be the minimum number of configurations in the roadmap for the previous statement to be true. Since n is the smallest such number, then every configuration of the roadmap appears exactly once as an intermediate point of the path connecting x to y. Note that it does not matter in what order the configurations are generated: the roadmap is permutation invariant. Since the samples are independent, it suffices to consider only the solutions where the path is generated in order and conclude that the probability of this occurring is then $\frac{P}{n!} > 0$. Algorithm 15, after running for n iterations, would have probability at least $\frac{P}{n!} > 0$ of succeeding.

This concludes the proof of the probabilistic completeness equivalence between algorithm 15 and PRM. It remains to show that the probability of failure for PRM decreases exponentially with the number of samples generated.

Define R^ℓ to be the ℓth iteration of R, i.e.,

$$(x_1, \ldots, x_\ell) \in R^\ell \Leftrightarrow x_1 R x_2 \cdots x_{\ell-1} R x_\ell.$$

Suppose PRM is probabilistically complete. For any query (x, y), there exists an ℓ such that a sequence of ℓ guesses is a path from x to y with probability $P > 0$. Let $S \subset R^\ell$ be the set of guesses which are length ℓ paths from x to y. The probability of choosing such a sequence of ℓ points is $\mu^\ell(S) = P > 0$ (it can be shown that S is measurable in $\mathcal{Q}_{\text{free}}^\ell$). The set S is decomposed into a union of disjoint rectangles, i.e.,

$$S = \bigcup_{i=1}^{\infty} A_1^i \times \cdots \times A_\ell^i.$$

Choose i such that

$$\mu^n\left(A_1^i \times \cdots A_\ell^i\right) = \prod_{j=1}^{\ell} \mu\left(A_j^i\right)$$

is maximized. Observe that it must be larger than zero. Let $x_1, \ldots, x_\ell$ be any set of points such that $x_j \in A_j^i$. It follows that $x R x_1 R \cdots R x_\ell R y$ by construction of S. Let

$$p = \min_j \mu\left(A_j^i\right).$$

The probability that PRM fails to find a path between x and y after generating n configurations is therefore bounded by the probability that no such $x_1, \ldots, x_\ell$ is contained in the configuration set. Let I_j be an indicator variable that witnesses the event of the configuration set containing a point from A_j^i.

$$Pr[(x, y) \text{ FAILURE}] \leq Pr\left[\bigvee_{j=1}^{\ell} I_j = 0\right]$$

$$\leq \sum_{j=1}^{\ell} Pr[I_j = 0]$$

$$= \sum_{j=1}^{\ell} \left(1 - \mu\left(A_j^i\right)\right)^n$$

$$\leq \ell(1 - p)^n \leq \ell e^{-pn}.$$

Finally, note that $\ell \geq 0$ and $p > 0$.   ∎

It is interesting to note that the symmetry and reflexivity properties of the local planner were never used in the proof. In particular, the proof will still hold for an asymmetric and irreflexive local planner. This is a natural way to incorporate the notion of time into PRM planning. Also, the sampling distribution is not necessarily uniform. The obtained bound is of the same form as the previous bounds and shows that probabilistic completeness ensures an inverse exponential bound on failure probability in terms of the number of configurations in the roadmap.

7.5   Beyond Basic Path Planning

Sampling-based planners are becoming powerful and this allows the solution of problems beyond the generalized movers' problem. Some instances are considered here.

7.5.1   Control-Based Planning

Control-based planning was initially introduced in the context of planners that used discretization [41]. The notion, however, extends naturally to sampling-based planners and the principal ideas are introduced here. This section, however, should be read in conjunction with the material introduced in chapters 11 and 12.

Consider a nonholonomic robot such as a carlike robot, or any type of system for which we are given a set of controls $\mathcal{U}$ and a well-behaved control function f, $f : \mathcal{Q} \times \mathcal{U} \rightarrow \mathcal{Q}$ that describes a method for propagating a robot state into the future. Many of the sampling-based planners that have been described in this chapter can be used with such systems.

In particular, when f is given, a simple way of generating new samples in the state space may be available, e.g., a new state can be obtained by sampling according to some distribution of values for the controls, and applying these to the system state via f. Several applications of random controls can yield a configuration far away from the original sample. When such a forward propagation of the system is relatively inexpensive, tree-based planners such as ESTs and RRTs can be directly applied. The sampling method of EST is purely a forward propagation method as it is explained in section 7.2.1. RRTs use steering to produce new configurations but do not require the system to achieve the configurations toward which it is steered (see section 7.2.2). In most cases when the above planners are applied to control-based planning a single tree is generated from the initial configuration. Planning finishes when the goal configuration is reached or approximated with a predefined accuracy.

EST and RRT planners have been used with success for several problems with robots that exhibit various kinematic and dynamic constraints [41, 192, 195, 196, 235, 249, 270–272, 350] as well as stability constraints [247, 248]. Examples of such problems are illustrated in figure 7.20. Time can also be accommodated as part of the state space of the robot. This allows the modeling of a dynamic workspace (e.g., [195]).

In some cases, it may be possible to solve for the set of controls that are required to travel between two given states either exactly or approximately. Then the application of PRM is possible. PRM has been applied successfully to path-planning for nonholonomic systems, such as carlike robots and tractor-trailer systems [379, 404, 405].

7.5.2   Multiple Robots

The multiple movers problem deals with path planning for many robots. A collision-free path from an initial configuration of the robots to a goal configuration of the robots implies that at every step there is no collision between a robot and an obstacle or between a robot and another robot. A solution to this problem, in addition to finding paths for the individual robots (which only guarantee that there are no collisions with the obstacles), must be able to coordinate these paths so that no two robots are in collision. This second requirement makes the problem significantly harder than in the case of a single robot. There are two classic approaches to the multiple robots problem: centralized and decoupled planning.

Centralized Planning

Centralized planning considers the different robots as if they were forming a single multibody robot and represents $\mathcal{Q}$ as the Cartesian product of the configuration spaces of all the robots. The dimensionality of $\mathcal{Q}$ is equal to the total number of

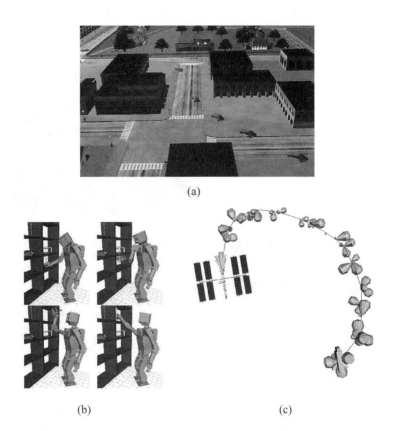

Figure 7.20 Control-based planning examples. (a) Car driving. (b) Humanoid robot. (c) Space shuttle docking at the space station—the yellow cones represent the plume of the shuttle that should not be directed toward the space station. ((a) From LaValle and Kuffner [272]; (b) from Kuffner [248]; (c) from Phillips, Kavraki, and Bedrossian [350].)

degrees of freedom of all the robots. Coordination of the robots is trivially achieved: a collision-free configuration in Q describes the configuration of each individual robot and ensures that no robot is in collision with some obstacle or some other robot. The difficulty of centralized planning arises from high the dimensionality of Q. As planners become more efficient in dealing with high-dimensional configuration spaces Q, harder problems with multiple robots can be solved. Figure 7.21 shows a workspace where six robots cooperate on a welding task [367].

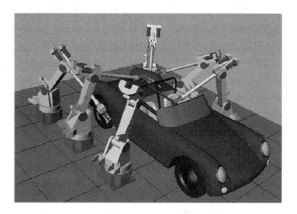

Figure 7.21   Multiple robots manipulating a car. (From Sánchez and Latombe [367].)

Decoupled Planning

Decoupled planning works in two stages. Initially, collision-free paths are computed for each robot individually, not taking into account the other robots but simply considering the obstacles of the workspace. In the second stage, coordination is achieved by computing the relative velocities of the robots along their individual paths that will avoid collision among them [219]. Decoupled planning does not increase the dimensionality of the configuration space. It is incomplete, however, even when the algorithms used in both of its stages are complete: it may be impossible to coordinate some of the paths generated during the first stage so that two different robots do not collide. Alternatively, in what is known as prioritized planning, robots are processed in some preassigned order and a robot is treated as a moving obstacle as soon as its path has been computed.

Planners for the Multiple Robots Problem

In principle, all sampling-based planners of this chapter can be adapted for multiple robots. Some key changes may be needed to retain good performance. For example, ESTs and RRTs can be used directly, as presented in section 7.2 but their performance can be improved with small modifications. A proposed scheme for connecting an existing configuration in the tree to a random configuration q_{rand} has been proposed [14]. Each robot is moved incrementally toward q_{rand}. The path is checked for collisions by adding one robot at a time and checking for collisions with the obstacles and with the previous robots. If a collision is found, then a new random configuration for the robot being added is generated. Although this local planner is more expensive than checking all robots simultaneously, it is considerably more effective in covering the space. The configuration returned by the call is the final configuration that was computed. In the case where no robot can move, no configuration is returned. It has been observed [14, 43] that this strategy avoids the problem of producing many configurations close to obstacles, a problem that arose from the direct application of EST and RRT algorithms to multiple robots.

The SRT [14, 43] algorithm presented in section 7.3 can be adapted to efficiently plan for multiple robots. SRT uses a prioritized approach for the computation of each edge on the roadmap and an incremental centralized approach for the computation of the configurations at the endpoints of the edges. An advantage of SRT is that it can be run in parallel to cut the cost of computation for solving planning problems involving many robots.

7.5.3 Manipulation Planning

Another important area of motion planning is manipulation planning [15, 16, 36, 170, 241–243, 335]. An example that involves an animated character manipulating different objects is given in figure 7.22. The objective is to move certain objects from some initial configuration to a goal configuration while avoiding collisions with the other objects and obstacles. Initially, the objects are static and at stable positions, e.g., resting against the obstacles or other stable objects. Since the objects cannot move autonomously, the robot must grasp the object and move it from one stable position to another, until it obtains the desired arrangement. A set of grasping positions at which the objects can be grasped by the robot is given to the planner.

Figure 7.22 Manipulation example. (Courtesy of J. C. Latombe.)

One approach to manipulation planning is to model the problem as fully dynamic and use control-based planning. This is expensive, however, and thus other approaches have been developed that make a distinction between the transit and transfer parts of the path [16, 264]. Transfer paths are defined as motions of the system while the robot grasps the object. Transit paths are defined as motions of the robot when it is not grasping an object as it moves from one grasp to the next. The manipulation planner is also responsible for computing regrasping operations. Fast planners are needed for all three subproblems.

Initial attempts to solve the manipulation problem [241, 243] for robots with many degrees of freedom proceed by finding a path for the object from q_{init} to q_{goal}. The planner then computes a series of transfer and transit paths for the robot that make it possible for the object to move along the path computed in the first stage. Variational dynamic programming [36] methods have also been used. A manipulation path is initially computed by assuming that the object and the robot move independently. Then an iterative process deforms the path to satisfy the constraints that the object can only move when it is in a proper grasp.

In [335] a two-level PRM is developed to handle manipulation planning. The first level of the PRM builds a manipulation graph, whose nodes represent stable placements of the manipulated objects while the edges represent transfer and transit actions. The second level of the PRM does the actual planning for the transfer and transit paths. The computation is made efficient by verifying that the edges are collision-free only if they are part of the final path. Otherwise, the local planner assigns a probability to the edge that expresses its belief that the edge is collision-free. The resulting planner, called FuzzyPRM, is yet another example of how sampling and connection strategies can be used in the context of PRMs. More advanced recent methods use several specialized roadmaps to address more complex problems and use manipulation planning as a vehicle to connect task level AI planning and motion planning [170].

Manipulation is a broad topic in itself that has also been addressed with techniques that do not fall under the general category of motion planning. For example, parts feeding often relies on nonprehensile manipulation. Nonprehensile manipulation exploits task mechanics to achieve a goal state without grasping and frequently allows accomplishing complex feeding tasks with few degrees of freedom. It may also enable a robot to move parts that are too large or heavy to be grasped and lifted. Pushing is one form of nonprehensile manipulation. Work on pushing originated in [311] where a simple rule is established to qualitatively determine the motion of a pushed object. A number of interesting results followed: among them were the development of a planning algorithm for a robot that tilts a tray containing a planar part of known shape to orient it to a desired orientation [146], the development of an algorithm to compute the shape of curved fences along a conveyor belt to reorient a given polygonal part [426], and the demonstration of a sequence of motions of a single articulated fence on a conveyor belt that achieves a goal orientation of an object [13]. A frictionless parallel-jaw gripper was used in [166] to orient polygonal parts. For any part P having an n-sided convex hull, there exists a sequence of $2n - 1$ squeezes achieving a single orientation of P (up to symmetries of the convex hull). The result has been generalized to planar parts having a piecewise algebraic convex hull [358]. It was later shown how to use a combination of squeeze and roll primitives to orient a polygonal part without changing the orientation of the gripper [317]. Last but not least, distributed manipulation systems provide another form of nonprehensile manipulation. These systems induce motions on objects through the application of many external forces and are realized typically on a flat surface. One way of implementing such forces is through the use of MicroElectoMechanical Systems (MEMS). Algorithms that position and orient parts based on identifying a finite number (depending on the number of vertices of the part) of distinct equilibrium configurations were given in [56]. Subsequent work showed that using a carefully selected actuators field, it is possible to position and orient parts in two stable equilibrium configurations [220]. Finally, a long standing conjecture was proven, namely that there exist actuators fields that can uniquely position and orient parts in a single step [55, 256, 399]. On the macroscopic scale it was shown that in-plane vibration can be used for closed-loop manipulation of objects using vision systems for feedback [363], that arrays of controllable airjets can manipulate paper [431] and that foot-sized discrete actuator arrays can handle heavier objects under various manipulation strategies [302].

7.5.4   Assembly Planning

An assembly operation is typically defined as a merging motion of pairwise-separated subassemblies into a new assembly. Two subassemblies are considered separated if they are arbitrarily far apart from each other. During the operation each assembly is treated as a single body and is not allowed to overlap with other subassemblies.

The assembly planning problem can be cast in a path-planning framework by considering one of the subassemblies that is to be merged as the robot and the other as the workspace. The objective then becomes to find a collision-free path for the moving subassembly to its final configuration. PRMs, ESTs, and RRTs have been used successfully to solve these problems [192]. Using planners for determining merging (or, equivalently, removal) paths for parts has important applications in the manufacturing cycle of new mechanical assemblies. E.g., when a new engine is designed, the CAD model of the engine is available. Using this model a planner can test the removal of various parts for maintainability purposes. Figure 7.2(d) shows such an example,

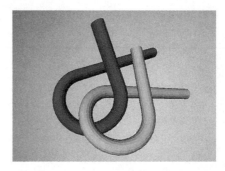

Figure 7.23   An example of assembly planning. The objective is to separate the two α-shaped pieces.   (From Amato et al. [18].)

while figure 7.23 shows an assembly that is frequently used to test how well planners can deal with the narrow passage problem.

It is worth noting that work on assembly planning has given rise to interesting analysis methods in robotics. Besides planning, researchers have tried to analytically determine the order in which the different parts of an assembly need to be assembled [230, 428, 429] by using the NonDirectional Blocking Graph (NDBG) [428], which represents all the blocking relations in a polygonal assembly.

7.5.5   Flexible Objects

Motion planning for flexible objects [22, 224, 255, 318] is an important problem as several applications could benefit from planners that account for the physical properties of the manipulated objects. For example, in industrial settings there is a need to handle sheets of metal, pipes that can bend, and cables. In assembly maintainability studies done with virtual prototyping, planning is used to compute a removal path for a part from an assembly, given only the CAD model of the assembly. The flexibility of the part needs to be considered as engineers use deformable parts to produce compact assemblies. In medical and surgical procedures, flexible catheters are inserted in human vessels. Accurate planning studies may help in choosing the size and properties of the catheter to be used. In computer-assisted pharmaceutical drug design, path-planning techniques are used to compute paths for drug molecules to their docking sites. In that context, the rigorous treatment of the physical properties of the drug molecule, expressed by its energy, is crucial for obtaining sequences that are of low energy and can thus be encountered in nature.

A major difficulty in planning for flexible objects stems from the fact that the configuration space is potentially of infinite dimension. So there is a need for geometric representations that approximate well the possible shapes of the flexible object and are still compact in terms of the number of parameters used. The energy of the object needs to be taken into consideration as a path must not only be collision-free but also energetically feasible. For example, if an elastic object is manipulated, care must be taken to not bend or stretch the object excessively and permanently deform it. This is achieved by keeping the elastic energy of the object below a predefined energy threshold. The computation of the energy is typically expensive [224]. Presently, there is no efficient way to relate the geometric representation of a flexible object to its flexibility/deformation properties except in specific cases [318, 319]. Collision checking is finally a significant bottleneck for path planning for flexible objects. In modern collision checking packages, some preprocessing of the robot is done to compute an internal representation that is used to speed up collision checking [168]. As the shape of the flexible robot changes continuously, such preprocessing is not possible and, as a result, collision checking is very expensive.

One approach to obtaining realistic (physical) paths for flexible objects is to create roadmaps of quasi-static nodes and then to connect the nodes using interpolating paths of low-energy configurations. Quasi-static configurations can be found by energy minimization or by physics-based simulation. Figure 7.24 shows a path for a thin elastic metal sheet. In the considered setting, two actuators control the deformation of the metal sheet by constraining the position of the two opposite sides of the sheet. The path has been computed by the application of PRM [255]. Configurations of the object in the roadmap are produced by first obtaining a low-energy random deformation and then a random configuration with that deformation. Any computed paths keep the elastic energy of the sheet below an energy minimum to avoid permanent deformations

Figure 7.24   An example of planning for flexible objects. The metal sheet needs to bend to go through the hole.   (From Kavraki [224].)

of the object. Local deformation fields over the volume of the object can be used to describe its deformation [255]. The choices for the local planner and the distance measure in the above framework are nontrivial.

Recently, it has been shown that, in certain cases, it is possible to obtain geometric representations for the flexible object that enforce certain physical properties of the flexible object. For example, in [318, 319] a low-dimensional representation is given for a three-dimensional curve that enforces the length of the curve to be constant. Hence there is no need for optimization procedures to maintain the the constant length constraint, which in general speeds up computation. PRM roadmaps of low-energy curves manipulated by actuators at their end points can then be constructed.

Finally, planning for flexible objects raises the issue of variable parametrization methods for the objects/robots [251]. It is sometimes necessary to change the parametrization over time to capture the shape of the object as accurately as possible, or to reduce the number of parameters of the problem, if the latter is feasible. When planning with a variable parametrization, rules for relating motion between different parametrizations must be established. In this case, the planner needs a mechanism for deciding how much and when to reparametrize. The energy of the system can also be seen as a heuristic that drives the exploration of tree sampling-based motion planners [251].

In summary, planning for flexible objects raises important questions and challenges to motion planning research. Planning for flexible objects in contact with obstacles remains a largely unexplored problem.

7.5.6   Biological Applications

Motion planning algorithms can also be applied to problems from computational structural biology [20, 23–25, 387, 408, 409, 433]. The problems in this domain are high-dimensional and of a complexity that tests the limits of current motion-planning algorithms. This section considers protein folding and protein-ligand docking. The first problem is a long-standing open problem in biochemistry. The second problem is central to understanding biomolecular interactions that regulate biochemical processes and can lead to the generation of new therapeutics. An example of folding is given in figure 7.2(g). Different three-dimensional representations of a widely targeted protein and a ligand are shown in figure 7.25.

A protein is a linear sequence, or polymer, of amino acid residues. The genome codes for twenty different residues give rise to a great variety of possible protein sequences, and a corresponding variety in three-dimensional structure and function. Proteins are broadly classified by their function: enzymes catalyze biochemical reactions; regulatory proteins modulate gene expression; peptide hormones and

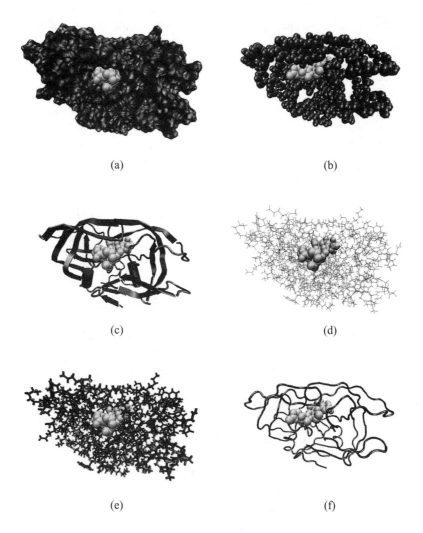

(a)   (b)

(c)   (d)

(e)   (f)

Figure 7.25   Docking examples. (a) HIV-I protease and docked ligand (PDB ID 4HVP), where the receptor (HIV-I protease) is rendered as a Connolly surface. The complex was obtained using x-ray crystallography. (b) Receptor—HIV-I protease—rendered with backbone atoms only. (c) Receptor rendered showing α-helices and β-sheets. (d) Receptor rendered as linkage. (e) Receptor rendered as stick model. (f) Receptor's backbone rendered as a tube. In all figures, the ligand is rendered using a sphere for each of its atoms and can be found close to the center of the HIV-I protease.

signaling proteins carry chemical messages both within and between cells; and structural proteins make up microfilaments and microtubules, which act as frameworks and molecular transport routes within cells, as well as macroscopic structures such as hair, claws, and silk.

Folding and Docking

A guiding principle in biochemistry is that molecular structure determines function. This is particularly evident in proteins where the biological function is strongly determined by the protein's ability to fold into a stable three-dimensional structure, also known as its native configuration. It is very important that proteins be able to reach and maintain their native configuration since failure to do so would render the protein nonfunctional. The pathway that the protein follows to reach its native configuration is hard to determine experimentally because the intermediate steps usually occur too rapidly to detect. The folding problem is concerned with trying to understand and characterize the sequence of motions followed by a protein to go from a disorganized, unfolded state to its highly ordered native configuration. A number of diseases result from the misfolding of a particular protein, so an understanding of how normal proteins fold may eventually aid medical researchers in understanding what goes wrong when a protein misfolds, and what medical intervention may ultimately be possible.

Docking is an equally important problem. The biological function of enzymatic and signaling proteins is often achieved by their ability to bind transiently to and react with smaller molecules, known as ligands. This binding (docking) usually takes place in a distinctive cleft in the protein's surface known as the binding pocket or active site. The ability of a receptor protein to dock a given prospective ligand depends on the geometric matching of the ligand and the binding pocket, as well as the presence of stabilizing chemical interactions between atoms of the ligand and atoms on the surface of the binding pocket. When the receptor succeeds in docking a ligand the free energy of the biomolecular complex is lower in the docked configuration than any other possible configuration of the complex. Many drugs act by blocking the active site of an enzyme or by binding to a signaling protein and either blocking it or enhancing its activity (see figure 7.25). Finding a new drug candidate starts by finding a compound that binds to a particular site on a protein's surface. The screening of a large number of potential ligands or drug candidates in the laboratory is very slow and expensive. Computational docking methods therefore offer substantial savings in both time and money to pharmaceutical researchers by providing promising leads from a database of hundreds of thousands of known compounds, given a particular receptor. Laboratory tests can then proceed only on those compounds predicted to dock well in simulation.

Several researchers have used sampling-based motion-planning techniques for protein folding and docking problems [20, 23–25, 387, 408, 409]. The notion of configuration space offers a layer of abstraction that allows for problems from other areas to be cast as motion-planning problems.

Application of PRM **Methods**

Any molecule can be seen as a collection of atoms and bonds between pairs of atoms. An underlying graph representation of a molecule can be constructed with atoms at the vertices and bonds on the edges. A common simplification that works for most molecules is to represent a cycle in the graph, which corresponds to a ring in the molecule, as a single special atom that is connected by bonds to the rest of the molecule. It follows then that the underlying graph is a tree. One atom, called the anchor, is chosen arbitrarily as the root of the tree. Thus the molecule is represented as a treelike articulated robot [433]. For each atom, information is kept about its mass, van der Waals radius, and other physical properties relevant to predicting interactions with other atoms. For each bond, information is kept about the bond length, which is the separation distance between the two atoms the bond connects; the bond angle, which is the angle between a given bond and the previous bond in the direction toward the anchor atom; and the set of possible torsional (or dihedral) angles, which expresses the possible rotations of the structure at one end of the bond with respect to the structure at the other end. A bond is fixed if its dihedral angle must remain constant, otherwise it is rotatable. A common assumption is to consider bond lengths and angles as fixed, with dihedral angles as the only variables. Figure 7.26 offers an

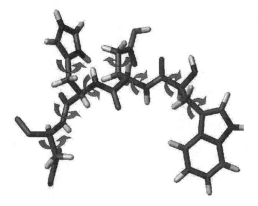

Figure 7.26 An example of a small molecule where arrows indicate rotatable bonds.

example for a small ligand. A small ligand may have 5–15 dihedral angles, while a protein has a few hundreds of dihedral angles. Robotics methodologies can be used to encode dihedral angles and efficiently compute molecular configurations [433].

Both folding and docking involve the exploration of a high-dimensional energy landscape for low-energy configurations or complexes. For a PRM roadmap that aims to explore the energy landscape of a small protein, node configuration can be generated by selecting uniformly at random values for the dihedral angles from their allowable range. Random configurations, however, do not all correspond to feasible configurations of the molecule that can be observed experimentally. The validity of a configuration is determined by the potential energy of the corresponding configuration, denoted E_{config}. The potential energy of a configuration depends on the properties of the atoms and the values of the dihedral angles and can be explicitly computed [387]. A configuration is considered feasible if its potential energy is below some threshold E_{max}. In addition to detecting unfavorable interactions, the use of an energy cutoff implicitly allows collision detection: most force fields include a term that imposes an exponential energetic penalty for overlapping atoms. In [21, 387], the following probability is used to add a configuration to the roadmap:

$$
Pr(\text{config is accepted}) =
\begin{cases}
0, & \text{if } E_{\text{config}} > E_{\text{max}} \\
\frac{E_{\text{max}} - E_{\text{config}}}{E_{\text{max}} - E_{\text{min}}}, & \text{if } E_{\text{min}} \leq E_{\text{config}} \leq E_{\text{max}} \\
1, & \text{if } E_{\text{config}} \leq E_{\text{min}}.
\end{cases}
$$

Selecting configurations as shown above results in denser sampling of low-energy configurations. For each configuration, a set of k closest neighbors is computed using either the Euclidean or least-root-mean-square distance as the metric. Neighboring configurations are connected by performing linear interpolation between the two configurations and checking that all the intermediate configurations correspond to feasible configurations. A weight is associated with each local path that reflects the difficulty of traversing the path. The probability of traversing a path is computed by using the energy of each intermediate configuration. Based on the ideas presented above, roadmaps have been constructed for exploring the energy landscape for the docking problem [387]. Also PRM roadmaps are used for tracing protein-folding or RNA-folding pathways [20, 21, 408] when the native configuration is known. Finally, for exploring the energy landscape of a protein a novel method influenced by PRM has been developed. Stochastic Roadmap Simulation [23–25] allows the simultaneous analysis of motion pathways and the computation of ensemble properties over the entire molecular energy landscape.

Computational structural biology offers challenging problems of unprecedented scale. Some promising solutions are currently influenced by a robotics methodology.

It is conceivable that, in the near future, we could see novel robotics planning methods inspired by biological problems.

Problems

1. You are given a rigid-body robot that can freely translate and rotate in an empty three-dimensional box. Quaternions are used to represent the configurations of the robot. Implement a procedure that generates random free configurations of the robot. Implement an efficient planner that connects two configurations.

2. Implement a procedure that determines if two polygons in a plane are in collision.

3. Implement an efficient local planner for four robots that move in the plane. There are no obstacles.

4. Define two functions to compute the distance between two configurations of three-dimensional rigid and articulated robots and discuss their advantages and disadvantages.

5. Implement the closest neighbors query using one of the distance functions defined in the previous problem and a grid-based approach.

6. Implement a basic PRM planner for a single robot operating in a two-dimensional Euclidean space. Assume that the robot and the obstacles are polygons.

7. Modify your implementation of PRM to include one of the sampling strategies discussed in section 7.1.3, e.g., Gaussian, bridge-test, and so on.

8. Implement one of the path-smoothing strategies discussed in section 7.1.2.

9. Implement a tree-based planner such as SBL or RRT for a point robot. Display the generated trees.

8 *Kalman Filtering*

HERETOFORE, WE have assumed that the planner has access either to an exact geometric description of its environment, or to a suite of sensors (e.g., sonars) that provide perfect information about the environment. In this chapter, we begin to consider cases for which the robot's knowledge of the world derives from measurements provided by imperfect, noisy sensors.

The Kalman filter is one of the most useful estimation tools available today. Loosely speaking, Kalman filtering provides a recursive method of estimating the state of a dynamical system in the presence of noise [215, 313]. A key feature of the Kalman filter is that it simultaneously maintains estimates of both the state vector ($\hat{x}$) and the estimate error covariance matrix (P), which is equivalent to saying that the output of a Kalman filter is a Gaussian probability density function (PDF) with mean $\hat{x}$ and covariance P. In the context of localization, the Kalman filter output is then a distribution of likely robot positions instead of a single position estimate. As such, the Kalman filter is a specific example of a more general technique known as probabilistic estimation. Some of more general probabilistic estimation techniques are presented in chapter 9.

We begin this chapter by presenting a conceptual overview of probabilistic estimation in section 8.1. Section 8.2 carefully derives the Kalman filter for linear systems. The approach taken here begins with a simplified version of the problem, then gradually adds complexity until the full Kalman filtering equations are reached. Section 8.3 describes the extended Kalman filter (EKF), which is a Kalman filtering variant that can be used on nonlinear systems. Examples that use the EKF for mobile robot localization are presented and discussed. Section 8.4 concludes the chapter by introducing the problem of simultaneous localization and mapping (SLAM) and showing how it can be solved using a Kalman filter.

8.1   Probabilistic Estimation

In this section, we introduce the concept of probabilistic estimation by considering the fundamental problem of estimating the location of a mobile robot. Probabilistic localization is a probabilistic algorithm: instead of maintaining a single hypothesis as to where in the world a robot might be, probabilistic localization maintains a *probability distribution* over the space of all such hypotheses. The probabilistic representation allows for the uncertainties that arise from uncertain motion models and noisy sensor readings to be accounted for in a principled way. The challenge is then to maintain a position probability density over all possible robot poses. Such a density can have arbitrary forms representing various kinds of information about the robot's position. For example, the robot can start with a uniform distribution representing that it is completely uncertain about its position, i.e., that the robot could be in any location with equal probability. It furthermore can contain multiple modes in the case of ambiguous situations. In the usual case, in which the robot is highly certain about its position, it consists of a unimodal distribution centered around the true position of the robot. This chapter discusses Kalman filtering, which is a form of probabilistic estimation where the estimate is assumed to be a Gaussian (unimodal) PDF. Other probabilistic estimation methods, such as Bayesian methods and particle filtering, can handle more general distributions and are discussed in chapter 9.

One crude method of mobile robot localization is achieved by simply integrating robot velocity commands from a known starting position. When the commands are executed perfectly and the robot starting position is perfectly known, this method gives a perfect estimate of position. Of course, perfect performance and knowledge are impossible to achieve in the real world. Errors between the velocity commands and the actual robot velocities will accumulate over time. In other words, as the robot moves, it is continuously losing information about its location. Eventually, the robot will lose so much information that the command integration estimate becomes meaningless for any practical purpose. A similar approach is to integrate robot velocity measurements reported by onboard odometry sensors. Figure 8.1 shows typical odometry data of a B21 robot as it is recorded with its odometry sensors. Note that the error in odometry quickly accumulates over time up to a rotational error of almost 45 degrees.

With just a few extensions, command integration can be thought of as a probabilistic estimation method. First, consider the robot starting location. Since, in the real world, the starting position cannot be perfectly known, it makes sense to represent the starting

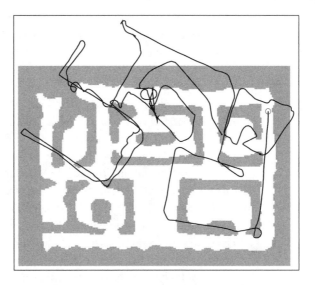

Figure 8.1   Odometry measurements of a B21 robot.

location as a PDF over the state space, i.e., the space of possible robot positions. If a good estimate of the starting position is available, the PDF will have a peak at that location, and the "sharpness" of the peak will represent the certainty of the initial estimate: the more certain the initial estimate, the higher and narrower the peak. Now the challenge is to propagate this PDF as the robot moves. The location of the peak is propagated by integrating the velocity commands just as before. However, since the commands are not executed perfectly, the estimate becomes more uncertain as time progresses and the peak of the PDF will become smaller and more spread out. The rate at which the peak spreads is determined by the amount of error (noise) in the velocity command: the greater the noise, the faster the peak spreads. Eventually, the peak will become so flat that the estimate provided by the PDF is meaningless.

It goes without saying that the goal of probabilistic estimation (or any estimation method, for that matter) is to provide a meaningful estimate of the system state, which in this case is the robot pose. The problem with the command integration method is that information is continually lost and no new information is ever gained. The solution to this is to inject new information into the system through the use of sensors that gather infomation about the environment. Consider, e.g., a robot equipped with a sensor capable of measuring the range and bearing to a landmark with a known location. Such a measurement adds new information to the system and can be used (at least partially) to compensate for the information that was lost by integrating. The new information can be represented as a PDF in sensor space, usually with a peak at the value of the sensor reading. As we have already described, knowledge about the robot location prior to the sensor measurement is described by a PDF in the state space. The crux of the probabilistic estimation problem is to merge these two distributions in a meaningful way.

Generally, any probabilistic estimation method can be thought of as a two–step process of *prediction* and *update*. Given an estimate of the system state in the form of a PDF, the prediction propagates the PDF according to robot commands together with a motion model for the robot. The update step then corrects the prediction by merging the predicted PDF with information collected by the sensors. The "new" estimate is given by the updated PDF, and the process is iterated. Note that in each iteration, the prediction step accounts for the information lost due to errors in the motion model while the update step incorporates information gained by the sensors.

The following sections describe the well-known Kalman filter, which is a specific probabilistic estimation technique. In Kalman filtering, the motion model is assumed to be a linear function of the state variables and the inputs (commands). The quantities measured by the sensors are assumed to be linear functions of the state variables. Errors in both the motion model and the sensor model are assumed to be zero-mean white Gaussian noise. Because of this simple form, it is possible to derive closed-form equations to perform the prediction and update steps, making Kalman filter implementation a straightforward process.

8.2   Linear Kalman Filtering

One reason that Kalman filtering has become such a popular estimation method is that it is extremely easy to implement for linear systems. The equations in section 8.2.5 can be implemented directly with little understanding of the underlying theory. This feature makes Kalman filtering useful and accessible to a broad range of potential users, but it does not mean that the underlying theory is unimportant. In fact, most modern applications of Kalman filtering employ substantial modifications of the original equations. For example, modifications are necessary to address nonlinear sensor models or non-Gaussian noise models in robot localization and mapping problems. Other modifications are often used to reduce computational complexity.

This section is intended to provide the reader with an understanding of the fundamentals of Kalman filtering for linear systems. The approach taken here is intuitive and uses basic facts from geometry and linear algebra to reconstruct Kalman's equations. Some knowledge of multivariate Gaussian distributions is assumed (see the statistics primer in appendix I for an overview). We begin with a simplified version of the Kalman filtering problem to illustrate the basic concept, then we incrementally add complexity until we arrive at the full Kalman equations. An example illustrating the application of the Kalman filter equations is presented, and the property of observability in linear systems is introduced. With the understanding provided here, the reader should be able to modify the Kalman filter to fit the needs of a specific estimation problem.

8.2.1   Overview

In order to apply Kalman filtering to the problem of robot localization, it is necessary to define equations that can be used to model the dynamics and sensors of the robot system. The vector x is used to denote the system (robot) state as it evolves through time. This chapter uses discrete time models, meaning that the continuously varying robot state is sampled at discrete, regularly spaced intervals of time to create the sequence $x(k)$, $k \in \{0, 1, 2, \ldots\}$. Specifically, if $x(0)$ is the value of the state at time $t = t_0$, then $x(k)$ is the value of the state at time $t_0 + Tk$, where T is defined to be the sampling time step.

For now we assume that the evolution of the robot state and the values measured by the robot sensors can be modeled as a linear dynamical discrete–time system:

$$(8.1) \quad x(k+1) = F(k)x(k) + G(k)u(k) + v(k)$$

$$(8.2) \quad y(k) = H(k)x(k) + w(k).$$

The vector $x(k) \in \mathbb{R}^n$ denotes the full system state. The vector $u(k) \in \mathbb{R}^m$ is used to represent the system input such as velocity commands, torques, or forces intentionally applied to the robot. The vector $y(k) \in \mathbb{R}^p$ is the system output and contains the values reported by the system sensors. The matrix $F(k) \in \mathbb{R}^{n \times n}$ encodes the dynamics of the system, and $G(k) \in \mathbb{R}^{n \times m}$ describes how the inputs drive the dynamics. The vector $v(k) \in \mathbb{R}^n$ is called the *process noise* and is assumed to be white Gaussian noise with zero mean and covariance matrix $V(k)$.[1] The process noise is used to account for unmodeled disturbances (such as slipping wheels) that affect the system dynamics. The matrix $H(k) \in \mathbb{R}^{p \times n}$ describes how state vectors are mapped into outputs. The *measurement noise* vector $w(k) \in \mathbb{R}^p$ is assumed to be white Gaussian noise with zero mean and covariance matrix $W(k)$. Here we assume that $H(k)$ is full row rank for all k, although it may not be square.

---

1. Here the term "white" means that the vector $v(k)$ is independent of $v(k-1)$ for all k. The properties of a Gaussian distribution, which is defined entirely by its mean vector and covariance matrix, is discussed in more detail later in this chapter.

The objective of Kalman filtering is to determine the "best" estimate of the state x at the kth time step given a previous estimate together with the known input $u(k)$ and output $y(k)$. In order to achieve this there are two separate difficulties that must be overcome. The first is the presence of the unknown and unmeasurable noise vectors $v(k)$ and $w(k)$. Hence, as its name implies, one task of the Kalman filter is to filter out these unwanted disturbances. The second difficulty is that the state in general cannot be directly observed from the outputs because $H(k)$ may not be invertible. This means that the state estimate must be reconstructed using the time history of the known signals $y(k)$ and $u(k)$ together with known parameters $F(k)$, $G(k)$, $H(k)$, $V(k)$, and $W(k)$.[2] A device that does this is called an observer. The Kalman filter is both an observer and a filter.

In this section we build up to Kalman's equations by first building an observer for a system with no measurement noise. Specifically, we derive the equations for a simple two-step observer using only a few simple facts from linear algebra. We then introduce the concept of using a multivariate Gaussian distribution as a state estimate, and we rederive the simple observer equations that use this kind of estimate. This leads naturally to the derivation of the Kalman filter equations for linear discrete time systems.

8.2.2   A Simple Observer

Here we consider a linear discrete time system with no noise:

(8.3) $x(k + 1) = F(k)x(k) + G(k)u(k)$

(8.4) $y(k) = H(k)x(k)$

Here, $H(k)$ is assumed to be full row rank at every k. The objective is to build an observer for this system, i.e., we would like to find a set of equations that allows us to reconstruct the state x. The observer we build will be recursive[3]: it will take the most recent estimate together with the most recent input u and output y, and then return the next estimate. If the observer works (and the assumptions are valid), then the estimate will converge to the actual value of x over time.

Before we begin deriving the necessary equations, we first introduce some notation to make the job of keeping track of the estimate easier. Given two integers k_1 and k_2

---

2. For this to be possible the pair (F, H) must be observable, a property which is discussed briefly in section 8.2.7. Observability is also discussed in the overview of linear time invariant control systems in appendix J. A more thorough discussion can be found in any good linear systems theory textbook, e.g., [214].

3. Note that the definition of *recursive* is subtly different from what is commonly found in computer science.

with $k_1 \geq k_2$, we use $\hat{x}(k_1 \mid k_2)$ to denote the value of the state estimate at time k_1 given the value of the output at all times up to k_2. The symbol $\hat{x}(k_1 \mid k_2)$ is pronounced "x hat at k-one given k-two." This notation may seem cumbersome at first, but its usefulness will soon become apparent.

Now the observer follows an intuitive two-step process. Given the current state estimate $\hat{x}(k \mid k)$, we first generate a prediction $\hat{x}(k+1 \mid k)$ by propagating the prior estimate according to the system dynamics in equation (8.3). We then correct the prediction based on the output $y(k+1)$ to generate the next estimate $\hat{x}(k+1 \mid k+1)$. We call these two steps the prediction and update steps, respectively.

For the prediction step, we simply substitute $\hat{x}(k \mid k)$ into equation (8.3) to get

(8.5) $\hat{x}(k+1 \mid k) = F(k)\hat{x}(k \mid k) + G(k)u(k).$

To perform the update, we first note that given the output $y(k+1)$, the system state is constrained to lie on the hyperplane

$$\Omega = \{x \in \mathbb{R}^n \mid H(k+1)x = y(k+1)\}.$$

Note that Ω is the set of states that are consistent with the measurement $y(k+1)$. For our simple observer, we choose the next estimate $\hat{x}(k+1 \mid k+1)$ to be the point in Ω that has the shortest distance to the prediction $\hat{x}(k+1 \mid k)$. This is an intuitive choice: we have some reason to believe that $\hat{x}(k+1 \mid k)$ is close to the actual state value, and we know that the actual state must be in Ω. So it makes sense to choose the update to be the point in Ω that is closest to $\hat{x}(k+1 \mid k)$. This choice of update is depicted graphically in figure 8.2. We can use algebra to find an expression for $\hat{x}(k+1 \mid k+1)$. Define the vector Δx to be the vector that points from $\hat{x}(k+1 \mid k)$ to $\hat{x}(k+1 \mid k+1)$, i.e.,

$$\Delta x = \hat{x}(k+1 \mid k+1) - \hat{x}(k+1 \mid k).$$

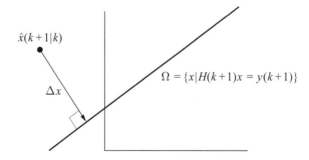

Figure 8.2 The set Ω corresponds to the states consistent with the current output $y(k+1)$. The corrected state lies in this set and is the state closest to the predicted estimate.

By our choice of $\hat{x}(k+1 \mid k+1)$, Δx is the shortest vector pointing from $\hat{x}(k+1 \mid k)$ to Ω. This means that Δx must be *orthogonal* to Ω by the standard inner product on $\mathbb{R}^n$, i.e., we must have $a^T \Delta x = 0$ for any a that is parallel to Ω.[4] Now we need two basic facts from linear algebra [398]:

1. A vector $a \in \mathbb{R}^n$ is parallel to Ω if and only if $H(k+1)a = 0$. The set of all such a is called the *null space* of $H(k+1)$ and is denoted by null($H(k+1)$).

2. A vector $b \in \mathbb{R}^n$ is orthogonal to every vector in the space null($H(k+1)$) if and only if b is in the *column space* of $H(k+1)^T$, where the column space of $H(k+1)^T$ is denoted column($H(k+1)^T$) and is defined to be the span of the columns of $H(k+1)^T$.

Note that any vector $b \in$ column($H(k+1)^T$) can be written as a weighted sum of the columns of $H(k+1)^T$, which is equivalent to saying that $b = H(k+1)^T \gamma$ for some $\gamma \in \mathbb{R}^p$. Combining these two facts, we see that in order to have Δx orthogonal to Ω, we must have

$$\Delta x = H(k+1)^T \gamma$$

for some vector γ in $\mathbb{R}^p$. Next, we will try to find γ.

Define the *innovation error* v to be the difference between the actual output $y(k+1)$ and the predicted output $H(k+1)\hat{x}(k+1 \mid k)$. In other words, v is the difference between what the sensors reported and what they would have reported if the prediction was correct. The larger the discrepancy between the actual and predicted measurements, the larger the necessary correction Δx will be. So for now we make the guess that γ can be written as a linear function of v, i.e., $\gamma = Kv$ for some $K \in \mathbb{R}^{p \times p}$. This yields the equation

$$\begin{aligned}\Delta x &= H(k+1)^T K v \\ &= H(k+1)^T K(y(k+1) - H(k+1)\hat{x}(k+1 \mid k)).\end{aligned}$$

If we can find a K such that $H(k+1)(\hat{x}(k+1 \mid k) + \Delta x)$ agrees with the measurement $y(k+1)$ (i.e., $(\hat{x}(k+1 \mid k) + \Delta x) \in \Omega$), then our guess is correct and we have an expression for $\hat{x}(k+1 \mid k+1)$. To find K, we start with the requirement that

(8.6) $$H(k+1)(\hat{x}(k+1 \mid k) + \Delta x) = y(k+1),$$

which implies that

$$H(k+1)\Delta x = y(k+1) - H(k+1)\hat{x}(k+1 \mid k) = v.$$

---

4. Technically, we must also define what we mean by "parallel." We say a vector a is *parallel* to a hyperplane Ω if $x + a \in \Omega$ for every $x \in \Omega$.

Substituting $\Delta x = H(k+1)^T K \nu$ yields

(8.7) $\qquad H(k+1)H(k+1)^T K \nu = \nu,$

which implies that $K = (H(k+1)H(k+1)^T)^{-1}$. Note that the matrix $H(k+1)H(k+1)^T$ is guaranteed to be invertible by the assumption that $H(k+1)$ is full row rank for all k. We were able to find a K that solves equation (8.7) meaning that for the choice $\Delta x = H^T K \nu$, equation (8.6) is satisfied. Our guess that Δx is a linear function of ν is then verified. As a result, we now have equations that fully express our simple two-step observer:

prediction:

$$\hat{x}(k+1 \mid k) = F(k)\hat{x}(k \mid k) + G(k)u(k)$$

update:

$$\hat{x}(k+1 \mid k+1) = \hat{x}(k+1 \mid k) + H^T(HH^T)^{-1}(y(k+1) - H\hat{x}(k+1 \mid k))$$

Note that in the update equation we have denoted $H(k+1)$ simply by H to keep the expression manageable.

It turns out that there are some problems with this observer. Our choice of the update is naive. Since the update is always perpendicular to the set Ω, only the component of the state that directly affects the current sensor reading is updated. Estimate errors in the direction parallel to Ω are never corrected. As a result, the estimate $\hat{x}$ will not in general converge to x. However, what is important is that the intuitive notions of prediction and correction are the same as those used in the Kalman filter. In the following discussion we follow this intuition toward Kalman's equations, and in the process we fix the problems associated with our simple observer.

8.2.3 Observing with Probability Distributions

The estimate produced by the simple observer discussed in the previous section is a vector. In contrast, the estimate produced by a Kalman filter is a multivariate Gaussian probability distribution over the state space. In addition to providing a vector estimate $\hat{x}(k \mid k)$, a Kalman filter also provides an estimate of the error covariance $P(k \mid k)$ associated with $\hat{x}(k \mid k)$. In this section, we advance the simple observer from the previous section one step toward Kalman's filter by augmenting it to provide a covariance estimate.

First we review some basic facts about multivariate Gaussian distributions. A more detailed discussion can be found in the statistics primer in appendix I. For $x \in \mathbb{R}^n$, a multivariate Gaussian distribution has a PDF of the form

(8.8) $$p(x) = \frac{1}{\sqrt{(2\pi)^n |P|}} e^{-\frac{1}{2}(x-\bar{x})^T P^{-1}(x-\bar{x})},$$

where $\bar{x}$ is a vector in $\mathbb{R}^n$ and P is a symmetric, positive definite $n \times n$ matrix. It is clear that $p(x)$ is entirely defined by $\bar{x}$ and P. Further, $E[x] = \bar{x}$ and $E[(x-\bar{x})(x-\bar{x})^T] = P$, so $\bar{x}$ and P are called the *mean vector* and *covariance matrix,* respectively. In the Kalman filter, we maintain a state estimate which will be the mean of a Gaussian distribution, so in the sequel we replace $\bar{x}$ with $\hat{x}$.

In this section, we consider linear discrete time systems with process noise but no measurement noise, i.e.,

(8.9) $$x(k + 1) = F(k)x(k) + G(k)u(k) + v(k)$$

(8.10) $$y(k) = H(k)x(k).$$

As before, $v(k) \in \mathbb{R}^n$ is assumed to be white noise chosen from a zero-mean Gaussian distribution with covariance matrix $V(k)$ and the matrix $H(k)$ is assumed to be full row rank for all k.

Here we follow the same basic steps of prediction and update that were used for the simple observer. The main difference is that this time we must generate both a state vector estimate $\hat{x}(k \mid k)$ and a covariance matrix estimate $P(k \mid k)$. Hence the prediction step will generate $\hat{x}(k + 1 \mid k)$ and $P(k + 1 \mid k)$, and the update step will generate the next estimate given by $\hat{x}(k + 1 \mid k + 1)$ and $P(k + 1 \mid k + 1)$.

The state vector prediction $\hat{x}(k + 1 \mid k)$ is found by substituting $\hat{x}(k \mid k)$ into equation (8.9). Since the expected value of $v(k)$ is zero, the resulting prediction is

(8.11) $$\hat{x}(k + 1 \mid k) = F(k)\hat{x}(k \mid k) + G(k)u(k).$$

To compute the predicted covariance matrix we start with the definition of the covariance matrix:

$$P(k + 1 \mid k) = E\left[(x(k + 1) - \hat{x}(k + 1 \mid k))(x(k + 1) - \hat{x}(k + 1 \mid k))^T\right]$$

Substituting $x(k + 1)$ from equation (8.9) and $\hat{x}(k + 1 \mid k)$ from equation (8.11), then multiplying the terms inside the expectation, yields

$$P(k + 1 \mid k) = E\left[F(k)(x(k) - \hat{x}(k \mid k))(x(k) - \hat{x}(k \mid k))^T F(k)^T \right. \\ \left. + 2F(k)(x(k) - \hat{x}(k \mid k))v(k)^T + v(k)v(k)^T\right].$$

The fact that $v(k)$ is independent of both $x(k)$ and $\hat{x}(k \mid k)$ implies that $E[(x(k) - \hat{x}(k \mid k))v(k)] = E[x(k) - \hat{x}(k \mid k)]E[v(k)]$, which is zero due to the fact that $v(k)$ is assumed to be zero mean. Using this fact together with the linearity property of the expectation yields

$$P(k+1\mid k) = F(k)E\left[(x(k) - \hat{x}(k\mid k))\,(x(k) - \hat{x}(k\mid k))^T\right]F(k)^T$$
$$+ E\left[v(k)v(k)^T\right].$$

The first expectation term in this equation matches the definition of the covariance matrix $P(k\mid k)$, while the second expectation term matches the definition of the covariance matrix $V(k)$. As a result we can write the prediction equation

(8.12) $$P(k+1\mid k) = F(k)P(k\mid k)F(k)^T + V(k).$$

To perform the update step, we choose $\hat{x}(k+1\mid k+1)$ to be the most likely point x in the set

$$\Omega = \{x \in \mathbb{R}^n \mid H(k+1)x = y(k+1)\}.$$

Hence, we look for $x \in \Omega$ that maximizes the Gaussian distribution defined by $\hat{x}(k+1\mid k)$ and $P(k+1\mid k)$, i.e.,

$$p(x) = \frac{1}{\sqrt{(2\pi)^n\,|P(k+1\mid k)|}}e^{-\frac{1}{2}(x-\hat{x}(k+1\mid k))^T P(k+1\mid k)^{-1}(x-\hat{x}(k+1\mid k))}.$$

Because the exponential is monotonically increasing, $p(x)$ is maximized when $(x - \hat{x}(k+1\mid k))^T P(k+1\mid k)^{-1}(x - \hat{x}(k+1\mid k))$ is minimized. With this in mind, we introduce a new notion of distance with the norm[5]

$$\|x\|_M^2 = x^T P(k+1\mid k)^{-1}x,$$

which is derived from the new inner product on $\mathbb{R}^n$,

$$\langle x_1, x_2 \rangle_M = x_1^T P(k+1\mid k)^{-1}x_2.$$

Define $\Delta x = \hat{x}(k+1\mid k+1) - \hat{x}(k+1\mid k)$. So we want to find $\hat{x}(k+1\mid k+1)$ such that

1. $\|\Delta x\|_M$ is minimized.

2. $(\hat{x}(k+1\mid k) + \Delta x) \in \Omega$.

The first condition means that the vector Δx is orthogonal to the hyperplane Ω with respect to the inner product $\langle \cdot, \cdot \rangle_M$. This notion is depicted graphically in figure 8.3. The ellipses in this figure represent sets of points that are equidistant to $\hat{x}(k+1\mid k)$ according to the $\|\cdot\|_M$ norm. With this notion of distance, choosing $\hat{x}(k+1\mid k+1)$ to be the closest point on Ω to $\hat{x}(k+1\mid k)$ is equivalent to choosing the point at which

---

5. $d_M(x, \hat{x}) = \|x - \hat{x}\|_M$ is called the *Mahalanobis distance* between x and $\hat{x}$. The Mahalanobis distance indicates how far away the point x is from the mean $\hat{x}$ in units of standard deviations.

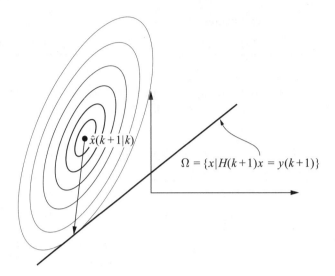

Figure 8.3 Correction determines the "closest" and "most likely" state on the set of states Ω.

one of the equidistant ellipses tangentially intersects Ω. The resulting Δx must be orthogonal to Ω, but our notion of orthogonality is skewed by $P(k+1 \,|\, k)^{-1}$. This means that we must have

$$a P(k+1 \,|\, k)^{-1}(\Delta x) = 0$$

for all $a \in \text{null}(H(k+1))$. In the remainder of this section, we simply denote $H(k+1)$ by H for brevity. Using the linear algebra facts presented earlier, this expression can only be true if $\Delta x \in \text{column}(P(k+1 \,|\, k)H^T)$, which means that

$$\Delta x = P(k+1 \,|\, k)H^T \gamma$$

for some $\gamma \in \mathbb{R}^p$. As in the case of the simple observer, we guess that γ can be expressed as a linear function of the innovation error $v = y(k+1) - Hx(k+1 \,|\, k)$, i.e.,

(8.13) $$\Delta x = P(k+1 \,|\, k)H^T K v$$

for some $K \in \mathbb{R}^{p \times p}$. Now we enforce $(\hat{x}(k+1 \,|\, k) + \Delta x) \in \Omega$, i.e.,

$$H(\hat{x}(k+1 \,|\, k) + \Delta x) = y(k+1),$$

which implies that $H \Delta x = v$. Substituting for Δx from equation (8.13) yields

$$HP(k+1 \,|\, k)H^T K v = v,$$

which means that we must have

$$K = \left(H P(k + 1 \mid k) H^T \right)^{-1}.$$

The resulting update equation for the state vector estimate is

(8.14) $\hat{x}(k + 1 \mid k + 1) = \hat{x}(k + 1 \mid k) + P(k + 1 \mid k) H^T \left(H P(k + 1 \mid k) H^T \right)^{-1} v.$

To ease notation, we define

$$R = P(k + 1 \mid k) H^T \left(H P(k + 1 \mid k) H^T \right)^{-1}$$

so that the update equation can be written simply

$$\hat{x}(k + 1 \mid k + 1) = \hat{x}(k + 1 \mid k) + R v.$$

To find the update equation for the covariance matrix estimate, we use the definition of the covariance matrix together with the update equation for the state vector estimate to get

(8.15) $P(k + 1 \mid k + 1) = P(k + 1 \mid k) - R H P(k + 1 \mid k).$

The details of this derivation are the subject of problem 7.
Summarizing the observer derived in this section:

prediction:

(8.16) $\hat{x}(k + 1 \mid k) = F(k)\hat{x}(k \mid k) + G(k)u(k)$

(8.17) $P(k + 1 \mid k) = F(k)P(k \mid k)F(k)^T + V(k)$

update:

(8.18) $\hat{x}(k + 1 \mid k + 1) = \hat{x}(k + 1 \mid k) + R v$

(8.19) $P(k + 1 \mid k + 1) = P(k + 1 \mid k) - R H P(k + 1 \mid k)$

where

(8.20) $v = y(k + 1) - H x(k + 1 \mid k)$

(8.21) $R = P(k + 1 \mid k) H^T \left(H P(k + 1 \mid k) H^T \right)^{-1}$

and H is shorthand for $H(k + 1)$.

As in the case of the simple observer, this observer also has some problems. Because we assumed no sensor noise, the update equations will cause the covariance matrix estimate to become singular. This makes sense: noiseless measurements mean that the uncertainty in the directions associated with the sensor measurements will be zero. But the singular covariance makes the resulting notions of Gaussian distribution and Mahalanobis distance meaningless since they rely on the inverse of P. Still, the intuition behind using and propagating a Gaussian distribution as a state estimate is in line with the intuition behind the Kalman filter in spite of this problem. In the next section, we advance this intuition one final step to derive the full Kalman filter equations.

8.2.4   The Kalman Filter

Consider the system described at the beginning of this chapter:

$$(8.22) \qquad x(k+1) = F(k)x(k) + G(k)u(k) + v(k)$$

$$(8.23) \qquad y(k) = H(k)x(k) + w(k)$$

The only difference between this system and the system in the previous section is that we have included the sensor noise term $w(k)$, a zero-mean white Gaussian random vector with covariance matrix $W(k)$.

Since the dynamic equation has not changed, the prediction step for the Kalman filter is identical to the prediction step for the observer defined in section 8.2.3. The addition of noise to the sensor equation significantly changes the update step, however. In the previous case, the output $y(k+1)$ constrained the next estimate $\hat{x}(k+1 \,|\, k+1)$ to lie in the hyperplane Ω. We knew exactly what the output $y(k+1)$ had to be, and we chose $\hat{x}(k+1 \,|\, k+1)$ to match it. As a result, we could use the algebraic equation $y(k+1) = H(k+1)\hat{x}(k+1 \,|\, k+1)$ to find $\hat{x}(k+1 \,|\, k+1)$. In the current case, there is no such algebraic constraint. We do not know exactly what the output should be; we only know that it is drawn from a Gaussian distribution in $\mathbb{R}^p$ with mean $y(k+1)$ and covariance matrix $W(k)$. Without this constraint, we cannot use the same algebraic approach to define $\hat{x}(k+1 \,|\, k+1)$. Instead, we will first look for the most likely output y^* given the prediction $(\hat{x}(k+1 \,|\, k), P(k+1 \,|\, k))$ together with the measured output $y(k+1)$. Once we have y^*, we can introduce the algebraic constraint $y^* = H(k+1)\hat{x}(k+1 \,|\, k+1)$ and proceed as before.

We begin to find y^* by projecting the prediction into output space. Using the output map $H(k+1)$ and the definition of covariance, we see that the state space distribution with mean $\hat{x}(k+1 \,|\, k)$ and covariance matrix $P(k+1 \,|\, k)$ projects into a Gaussian distribution in the output space ($\mathbb{R}^p$) with mean

$$\hat{y}(k+1) = H(k+1)\hat{x}(k+1 \,|\, k)$$

and covariance matrix

$$\begin{aligned}
\widehat{W} &= E\big[(\hat{y}(k+1) - y(k+1))\,(\hat{y}(k+1) - y(k+1))^T\big] \\
&= E\big[H(k+1)\big(\hat{x}(k+1 \,|\, k) - x(k+1)\big)\big(\hat{x}(k+1) \\
&\qquad - x(k+1)\big)^T H(k+1)^T\big] \\
&= H(k+1)P(k+1 \,|\, k)H(k+1)^T.
\end{aligned}$$

The most likely output y^* is then defined to be the most likely point in the output space $\mathbb{R}^p$ given the Gaussian distribution that results from projecting the prediction and the Gaussian distribution that results from taking the measurement. The projected prediction and output distributions have mean-covariance pairs $(\hat{y}, \widehat{W})$ and $(y(k+1), W(k+1))$, respectively. Since these distributions are independent, y^* will be the peak of the function that results from taking their product. Fortunately the product of two Gaussian distributions is also Gaussian and the result can be obtained using a well-known formula [389]. We summarize the required result as a theorem:

THEOREM 8.2.1 (Product of Gaussians)   *The product of two Gaussian distributions with mean-covariance pairs (z_1, C_1) and (z_2, C_2) is proportional to a third Gaussian with mean vector*

$$z_3 = z_1 + Q(z_2 - z_1)$$

and covariance matrix

$$C_3 = C_1 - QC_1,$$

where

$$Q = C_1(C_1 + C_2)^{-1}.$$

To sketch the proof of this theorem, we use the property that the product of two exponentials is the exponential of the sum of the exponents. A clever reordering of the terms in the resulting sum yields the result. See problem 8 for the details of the proof.

Applying theorem 8.2.1,

$$y^* = \hat{y} + \widehat{W}(\widehat{W} + W(k+1))^{-1}(\hat{y} - y(k+1)).$$

Now that we have found the most likely output y^*, we can define $\Omega^* = \{x \in \mathbb{R}^n \,|\, H(k+1)x = y^*\}$ and, as in the previous section, proceed to find the $\Delta x = \hat{x}(k+1\,|\,k+1) - \hat{x}(k+1\,|\,k)$ that minimizes $\|\Delta x\|_M$ while satisfying $\hat{x}(k+1\,|\,k+1) \in \Omega^*$ (see figure 8.4). Using H to denote $H(k+1)$, we get the result

$$
\begin{aligned}
\Delta x &= P(k+1\,|\,k)H^T \left(HP(k+1\,|\,k)H^T \right)^{-1} \\
&\quad (y^* - Hx(k+1\,|\,k)) \\
&= P(k+1\,|\,k)H^T \left(HP(k+1\,|\,k)H^T + W(k) \right)^{-1} v,
\end{aligned}
$$

where, as before, $v = y(k+1) - Hx(k+1\,|\,k)$ is the innovation error. Defining

$$R = P(k+1\,|\,k)H^T \left(HP(k+1\,|\,k)H^T + W(k+1) \right)^{-1},$$

we can write the state vector estimate update equation

(8.24) $\hat{x}(k+1\,|\,k+1) = \hat{x}(k+1\,|\,k) + Rv.$

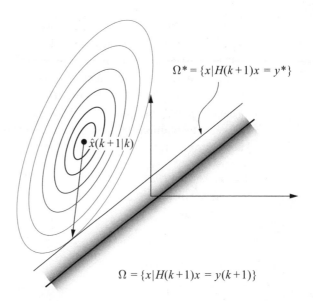

$\Omega^* = \{x \mid H(k+1)x = y^*\}$

$\hat{x}(k+1 \mid k)$

$\Omega = \{x \mid H(k+1)x = y(k+1)\}$

Figure 8.4   The sensor noise distribution is projected into the state space and is an extruded Gaussian centered on the states consistent with the current sensor reading. The most likely output y^* is determined by multiplying the Gaussian distribution that results from the measurement $y(k+1)$ with the Gaussian distribution that results from projecting the prediction into the output space. This then corresponds to a set of states Ω^*. The update is the point on Ω^* that is closest to the prediction $\hat{x}(k+1 \mid k)$ in the sense of Mahalanobis distance.

To find the update equation for the covariance matrix estimate, we use the definition of the covariance matrix together with the update equation for the state vector estimate to get

$$(8.25) \qquad P(k+1 \mid k+1) = P(k+1 \mid k) - RHP(k+1 \mid k).$$

8.2.5   Kalman Filter Summary

The Kalman filter equations are summarized as follows:
prediction:

$$(8.26) \qquad \hat{x}(k+1 \mid k) = F(k)\hat{x}(k \mid k) + G(k)u(k)$$

$$(8.27) \qquad P(k+1 \mid k) = F(k)P(k \mid k)F(k)^T + V(k)$$

update:

(8.28) $\hat{x}(k+1\,|\,k+1) = \hat{x}(k+1\,|\,k) + Rv$

(8.29) $P(k+1\,|\,k+1) = P(k+1\,|\,k) - RH(k+1)P(k+1\,|\,k)$

where

(8.30) $v = y(k+1) - H(k+1)x(k+1\,|\,k)$

(8.31) $S = H(k+1)P(k+1\,|\,k)H(k+1)^T + W(k+1)$

(8.32) $R = P(k+1\,|\,k)H(k+1)^T S^{-1}.$

These equations provide the optimal estimate of x in the sense that the expected value of the error between $x(k)$ and $\hat{x}(k\,|\,k)$ is minimized at every k. One can view R as the weighting factor that takes into account the relationship between the accuracy of the predicted estimate and the measurement noise. If R is "large," then the sensor readings are more believable than the prediction and the Kalman filter weights the sensor reading highly when computing the updated estimate. If R is "small," then the sensor readings are not as believable and, as a result, they do not have as much influence in the update step.

In this chapter we have chosen to present the derivation of the Kalman filter equations as an optimization problem because we believe that to be an intuitive approach. It is important to note, however, that the state and covariance estimates that result from the use of these equations are not only the "best" estimates, they are also the "correct" estimates. If the estimate at time k is Gaussian and described by $(\hat{x}(k\,|\,k), P(k\,|\,k))$, then the correct distribution at time $k+1$ (i.e., the *posterior* distribution) is in fact also Gaussian and is described by $(\hat{x}(k+1\,|\,k+1), P(k+1\,|\,k+1))$.

If we allow the noise terms $v(k)$ and $w(k)$ to have non-Gaussian distributions, then these equations still provide the best *linear* estimator, but there may be nonlinear estimators that do a better job.

8.2.6   Example: Kalman Filter for Dead Reckoning

In mobile robotics, the term *dead reckoning* typically refers to a position estimate achieved by integrating odometry measurements. Here we present an example of a more sophisticated form of dead reckoning where a Kalman filter is used to fuse the robot commands (inputs) with measurements from odometry sensors.

Consider a mobile robot constrained to move along a straight line. The robot state is defined to be $x = [x_r, \ v_r]^T$ where x_r and v_r are the robot position and velocity, respectively. The input u is a real–valued force applied to the robot. Newton's law states that $\frac{dv_r}{dt} = \frac{u}{m}$, where m is the mass of the robot. This can be approximated by the discrete time equation

$$\frac{v_r(k+1) - v_r(k)}{T} = \frac{u(k)}{m},$$

where T is the sampling rate (in seconds) of the discretization. So then the discrete time state equation can be written as

(8.33)
$$x(k+1) = \begin{bmatrix} 1 & T \\ 0 & 1 \end{bmatrix} x(k) + \begin{bmatrix} 0 \\ \frac{T}{m} \end{bmatrix} u(k) + v(k)$$
$$\stackrel{\triangle}{=} Fx(k) + Gu(k) + v(k),$$

where the process noise term $v(k)$ is used to account for errors that arise from unmodeled sources such as discretization and friction. The vector $v(k)$ is assumed to be zero-mean white Gaussian noise with covariance matrix V.

We assume that the robot is equipped with a sensor that measures velocity. We also assume that the error in this measurement is well modeled as zero-mean white Gaussian noise with known variance W. Then the output $y(k)$ can be written

(8.34)
$$y(k+1) = \begin{bmatrix} 0 & 1 \end{bmatrix} x(k) + w(k)$$
$$\stackrel{\triangle}{=} Hx(k) + w(k),$$

where w is the noise term.

Now the Kalman filter can be applied using the sequence of equations listed in section 8.2.5. We simulated this example in MATLAB using the parameters $m = 1$, $W = .5$, $T = 0.5$, and

$$V = \begin{bmatrix} 0.2 & 0.05 \\ 0.05 & 0.1 \end{bmatrix}.$$

Assume that the input at time k is known to be $u(k) = 0$, and assume an initial state estimate of $\hat{x}(k \mid k) = [2, \ 4]^T$ and an initial covariance estimate of

$$P(k \mid k) = \begin{bmatrix} 1 & 0 \\ 0 & 2 \end{bmatrix}.$$

Further, assume that the (unknown) value of the actual state is $x(k+1) = [1.8, \ 2]^T$. The sequence of prediction, combining prediction with measurement in output space, and update are depicted graphically in figures 8.5, 8.6, and 8.7. Here, the two-dimensional Gaussian distributions that result from $\hat{x}$ and P are represented by confidence ellipses. Specifically, these ellipses are chosen so that the probability that the actual value of the state x is contained within the ellipse is 0.95.

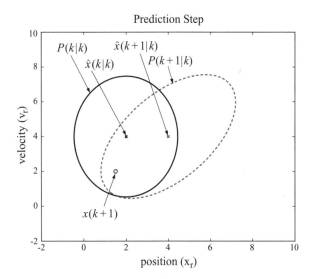

Figure 8.5   The initial estimate $\hat{x}(k \mid k)$ has an uncertainty $P(k \mid k)$ which grows to $P(k+1 \mid k)$ when the robot moves, reflecting the increase in uncertainty. This increase in uncertainty is depicted by plotting the 0.95 confidence ellipses.

8.2.7   Observability in Linear Systems

Note that in the update step of the previous example, the updated ellipse is "squished" significantly in the vertical direction. This squishing corresponds to the information gained from the velocity measurement. For this particular example, each iteration of the Kalman filter will reflect a gain of information in the velocity direction and a loss of information in the position direction. As a result, the expected error on the position estimate will grow monotonically without bound. This failure is not the fault of the Kalman filter, which is guaranteed to provide the best possible estimate. The problem instead lies with the system itself; specifically, the system dynamics and output equations do not interact in a way that allows the state to be recovered from the available outputs. In other words, the system in the example fails to be observable.

Loosely speaking, a system is said to be *observable* if the full state can be reconstructed by observing the input u and the output y over some period of time (see appendix J for a discussion of observability in linear systems.) For linear systems where the system matrices $F(k)$ and $H(k)$ do not vary with k, there is a simple test to determine observability:

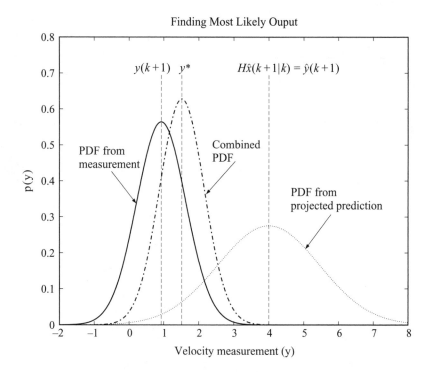

Figure 8.6 Measurements and predictions are then merged. The PDF plotted with the dot-dashed line results from the combination of the measurement PDF and the PDF of the prediction projected into output space, where the combination is computed using theorem 8.2.1. The most likely output y^* is the value at which this combined distribution reaches its peak.

THEOREM 8.2.2 (Observability Test)   *The linear time-invariant discrete time system*

$$x(k + 1) = Fx(k) + Gu(k) + v(k)$$
$$y(k) = Hx(k) + w(k)$$

is observable if and only if the observability matrix

$$Q = \begin{bmatrix} H \\ HF \\ HF^2 \\ \vdots \\ HF^{(n-1)} \end{bmatrix}$$

has rank n.

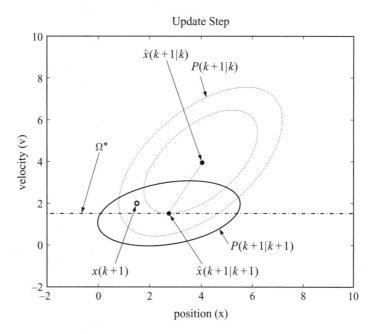

Figure 8.7 The updated estimate $\hat{x}(k+1\,|\,k+1)$ is the point on Ω^* that is closest to $\hat{x}(k+1\,|\,k)$ in terms of Mahalanobis distance. The smaller of the two dotted ellipses is the smallest ellipse that intersects Ω^*, hence $\hat{x}(k+1\,|\,k+1)$ is defined by this intersection. Note that the line Ω^* represents the set of states that are consistent with the most likely output y^* that was found in figure 8.6.

For any observable linear system, the estimate provided by the Kalman filter is guaranteed to converge in the sense that the expected error between the actual and estimated state will be bounded for all time.

8.3 Extended Kalman Filter

The Kalman filter is a powerful tool for linear systems, but many systems encountered in practice are nonlinear. Consider the system

$$(8.35) \quad x(k+1) = f(x(k), u(k), k) + v(k)$$

$$(8.36) \quad y(k) = h(x(k), k) + w(k),$$

where x, y, u, v, and w are as before and

$$f : \mathbb{R}^n \times \mathbb{R}^m \times \mathbb{Z}^+ \to \mathbb{R}^n$$

and

$$h : \mathbb{R}^n \times \mathbb{Z}^+ \rightarrow \mathbb{R}^p$$

are both continuously differentiable in $x(k)$. One approach to state estimation for systems of this type is to linearize the equations about the current estimate and then apply Kalman's equations using the resulting approximation. This formulation is called the extended Kalman filtering. The EKF equations are:

  prediction:

(8.37) $\hat{x}(k+1 \,|\, k) = f(\hat{x}(k \,|\, k), u(k), k)$

(8.38) $P(k+1 \,|\, k) = F(k)P(k \,|\, k)F(k)^T + V(k)$

  where

(8.39) $F(k) = \left. \dfrac{\partial f}{\partial x} \right|_{x=\hat{x}(k \,|\, k)} = \begin{bmatrix} \frac{\partial f_1}{\partial x_1} & \frac{\partial f_1}{\partial x_2} & \cdots & \frac{\partial f_1}{\partial x_n} \\ \frac{\partial f_2}{\partial x_1} & \frac{\partial f_2}{\partial x_2} & \cdots & \frac{\partial f_2}{\partial x_n} \\ \vdots & \vdots & \ddots & \vdots \\ \frac{\partial f_n}{\partial x_1} & \frac{\partial f_n}{\partial x_2} & \cdots & \frac{\partial f_n}{\partial x_n} \end{bmatrix}_{x=\hat{x}(k \,|\, k)}.$

  update:

(8.40) $\hat{x}(k+1 \,|\, k+1) = \hat{x}(k+1 \,|\, k) + Rv$

(8.41) $P(k+1 \,|\, k+1) = P(k+1 \,|\, k) - RH(k+1)P(k+1 \,|\, k)$

  where

(8.42) $v = y(k+1) - h(x(k+1 \,|\, k), k+1)$

(8.43) $S = H(k+1)P(k+1 \,|\, k)H(k+1)^T + W(k+1)$

(8.44) $R = P(k+1 \,|\, k)H(k+1)^T S^{-1}$

and

(8.45) $H(k+1) = \left. \dfrac{\partial h}{\partial x} \right|_{x=\hat{x}(k+1 \,|\, k)} = \begin{bmatrix} \frac{\partial h_1}{\partial x_1} & \frac{\partial h_1}{\partial x_2} & \cdots & \frac{\partial h_1}{\partial x_n} \\ \frac{\partial h_2}{\partial x_1} & \frac{\partial h_2}{\partial x_2} & \cdots & \frac{\partial h_2}{\partial x_n} \\ \vdots & \vdots & \ddots & \vdots \\ \frac{\partial h_p}{\partial x_1} & \frac{\partial h_p}{\partial x_2} & \cdots & \frac{\partial h_p}{\partial x_n} \end{bmatrix}_{x=\hat{x}(k+1 \,|\, k)}.$

8.3.1   EKF for Range and Bearing Localization

The EKF is well suited to the problem of localizing a mobile robot equipped with sensors that can detect range and bearing to previously mapped landmarks in the environment [278]. Consider a robot whose state at time k is given by $x(k) = [x_r(k), y_r(k), \theta_r(k)]^T$, where $(x_r(k), y_r(k))$ denotes its position in the plane and $\theta(k)$ denotes its orientation. The input is $u(k) = [u_1(k), u_2(k)]^T$, where $u_1(k)$ and $u_2(k)$ denote the forward and angular velocities of the robot, respectively. The process model for this robot is nonlinear, i.e.,

$$x(k+1) = \begin{bmatrix} \cos\theta_r(k)u_1(k) + x_r(k) \\ \sin\theta_r(k)u_1(k) + y_r(k) \\ u_2(k) + \theta_r(k) \end{bmatrix} + v(k),$$

where $v(k)$ is a random vector from a Gaussian distribution whose mean is zero and covariance is $V(k)$.

The robot is equipped with sensors that can measure the range and bearing to certain landmarks in the environment. Assume that the free space is populated with n_ℓ landmarks whose locations are known to be $(x_{\ell i}, y_{\ell i})$, $i = 1, 2, \ldots, n_\ell$. At any time k, the robot can only see the subset of landmarks that is within the range of its sensors, so the number of measurements taken varies with k. The number of measurements taken at the kth timestep is denoted by $p(k)$. Each measurement has two components, a range component and a bearing component. We also assume for now that for each measurement, we somehow know which landmark was observed. We introduce the association map $a: \{1, 2, \ldots, p(k)\} \to \{1, 2, \ldots, n_\ell\}$ which is defined such that the ith measurement at time k corresponds to the $a(i)$th landmark. The output equation for this system is then given as

$$y(k) = \begin{bmatrix} h_1(x(k), a(1)) \\ h_2(x(k), a(2)) \\ \vdots \\ h_{p(k)}(x(k), a(p(k))) \end{bmatrix} + \begin{bmatrix} w_1(k) \\ w_2(k) \\ \vdots \\ w_{p(k)}(k) \end{bmatrix},$$

where, for $i = 1, 2, \ldots, p(k)$,

$$h_j(x(k), j) = \begin{bmatrix} \sqrt{(x_r(k) - x_{\ell j})^2 + (y_r(k) - y_{\ell j})^2} \\ \mathrm{atan2}(y_r(k) - y_{\ell j}, x_r(k) - x_{\ell j}) - \theta_r(k) \end{bmatrix}$$

and $w_i(k) \in \mathbb{R}^2$ is a random vector taken from a Gaussian distribution with zero mean and covariance matrix $W_i(k)$.

In order to linearize, we differentiate the process and sensor models with

$$F(k) = \left. \frac{\partial f}{\partial x} \right|_{x=\hat{x}(k|k)}$$

and

$$H(k+1) = \begin{bmatrix} H_1(k+1, a(1)) \\ H_2(k+1, a(2)) \\ \vdots \\ H_{p(k+1)}(k+1, a(p(k+1))) \end{bmatrix} = \begin{bmatrix} \frac{\partial h_1}{\partial x}\Big|_{x=\hat{x}(k+1|k)} \\ \frac{\partial h_2}{\partial x}\Big|_{x=\hat{x}(k+1|k)} \\ \vdots \\ \frac{\partial h_{p(k+1)}}{\partial x}\Big|_{x=\hat{x}(k+1|k)} \end{bmatrix}$$

resulting in

$$F = \begin{bmatrix} 1 & 0 & -\sin\theta_r(k)u_1(k) \\ 0 & 1 & \cos\theta_r(k)u_1(k) \\ 0 & 0 & 1 \end{bmatrix}$$

and

$$H_i(k+1, j) =$$

$$\begin{bmatrix} \dfrac{(\hat{x}_r(k+1|k)-x_{\ell j})}{\sqrt{(\hat{x}_r(k+1|k)-x_{\ell j})^2+(\hat{y}_r(k+1|k)-y_{\ell j})^2}} & \dfrac{(\hat{y}_r(k+1|k)-y_{\ell j})}{\sqrt{(\hat{x}_r(k+1|k)-x_{\ell j})^2+(\hat{y}_r(k+1|k)-y_{\ell j})^2}} & 0 \\[3ex] \dfrac{-(\hat{y}_r(k+1|k)-y_{\ell j})}{1+\left(\frac{\hat{y}_r(k+1|k)-y_{\ell j}}{\hat{x}_r(k+1|k)-x_{\ell j}}\right)^2 (\hat{x}_r(k+1|k)-x_{\ell j})^2} & \dfrac{1}{1+\left(\frac{\hat{y}_r(k+1|k)-y_{\ell j}}{\hat{x}_r(k+1|k)-x_{\ell j}}\right)^2 (\hat{x}_r(k+1|k)-x_{\ell j})} & -1 \end{bmatrix}.$$

With these matrices in hand, we can use the linearized Kalman filter equations to estimate the robot state.

8.3.2   Data Association

The solution presented in the previous section glosses over one very important aspect of localization: it assumes that each measurement is automatically associated with the correct landmark. In practice, landmarks have similar properties which make them good features but often make them difficult to distinguish one from another. When this happens, we must address the problem of *data association,* which is the question of which landmark corresponds to a particular measurement. This is equivalent to finding the association map used in the previous section.

The basic idea used for data association is as follows. Consider the ith measurement $y_i(k+1)$. For each landmark in the map, we compute the innovation v_{ij} which is defined to be the difference between the actual measurement $y_i(k+1)$ and the measurement that we would expect if $y_i(k+1)$ corresponded to the jth landmark and the prediction $\hat{x}(k+1|k)$ was correct. This means that

$$v_{ij} \overset{\triangle}{=} y_i(k+1) - h_i(\hat{x}(k+1|k), j).$$

The smaller the innovation v_{ij}, the more likely it is that the ith measurement corresponds to the jth landmark. We then can make a good guess of which landmark corresponds to the measurement by choosing the landmark that yields the smallest innovation. However, the notion of size must be weighted by the uncertainties in the predictions and measurements. Fortunately, these uncertainties are encoded in the matrix S from the Kalman filter update equation (8.31), and S can be used to create a Mahalanobis norm for v_{ij} which indicates the size of the innovation in units of standard deviations. We write this measure of the innovation as

$$\chi_{ij}{}^2 = v_{ij}^T S_{ij}^{-1} v_{ij},$$

where

$$S_{ij} = H_i(k+1, j)P(k+1 \mid k)H_i(k+1, j)^T + W_i(k+1).$$

We then can build the data association function $a()$ by setting $a(i)$ equal to the value of j that minimizes χ_{ij}.

Figure 8.8 contains an example of localization using an extended Kalman filter. The actual path, i.e., ground truth, is displayed along with estimates of the robot's location and its uncertainty of that estimate as the robot moves along the path. Note that the estimate converges to the actual path as the robot moves along the path and as more measurements are taken. Also, note that the uncertainty of the estimate considerably decreases as well.

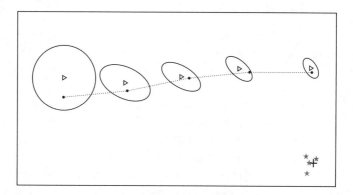

Figure 8.8 The dotted line displays the actual path of the Nomad Scout mobile robot. The triangles and ellipses correspond to the estimated location and variance, respectively, of the robot's location. The stars correspond to the measured location of the beacon, which also has an associated variance (due to noise) which is not displayed.

8.3.3   EKF for Range-Only Localization

The EKF solution to the problem of localization using range-only sensors is a trivial extension to the range and bearing case. The only difference is that the output equation will not contain any bearing information, so we simply remove the rows from h and H that correspond to the bearing measurements. The EKF equations are then applied in the usual manner.

8.4   Kalman Filter for SLAM

In this section we introduce the use of the Kalman filter to solve the problem SLAM which has been an active topic of research in recent years (see, e.g., [99, 128, 321, 390]). We begin with a very simple case where the robot is able to measure the relative displacement between itself and a number of fixed landmarks. The simple example also assumes that each sensor reading is automatically associated with the correct landmark so that the robot does not have to determine which landmark corresponds to any given measurement. After using this simple example to demonstrate the basic concept of Kalman filter-based SLAM, we present a more realistic example where the robot measures range and bearing to fixed landmarks and the data association problem is not automatically solved.

8.4.1   Simple SLAM

One common approach to solving the SLAM problem is to use a Kalman filter to simultaneously estimate the position of a moving vehicle along with the positions of landmarks seen by the vehicle. This technique was originally suggested by Smith, Self, and Cheeseman [390]. Here we present the most basic example of this technique: we assume an omnidirectional motion model for the vehicle and we assume that the vehicle has sensors capable of uniquely identifying each landmark and providing a measurement of the relative displacement between the vehicle and the landmark. We assume that the vehicle's sensor can see every landmark at every instant of time.

We first define the state to be the location of the vehicle (x_r, y_r) together with the locations of each of the landmarks, $(x_{\ell i}, y_{\ell i})$, $i = 1, 2, \ldots, n_\ell$, where n_ℓ is the total number of landmarks. In other words,

$$x = [x_r \quad y_r \quad x_{\ell 1} \quad y_{\ell 1} \quad x_{\ell 2} \quad y_{\ell 2} \quad \cdots \quad x_{\ell n_\ell} \quad y_{\ell n_\ell}]^T.$$

We assume that the control inputs are u_x and u_y, the vehicle velocities in the x- and y-directions, respectively. We model the errors associated with this motion with the random vector $v_r(k) = [v_{rx}(k), v_{ry}(k)]^T$, which is zero-mean white Gaussian noise with covariance matrix $V_r(k)$. The landmarks do not move, so the resulting dynamic equations for the system are

$$x(k+1) = x(k) + \begin{bmatrix} 1 & 0 \\ 0 & 1 \\ 0 & 0 \\ 0 & 0 \\ \vdots & \vdots \\ 0 & 0 \end{bmatrix} \begin{bmatrix} u_x(k) \\ u_y(k) \end{bmatrix} + \begin{bmatrix} v_{rx}(k) \\ v_{ry}(k) \\ 0 \\ 0 \\ \vdots \\ 0 \end{bmatrix}.$$

This equation can clearly be written in the form

$$x(k+1) = Fx(k) + Gu(k) + v(k),$$

where $v(k)$ is a zero-mean white Gaussian noise with covariance matrix

$$V(k) = \begin{bmatrix} V_r(k) & 0 & \cdots & 0 \\ 0 & 0 & \cdots & 0 \\ \vdots & \vdots & \ddots & \vdots \\ 0 & 0 & \cdots & 0 \end{bmatrix}.$$

The measurement to the ith landmark is the position of the landmark relative to the vehicle plus some noise, i.e.,

$$y_i(k) = \begin{bmatrix} x_{\ell i}(k) - x_r(k) \\ y_{\ell i}(k) - y_r(k) \end{bmatrix} + w_i(k),$$

where $w_i(k)$ is an independently distributed Gaussian random vector with covariance matrix $W_i(k)$. Note the $y_i(k)$ is a linear function of the system state $x(k)$. Specifically, we can write

$$y_i(k) = H_i x(k) + w_i(k),$$

where

$$H_i = \begin{bmatrix} -1 & 0 & 0 & \cdots & 0 & 1 & 0 & 0 & \cdots & 0 \\ 0 & -1 & 0 & \cdots & 0 & 0 & 1 & 0 & \cdots & 0 \end{bmatrix}.$$

The first row of H has a -1 in the first column that to corresponds x_r and a 1 in the $(2i+1)$th column that corresponds to $x_{\ell i}$, and zeros everywhere else. Similarly, the second row is all zeros except for a -1 in the second column and a 1 in the $(2i+2)$th column.

With this notation, we can stack all of the measurements together to create one big measurement vector $y = [y_1, y_2, \ldots, y_{n_\ell}]^T$ which gives the measurement equation

$$y(k) = Hx(k) + w(k),$$

where

$$H = \begin{bmatrix} H_1 \\ H_2 \\ \vdots \\ H_{n_\ell} \end{bmatrix}, \quad \text{and} \quad w(k) = \begin{bmatrix} w_1(k) \\ w_2(k) \\ \vdots \\ w_{n_\ell}(k) \end{bmatrix},$$

and the covariance matrix associated with $w(k)$ is

$$W(k) = \begin{bmatrix} W_1(k) & 0 & \cdots & 0 \\ 0 & W_2(k) & \ddots & \vdots \\ \vdots & \ddots & \ddots & 0 \\ 0 & \cdots & 0 & W_{n_\ell} \end{bmatrix},$$

where $W_i(k)$ is the covariance matrix associated with $w_i(k)$. The problem has now been put into a form suitable for the Kalman filtering equations in section 8.2.5. Kalman estimates of the system state x provide estimates of both vehicle and landmark locations, hence solving the SLAM problem.

8.4.2   Range and Bearing SLAM

Now we consider the SLAM problem for a mobile robot whose inputs are forward velocity and angular velocity and whose measurements are range and bearing readings. In a sense, we are combining the range-bearing localization approach from section 8.3.1 with the SLAM approach described above in section 8.4.1. The difference is that the number of columns in the H matrix is the same as the number of rows in the state vector. Moreover, the H matrix now contains partial derivatives of the measurement equations with respect to the state.

The measurement equations are the same as in the range and bearing localization example, i.e.,

(8.46) $$y_i(k) = \begin{bmatrix} \sqrt{(x_{\ell i}(k) - x_r(k))^2 + (y_{\ell i}(k) - y_r(k))^2} \\ \text{atan2}((y_{\ell i}(k) - y_r(k)), (x_{\ell i}(k) - x_r(k)) - \theta_r(k)) \end{bmatrix} + w_i(k).$$

The first three columns of the H matrix will be fairly dense since the planar location of the robot is part of both measurement equations. The columns to the right will be sparse as in the last example of EKF SLAM since the measurement of each landmark is only a function of the robot position and that landmark's position.

$$
(8.47) \quad H_i = \begin{bmatrix} \frac{\partial y_i}{\partial x_r} \\ \frac{\partial y_i}{\partial y_r} \end{bmatrix}
$$

$$
= \begin{bmatrix} \frac{-x_{\ell i}(k)+x_r(k)}{\rho_i} & \frac{-y_{\ell i}(k)+y_r(k)}{\rho_i} & 0 & \cdots & \frac{x_{\ell i}(k)-x_r(k)}{\rho_i} & \frac{y_{\ell i}(k)-y_r(k)}{\rho_i} & \cdots \\ \frac{y_{\ell i}(k)-y_r(k)}{\rho_i^2} & \frac{-x_{\ell i}(k)+x_r(k)}{\rho_i^2} & -1 & \cdots & \frac{-y_{\ell i}(k)+y_r(k)}{\rho_i^2} & \frac{x_{\ell i}(k)-x_r(k)}{\rho_i^2} & \cdots \end{bmatrix}
$$

where ρ_i is the range of the landmark as given in the measurement equation. Now, we substitute the modified H matrix into the previously defined framework for Kalman filter SLAM.

Again, we have the problem of data association, i.e., we must determine which landmark corresponds to each measurement. We also have to determine when a new landmark has been encountered. Once again, we use the Mahalanobis distance metric to compare the ith measurement with the measurement prediction for the jth landmark, i.e.,

$$
(8.48) \quad \chi_{ij}^2 = (y_i(k) - h(k)_j)^T S_{ij}(y_i(k) - h(k)_j).
$$

Once the χ_{ij} has been calculated for each combination of landmarks and measurements, the minimum is checked against an acceptance threshold to assure that the match is likely enough. If the minimum χ is above a high threshold, then the measurement is not likely to have come from any existing landmark. Therefore, we have an indication that a new landmark should be initialized and added to the map.

Problems

1. The methods presented in this chapter generally assume that the noise is modeled as a zero-mean white Gaussian random process and that the noise enters the system dynamics and measurement equations through addition. It is important to understand that this is always an approximation; real systems never contain such nice noise. For each of the noise sources listed below, briefly describe how the zero-mean white Gaussian noise assumption used in the Kalman filter fails:

   (a) distance measurements;
   (b) bearing measurements;
   (c) odometry error in a differentially steered wheeled robot due to a mismatch in wheel size;
   (d) odometry error in a wheeled robot due to wheel slippage;
   (e) sonar errors due to multipath reflections;
   (f) temperature dependent drift in a rate gyro.

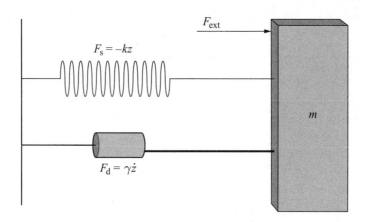

Figure 8.9   Mass–spring–damper.

2. The mass–spring–damper system shown in figure 8.9 with mass m, spring constant k, and damping coefficient γ can be modeled by the second order differential equation

    $$m\ddot{z} + \gamma\dot{z} + kz = 0.$$

    Define the system state to be $x = [z \ \dot{z}]^T$. In discrete time with sampling time step T, the derivative $\frac{d}{dt}$ can be approximated as

    $$\dot{x}(t)\big|_{t=kT} \approx \frac{x(k+1) - x(k)}{T}.$$

    Use this approximation to write the mass–spring–damper system as a linear discrete time system in state space.

3. Consider the linear time invariant, noise-free, zero-input, single-output discrete time system

    $$x(k+1) = Fx(k); \quad y(k) = Hx(k),$$

    where $x(k) \in \mathbb{R}^2$.

    (a) Describe the set of states $x(k)$ that are possible given the measurement $y(k)$.

    (b) Describe the set of states $x(k+1)$ that are possible given the measurement $y(k)$ (but not $y(k+1)$).

    (c) Given the measurement $y(k+1)$, under what condition will the set of possible $x(k+1)$ be a single point? How does this relate to the result in theorem 8.2.2?

4. Let $\hat{x} \in \mathbb{R}^n$, $y \in \mathbb{R}^p$, and $H \in \mathbb{R}^{p \times n}$. Define the hyperplane $\Omega = \{x \in \mathbb{R}^n \mid Hx = y\}$. Let P be a positive definite matrix and define the norm $\|x\|_P = x^T P x$. Show that, with respect to this norm, the shortest vector Δx that satisfies $H(x + \Delta x) = y$ must be orthogonal to Ω, i.e., $\Delta x^T P a$ must be zero for every a that is parallel to Ω.

5. Using the definitions from problem4, show that null(H) is parallel to Ω.

6. Show that for a matrix A, null(A) is orthogonal to column(A^T).

7. The solutions to the following sequence of problems combine to form the derivation of equation (8.15). To make the expressions more manageable, we introduce the notation $x_k = x(k)$, $\hat{x}_k = \hat{x}(k \mid k)$, $\hat{x}^-_{k+1} = \hat{x}(k+1 \mid k)$, $P_k = P(k \mid k)$, and $P^-_{k+1} = P(k+1 \mid k)$. We also denote the innovation as $\nu = \nu(k+1) = y(k+1) - H(k)x_k$.

   (a) Starting with the definition of P_k,

   $$P_k = E\left[(x_{k+1} - \hat{x}_{k+1})(x_{k+1} - \hat{x}_{k+1})^T\right],$$

   show that

   $$P_k = E\left[(x_{k+1} - \hat{x}^-_{k+1})(x_{k+1} - \hat{x}^-_{k+1})^T \right.$$
   $$\left. - 2R\nu(x_{k+1} - \hat{x}^-_{k+1})^T - R\nu(R\nu)^T\right].$$

   (b) Continue to show that

   $$P_k = P^-_{k+1} - 2RHP^-_{k+1} + RHP^-_{k+1}H^T R^T.$$

   (c) Next show that

   $$P_k = P^-_{k+1} - 2RHP^-_{k+1}H^T R^T + RHP^-_{k+1}H^T R^T.$$

   Equation (8.15) follows trivially from this last expression.

8. The solutions to the following sequence of problems combine to form the proof of theorem 8.2.1. Consider two multivariate Gaussian distributions with mean-covariance pairs (z_1, C_1) and (z_2, C_2).

   (a) Show that the product of these two Gaussian distributions is proportional to

   $$e^{-\frac{1}{2}(z^T(C_1^{-1}+C_2^{-1})z - 2z^T(C_1^{-1}z_1+C_2^{-1}z_2) + z_1^T C_1^{-1}z_1 + z_2^T C_2^{-1}z_2)}.$$

   (b) Consider the term in the exponential of a Gaussian with mean-covariance pair (z_3, C_3). By equating the terms that are quadratic in z, show that

   $$C_3 = C_1 - QC_1,$$

   where $Q = C_1(C_1 + C_2)^{-1}$.

   (c) By equating the terms that are linear in z, show that

   $$z_3 = z_1 + Q(z_2 - z_1).$$

9. Consider the nonlinear discrete time system

$$x(k+1) = \begin{bmatrix} x_1(k) + Tu_1(k)cos(x_3(k)) \\ x_2(k) + Tu_1(k)sin(x_3(k)) \\ x_3(k) + Tu_2(k) \end{bmatrix}; \qquad y(k) = x_1(k) + w(k),$$

where $v(k)$ is Gaussian white noise with zero mean and variance 0.5. Suppose the estimate $\hat{x}(1|1) = [1\,0.5\,\frac{\pi}{4}]^T$, the input $u(1) = [3\,\pi]$, and the time step $T = 0.25$. Also suppose that the covariance estimate $P(1|1)$ is the 3×3 identity matrix. Using the extended Kalman filter formulation,

(a) compute the predicted estimate and covariance, $\hat{x}(2|1)$ and $P(2|1)$.

(b) given the measurement $y(2) = 1.7$, compute $\hat{x}(2|2)$ and $P(2|2)$.

10. Consider the system of problem 9 with noise added to the inputs: $u_1(k)$ is replaced by $u_1(k) + s_1(k)$ and $u_2(k)$ is replaced by $u_2(k) + s_2(k)$, where s_1 and s_2 are Gaussian white noise with variances σ_1^2 and σ_2^2 respectively. Given an estimate $\hat{x}(k\,|\,k)$, state the equation used to find the estimate covariance prediction $P(k+1\,|\,k)$.

11. Consider the system given by equations (8.1) and (8.2) with all of the standard assumptions. Show that if the initial estimate $\hat{x}(0|0)$ is such that the expected value of the intitial estimate error $E[x(0) - \hat{x}(0|0)] = 0$, then the expected value of the error of the estimate provided by the Kalman filter remains zero for all k.

9 *Bayesian Methods*

OPERATING IN the real world, robots lack the perfect sensors and deterministic actions of many artificial worlds. Rather, robots are faced with various kinds of uncertainty. In this chapter we continue to discuss probabilistic frameworks for typical fundamental tasks of mobile robots such as localization, mapping, and simultaneous localization and mapping (SLAM). While the methods presented in this chapter employ the same iterative prediction-update process that is used in the Kalman filter (see chapter 8), they do not rely on the restrictive assumptions required by the Kalman filter. The methods described here can use nonlinear models for both robot motion and sensing. Most important, the resulting estimate may be an arbitrary distribution instead of a Gaussian. Throughout this chapter we present the key ideas of successful techniques together with a derivation of their mathematical foundations. We will also discuss ways to efficiently implement these approaches, since the capability to represent arbitrary distributions can lead to higher computational demands compared to Kalman filters.

9.1 Localization

In the previous chapter, we achieved localization by maintaining a distribution of the robot by iteratively estimating the mean and covariance matrix of a Gaussian distribution. This way of representing a belief about the location of the robot assumes that there is no "ambiguity" in the sense that the distribution is always unimodal or more specifically a Gaussian. One form of localization in which this assumption is often met is *position tracking,* which assumes that the initial configuration of the robot is (approximately) known and whose task is to keep track of the robot's location while it is moving through the environment. If the robot's configuration is approximately known and if there is only a small region of uncertainty around the true location of the robot, the observations of the robot can usually be associated uniquely with the corresponding features in its map. Consider, e.g., that a robot knows its location up to a few centimeters. If it detects a door, it can use this observation to accurately compute its location given the door stored in its map of the environment. If, however, the uncertainty is high and the robot knows its location only up to several meters, there might be multiple doors in the map that its current observation can correspond to. Accordingly, the situation is ambiguous and a single Gaussian obviously cannot appropriately represent the robot's belief about its location.

In this chapter, we consider a form of position estimation where the robot may have ambiguity, i.e., the belief about its location can be modeled by a multimodal distribution. The techniques described in this chapter are able to deal with a more complex version of localization called *global localization*. Here the robot has to estimate its location under global uncertainty as it is not given its initial location. The techniques can also solve the most complex problem of robot localization, the so-called *kidnapped robot problem*. The kidnapped robot problem, or the relocalization problem, is more complicated than the global localization problem because the robot has generated a false belief of its most likely location which it must identify and "unlearn" before it can relocalize.

9.1.1   The Basic Idea of Probabilistic Localization

Before we delve into mathematical detail, let us illustrate the basic concepts with a simple example. Consider the environment depicted in figure 9.1. For the sake of simplicity, assume that the space of robot locations is one-dimensional, i.e., the robot can only move horizontally. Now suppose the robot is switched on somewhere in this environment to start its operation, but it is not told its location. Probabilistic localization represents this state of uncertainty by a *uniform distribution* over all locations, as shown by the graph in the top diagram in figure 9.1. Now assume the robot queries its sensors and finds out that it is next to a door. Probabilistic localization modifies the belief by raising the probability for locations next to doors, and lowering it elsewhere. This is illustrated in the second diagram in figure 9.1. Notice that the resulting belief is multimodal, reflecting the fact that the available information is insufficient to uniquely derive the robot's configuration. Also note that locations not close to a door still possess nonzero probability. This is because sensor readings are noisy, and a single sight of a door is typically insufficient to exclude the possibility of not being next to a door.

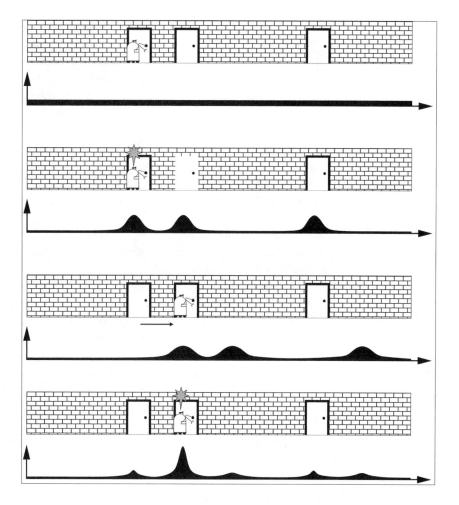

Figure 9.1   The basic idea of probabilistic localization: a mobile robot during global localization.

Now the robot advances to the next door. Probabilistic localization incorporates this information by propagating the belief distribution accordingly. To account for the inherent noise in robot motion, which in this situation inevitably leads to a loss of information, the new belief is smoother (and less certain) than the previous one. This is visualized in the third diagram in figure 9.1. Finally, the robot senses a second time, and again finds itself next to a door. This observation is combined with the current (nonuniform) belief, which leads to the final belief shown in the bottom diagram in figure 9.1. At this point, "most" of the probability is centered around a single location. The robot is now quite certain about its location.

Note that the final belief includes five different peaks given our sequence of two observations and one motion. The four smaller peaks correspond to the four cases in which the robot could only once explain its two observations given the map of the environment. At the location of the highest peak, which is in front of the second door at the true location of the robot, the robot has correctly identified a door twice. All other locations have small probabilities, since the robot could not explain its observations using its map.

Note that in this example the robot did not have an erroneous measurement. A false-positive detection of a door would lead to a situation in which the highest peak does not correspond to the true location of the robot. If, however, the robot knows about potential measurement errors, it would not become overly confident by just a few observations. One of the key features of probabilistic localization is that it uses the sensory information obtained to compute a belief that most accurately reflects the uncertainty about the configuration of the robot, given the knowledge about the behavior of the sensors of the robot.

Moreover, if the doors were uniquely identifiable by the robot, a Kalman filter would be sufficient for global localization. Since the robot is not able to identify the door it has sensed, it cannot associate an observation of a door uniquely to the doors given in its map. This problem is well-known as the *data association* problem. If the data association is known, Kalman filters can in fact be sufficient. Without knowing how to associate measurements to features in the map, the resulting beliefs will be inherently multimodal due to the resulting ambiguities. The strength of probabilistic localization lies in its capability to allow the representation of arbitrary distributions that are much more flexible than Gaussians. Put another way, probabilistic localization can also be applied when the data association is unknown or when the robot's motion models or sensor models are highly nonlinear.

9.1.2   Probabilistic Localization as Recursive Bayesian Filtering

Let X be the state space for the robot. We want to estimate the state $x \in X$ of the robot, which essentially is its configuration given as its position and orientation. In probabilistic localization the robot estimates at every time step k the conditional probability $P(x(k) \mid u(0 : k-1), y(1 : k))$ over all possible configurations given the sensor information $y(1 : k)$ it gathered about the environment and the movements $u(0 : k-1)$ carried out. The term $y(1 : k)$ denotes all observations obtained in the time steps $1, \ldots, k$. The notation $y(1 : k)$ "unfolds" to $y(1), y(2), \ldots, y(k)$ and $u(0 : k-1)$ unfolds in a similar fashion. The term $P(x(k) \mid u(1 : k-1), y(1 : k))$ is usually called the *posterior probability* (or simply *posterior*) [347]. Note that when u and y are written side by side, we assume that data have arrived in a synchronized way, i.e., in the form $u(0), y(1), \ldots, u(k-1), y(k)$. This assumption makes the derivation of probabilistic localization easier, but our algorithms can easily be extended to data streams that are not synchronized.

Note that in the previous chapter, $u(k)$ was simply the control input. In this chapter, we denote it as movements and do not rely on a specific interpretation of $u(k)$. It can represent the commanded velocities, the odometry measurements, or the result of filtering and fusing commanded velocities and odometry measurements, as described in the previous chapter.

Throughout this section we will assume that the robot is also given a model or map m of the environment. In principle we have to add this model as background knowledge in every term. However, for the sake of simplicity we will skip m in the equations below and assume that it is given as background knowledge. The heart of probabilistic localization is the following equation which tells us how to use the sensory input to update the most recent estimate (the *prior*) to obtain a new estimate (the posterior):

$$P(x(k) \mid u(0:k-1), y(1:k))$$
$$= \eta(k)\, P(y(k) \mid x(k)) \sum_{x(k-1)\in X} (P(x(k) \mid u(k-1), x(k-1))$$
(9.1)
$$P(x(k-1) \mid u(0:k-2), y(1:k-1)))$$

As mentioned above, the term $P(x(k) \mid u(0:k-1), y(1:k))$ is the posterior about the location of the robot at time k given the input data gathered so far. The term $P(x(k-1) \mid u(0:k-2), y(1:k-1)))$, in contrast, is denoted as the *prior* as it quantifies the probability that the robot is at location $x(k-1)$ before the integration of $u(k-1)$ and $y(k)$. The term $P(y(k) \mid x(k))$ is called the *observation model* which specifies the likelihood of the measurement $y(k)$ given the robot is at location $x(k)$. The term $P(x(k) \mid u(k-1), x(k-1))$ represents the *motion model* and can be regarded as a transition probability. It specifies the likelihood that the movement action $u(k-1)$ carried out at location $x(k-1)$ carries the robot to the location $x(k)$. Finally, $\eta(k)$ is a *normalization constant* that ensures that the left-hand side of this equation sums up to one over all $x(k)$. Note that (9.1) effectively accomplishes a combination of both the prediction and update steps of the Kalman filter.

Equation (9.1) is a special case of the following general equation for recursive Bayesian filtering.

$$P(x(k) \mid u(0:k-1), y(1:k))$$
$$= \eta(k)\, P(y(k) \mid x(k)) \int_X (P(x(k) \mid u(k-1), x(k-1))$$
(9.2)
$$P(x(k-1) \mid u(0:k-2), y(1:k-1)))\, dx(k-1)$$

Whereas (9.1) assumes discrete state spaces, (9.2) deals with continuous state spaces. Additionally, (9.2) can be shown to be a generalization of Kalman filtering. In this context the term $P(x(k) \mid u(k-1), x(k-1))$ is a generalization of (8.1) (see chapter 8) to arbitrary and nonlinear noise. Similarly, the term $P(y(k) \mid x(k))$ can handle arbitrary and nonlinear noise in the measurements. Finally, the posterior $P(x(k) \mid u(0:k-1),$ $y(1:k))$ generalizes the belief representation of Kalman filters from Gaussians to arbitrary probability density functions (PDFs).

Note the recursive character of probabilistic localization. The belief at time k is computed out of the posterior at time $k-1$ by incorporating two quantities, namely $P(y(k) \mid x(k))$ and $P(x(k) \mid u(k-1), x(k-1))$. Obviously, both the motion model and the observation model are the crucial components of probabilistic localization. Further below we will describe typical realizations of these models. We will also discuss different ways to represent the posterior $P(x(k) \mid u(k-1), x(k-1))$ and describe how to update the posterior given these representations.

Independent of the specific representation, the update of the belief is generally carried out in two different steps. The two steps are the *prediction step* and the *update step* which are joined together by (9.1). Note that the separation of a filtering process into these two steps is common in the context of Kalman filtering. The prediction step is applied whenever the belief has to be updated because of an odometry measurement $u(k-1)$. Suppose $u(0:k-2)$ and $y(1:k-1)$ are the data obtained thus far and $P(x(k-1) \mid u(0:k-2), y(1:k-1)))$ is the current belief about the configuration of the robot. Then we obtain the resulting belief $P(x(k) \mid u(0:k-1), y(1:k-1))$ by integrating over all possible previous configurations $x(k-1)$. For each such $x(k-1)$ we multiply $P(x(k-1) \mid u(0:k-2), y(1:k-1))$ by the probability $P(x(k) \mid u(k-1), x(k-1))$ that the measured motion action $u(k-1)$ has carried the robot from $x(k-1)$ to $x(k)$ and compute $P(x(k) \mid u(0:k-1), y(1:k-1))$ as the sum over all these values, i.e.,

$$
\begin{aligned}
P(x(k) &\mid u(0:k-1), y(1:k-1)) \\
&= \sum_{x(k-1)\in X} (P(x(k) \mid u(k-1), x(k-1)) \\
&\qquad\qquad P(x(k-1) \mid u(0:k-2), y(1:k-1)))
\end{aligned}
$$

(9.3)

Note that this operation basically corresponds to the step depicted in the third diagram of figure 9.1.

The update step is carried out whenever the robot perceives a measurement $y(k)$ with information about its environment. Suppose the current belief of the robot is $P(x(k) \mid u(0:k-1), y(1:k-1))$. In the update step we simply multiply for each configuration $x(k)$ the current value $P(x(k) \mid u(0:k-1), y(1:k-1))$ with the likelihood of $P(y(k) \mid x(k))$ that the robot perceives $y(k)$ given the map of the environment and given that the robot's configuration is $x(k)$. Additionally, we multiply each value with a normalization constant that ensures that $P(x(k) \mid u(0:k-1), y(1:k))$ sums up to one over all $x(k)$, i.e.,

$$P(x(k) \mid u(0:k-1), y(1:k))$$

(9.4)
$$= \eta(k) \, P(y(k) \mid x(k)) \, P(x(k) \mid u(0:k-1), y(1:k-1)).$$

According to Bayes rule, the constant $\eta(k)$ is given as

(9.5) $$\eta(k) = P(y(k) \mid u(0:k-1), y(1:k-1))^{-1},$$

which generally is hard to compute. This is mainly because the dependency between consecutive measurements without any information about the location of the robot in general is hard to determine. However, if we apply the law of total probability, we can sum over all locations $x(k)$ and transform (9.5) to

(9.6) $$\eta(k) = \left[\sum_{x(k) \in X} P(y(k) \mid x(k)) \, P(x(k) \mid u(0:k-1), y(1:k-1)) \right]^{-1}.$$

Obviously, we now can compute $\eta(k)$ using the terms that are already contained in (9.4). As we will see later, using this equation the normalization constant can be computed on the fly while integrating $y(k)$ into the current belief.

To summarize, we have the following equations that completely describe the two individual steps of recursive Bayesian filtering and that correspond to the prediction and update steps also found in the Kalman filter:

prediction:

$$P(x(k) \mid u(0:k-1), y(1:k-1))$$
$$= \sum_{x(k-1) \in X} \left(P(x(k) \mid u(k-1), x(k-1)) \right.$$
$$\left. P(x(k-1) \mid u(0:k-2), y(1:k-1)) \right)$$

update:

$$\eta(k)$$
$$= \left[\sum_{x(k) \in X} P(y(k) \mid x(k)) \, P(x(k) \mid u(0:k-1), y(1:k-1)) \right]^{-1}$$
$$P(x(k) \mid u(0:k-1), y(1:k))$$
$$= \eta(k) \, P(y(k) \mid x(k)) \, P(x(k) \mid u(0:k-1), y(1:k-1)).$$

In probabilistic localization the initial belief $P(x(0))$, reflects the prior knowledge about the initial configuration of the robot. This distribution can be initialized arbitrarily, but in practice two cases prevail. If the configuration of the robot relative to its map is entirely unknown, $P(x(0))$ is usually uniformly distributed, or if the initial state of the robot would be known up to a slight uncertainty, one would initialize $P(x(0))$ using a narrow Gaussian distribution centered at the robot's believed configuration.

The reader may notice that the principle of probabilistic localization leaves open

1. how the belief $P(x)$ is represented as well as

2. how the conditional probabilities $P(x(k) \mid u(k-1), x(k-1))$ and $P(y(k) \mid x(k))$ are computed.

Accordingly, existing approaches to probabilistic localization mainly differ in the representation of the belief and the way the perceptual and motion models are represented. After a derivation of the equation for probabilistic localization in the following subsection, we will discuss different ways to represent the posterior. As we will see, the representation of the posterior has a serious impact on the efficiency of probabilistic localization and the type of situations that can be accommodated with probabilistic localization.

9.1.3   Derivation of Probabilistic Localization

When computing $P(x(k) \mid u(0 : k-1), y(1 : k))$, we distinguish two cases, depending on whether the most recent data item is an odometry reading or a sensor measurement.[1] Let us first consider how to incorporate the most recent data item, namely a sensor measurement $y(k)$ the robot uses to gather information about its environment. If we apply Bayes rule considering $y(1 : k-1)$ and $u(0 : k-1)$ as background knowledge, we obtain

$$(9.7) \quad
\begin{aligned}
&P(x(k) \mid u(0 : k-1), y(1 : k)) \\
&= \frac{P(y(k) \mid u(0 : k-1), y(1:k-1), x(k)) P(x(k) \mid u(0:k-1), y(1:k-1))}{P(y(k) \mid u(0:k-1), y(1:k-1))}.
\end{aligned}$$

First consider the left term $P(y(k) \mid u(k-1), y(1:k-1), x(k))$ in the numerator. This term represents the likelihood of the most recent measurement $y(k)$ given all previous measurements and given that the configuration $x(k)$ of the robot at time k is known. In recursive Bayesian filtering, one generally makes the assumption that, once the state $x(k)$ is known, the measurement $y(k)$ is independent of all previous measurements and controls. Given this assumption we can simply remove $y(1 : k-1)$

---

1. These two cases are analogous to the Kalman prediction and update steps, respectively.

and $u(0 : k - 1)$ from this term. Accordingly, we simplify (9.7) to

(9.8)

$$P(x(k) \,|\, u(0 : k - 1), y(1 : k))$$
$$= \frac{P(y(k) \,|\, x(k)) \, P(x(k) \,|\, u(0 : k - 1), y(1 : k - 1))}{P(y(k) \,|\, u(0 : k - 1), y(1 : k - 1))}.$$

Observe that the denominator is a normalizer that does not depend on the configuration of the robot. It simply ensures that the left-hand side of (9.8) sums up to one over all $x(k)$. Accordingly, we can replace the denominator by a normalization constant $\eta(k)$ which is the same for all $x(k)$. This leads to

(9.9)

$$P(x(k) \,|\, u(0 : k - 1), y(1 : k))$$
$$= \eta(k) \, P(y(k) \,|\, x(k)) \, P(x(k) \,|\, u(0 : k - 1), y(1 : k - 1)).$$

To see how to incorporate the motions of the robot into the belief we next consider the rightmost term $P(x(k) \,|\, u(0 : k - 1), y(1 : k - 1))$ in this equation. If we use the law of total probability we derive

$$P(x(k) \,|\, u(0 : k - 1), y(1 : k - 1))$$
$$= \sum_{x(k-1) \in X} \Big[P(x(k) \,|\, u(0 : k - 1), y(1 : k - 1), x(k - 1))$$

(9.10)

$$P(x(k - 1) \,|\, u(0 : k - 1), y(1 : k - 1)) \Big].$$

To simplify the lefthand term in the sum we again make an independence assumption. We assume that $x(k)$ is independent of the measurements $y(1 : k - 1)$ and the movements $u(0 : k - 2)$ obtained and carried out before the robot arrived at $x(k - 1)$ given we know $x(k - 1)$. Rather the likelihood of being at $x(k)$ only depends on $x(k - 1)$ and the most recent movement $u(k - 1)$, i.e.,

$$P(x(k) \,|\, u(0 : k - 1), y(1 : k - 1), x(k - 1))$$

(9.11)

$$= P(x(k) \,|\, u(k - 1), x(k - 1))$$

Thus we simplify (9.10) to

$$P(x(k) \,|\, u(0 : k - 1), y(1 : k - 1))$$
$$= \sum_{x(k-1) \in X} (P(x(k) \,|\, u(k - 1), x(k - 1))$$

(9.12)

$$\cdot P(x(k - 1) \,|\, u(0 : k - 1), y(1 : k - 1))).$$

Now consider the second factor $P(x(k - 1) \,|\, u(0 : k - 1), y(1 : k - 1))$ in the sum. This term specifies the probability that the robot's configuration at time $k - 1$ is $x(k - 1)$ given the motions $u(0 : k - 1)$ and given the observations $y(1 : k - 1)$.

According to our terminology, the motion $u(k-1)$ is carried out at time step $k-1$ so that $u(k-1)$ carries the robot away from $x(k-1)$. Since we have no information about $x(k)$ and under the assumption that the time that elapses between consecutive measurements is small, we can in fact conclude that the information that the robot has moved after it was at $x(k-1)$ does not provide any information about $x(k-1)$. Note that this is not true in general. Suppose the environment of the robot consists of two rooms, a small and a large room, and that there is no door between these two rooms. Furthermore suppose the robot moved a distance that is larger than the diameter of the small room. After that movement the probability that the robot is in the larger room must exceed the probability that the robot is in the smaller room. If, however, the time intervals between consecutive measurements are small, each movement can only represent a small distance and the fact that the robot has moved a few inches away from its current location $x(k-1)$ carries almost no information about $x(k-1)$ given we do not know $x(k)$. Under this assumption we therefore can conclude that $u(k-1)$ does not provide information about $x(k-1)$ if we have no information about $x(k)$. Thus, we assume that $x(k-1)$ is independent of $u(k-1)$ in the term $P(x(k-1) \mid u(0:k-1), y(1:k-1))$. Thus we obtain

$$P(x(k-1) \mid u(0:k-1), y(1:k-1))$$
$$(9.13) \qquad = P(x(k-1) \mid u(0:k-2), y(1:k-1))$$

and simplify (9.12) to

$$P(x(k) \mid u(0:k-1), y(1:k-1))$$
$$= \sum_{x(k-1)\in X} (P(x(k) \mid u(k-1), x(k-1))$$
$$(9.14) \qquad \qquad \cdot P(x(k-1) \mid u(0:k-2), y(1:k-1))).$$

If we now substitute this result into (9.9) we obtain

$$P(x(k) \mid u(0:k-1), y(1:k))$$
$$= \eta(k)\, P(y(k) \mid x(k)) \sum_{x(k-1)\in X} (P(x(k) \mid u(k-1), x(k-1))$$
$$(9.15) \qquad P(x(k-1) \mid u(0:k-2), y(1:k-1)))$$

which directly corresponds to the (9.1).

9.1.4   Representations of the Posterior

As mentioned above, the probabilistic formulation leaves open how the posterior is represented. In principle, there are various ways to represent the posterior. Mathematically speaking, P is a function $P : X \to \mathbb{R}$. When X is continuous (or infinite), P lives in an infinitely dimensional space. Of course it is impossible to arbitrarily represent an infinitely dimensional map. Thus we have to be content with a finite approximation. Throughout this section we discuss three different approaches: Kalman filters, discrete approximations, and particle filters.

Kalman Filters

The previous chapter covered a common approach to the representation of the belief $P(x(k) | u(0:k-1), y(1:k))$ as Extended Kalman filters (EKFs) [215, 390]. In this case, the posterior is represented using a unimodal Gaussian distribution. Many successful applications of Kalman filters for mobile robot localization have been demonstrated [28, 179, 276, 371]. One advantage of Kalman filtering is that it can be implemented quite efficiently and that it works well in high-dimensional state spaces. Additionally, Kalman filters provide a floating-point resolution and in this way allow highly accurate estimates.

Unfortunately, Kalman filters are only optimal for systems whose behavior is governed by the linear equations given by (8.1) and (8.2). Since Kalman filters use Gaussian distributions, they cannot appropriately represent beliefs that correspond to ambiguous situations as they appear, e.g., in the context of global localization. As a result, localization approaches using Kalman filters typically require that the starting location of the robot is known or that unique landmarks are given so that there is no data association problem. To overcome the limitations of Kalman filters, recent extensions of this approach have been developed. For example, Jensfeld and Christensen [209] use a mixture of Gaussians to represent the belief about the location of the robot. They also present techniques to update this mixture based on sensory input and robot motions. In this chapter, we consider two alternative nonparametric state representations—discrete grids and samples—to bypass the Gaussian assumption.

Discrete Approximations

An alternative form to represent $P(x(k))$ is to use a discrete approximation of the configuration space. Independent of the structure of the discretization all approaches store in each element of their discrete structure the probability that the robot is at the location that corresponds to this element. In practice, one mainly finds topological and geometric discretizations. In the first case, the configuration space is separated according to the topological structure of the environment. Many systems that exploit topological structures use individual states for junctions, doorways and rooms and four possible headings. Several systems [213, 339, 386] follow this approach and perform probabilistic localization for landmark-based corridor navigation. Choset and Nagatani exploit the topology of the generalized Voronoi diagram (GVD) [108], described in chapter 5. The advantage of a topological representation lies in its compactness, because only a limited number of states need to be considered. Its disadvantage, on the other hand, is limited accuracy with respect to position and orientation. To achieve more accurate estimates, Burgard, Fox, and coworkers [83, 158] use a fine-grained grid to represent the posterior. Throughout this section we will give a detailed description of this grid-based technique.

If we assume that the configuration of the robot is $SE(2)$ and thus a configuration is represented by a three-dimensional random vector consisting of the (x_r, y_r)-position and the orientation θ_r of the vehicle, our grid needs to be three-dimensional. Whereas the first two dimensions are used for the position of the vehicle, the third dimension is used for its orientation. Figure 9.2 shows the structure of the grid for this kind of representation. In practical applications of this technique, spatial resolution of 10 to 30 cm and an angular resolutions of 2 to 10 degrees turned out to be sufficient for robust and accurate localization of mobile robots [81, 177].

To compute the posterior $P(x(k) \mid u(0 : k-1), y(1 : k))$ represented by a grid we follow the procedure summarized in Algorithm 16. Given $u(0 : k-1)$, $y(1 : k)$ and an initial belief $P(x(0))$, we carry out k loops. In the every round i we integrate $u(i-1)$ and $y(i)$. The first step (i.e., the prediction step) incorporates the movement $u(i-1)$, which means that we have to recompute the grid according to the motion model $P(x \mid u(i-1), x')$. In principle, this involves integrating over all possible prior states of the robot, which would result in an $O(N^2)$ complexity, where N is the number of states represented by the grid. One way to reduce the complexity of this operation is to

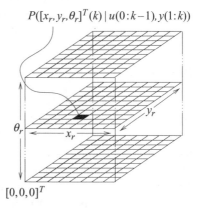

Figure 9.2 Grid-based representation of the state space.

Algorithm 16 Probabilistic localization for discrete state spaces

---

Input: Sequence of measurements $y(1 : k)$ and movements $u(0 : k - 1)$ and initial belief $P(x(0))$

Output: A posterior $P(x(k) \,|\, u(0 : k - 1), y(1 : k))$ about the configuration of the robot at time step k

---

1: $P(x) \leftarrow P(x(0))$

2: **for** $i \leftarrow 1$ to k **do**

3:      **for all** states $x \in X$ **do**

4:          $P'(x) \leftarrow \sum_{x' \in X} P(x \,|\, u(i - 1), x') \cdot P(x')$

5:      **end for**

6:      $\eta \leftarrow 0$

7:      **for all** states $x \in X$ **do**

8:          $P(x) \leftarrow P(y(i) \,|\, x) \cdot P'(x)$

9:          $\eta \leftarrow \eta + P(x)$

10:      **end for**

11:      **for all** states $x \in X$ **do**

12:          $P(x) \leftarrow P(x)/\eta$

13:      **end for**

14: **end for**

---

limit the number of predecessor states summed over. In a successful implementation of the grid-based representation Fox, Burgard and Thrun [158] applied the following approach to approximate the integration over all potential previous states: First, all grid cells are shifted according to the motion $u(i - 1)$ carried out by the robot, and then the whole grid is convolved using a bounded Gaussian kernel that corresponds to the uncertainty of $u(i - 1)$. Whenever the robot has moved, we can easily compute for every (x, y)-plane of the grid the offsets Δx and Δy by which each cell has to be shifted according to $u(i - 1)$ (and assuming there are no odometry errors). Please note that Δx and Δy both depend on the angle θ that the corresponding plane in the grid represents. The convolution operation can be carried out efficiently using a separable kernel. This involves convolving independently over the individual dimensions of the grid. To realize this, one usually introduces a one-dimensional array P' that stores the intermediate results of this computation. In the case of the x-dimension we proceed as follows for all x:

$$
(9.16) \qquad
\begin{aligned}
P'((x, y, \theta)) ={} & 0.5 \cdot P((x, y, \theta)) \\
& + 0.25 \cdot (P((x - 1, y, \theta)) + P((x + 1, y, \theta)))
\end{aligned}
$$

Special care has to be taken at the borders of the grid. In this case only one neighboring cell is given and one chooses the coefficients $\frac{2}{3}$ for the cell itself and $\frac{1}{3}$ for the neighbor cell ($x+1$ or $x-1$ depending on where one is in the grid). Whenever $P'((x, y, \theta))$ has been computed for all x and a given pair of y and θ, the results are then stored back in the original cells. Similarly we proceed with all y and θ. To correctly model the uncertainty introduced by the motion $u(i-1)$ the convolution process can be repeated appropriately. The second step (i.e., the update step) of Algorithm 16 integrates the observation $y(i)$ into the grid. To achieve this, we simply multiply every grid cell by the likelihood of the observation $y(i)$, given the robot has the configuration corresponding to that particular cell. Afterward the whole grid is normalized.

Algorithm 16 can also be used for incremental filtering. If the initial belief is set to the output obtained from the preceeding application of the algorithm and if all measurements and movements since this point in time are given as input, the algorithm incrementally computes the corresponding posterior. Please also note that Algorithm 16 can easily be extended to situations in which the movements and the environment measurements do not arrive in an alternating and fixed scheme.

One important aspect of all state estimation procedures is the extraction of relevant statistics such as the mean and the mode. These parameters are important whenever the robot has to generate actions based on the current belief about its state. For example, this can be the next motion command to enter a specific room. Both the mode and the mean can be determined efficiently given a grid-based approximation. The x- and y-coordinates of the mean ($\bar{x}$ and $\bar{y}$) can be computed by computing the weighted sums $i = 1, \ldots, N$ over all cells of the grid. To compute the angle mean we use the following equation:

$$(9.17) \quad \hat{\phi} = \text{atan2}\left(\sum_{i=0}^{N-1} P(i) \cdot \sin\phi(i), \sum_{i=0}^{N-1} P(i) \cdot \cos\phi(i) \right)$$

Unfortunately, the mean has the disadvantage that the resulting values can lack any useful meaning, especially in the context of multimodal distributions. For example, the mean of a bimodal distribution might lie within an obstacle so that no meaningful commands can be generated. An alternative statistic is the mode of the distribution which, given a grid-based approximation, can be computed efficiently by a simple maximum operation. Compared to the mean, the mode has the advantage that it generally corresponds to a possible location of the vehicle. However, the locations of subsequent modes can differ largely so that the estimates are not as continuous as if we choose the mean. Whereas the mean automatically yields estimates at subgrid-resolution accuracy, we can obtain the same for the mode by averaging over a small region around the cell containing the maximum probability [82].

To illustrate an application example of a grid-based representation, consider the map and the data set depicted in the left image of figure 9.3. The map, which was generated using the system described by Buhmann and coworkers [73], corresponds to the environment of the AAAI '94 mobile robot competition. The size of this environment is 31 by 22 m. The right image of the same figure depicts the path of the B21 robot Rhino [415] along with measurements of the twenty four ultrasound sensors obtained as the robot moved through the competition arena. Here we use this sensor information to globally localize the robot from scratch. The time required to process this data on a 400 MHz Pentium II is 80 seconds, using a position probability grid with a spatial resolution of 15 cm and an angular resolution of 3 degrees.

The right image of figure 9.3 also marks the points in time when the robot perceived the fifth (A), eighteenth (B), and twenty-fourth (C) sensor sweep. The posteriors during global localization at these three points in time are illustrated in figure 9.4. The figures show the belief of the robot projected onto the (x, y)-plane by plotting

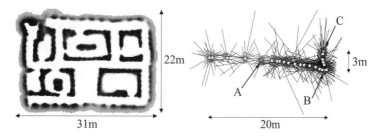

Figure 9.3   Occupancy grid map of the 1994 AAAI mobile robot competition arena (left) and data set recorded in this (right). It includes the odometry information and the ultrasound measurements. Point *A* is after five steps, *B* is after eighteen, and *C* is after twenty-four.

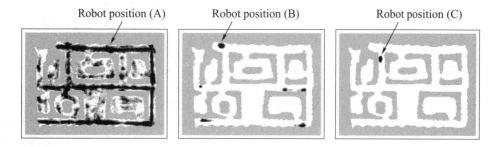

Figure 9.4   Density plots after incorporating five, eighteen, and twenty-four sonar scans (the darker locations are more likely).

for each (x, y)-position the maximum probability over all possible orientations. More likely locations are darker and, for illustration purposes, the left and middle images use a logarithmic scale in intensity. The leftmost image of figure 9.4 shows the belief state after integrating five sensor sweeps (i.e. when the robot is at step A on its path). At this point in time, all the robot knows is that it is likely in one of the corridors of the environment. After integrating eighteen sweeps of the ultrasound sensors (at step B) the robot is almost certain that it is at the end of a corridor (see center image of figure 9.4). After incorporating twenty-four scans (step C) the robot has determined its location uniquely. This is represented by the unique peak containing 99% of the whole probability mass (see rightmost image of figure 9.4).

Although the grid-based approach has the advantage that it provides a well-understood approximation of the true distribution and that important statistics such as the mean and the mode can be easily assessed, it has certain disadvantages. First, the number of grid cells grows exponentially in the number of dimensions and therefore limits the application of this approach to low-dimensional state spaces. Additionally, the approach uses a rigid grid. If the whole probability mass is concentrated on a unique peak, most of the states in the grid are useless and approaches that focus the processing time on regions of high likelihood are preferable. One method to dynamically adapt the number of states that have to be updated is the selective updating scheme [158]. Burgard, Derr, Fox, and Cremers [82] use a tree structure and store only cells whose probability exceeds a certain threshold. In this way, memory and computational requirements can be adapted to the complexity of the posterior.

Particle Filters

An alternative and efficient way of representing and maintaining probability densities is the particle filter. The key idea of particle filters is to represent the posterior by a set $\mathcal{M}$ of N samples. Each sample consists of a pair (x, ω) containing a state vector x of the underlying system and a weighting factor ω, i.e., $\mathcal{M} = (X, [0, 1])^N$. The latter is used to store the importance of the corresponding particle. The posterior is represented by the distribution of the samples and their importance factors. In the past a variety of different particle filter algorithms have been developed and many variants have been applied with great success to various application domains [97,125,138,157,167,201,218]. Algorithm 17 describes a particle filter algorithm that uses *sequential importance sampling with resampling* [29] to implement the update step. This algorithm follows a survival of the fittest scheme. Whenever a new measurement $y(k)$ arrives, the weight ω of a particle (x, ω) is computed as the likelihood $p(y(k) \mid x)$ of this observation given the system is in state x. After computing

---

Algorithm 17 Probabilistic localization using a particle filter

---

Input: Sequence of measurements $y(1:k)$ and movements $u(0:k-1)$ and set $\mathcal{M}$ of N samples (x_j, ω_j) corresponding to the initial belief $P(x)$

Output: A posterior $P(x(k) \mid u(0:k-1), y(1:k))$ about the configuration of the robot at time step k represented by $\mathcal{M}$.

1: **for** $i \leftarrow 1$ to k **do**
2:     **for** $j \leftarrow 1$ to N **do**
3:         compute a new state x by sampling according to $P(x \mid u(i-1), x_j)$.
4:         $x_j \leftarrow x$
5:     **end for**
6:     $\eta \leftarrow 0$
7:     **for** $j \leftarrow 1$ to N **do**
8:         $w_j = P(y(i) \mid x_j)$
9:         $\eta = \eta + w_j$
10:     **end for**
11:     **for** $j \leftarrow 1$ to N **do**
12:         $w_j = \eta^{-1} \cdot w_j$
13:     **end for**
14:     $\mathcal{M} = resample(\mathcal{M})$
15: **end for**

---

the weights, a so-called resampling procedure is applied. We draw N samples with replacement from $\mathcal{M}$ such that each sample in $\mathcal{M}$ is selected with a probability that is proportional to its weight ω. Accordingly, samples with greater weights survive with higher likelihood than samples with values of small importance. In principle, there are many ways of achieving this. One popular approach (see also [29]) is described by algorithm 18. In this algorithm, the procedure *rand(I)* draws a random value from the interval I according to a uniform distribution. The major advantage of this algorithm is that the whole resampling process is carried out in $O(N)$ steps. One alternative technique is the one used by Isard and Blake [201]. This approach relies on binary search to select a sample and thus requires $O(N \log N)$ steps.

We also need to describe the prediction step that we use to incorporate the motions of the robot into the sample set. Throughout this chapter we assume that incremental motions of a robot between two configurations x_1 and x_2 are encoded by the three parameters α, β, and d (see figure 9.5). Here α is an initial rotation in x_1 toward x_2, d is the distance to be traveled from x_1 to x_2, and β is the final rotation carried out at the location x_2 to reach the orientation of the robot in x_2. Since the motions carried

---

Algorithm 18 The procedure *resample*($\mathcal{M}$)

Input: Set $\mathcal{M}$ of N samples

Output: Set $\mathcal{M}'$ of N samples obtained by importance resampling from $\mathcal{M}$

1: $\mathcal{M}' \leftarrow \emptyset$
2: $\Delta \leftarrow rand((0; N^{-1}])$
3: $c \leftarrow \omega_0$
4: $i \leftarrow 0$
5: **for** $j \leftarrow 0$ to $N - 1$ **do**
6:    $u \leftarrow \Delta + j \cdot N^{-1}$
7:   **while** $u > c$ **do**
8:     $i \leftarrow i + 1$
9:     $c \leftarrow c + \omega_i$
10:   **end while**
11:    $\mathcal{M}' \leftarrow \mathcal{M}' \cup \{(x_i, N^{-1})\}$
12: **end for**

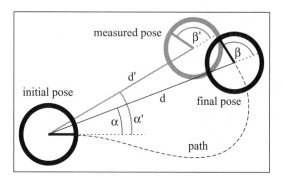

Figure 9.5 The parameters α, β, and d specifying any incremental motion of a robot in the (x, y, θ)-space.

out by the robot are not deterministic, we need to cope with potential errors when we compute new locations for samples. We proceed as follows: Whenever we compute the new configuration for a sample after a movement $u(i - 1)$, we incorporate the possible deviations from the values of α, β, and d. Throughout this section, we assume Gaussian noise in these values and compute the new location of a sample according to values α', β', and d' that deviate from the measured values α, β, and d according to Gaussian distributions. If we denote the robot state as $x = [x_r, y_r, \theta_r]$

and the ith particle by x^i, then the motion model is implemented by assigning

$$(9.18) \qquad x^i = x^i + \begin{bmatrix} d' \cos \left(\theta_r^i + \alpha' \right) \\ d' \sin \left(\theta_r^i + \alpha' \right) \\ \alpha' + \beta' \end{bmatrix}$$

for each particle in the collection, where

$$(9.19) \qquad \alpha' = \alpha + \alpha \cdot \mathrm{norm}(\sigma_1) + d \cdot \mathrm{norm}(\sigma_2)$$

$$(9.20) \qquad \beta' = \beta + \beta \cdot \mathrm{norm}(\sigma_3) + d \cdot \mathrm{norm}(\sigma_4)$$

$$(9.21) \qquad d' = d + d \cdot \mathrm{norm}(\sigma_5) + (\alpha + \beta) \cdot \mathrm{norm}(\sigma_6).$$

Here $\mathrm{norm}(\sigma)$ is a random number generator that outputs random numbers according to a normal distribution with mean 0 and standard deviation σ. The standard deviations σ_i are parameters that describe the influence of the translation d and the rotations α and β on the potential errors. In this model the errors in all three values depend on the rotations and the translation carried out. Note that the σ_i can be learned by generating a statistic about typical deviations of the actual movements from the values α, β, and d.

Figure 9.6 illustrates an application of this motion model to a sample set in which all samples are concentrated in a single state. The line depicts the path taken by the robot and the sample sets illustrate the belief about the robot's configurations at

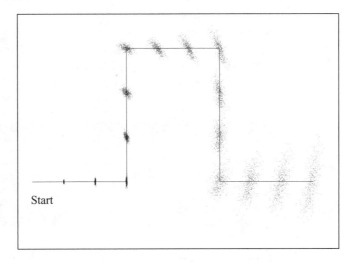

Figure 9.6   Sample-based approximation of the belief of the robot after repeatedly executing motion commands. In this example, the robot did not perceive the environment while it was updating the sample set.

certain points in time. In this particular example we incorporated no observations of the environment into the sample set. As can be seen from the figure, the robot's uncertainty grows indefinitely while it is moving, and the overall distribution is slightly bent.

Note that the motion model described above does not incorporate any information about the environment. Accordingly, samples might end up inside obstacles in the map of the environment. One advantage of sample-based approaches, however, lies in the fact that such environmental information can easily be incorporated. To avoid that samples move through obstacles we can simply reject such values for α', β', and d'.

To extract the mean of a posterior represented by N samples, we can proceed in a similar way as for the grid-based representation. We simply average over all samples of the distribution. In the case that not all importance factors are equal we use the normalized importance factors as weighting factors. When computing the mode, we distinguish two different situations. If we compute the mode just before the resampling step, we can simply select that sample with the highest importance factor, which requires $O(N)$ steps. After resampling, however, the mode cannot be computed as easily, because the actual form of the posterior is only encoded in the density of the samples. One popular approach to approximate the mode is to use kd-trees [44], which, however, requires $O(N \log N)$ steps. Alternatively, one can compute a histogram based on a coarse discretization of the state space. In this case, the space requirements are similar to the grid-based approach, but the mode can be extracted in $O(N)$ steps.

Figure 9.7 shows a particle filter in action. This example is based on the ultrasound and odometry data obtained while the robot traveled along the path depicted in figure 8.1 in chapter 8. To achieve global localization we initialized the filter by selecting the initial set of particles from a uniform distribution over the free space in the environment (see left image of figure 9.7). After incorporating ten ultrasound

Figure 9.7 Global localization using a particle filter with 10,000 samples. The left image shows the initial distribution. The middle image shows the distribution after incorporating ten ultrasound beams. The right image shows a typical situation (here after 65 steps) when the location of the robot has been identified.

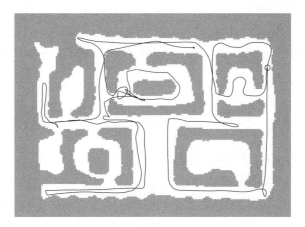

Figure 9.8 Trajectory of the robot obtained by applying a particle filter for tracking the configuration of the robot using the data gathered along the path depicted in figure 8.1 in chapter 8.

measurements we obtain the particle set depicted in the middle image of figure 9.7. After incorporating 65 measurements, the particle filter has converged to configurations close to the true location of the robot. The resulting density is depicted in the right image of figure 9.7.

Figure 9.8 shows the path of a robot as it is estimated by a particle filter. Here the particle filter was initialized using a Gaussian distribution and therefore was just tracking the location of the robot. Again, the odometry data used as input are shown in figure 8.1. Additionally, the filter used the data of the 24 ultrasound sensors. As can be seen, the particle filter can robustly track the position of the robot, although ultrasound measurements are noisy and there are larger errors in odometry.

As the global localization example illustrates, particle filters typically converge to the most likely estimate, i.e., after a certain period of time all of the particles usually cluster around the true configuration of the system. While this is desired in most situations, it also can be disadvantageous. This is especially true if a failure occurs that is not modeled in the motion model. One typical scenario in which such unpredicted localization errors frequently occur is the RoboCup environment [237]. There are certain conditions under which a referee removes a player from the soccer field. After a short period of time the player is then placed back onto the field. The problem that has to solved by the robot in such a case is usually denoted as the "kidnapped robot problem" [145]. To deal with such situations, or more generally, with situations in which the estimation process fails, the robot requires techniques to detect localization failures and to initiate a global localization. One approach is to modify the motion model and to choose random configurations for a certain fraction of the samples [156]. Alternatively, one can monitor the average observation likelihood $\tilde{p} = N^{-1} \cdot \sum_{j=1}^{N} P(y(i) \mid x_j)$ of all samples. For example, Burgard et al. [82] restart a global localization if this value falls below a certain threshold. Lenser and Veloso [275] adjust the number of samples with randomly chosen locations according to the value of $\tilde{p}$. Gutmann and Fox [176] additionally smooth $\tilde{p}$ to be more robust against short-term changes of $\tilde{p}$.

9.1.5   Sensor Models

One of the crucial aspects of probabilistic localization is how the likelihood of the robot's sensor measurements is computed. In particular, we are interested in the quantity $P(y \mid x)$, which represents the likelihood of measuring y given x is the location of the system. Throughout this chapter, we denote the way in which we compute this quantity as the sensor model. Obviously, a good sensor model largely depends on the type of sensor that is used for localization. Additionally, it also may depend on the environment. For example, it might exploit particular features of the environment, such as landmarks. Finally, it also depends on the way the environment is represented, i.e., on the type of the map.

In this subsection we describe a sensor model that captures several of the physical properties of frequently used proximity sensors such as ultrasound or laser range scanners. To motivate this sensor model let us first investigate a typical scan obtained with the 24 ultrasound sensors of a B21 robot. One such scan is shown in figure 9.9. In this figure the objects in the environment are shown in light gray. The dark lines indicate the central axis of 24 ultrasound beams as they are obtained at the corresponding location in this environment. As can be seen from the figure, most of the measurements are quite accurate. For example, the beams 0, 2, 3, 4, 10, 13, and 15 quite accurately correspond to the distance to the nearest obstacle in the measurement direction. Other beams, such as 1, 12, and 14, are shorter than the distance to the nearest obstacle. In this particular situation, the measurement 1 resulted from a crosstalk: the sensor received a sound signal emitted by another sensor [61]. The other two short measurements (12 and 14) were caused by objects not contained in the map. Whereas beam 12 was reflected by a person entering the room, the cone of beam 14 was echoed by a refrigerator installed in the niche. Furthermore, some of the measurements, such as 18, 19, and 20, appear to be quite random. They apparently pass through a bookshelf and appear to be echoed by an unmodeled object behind it. Finally, the sensors 6, 8, and 23 never received an echo and therefore report a maximum range reading.

The sensor model that we describe in the remainder of this subsection is designed to capture the noise and error characteristics of many active range sensors. It can model

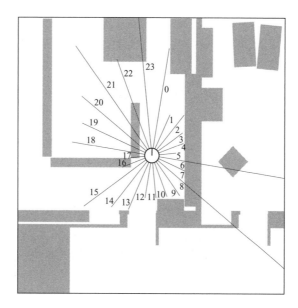

Figure 9.9   Ultrasound scan perceived with a B21 robot in an office environment.

the accuracy of the sensor whenever the beam hits the nearest object in the direction of the measurement. Additionally, it represents random measurements. It furthermore provides means to model objects not contained in the map and to represent the effects of crosstalk between different sensors. Finally, it incorporates a technique to deal with detection errors in which the sensor reports a maximum range measurement. The model has been applied successfully in the past for mobile robot localization with proximity sensors such as ultrasound sensors and laser range finders. In 1997 and 1998 [81] the mobile robots Rhino and Minerva operated several weeks in populated museum environments using the sensor model described here for localization with laser range scanners.

Throughout this subsection we assume that range sensors have a limited numerical resolution, i.e., the information they provide is discrete. Accordingly, we consider a discrete set of distances $d_0, \ldots, d_{n-1}$ where d_{n-1} corresponds to the maximum distance that can be measured. We also assume that the size of the ranges $\Delta = \Delta_{i+1} = d_{i+1} - d_i$ is the same for all i. In principle, the distribution $P(y \mid x)$ that y is observed given the state of the system is x can be specified by a histogram that stores in each of its n bins the likelihood that y is d_i, $i = 0, 1, \ldots n - 1$. Obviously, storing an individual histogram for sufficiently large number of potential states would consume too much space.

The key idea of the model that we describe here is to compute $P(y \mid x)$ based on the distance $d(x)$ to the closest obstacle in the map within the perceptual field of the sensor. If the environment is represented geometrically, i.e., by an evidence grid or by geometric primitives such as polygons or lines, the expected distance $d(x)$ can be computed efficiently using ray-casting. It is natural to assume that

(9.22) $$P(y \mid x) = P(y \mid d(x)).$$

Accordingly, it suffices to determine the expected distance d for the given measuring direction at the current location of the vehicle and then compute $P(y \mid d)$.

The quantity $P(y \mid d)$ is calculated as follows. According to the different situations identified in the scan depicted in figure 9.9, we distinguish the following four situations:

1. *The nearest object in the direction of the beam is detected.* The actual measurement depends on the accuracy of the underlying sensor. Typically, the likelihood of the measurement y is then well-approximated by a Gaussian distribution $\mathcal{N}(y, d, \sigma)$, where d is the true distance to the object and the variance σ depends on the accuracy of the sensor; it is higher for ultrasound sensors than for laser range scanners.

2. *An object not contained in the map reflects the beam, or there is crosstalk.* The sensor will report a distance that is shorter than the expected distance. In our model, we represent this by an exponential distribution, e.g., $\lambda e^{-\lambda y}$.

3. *The sensor produces a random measurement.* As mentioned above, there are situations in which the sensor provides a random measurement that cannot be explained given the current map. We model these types of measurements by a uniform distribution over the possible distances reported by the sensor, represented by a constant γ.

4. *The sensor reports a maximum range reading.* In some situations, time-of-flight sensors such as ultrasounds or laser range scanners, or intensity-based sensors such as those based on infrared light fail to detect the beam reflected by an object. If no random measurement is obtained and if no crosstalk happens, the sensor may report a maximum range measurement. The likelihood of this event is represented by a constant δ.

Since we do not know which situation is given, our distribution needs to represent all the different cases. Accordingly, the distribution $P(y \mid d)$ is computed based on a mixture of the four different densities that correspond to the individual situations. Suppose d_i is the expected distance in a particular measurement direction at a given position in the environment. Then we determine a complete histogram h_i containing in each bin $h_{i,j}$ the likelihood of each possible measurement d_j the robot can obtain. If $j < n - 1$, i.e., the actual measurement is not a maximum range measurement, $h_{i,j}$ is computed based on a mixture of the three densities representing the situations 1, 2, and 3 described above. If, however, $j = n - 1$, then we have a single value representing the likelihood of maximum range measurements.

$$(9.23) \quad h_{i,j} = \begin{cases} \int_{d_j - \frac{\Delta}{2}}^{d_j + \frac{\Delta}{2}} \alpha_i \cdot \mathcal{N}(y, d_i, \sigma_i) + \beta_i \cdot \lambda_i e^{-\lambda_i y} + \gamma_i \ dy & \text{if } j < n - 1 \\ \delta_i & \text{if } j = n - 1 \end{cases}$$

The values of the parameters σ_i, α_i, β_i, γ_i and δ_i have to be chosen appropriately such that each $h_{i,j}$ reflects the correct likelihood. Thereby, we have to consider the constraint that

$$(9.24) \quad \sum_{j=0}^{n-1} h_{i,j} = 1$$

for each histogram h_i. A typical approach to determine the optimal values for the different parameters of each histogram is log-likelihood maximization. Given a set of actual measurements for a given expected distance d_i this approach seeks to determine those values for the parameters that maximize the sum of the log likelihoods $\log P(d \mid d_i)$ over all measurements in the data set.

Figures 9.10 and 9.11 show the resulting maximum-likelihood approximations for two different expected distances. Whereas figure 9.10 plots the histograms for ultrasound data, figure 9.11 plots the histograms obtained for laser range data. In all plots, the data are shown as boxes and the approximation is shown as dots. As can

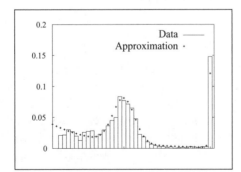
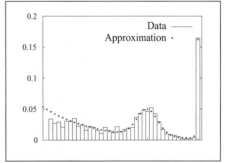

Figure 9.10   Histograms (data and maximum-likelihood approximation) for data sets obtained with ultrasound sensors and for two different expected distances.

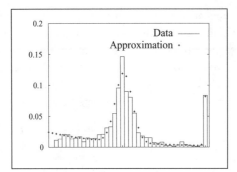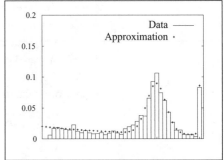

Figure 9.11   Histograms (data and maximum-likelihood approximation) for data sets obtained with a laser range finder and for two different expected distances.

be seen, our model is quite accurate and reflects well the properties of the real data. It is furthermore obvious that the laser range finder yields much more accurate data than the ultrasound sensor. This is represented by the fact that the Gaussians have a lower variance in both histograms. Please also note that the variances of the Gaussians do not depend solely on the accuracy of the sensor. They also encode the uncertainty of the map. For example, if dynamic objects such as chairs or tables are slightly moved, the error in the measurements increases, which results in a higher uncertainty of the sensor model.

Range scans obtained with laser range scanners typically consist of multiple measurements. Some robots are also equipped with arrays of ultrasound sensors which provide several measurements $y_1, y_2, \ldots y_m$ at a time. In practice it is often assumed that these measurements are independent, i.e, that

$$(9.25) \qquad P(y_1, y_2, \ldots, y_m \mid x) = \prod_{i=1}^{I} P(y_i \mid x).$$

In general, however, this assumption is not always justified, e.g., if the environment contains objects that are not included in the map. In such a situation, the knowledge that one measurement is shorter than expected raises the probability that a neighboring beam of the same scan that intercepts a region close to the first measurement is also shorter. A popular solution to this problem is to use only a subset of all beams of a scan. For example, Fox et al. [158] typically used only 60 beams of the 181 beams of a SICK PLS laser range scan.

Finally, we want to describe some aspects that might be important when implementing this model. The first disadvantage of this model is that one has to perform a ray-casting operation for every potential state of the system. In a grid-based representation of the state space this involves a ray-casting operation for every cell of the grid. If a sample-based representation is used, the same operation has to be performed for every sample in the sample set. One approach to reduce the computation time is to precompute all expected distances and to store them in a large lookup table. Given an appropriate discretization of the state space, the individual entries of this table can be accessed in constant time. If, furthermore, one limits the resolution of the range data such that each beam can have no more than 256 values, only one byte is needed for each entry of the table. In this case, also, the number of histograms for the computation of $P(y \mid d)$ is limited. We only need 256 histograms with 256 bins each.

A further disadvantage of this approach is that the likelihood function sometimes lacks smoothness with respect to slight changes of the locations of the robot. For example, consider a situation in which a laser range finder points into a doorway. If we move the robot slightly, the beam might hit the adjacent wall. Alternatively, consider a beam that hits a wall at a small angle. Slight changes in the orientation of the robot in this case have a high influence on the measured distance. In contrast to that, if the beam is almost perpendicular to a wall, slight changes of the orientation of the robot have almost no influence on the measured distance. One way to solve this problem is to also consider the variance of the expected measurement. This variance can also be computed beforehand by integrating over the local neighborhood of the state x. Given an appropriate discretization of the variances, the histograms then have to be learned for each pair of expected distances and variance. Both techniques, the compact representation of expected distances and the integration of the variance of the expected distance with respect to slight changes in the location of the robot, have been used successfully in Rhino and Minerva [81, 158, 414].

Note that several alternative models have been proposed in the past. The goal of all these models is to provide robust and accurate estimates of the location of the robot. For example, Moravec [324] and Elfes [143], who introduced occupancy grid maps, also presented a probabilistic technique to compute the likelihood of ultrasound measurements given such a map and the position of the robot. Yamauchi and Langley [430] compared local maps built from the most recent measurements with a global map. The sensor models proposed by Konolige [245] and Thrun [412] are more efficient than the model presented here since they avoid the ray-casting operation and only consider the endpoint of each beam. Finally, Simmons and Koenig [386] extracted doorways and corridor junction types out of local grid maps and compared this information to landmarks stored in a topological representation of the environment.

9.2 Mapping

In chapter 8 we learned how to use a Kalman filter for acquiring a map of the environment. The assumption there was that the robot can identify landmarks in the environment and that the posterior about the location of the robot and the landmarks can be represented by a Gaussian distribution. In this section we consider probabilistic forms of mapping that—similarly to probabilistic localization—allow representation of arbitrary posteriors about the state of the environment and the location of the robot during mapping.

To map an environment, a robot has to cope with two types of sensor noise: noise in perception (e.g., range measurements), and noise in odometry (e.g., wheel encoders). Because of the latter, the problem of mapping creates an inherent localization problem. The mobile robot mapping problem is therefore often referred to as the *concurrent mapping and localization problem (CML)* [277], or as the *simultaneous localization and mapping problem* [99, 128] (see also chapter 8). As in chapter 8 we will use the acronym SLAM when referring to the latter. In fact, errors in odometry render the errors of individual features in the map dependent even if the measurement noise is independent, which suggests that SLAM is a high-dimensional statistical estimation problem, often with tens of thousands of dimensions. In this chapter we approach this problem in two steps. First we concentrate on the question of how to build maps given the location of the robot is known. Afterward we relax this assumption and describe a recently developed technique for SLAM.

9.2.1 Mapping with Known Locations of the Robot

A very popular, probabilistic approach to represent the environment is the so-called occupancy probability grid pioneered by Elfes and Moravec in the 80s [325]. Occupancy probability grids are approximative. Each cell m_l of such a two-dimensional grid m stores the probability $P(m_l \mid x(1:k), y(1:k))$ that the place in the environment corresponding to m_l is occupied given the observations $y(1:k) = y(1), \ldots, y(k)$ and all locations of the robot $x(1:k) = x(1), \ldots, x(k)$ at the corresponding points in time. Because of their probabilistic nature, occupancy probability grids can be updated easily based on sensory input.

Occupancy probability grids seek to find the map m that maximizes $P(m \mid x(1:k), y(1:k))$. If we apply Bayes rule using $x(1:k)$ and $y(1:k-1)$ as background knowledge, we obtain

$$(9.26) \quad \begin{aligned} &P(m \mid x(1:k), y(1:k)) \\ &= \frac{P(y(k) \mid m, x(1:k), y(1:k-1)) \, P(m \mid x(1:k), y(1:k-1))}{P(y(k) \mid x(1:k), y(1:k-1))}. \end{aligned}$$

If we assume that $y(k)$ is independent from $x(1 : k - 1)$ and $y(1 : k - 1)$ given we know m, then the right side of this equation can be simplified to

$$(9.27) \quad \begin{aligned} & P(m \mid x(1:k), y(1:k)) \\ & = \frac{P(y(k) \mid m, x(k)) \, P(m \mid x(1:k), y(1:k-1))}{P(y(k) \mid x(1:k), y(1:k-1))}. \end{aligned}$$

We now again apply Bayes rule to determine

$$(9.28) \quad P(y(k) \mid m, x(k)) = \frac{P(m \mid x(k), y(k)) \, P(y(k) \mid x(k))}{P(m \mid x(k))}.$$

If we insert (9.28) into (9.27) and since $x(k)$ does not carry any information about m if there is no observation $y(k)$, we obtain

$$(9.29) \quad \begin{aligned} & P(m \mid x(1:k), y(1:k)) \\ & = \frac{P(m \mid x(k), y(k)) \, P(y(k) \mid x(k)) \, P(m \mid x(1:k-1), y(1:k-1))}{P(m) \, P(y(k) \mid x(1:k), y(1:k-1))}. \end{aligned}$$

If we exploit the fact that each m_l is a binary variable, we derive the following equation for the posterior probability that all cells of m are free in an analogous way.

$$(9.30) \quad \begin{aligned} & P(\neg m \mid x(1:k), y(1:k)) \\ & = \frac{P(\neg m \mid x(k), y(k)) \, P(y(k) \mid x(k)) \, P(\neg m \mid x(1:k-1), y(1:k-1))}{P(\neg m) \, P(y(k) \mid x(1:k), y(1:k-1))}, \end{aligned}$$

where $\neg m$ denotes the complement of m. By dividing (9.29) by (9.30), we obtain

$$(9.31) \quad \begin{aligned} & \frac{P(m \mid x(1:k), y(1:k))}{P(\neg m \mid x(1:k), y(1:k))} \\ & = \frac{P(m \mid x(k), y(k)) \, P(\neg m) \, P(m \mid x(1:k-1), y(1:k-1))}{P(\neg m \mid x(k), y(k)) \, P(m) \, P(\neg m \mid x(1:k-1), y(1:k-1))}. \end{aligned}$$

Finally, we use the fact that $P(\neg A) = 1 - P(A)$ and obtain the following equation:

$$(9.32) \quad \begin{aligned} & \frac{P(m \mid x(1:k), y(1:k))}{1 - P(m \mid x(1:k), y(1:k))} \\ & = \frac{P(m \mid x(k), y(k))}{1 - P(m \mid x(k), y(k))} \frac{1 - P(m)}{P(m)} \frac{P(m \mid x(1:k-1), y(1:k-1))}{1 - P(m \mid x(1:k-1), y(1:k-1))} \end{aligned}$$

If we define

$$(9.33) \quad \text{Odds}(x) = \frac{P(x)}{1 - P(x)},$$

(9.32) turns into

$$\text{Odds}(m \mid x(1:k), y(1:k))$$

(9.34)
$$= \frac{\text{Odds}(m \mid x(k), y(k))\ \text{Odds}(m \mid x(1:k-1), y(1:k-1))}{\text{Odds}(m)}.$$

The corresponding log Odds representation of (9.34) is

$$\log \text{Odds}(m \mid x(1:k), y(1:k))$$
$$= \log \text{Odds}(m \mid x(k), y(k)) - \log \text{Odds}(m)$$

(9.35)
$$+ \log \text{Odds}(m \mid x(1:k-1), y(1:k-1)).$$

Please note that this equation also has a recursive structure similar to that of the recursive Bayesian filtering scheme described in Section 9.1. To incorporate a new scan into a given map we multiply its Odds ratio with the Odds ratio of a local map constructed from the most recent scan and divide it by the Odds ratio of the prior. Often it is assumed that the prior probability of m is 0.5. In this case the prior can be canceled so that (9.35) simplifies to

$$\log \text{Odds}(m \mid x(1:k), y(1:k))$$
$$= \log \text{Odds}(m \mid x(k), y(k))$$

(9.36)
$$+ \log \text{Odds}(m \mid x(1:k-1), y(1:k-1)).$$

To recover the occupancy probability from the Odds representation given in (9.34), we use the following law which can easily be derived from (9.33).

(9.37)
$$P(x) = \frac{\text{Odds}(x)}{1 + \text{Odds}(x)}$$

(9.38)
$$= \left[1 + \frac{1}{\text{Odds}(x)} \right]^{-1}$$

This leads to

$$P(m \mid x(1:k), y(1:k))$$

$$= \left[1 + \frac{\text{Odds}(m)}{\text{Odds}(m \mid x(k), y(k))\ \text{Odds}(m \mid x(1:k-1), y(1:k-1))} \right]^{-1}$$

$$= \left[1 + \frac{1 - P(m \mid x(k), y(k))}{P(m \mid x(k), y(k))} \frac{P(m)}{1 - P(m)} \right.$$

(9.39)
$$\left. \frac{1 - P(m \mid x(1:k-1), y(1:k-1))}{P(m \mid x(1:k-1), y(1:k-1))} \right]^{-1}.$$

Algorithm 19 Occupancy grid mapping with known locations

Input: Sequence of measurements $y(1 : k)$ and corresponding positions $x(1 : k)$ and an initial belief $P_0(m)$ that the cells in the map are occupied

Output: Posterior $P_m = P(m \mid x(1 : k), y(1 : k))$ that the cells in the map are occupied

---

1: $P_m \leftarrow P_0(m)$
2: **for** $i \leftarrow 1$ to k **do**
3: $\quad P_m \leftarrow \left[1 + \dfrac{1 - P(m \mid x(i), y(i))}{P(m \mid x(i), y(i))} \dfrac{P(m)}{1 - P(m)} \dfrac{1 - P_m}{P_m} \right]^{-1}$
4: **end for**

Algorithm 19 uses the recursive nature of (9.39) to compute the posterior $P(m \mid x(1 : k), y(1 : k))$. It receives as input the sequence of measurements $y(1 : k)$ and the corresponding locations of the robot $x(1 : k)$, as well as the initial probability $P_0(m)$ about the occupancy probability of the cells in the map. Typically, $P_0(m)$ will be initialized with the prior probability $P(m)$. If one wants to apply Algorithm 19 to multiple sequences of measurements, $P_0(m)$ can also be initialized with the output obtained from the previous application of the algorithm.

It remains to describe how we actually compute $P(m \mid x(k), y(k))$. Several techniques for determining this quantity have been presented. Whereas Moravec and Elfes [144, 325] used a probabilistic model to compute this quantity for ultrasound measurements, Thrun [411] applied a neural network to learn the appropriate interpretation of the measurements obtained with sonar sensors. The map depicted in the right image of figure 5.1 in chapter 5 has been computed with Thrun's approach. In this chapter we present a model that can be regarded as an approximate version of the approach described by Elfes [144].

One key assumption of occupancy probability grid-mapping techniques is that the individual cells of the map m can be considered independently. Accordingly, the posterior probability of m is computed as

$$(9.40) \quad P(m) = \prod_l P(m_l).$$

The advantage of this approach is that it suffices to describe how to update a single cell upon sensory input. Given this assumption, all we need to specify is the quantity $P(m_l \mid x(k), y(k))$ which is the probability that cell m_l is occupied given the measurement $y(k)$ and the state $x(k)$ of the robot.

The model $P(m_l \mid x(k), y(k))$ described here considers for each cell m_l the difference between the measured distance $y(k)$ and distance of m_l from $x(k)$. In the case of

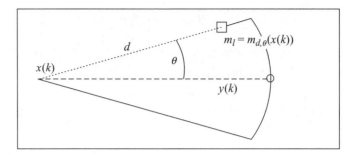

Figure 9.12   The occupancy probability of a cell $m_l = m_{d,\theta}(x(k))$ depends on the distance d to $x(k)$ and the angle θ to the optical axis of the cone.

ultrasound sensors the signal is typically emitted in a cone. To compute the occupancy probability of a cell m_l we therefore also consider the angle θ between the optical axis of the sensor and the ray going through m_l and $x(k)$ (see figure 9.12). The occupancy probability $P(m_l \mid x(k), y(k)) = P(m_{d,\theta}(x(k)) \mid y(k), x(k))$ of m_l is then computed using the following function, which can be regarded as an approximation of the mixture of Gaussians and linear functions applied by Elfes [144].

$$P(m_{d,\theta}(x(k)) \mid y(k), x(k)) = P(m_{d,\theta}(x(k)))$$

(9.41)
$$+ \begin{cases} -s(y(k), \theta) & d < y(k) - d_1 \\ -s(y(k), \theta) + \frac{s(y(k),\theta)}{d_1}(d - y(k) + d_1) & d < y(k) + d_1 \\ s(y(k), \theta) & d < y(k) + d_2 \\ s(y(k), \theta) - \frac{s(y(k),\theta)}{d_3 - d_2}(d - y(k) - d_2) & d < y(k) + d_3 \\ 0 & \text{otherwise.} \end{cases}$$

In this definition $s(y(k), \theta)$ is a function that computes the deviation of the occupancy probability from the prior occupancy probability $P(m)$ given the measured distance $y(k)$ and the angle θ between the cell, the sensor, and its optical axis. A common choice for $s(y(k), \theta)$ is a product of a linear function $g(y(k))$ and a Gaussian $\mathcal{N}(0, \sigma_\theta)$:

(9.42) $s(y(k), \theta) = g(y(k))\, \mathcal{N}(0, \sigma_\theta)$

figure 9.13 plots these two components as they are used in the examples shown in this section. The variance σ_θ of the Gaussian is 0.05. Figure 9.14 plots $s(y(k), \theta)$ for $y(k) \in [0; 3m]$ and $\theta \in [-\frac{\pi}{24}; \frac{\pi}{24}]$. This angular range is identical to the opening angle of 15 degrees of the ultrasound sensors used to acquire the data of the examples presented here.

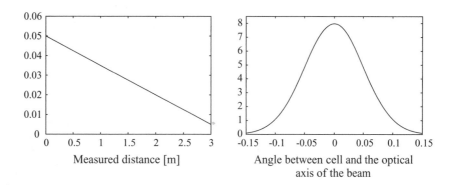

Figure 9.13 Functions used to compute the function $s(y(k), \theta)$: linear function (left) and Gaussian (right).

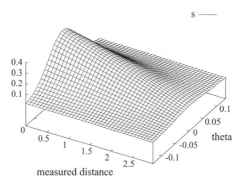

Figure 9.14 Function $s(y(k), \theta)$ used to model the deviation from the prior for cells in the sensor cone.

The constants d_1, d_2, and d_3 in (9.41) specify the interval in which the different linear functions of the piecewise linear approximation are valid (see also figure 9.15). The occupancy probability of cells lying between $x(k)$ and the arc from which the signal was reflected must be smaller than the prior probability for occupancy. In our model the occupancy probability of cells with $d < y(k) - d_1$ therefore is computed as $P(m_l) - s(y(k), \theta)$. The occupancy probability of cells whose distance to $x(k)$ is close to $y(k)$, i.e., for which $y(k) - d_1 \leq d < y(k) + d_1$, is computed by a linear function that increases with d. If a beam ends in a cell it is commonly assumed that the world is also occupied at least for a certain range behind that cell. In our model the occupancy probability therefore stays at a high but constant level $P(m_l) + s(y(k), \theta)$ for all cells whose distance lies between $y(k) + d_1$ and $y(k) + d_2$. Accordingly, the constants

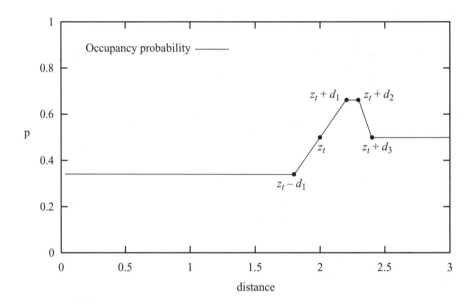

Figure 9.15   Probability $P(m_{d,\theta}(x(k))$ that a cell on the optical axis of the sensor ($\theta = 0$) is occupied depending on the distance of that cell from the sensor. The measured distance is $2m$.

d_1 and d_2 encode the average depth of obstacles. For distances d with $y(k) + d_2 \leq d < y(k) + d_3$, we assume that the occupancy probability linearly decreases to the prior occupancy probability $P(m_l)$. Finally, for cells $m_l = m_{d,\theta}(x(k))$ whose distance from $x(k)$ exceeds $y(k) + d_3$ we can safely assume that $P(m_{d,\theta}(x(k)) \mid y(k), x(k))$ equals the prior probability $P(m_l)$, since $y(k)$ does not give us any information about such cells. Figure 9.15 plots $P(m_{d,\theta}(x(k)) \mid y(k), x(k))$ for d ranging from $0\ m$ to $3\ m$ given that $y(k)$ is $2\ m$ and that θ is 0 degrees. In this case, the value of $s(y(k), \theta)$ is approximately 0.16. We additionally assume that the prior probability $P(m_l) = P(m_{d,\theta}(x(k))) = 0.5$ for all d, θ, and $x(k)$.

Figure 9.16 shows three-dimensional plots of the resulting occupancy probabilities for measurements of $2.0\ m$ and $2.5\ m$. In both plots the optical axis of the sensor cone is identical to the x-axis and the sensor is located in the origin of the coordinate frame. As can be seen from the figure, the occupancy probability is high for cells whose distance to $x(k)$ is close to $y(k)$. It decreases for cells with distance $y(k)$ from $x(k)$ and with increasing values of θ. Furthermore, it stays constant for cells that are immediately behind a cell that might have reflected the beam and linearly decreases to the prior probability afterward. For cells that are covered by the beam but did not reflect it, the occupancy probability is decreased.

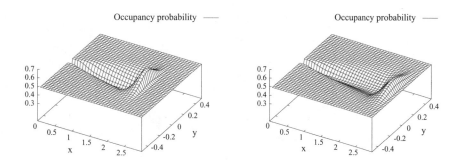

Figure 9.16   Local occupancy probability grids for a single ultrasound measurement of $y(k) =$ 2.0m (left) and $y(k) = 2.5m$ (right).

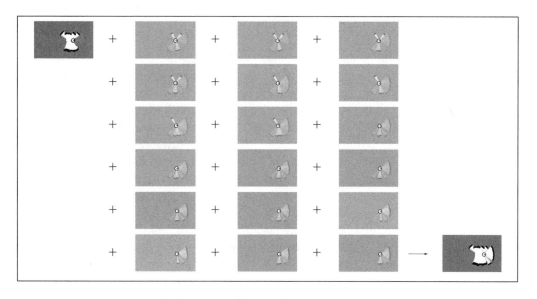

Figure 9.17   Incremental mapping in a corridor environment. The upper left image shows the initial map and the lower right image contains the resulting map. The maps in the three middle columns are the local maps built from the individual ultrasound scans perceived by the robot. Throughout this process, measurements beyond a 2.5 m radius have not been considered.

Figure 9.17 shows our sensor model in action for a sequence of measurements recorded with a B21r robot in a corridor environment. The upper-left corner shows a map that has been built from a sequence of ultrasound scans. Afterward the robot perceived a series of eighteen ultrasound scans each consisting of 24 measurements.

Figure 9.18   Occupancy probability map for the corridor of the Autonomous Intelligent Systems Lab at the University of Freiburg (left) and the corresponding maximum-likelihood map (right).

The occupancy probabilities for these eighteen scans are depicted in the three columns in the center of this figure. Note that during mapping we did not use measurements whose distance exceeded $2.5m$. The occupancy probability grid obtained by integrating the individual observations into the initial map is shown in the lower-right corner of this figure. As can be seen, the belief converges to an accurate representation of the corridor structure although the individual measurements show a high amount of uncertainty, as is usually the case for ultrasound sensors.

The left image of figure 9.18 shows the occupancy probabilities of the corridor environment obtained after incorporating all measurements of the data set used here. The map represents a 17 m long and 11 m wide part of a corridor environment including three rooms. The right image shows the corresponding maximum-likelihood map. This map is obtained from the occupancy probability grid by a simple clipping operation with a threshold of 0.5. The gray areas of the maximum-likelihood map correspond to cells that have not been sensed by the robot.

Let us briefly discuss some aspects that might be relevant to potential improvements of the models described here. The strongest restriction results from the assumption that all cells of the grid are considered independently. This independence assumption decomposes the high-dimensional state estimation problem into a set of onedimensional estimation problems. The independency of individual cells, however, is usually not justified in practice. For example, if the robot detects a door, then particular cells in the neighborhood need to be updated according to the specific shape of the door. Accordingly, techniques considering the individual cells of a grid independently might produce suboptimal solutions. One technique that addresses this problem has recently been presented by Thrun [413].

Additionally, occupancy probability grid maps assume that the environment has a binary structure, i.e., that every cell is either occupied or free. Occupancy probabilities cannot correctly represent situations in which a cell is only partly covered by an obstacle. Finally, most of the techniques, as well as our model, assume that the individual beams of the sensors can be considered independently when updating a map. This assumption also is not justified in practice, since neighboring beams of a scan often yield similar values. Accordingly, a robot ignoring this might become overly confident of the state of the environment.

9.2.2   Bayesian Simultaneous Localization and Mapping

In the previous subsection, we assumed that the robot always knows its position while it is mapping the environment. This assumption, however, is typically not justified, especially when a robot has to rely on its onboard sensors to determine its position due to the lack of a global positioning system, active beacons, or predefined landmarks. In such a situation, mapping turns into the so-called chicken and egg problem. Without a map the robot cannot determine its own position and without knowledge about its own position the robot cannot compute what its environment looks like. This is why this problem is often denoted as the SLAM problem (see also chapter 8).

In the past, research in the area of SLAM has led to two different types of approaches, each of which has its advantages and disadvantages [412]. The first class contains algorithms relying on the EKF to estimate joint posteriors over maps and robot locations [100, 128, 277]. These approaches provide a sound mathematical framework (see also chapter 8). However, they mainly have been applied in situations in which the environment contains predefined landmarks.

The second class of techniques considers the SLAM problem as a global optimization problem. For example, Lu and Milios [299] consider robot locations as random variables and derive constraints between locations from distances between overlapping range measurements and from odometry measurements. The constraints can be regarded as links in a network of springs, whose energy is to be minimized. Other approaches apply Dempster, Laird and Rubin's *expectation maximization,* or *EM* algorithm [127] to compute the maximum-likelihood estimate for the map and the locations of the robot. Examples of these kind of techniques can be found in [84, 126, 382, 417]. EM-based techniques have been applied successfully to mapping large cyclic environments with highly ambiguous features. However, they are inherently batch algorithms, requiring multiple passes through the entire data set. As a consequence, they usually cannot be applied when a robot has to map its environment online, i.e., while it is exploring it.

In probabilistic terms the problem of SLAM is to find the map and the robot positions which yield the best interpretation of the data gathered by the robot. As in

Section 9.1 the data consist of a stream of odometry measurements $u(0 : k - 1)$ and perceptions of the environment $y(1 : k)$. According to Thrun [412], the mapping problem can be phrased as recursive Bayesian filtering for estimating the robot positions along with a map of the environment:

$$P(x(1 : k), m \,|\, u(0 : k - 1), y(1 : k)) = \alpha \; P(y(k) \,|\, x(k), m)$$

$$\int \bigg(P(x(k) \,|\, u(k - 1), x(k - 1))$$

(9.43) $$P(x(1 : k - 1), m \,|\, u(0 : k - 2), y(1 : k - 1)) \bigg) dx(1 : k - 1).$$

As in probabilistic localization (see Section 9.1) we assume that the odometry measurements are governed by a so-called probabilistic motion model $P(x(k) \,|\, x(k - 1), u(k - 1))$ which specifies the likelihood that the robot is at $x(k)$ given that it previously was at $x(k - 1)$ and the motion $u(k - 1)$ was measured. On the other hand, the observations follow the sensor model $P(y(k) \,|\, x(k), m)$, which defines for every possible location $x(k)$ in the environment the likelihood of the observation $y(k)$ given the map m.

Unfortunately, estimating the full posterior in (9.43) is not tractable in general. One approach is to apply incremental scan matching [180, 182, 365, 424]. The general idea of such approaches can be summarized as follows. At any point $k - 1$ in time, the robot is given an estimate of its location $\hat{x}(k - 1)$ and a map $\hat{m}(\hat{x}(1 : k - 1), y(1 : k - 1))$. After moving and taking a new measurement $y(k)$, the robot determines the most likely new location $\hat{x}(k)$ such that

$$\hat{x}(k) = \underset{x(k)}{\operatorname{argmax}} \big\{ P(y(k) \,|\, x(k), \hat{m}(\hat{x}(1 : k - 1), y(1 : k - 1)))$$

(9.44) $$P(x(k) \,|\, u(k - 1), \hat{x}(k - 1)) \big\}.$$

It does this by trading off the consistency of the measurement with the map [first term on the right-hand side in (9.44)] and the consistency of the new location with the control action and the previous location [second term on the right-hand side in (9.44)]. The map is then extended by the new measurement $y(k)$, using the location $\hat{x}(k)$ as the location at which this measurement was taken. Popular techniques to determine $\hat{x}(k)$ in the context of laser range scans are the iterative-closest-point algorithm [46] or variants thereof.

The key limitation of scan-matching approaches lies in the greedy maximization step. Once the location $x(k)$ at time k has been computed it is not revised afterward so that the robot cannot recover from registration errors. Although scan matching techniques have been proven to be able to correct enormous errors in odometry, the resulting maps often are globally inconsistent. As an example consider figure 9.19

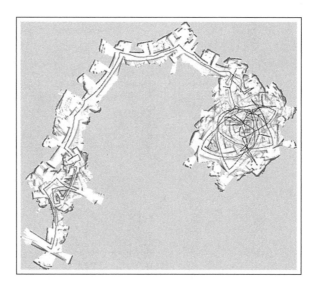

Figure 9.19 Map generated from raw odometry and laser range data gathered with a B21r robot.

which shows a map generated from raw odometry and laser range data obtained with a B21r robot. As can be seen from the figure, the robot suffers from serious errors in odometry so that the resulting map is useless without any correction. The size of this environment is 28 m × 28 m. When recording the data the robot traveled 491 m with an average speed of 0.19 m/s. Figure 9.20 shows the map created with the scan matching system presented by Hähnel, Schulz, and Burgard [182]. Although the local structures of the map appear to be very accurate, the map is globally inconsistent. For example, many structures like walls, doors, and such can be found several times and with a small offset between them.

To overcome this problem, alternative approaches have been developed. The key idea of these techniques is to maintain a posterior about the position of the vehicle. Whereas Gutmann and Konolige [178] used a discrete and grid-based approximation of the belief about the robots location, Thrun [412] applied a particle filter for this purpose. However, both approaches only maintain a single map and revise previous decisions whenever the robot closes a loop and returns to a previously visited place.

More recently, Murphy and coworkers [137,329] have proposed *Rao-Blackwellized particle filtering* as an efficient means to maintain multiple hypotheses during mapping. The key idea of this approach can be understood more easily when one considers the graphical model depicted in figure 9.21. If we know the map, the overall problem is transformed into a localization problem where the task is to estimate the location

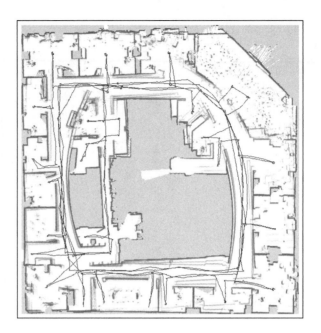

Figure 9.20   Map obtained by applying a scan-matching approach to the same data used in figure 9.19.

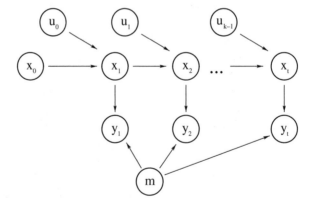

Figure 9.21   Graphical model of incremental probabilistic SLAM.

of the robot at each point in time. If, however, the locations are known, it remains solely to compute the map. Note that the knowledge of $x(1:k)$ is sufficient to figure out what the environment looks like, whereas $x(0)$ only determines the location of the map. Thus, if $x(1:k)$ is known but $x(0)$ is unknown, the robot can estimate its position relative to the map, but it cannot determine the location of the map.

The application of Rao-Blackwellized particle filtering to mapping is motivated by the observation that once the path $x(1:k)$ of the robot is known, the maximum-likelihood map can be computed analytically, e.g., using the method described in subsection 9.2.1. Therefore, the goal of SLAM with Rao-Blackwellized particle filters is to estimate the path of the robot using a particle filter and to analytically compute the map corresponding to that path. In practice this means that we use a set of particles to represent a posterior about potential paths of the robot. To each of these paths we associate an individual map that is computed based on the hypothesis that this path corresponds to the true path of the robot. The importance weight of a sample is proportional to the likelihood of the most recent observation given the map, which is computed based on the previous observations and the path of the robot according to that particular particle.

Note that Rao-Blackwellized particle filtering [136, 137] is a general technique to reduce the size of high-dimensional state estimation problems by marginalizing out parts of the state space. In this section we use this technique to develop an efficient solution to the SLAM problem.

Let us again consider the posterior $P(x(1:k), m \mid u(0:k-1), y(1:k))$ we want to estimate. If we apply the chain rule of probability theory, we obtain

$$
\begin{aligned}
&P(x(1:k), m \mid u(0:k-1), y(1:k)) \\
&\quad = P(m \mid x(1:k), y(1:k), u(0:k-1)) \\
&\qquad P(x(1:k) \mid y(1:k), u(0:k-1)).
\end{aligned}
$$

(9.45)

Obviously, we can safely assume that m is independent of $u(0:k-1)$ once we know the locations $x(1:k)$ of the robot, i.e.,

(9.46) $P(m \mid x(1:k), y(1:k), u(0:k-1)) = P(m \mid x(1:k), y(1:k)).$

This leads to

$$
\begin{aligned}
&P(x(1:k), m \mid u(0:k-1), y(1:k)) \\
&\quad = P(m \mid x(1:k), y(1:k)) \, P(x(1:k) \mid y(1:k), u(0:k-1)).
\end{aligned}
$$

(9.47)

In the previous section we saw that we can efficiently compute the posterior $P(m \mid x(1:k), y(1:k))$ for m given we know $x(1:k)$ and $y(1:k)$. Thus, all we need to do is to sample $P(x(1:k) \mid y(1:k), u(0:k-1))$ using a particle filter and compute for each particle the map that is associated to it.

We proceed as follows. Suppose $\mathcal{M}$ is a set of particles that represents the posterior about potential paths of the robot. In the beginning we assume that each particle starts at $[0, 0, 0]^T$, i.e., the robot is located at the origin of the coordinate system and its heading is 0. Let us furthermore denote the path associated with the jth particle by $h^{(j)}(1 : k)$. As described above, once the path of the robot is known, we can directly compute the most likely map for that particle:

$$(9.48) \qquad m^{(j)}(1 : k - 1) = \underset{m}{\arg\max} \, P(m \mid h^{(j)}(1 : k), y(1 : k - 1))$$

Whenever an odometry measurement $u(i-1)$ is obtained, we proceed in the same way as we do in probabilistic localization. For each sample we compute the next location $x = x'_j$ of its path by sampling from $P(x \mid x_j, u(i-1))$. Note that—as in probabilistic localization—we in principle had to sample from $P(x \mid x_j, u(i-1), m^{(j)}(1 : k - 1))$, i.e., we also had to consider the map $m^{(j)}(1 : k - 1)$ associated with each sample. In practice, however, the map is often ignored for reasons of efficiency since computing $P(x \mid x_j, u(i - 1), m^{(j)}(1 : k - 1))$ typically involves a time-consuming ray-casting operation in $m^{(j)}(1 : k-1)$. Once we have computed for the jth particle both the map $m^{(j)}(1 : k - 1)$ and the new location x'_j, we are ready to compute the likelihood of the observation $y(k)$ and to use the resulting quantity as an importance weight during the resampling step. As a sensor model we can, e.g., choose the model described in section 9.1. The overall approach is realized by algorithm 20. Note that, according to the recursive structure of the problem, this algorithm can easily be extended for multiple sequences of sensory input. To do so, one simply has to ensure that the initialization is carried out using the output obtained from the previous application of the algorithm.

Please note that two aspects of this algorithm need to be implemented carefully to obtain the desired efficiency and convergence. If the map $m^{(j)}(1 : k - 1)$ is computed from scratch in every round, the resulting algorithm will be quadratic in k. On the other hand, maintaining a complete map for each individual particle (which also needs to be updated in each round) is inefficient with respect to memory. Additionally, it involves a time-consuming operation if the map associated with a sample has to be copied once for each of its successors in the resampling step. A popular approach to overcome this problem is to use treelike structures such as those proposed by Montemerlo et al [321], as well as by Parr and Eliazar [346]. Since typically many of the particles have larger parts of their history in common, the maps associated with the particles can efficiently be represented using trees.

An alternative although approximative approach to compute the maps associated with the individual samples is based on the observation that, in order to compute $P(y(k) \mid x_j, m^{(j)}(1 : k - 1))$, we only need to determine the part of the map $m^{(j)}(1 : k - 1)$ that is covered by $y(k)$. If we furthermore use only a limited number of

---

Algorithm 20 Simultaneous localization and mapping using Rao-Blackwellized particle filtering

---

Input: Sequence of measurements $y(1:k)$ and movements $u(0:k-1)$ and set $\mathcal{M}$ of N samples (x_j, ω_j)

Output: Posterior $P(x(1:k), m \mid u(0:k-1), y(1:k))$ represented by $\mathcal{M}$ about the path of the robot at time and the map

---

1: **for** $j \leftarrow 1$ to N **do**
2:    $x_j \leftarrow (0, 0, 0)$
3: **end for**
4: **for** $i \leftarrow 1$ to k **do**
5:    **for** $j \leftarrow 1$ to N **do**
6:      compute a new state x by sampling according to $P(x \mid u(i-1), x_j)$.
7:      $x_j \leftarrow x$
8:    **end for**
9:    $\eta \leftarrow 0$
10:   **for** $j \leftarrow 1$ to N **do**
11:     $w_j = P(y(i) \mid x_j, m^{(j)}(1:i-1)))$
12:     $\eta = \eta + w_j$
13:   **end for**
14:   **for** $j \leftarrow 1$ to N **do**
15:     $w_j = \eta^{-1} \cdot w_j$
16:   **end for**
17:   $\mathcal{M} = resample(\mathcal{M})$
18: **end for**

---

measurements from the history $h^{(j)}(1:k)$ we obtain an approximation of $m^{(j)}(1:k-1)$ that, if, additionally, spatial indices are used to compute the set of relevant scans from $h^{(j)}(1:k)$, can be computed in constant time. Thus, the overall complexity is constant for each particle. This approach has been successfully applied by Hähnel, Burgard, Fox, and Thrun [181] and has been used for the examples presented here.

A further aspect which turns out to be crucial to the success of the overall approach is the limitation of the number of particles that are needed. Since each particle possesses an individual map, the memory required by using Rao-Blackwellized filtering can be quite high, especially if many samples are needed to appropriately represent the posterior. One technique to reduce the number of necessary samples has recently been developed by Hähnel et al. [181]. In their approach consecutive laser range scans are converted into highly accurate odometry measurements. This way the uncertainty in the location of the robot is reduced so that fewer samples are needed to represent the posterior.

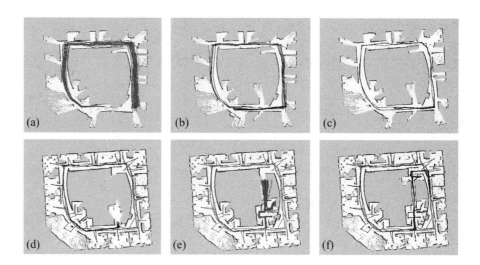

Figure 9.22 Sequence of maps corresponding to the particle with the highest accumulated importance weight during Rao-Blackwellized particle filtering of a large cyclic environment.

Figure 9.22 shows a Rao-Blackwellized particle filter for simultaneous localization and mapping in action. The individual figures illustrate the posterior about the robot's location as well as the map associated with the particle with the maximum accumulated importance factor. Image (a) shows the belief of the robot just before the robot is closing a loop. The second image (b) depicts the belief some steps later after the robot has closed the loop. As can be seen, the belief is more peaked due to the fact that particles whose observations do not match to their maps quickly die out when a loop is closed. Picture (c) shows a situation when the robot has moved around the loop for a second time. Please note that all figures also show the paths of all particles. A low number of different paths indicates that at the corresponding point in time, already many particles have a common history. In the situation depicted in image (d) the robot has visited all rooms in the building and enters a new corridor which imposes the task of closing another loop. The belief shortly before the robot closes this second loop is depicted in image (e). Image (f) shows the map and the particle histories after the robot finished its task. The resulting map is illustrated in figure 9.23.

After they have been demonstrated to be an efficient means for SLAM [137, 329], Rao-Blackwellized particle filters have been used with great success to learn large-scale maps of different types of environments. For example, Montemerlo et al. [321] have applied this technique to landmark-based mapping in which the locations of the individual landmarks are represented by Gaussians. In a more recent work [320] Montemerlo and Thrun extended this work to landmark-based mapping with uncertain

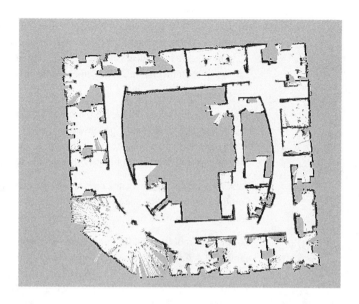

Figure 9.23 Resulting map obtained with Rao-Blackwellized particle filtering for SLAM.

data association. Additionally, this technique has been applied successfully to the simultaneous estimation of states of dynamic objects and the robot's locations [32, 322]. As mentioned above, new results present optimizations of this technique that allow the efficient application of Rao-Blackwellized particle filtering to SLAM with raw laser range scans [181, 346].

Problems

1. Prove the following variant of Bayes rule with background knowledge E:

$$P(A \mid B, E) = \frac{P(B \mid A, E) P(A \mid E)}{P(B \mid E)}$$

2. Use (9.2) to rederive the equations for the Kalman filter.

3. Implement the sensor model for mobile robot localization using data from proximity sensors described in Subsection 9.1.5. Proceed in the following steps.

   (a) Generate a model of your robot's environment and implement a function that takes as input the location and sensing direction of a sensor and that generates as output the expected distance to the next obstacle in the direction of the measurement.

(b) Use (9.23) to generate for each of a discrete set of expected distances a histogram whose values represent the likelihood of the corresponding measured distance given that expected distance. Choose a discretization of 3 inches. To ensure that the values $h_{i,j}$ of each histogram sum up to one over all j for each i, choose appropriate values for δ_i.

4. Implement a motion model for sample-based mobile robot localization according to (9.19), (9.20), and (9.21). Realize a procedure that continuously reads odometry measurements from your robot and propagates a set of samples according to your motion model based on this input. Your program should generate sample sets similar to those shown in figure 9.6.

5. Implement probabilistic localization based on a particle filter using Algorithms 17 and 18. Combine the procedures developed in the previous two assignments and apply your algorithm to data obtained from your robot's odometry and from its proximity sensors.

6. Consider a robot that resides in a circular world consisting of ten different places that are numbered counterclockwise. The robot is unable to sense the number of its present place directly. However, places 0, 3, and 6 contain a distinct landmark, whereas all other places do not. All three of these landmarks look alike. The likelihood that the robot observes the landmark given it is in one of these places is 0.8. For all other places, the likelihood of observing the landmark is 0.4. For each place on the circle compute the probability that the robot is in that place given that the following sequence of actions is carried out deterministically and the following sequence of observations is obtained: The robot detects a landmark, moves 3 grid cells counterclockwise and detects a landmark, and then moves 4 grid cells counterclockwise and finally perceives no landmark.

7. Implement a program that simulates the world described in Question 6. Assume that the actions are nondeterministic: When moving from one place to a neighboring place, the robot succeeds with probability 0.8 but stays in the same place with probability 0.2. Run your algorithm to calculate the posterior probability over places (a) under the deterministic action model and (b) the non-deterministic action model.

8. Argue why, once the location $x(k)$ of the robot and the map m is known, the measurement $y(k)$ is independent of all previous measurements $y(1:k-1)$. Show why this independence does *not* hold if the map m is not given.

9. What happens if you apply the particle filter algorithm to a robot whose sensor is almost perfect? For example, what happens when the robot uses (almost) noise-free range sensors? Hint: For near-perfect sensors, the likelihood-function $P(y\,|\,x)$ will be extremely peaked, i.e., it will be almost zero for all measurements that are slightly off the correct noise-free value. How does the accuracy of the sensor affect the number of particles needed?

10. Implement a motion model for the particle filter algorithm which takes into account that the robot cannot move through obstacles. Run a simulation in which the robot starts in the center of an empty and quadratic room without doors. Suppose the robot moves forward d meters, then turns left and again moves d meters, where d is the width of the room.

11. Implement the circular world described in Question 6 using a particle filter algorithm.

12. Consider a robot that has to use its camera for localization. Discuss what models could be used to compute the likelihood $P(y \mid x)$ for vision-based robot localization.

13. Suppose the robot has only two kinds of observations y_1 and y_2. Furthermore, suppose that the occupancy probability of a particular cell covered by an observation of type y_1 is 0.8 and for observations of type y_2 is $1/3$. To which value will the occupancy probability of m_i converge if only one out of three measurements is y_1 and all others are y_2? *Hint:* Use (9.35) and calculate to which value the occupancy probability of m_i will converge if the number of measurements goes to infinity.

14. Discuss how the sensor model for $P(m_l \mid x(k), y(k))$ for occupancy grids changes if more accurate sensors are used and if the opening angle of the sonar cone becomes smaller.

15. Consider an environment which has the shape of a snailshell as depicted below. What will happen if you apply the Rao-Blackwellized particle filtering algorithm in such an environment? Consider robots with accurate/weak odometry and with accurate/noisy sensors.

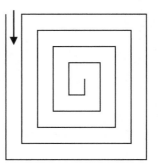

10 *Robot Dynamics*

WE HAVE SEEN THAT A path specifies the set of configurations a robot achieves as it moves from one configuration to another, and thus path planning (e.g., finding collision-free paths) is a kinematic/geometric problem. A path is not a complete description of the motion of a robot system, however, as the timing of the motion is not specified. A trajectory is a path plus a specification of the time at which each configuration is achieved. Trajectory planning is not only a geometric problem, but also a dynamic problem. Finding feasible trajectories of a system obeying dynamics requires knowledge of the masses and inertias of the system, actuator limits, and forces such as gravity and friction. Since we are now dealing with system dynamics, we can pose optimal control problems such as finding minimum-time or minimum-energy motions.

Since trajectory planning requires a full dynamic model of the robotic system, in section 10.1 we review the Lagrangian approach to deriving equations of motion for a mechanical system such as a robot arm. Section 10.2 explores the structure of the equations of motion and gives standard forms for writing them. In section 10.3 we consider systems subject to velocity constraints, such as those imposed when maintaining rolling and sliding contacts. Finally, section 10.4 studies the particular case of a rotating and translating rigid body.

10.1 Lagrangian Dynamics

The equations of motion for a mechanical system can be generated in a variety of ways. While all are equivalent (if correctly done!), the number of computations and the size of the resulting expressions may vary. In this chapter we use a Lagrangian formulation, which is based on the kinetic and potential energy of the system. Lagrange's equations provide a straightforward recipe, amenable to computer implementation (using, e.g., Mathematica or Maple), for calculating equations of motion for many robotic systems.

Let $q = [q_1, \ldots, q_{n_Q}]^T \in \mathbb{R}^{n_Q}$ be a vector of *generalized coordinates* representing the configuration of the system on the n_Q-dimensional configuration space, and let $u = [u_1, \ldots, u_{n_Q}]^T \in \mathbb{R}^{n_Q}$ be the vector of *generalized forces* acting on the generalized coordinates. For example, for a robot arm, the generalized coordinates would typically be the joint angles for revolute joints and the joint translations for prismatic joints, and the generalized forces would be torques about the joints and forces along the joints, respectively.

The *Lagrangian L* of a mechanical system is written as the kinetic energy minus the potential energy

$$L(q, \dot{q}) = K(q, \dot{q}) - V(q),$$

where K is the kinetic energy, a function of the configuration and the velocity, and V is the potential energy, a function of the configuration only. The Lagrangian equations of motion, also known as the Euler-Lagrange equations, can be written

$$\frac{d}{dt}\frac{\partial L}{\partial \dot{q}_i} - \frac{\partial L}{\partial q_i} = u_i, \quad i = 1 \ldots n_Q,$$

or simply

(10.1)
$$\frac{d}{dt}\frac{\partial L}{\partial \dot{q}} - \frac{\partial L}{\partial q} = u.$$

A derivation of these equations can be found in many dynamics textbooks.

EXAMPLE 10.1.1   *Consider a planar body described by the generalized coordinates* $q = [q_1, q_2, q_3]^T \in \mathbb{R}^2 \times S^1$, *where* $(q_1, q_2) \in \mathbb{R}^2$ *specify the location of the center of mass of the body in the plane and* $q_3 \in S^1$ *specifies the orientation of the body (figure 10.1). The generalized forces* $u = [u_1, u_2, u_3]^T \in \mathbb{R}^3$ *are the linear forces* (u_1, u_2) *through the center of mass and the torque* u_3 *about the center of mass. A gravitational acceleration* $a_g \geq 0$ *acts in the* $-q_2$ *direction,* $[0, -a_g, 0]^T$. *The mass of the planar body is* m *and* I *is the scalar inertia about an axis through the center of mass and out of the page.*

The Lagrangian for this system is

$$L = \frac{1}{2}m\left(\dot{q}_1^2 + \dot{q}_2^2\right) + \frac{1}{2}I\dot{q}_3^2 - ma_gq_2,$$

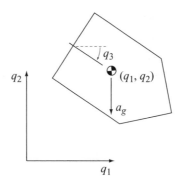

Figure 10.1   A planar body.

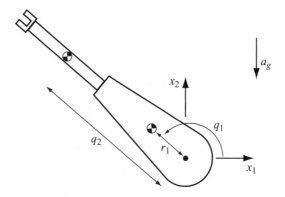

Figure 10.2   The RP manipulator.

i.e., the sum of the linear and angular kinetic energies minus the potential energy. Applying (10.1), we get

$$(10.2) \quad u_1 = \frac{d}{dt}\frac{\partial L}{\partial \dot{q}_1} - \frac{\partial L}{\partial q_1} = \frac{d}{dt}(m\dot{q}_1) - 0 = m\ddot{q}_1$$

$$(10.3) \quad u_2 = \frac{d}{dt}\frac{\partial L}{\partial \dot{q}_2} - \frac{\partial L}{\partial q_2} = \frac{d}{dt}(m\dot{q}_2) - ma_g = m\ddot{q}_2 - ma_g$$

$$(10.4) \quad u_3 = \frac{d}{dt}\frac{\partial L}{\partial \dot{q}_3} - \frac{\partial L}{\partial q_3} = \frac{d}{dt}(I\dot{q}_3) - 0 = I\ddot{q}_3.$$

EXAMPLE 10.1.2   *Figure 10.2 shows a robot arm consisting of one revolute joint and one prismatic (translational) joint. This type of robot is called an RP manipulator.*

The configuration of the robot is $[q_1, q_2]^T$, *where* q_1 *gives the angle of the first joint from a world frame* x_1-*axis, and* $q_2 > 0$ *gives the distance of the center of mass of the second link from the first joint. The center of mass of the first link is a distance* r_1 *from the first joint. The first link has mass* m_1 *and inertia* I_1 *about the center of mass, and the second link has mass* m_2 *with inertia* I_2 *about the center of mass. A gravitational acceleration* $a_g \geq 0$ *acts in the* $-x_2$ *direction of a world frame. To derive the Lagrangian for this RP arm, we will consider the two links' contributions independently.*

The kinetic energy of the first link can be expressed as

$$K_1(q, \dot{q}) = \frac{1}{2} m_1 v_1^2 + \frac{1}{2} I_1 \omega_1^2,$$

where v_1 *and* ω_1 *are the linear velocity of the link center of mass and angular velocity of the link, respectively. We have*

$$v_1 = r_1 \dot{q}_1$$

$$\omega_1 = \dot{q}_1$$

yielding the expression

$$K_1(q, \dot{q}) = \frac{1}{2} m_1 r_1^2 \dot{q}_1^2 + \frac{1}{2} I_1 \dot{q}_1^2.$$

The potential energy of the first link is

$$V_1(q) = m_1 a_g r_1 \sin q_1.$$

The kinetic energy of the second link can be expressed as

$$K_2(q, \dot{q}) = \frac{1}{2} m_2 v_2^2 + \frac{1}{2} I_2 \omega_2^2,$$

where v_2 *and* ω_2 *are the linear velocity of the link center of mass and angular velocity of the link, respectively. We have*

$$v_2 = \sqrt{\dot{q}_2^2 + (q_2 \dot{q}_1)^2}$$

$$\omega_2 = \dot{q}_1$$

yielding

$$K_2(q, \dot{q}) = \frac{1}{2} m_2 \left(\dot{q}_2^2 + (q_2 \dot{q}_1)^2 \right) + \frac{1}{2} I_2 \dot{q}_1^2.$$

The potential energy of the second link is

$$V_2(q) = m_2 a_g q_2 \sin q_1.$$

The Lagrangian for the system is $L = K_1 + K_2 - V_1 - V_2$:

$$L = \frac{1}{2}\left(\left(I_1 + I_2 + m_1 r_1^2 + m_2 q_2^2\right)\dot{q}_1^2 + m_2 \dot{q}_2^2\right) - a_g \sin q_1 (m_1 r_1 + m_2 q_2).$$

Applying Lagrange's equations, we get

(10.5)
$$\begin{aligned} u_1 &= \left(I_1 + I_2 + m_1 r_1^2 + m_2 q_2^2\right)\ddot{q}_1 + 2m_2 q_2 \dot{q}_1 \dot{q}_2 \\ &\quad + a_g(m_1 r_1 + m_2 q_2)\cos q_1 \end{aligned}$$

(10.6)
$$u_2 = m_2 \ddot{q}_2 - m_2 q_2 \dot{q}_1^2 + a_g m_2 \sin q_1.$$

10.2 Standard Forms for Dynamics

Let's take a closer look at equations (10.5) and (10.6), since they display much of the interesting structure of many second-order mechanical systems. On the right-hand side of each equation, there is a term depending on the second derivatives of the configuration variables, a term quadratic in the first derivatives of the configuration variables, and a term depending only on the configuration variables. These terms can be collected to write the dynamics in the following standard form:

(10.7)
$$u = M(q)\ddot{q} + C(q, \dot{q})\dot{q} + g(q),$$

where $C(q, \dot{q})\dot{q} \in \mathbb{R}^{n_Q}$ is a vector of velocity product terms with the $n_Q \times n_Q$ matrix $C(q, \dot{q})$ linear in $\dot{q}$, $g(q) \in \mathbb{R}^{n_Q}$ is a vector of gravitational forces, and $M(q)$ is an $n_Q \times n_Q$ symmetric, positive definite *mass* or *inertia matrix*. A matrix M is symmetric if $M_{ij} = M_{ji}$, where M_{ij} is the entry in the ith row and jth column of M. A matrix M is positive definite if $v^T M v > 0$ holds for any nonzero vector v. This is true if the determinant and trace (the sum of diagonal elements) of M are positive.

Equations (10.5) and (10.6) for the RP manipulator can be written in the standard form of equation (10.7), where

$$M(q) = \begin{bmatrix} I_1 + I_2 + m_1 r_1^2 + m_2 q_2^2 & 0 \\ 0 & m_2 \end{bmatrix},$$

$$C(q, \dot{q})\dot{q} = \begin{bmatrix} 2m_2 q_2 \dot{q}_1 \dot{q}_2 \\ -m_2 q_2 \dot{q}_1^2 \end{bmatrix}, \quad g(q) = \begin{bmatrix} a_g(m_1 r_1 + m_2 q_2)\cos q_1 \\ a_g m_2 \sin q_1 \end{bmatrix}.$$

The standard form (10.7) is compact, but the term $C(q, \dot{q})\dot{q}$ masks the fact that these velocity product terms can be derived from the inertia matrix $M(q)$. If we consider the individual components of this vector, $C(q, \dot{q})\dot{q} = [c_1(q, \dot{q}), \ldots, c_{n_Q}(q, \dot{q})]^T$, we find that

(10.8) $$c_i(q, \dot{q}) = \sum_{j=1}^{n_Q} \left(\sum_{k=1}^{n_Q} \Gamma^i_{jk}(q) \dot{q}_j \dot{q}_k \right)$$

where

(10.9) $$\Gamma^i_{jk}(q) = \frac{1}{2} \left(\frac{\partial M_{ij}(q)}{\partial q_k} + \frac{\partial M_{ik}(q)}{\partial q_j} - \frac{\partial M_{kj}(q)}{\partial q_i} \right).$$

The n_Q^3 scalars $\Gamma^i_{jk}(q)$ are known as the *Christoffel symbols* of the inertia matrix $M(q)$. In equation (10.8), squared velocity terms (where $j = k$) are known as *centrifugal* terms, and velocity product terms where $j \neq k$ are known as *Coriolis* terms. For example, the centrifugal term $-m_2 q_2 \dot{q}_1^2$ in the RP arm of example 10.1.2 indicates that the linear actuator at the prismatic joint must apply a force to keep the joint stationary as the revolute joint rotates. The Coriolis term $2m_2 q_2 \dot{q}_1 \dot{q}_2$ indicates that the actuator at the revolute joint must apply a torque for the two joints to move at constant velocities. This is because the inertia of the robot about the first joint is changing as the second joint extends or retracts, so the angular momentum is also changing, implying a torque at the first joint.

Although there are many ways to write the *Coriolis matrix* $C(q, \dot{q})$ as a function of the Christoffel symbols, one common choice is

$$C_{ij}(q, \dot{q}) = \sum_{k=1}^{n_Q} \Gamma^k_{ij}(q) \dot{q}_k.$$

Velocity product terms arise due to the noninertial reference frames implicit in the generalized coordinates q. The unforced motions (when $u - g(q) = 0$) are not "straight lines" in this choice of coordinates, and the Christoffel symbols carry geometric information on how unforced motions "bend" in this choice of coordinates. For example, if we represent the configuration of a point mass m in the plane by standard Cartesian (x, y) coordinates, unforced motions are straight lines in these inertial coordinates, and the Christoffel symbols are zero. If we represent the configuration by polar coordinates $[q_1, q_2]^T = [r, \theta]^T$, however, unforced motions are not straight lines in this choice of coordinates, and we find $\Gamma^1_{22} = -mq_1$, $\Gamma^2_{12} = \Gamma^2_{21} = mq_1$ (see problem 1). The geometry of the dynamics of mechanical systems is discussed further in chapter 12.

The main point is that the equations of motion (10.7) depend on the choice of coordinates q. For this reason, neither $M(q)\ddot{q}$ nor $C(q, \dot{q})\dot{q}$ *individually* should be thought of as a generalized force; only their sum is a force.

When we wish to emphasize the dependence of the velocity product terms on the Christoffel symbols, which in turn are determined by the inertia matrix, we write

$$u = M(q)\ddot{q} + \begin{bmatrix} \dot{q}^T \Gamma^1(q)\dot{q} \\ \vdots \\ \dot{q}^T \Gamma^{n_Q}(q)\dot{q} \end{bmatrix} + g(q),$$

where $\Gamma^i(q)$ is the $n_Q \times n_Q$ symmetric matrix with elements $\Gamma^i_{jk}(q)$, $j, k = 1 \ldots n_Q$. We write this more compactly as

(10.10) $u = M(q)\ddot{q} + \dot{q}^T \Gamma(q)\dot{q} + g(q).$

Conceptually, $\Gamma(q) \in \mathbb{R}^{n_Q \times n_Q \times n_Q}$ can be viewed as an n_Q-dimensional column vector, where each element of the "vector" is a matrix $\Gamma^i(q)$, as shown in the following example.

EXAMPLE 10.2.1   *For the RP arm of Example 10.1.2, there are $n_Q^3 = 2^3 = 8$ Christoffel symbols. The only nonzero Christoffel symbols are $\Gamma^1_{12} = \Gamma^1_{21} = m_2 q_2$ and $\Gamma^2_{11} = -m_2 q_2$. The Coriolis and centrifugal terms $\dot{q}^T \Gamma(q)\dot{q}$ can be calculated as follows:*

$$\dot{q}^T \Gamma(q)\dot{q} = [\dot{q}_1 \ \dot{q}_2] \begin{bmatrix} \Gamma^1(q) \\ \Gamma^2(q) \end{bmatrix} \begin{bmatrix} \dot{q}_1 \\ \dot{q}_2 \end{bmatrix}$$

$$= [\dot{q}_1 \ \dot{q}_2] \begin{bmatrix} \begin{bmatrix} 0 & m_2 q_2 \\ m_2 q_2 & 0 \end{bmatrix} \\ \begin{bmatrix} -m_2 q_2 & 0 \\ 0 & 0 \end{bmatrix} \end{bmatrix} \begin{bmatrix} \dot{q}_1 \\ \dot{q}_2 \end{bmatrix}$$

$$= \begin{bmatrix} 2m_2 q_2 \dot{q}_1 \dot{q}_2 \\ -m_2 q_2 \dot{q}_1^2 \end{bmatrix}.$$

The dynamics described by equation (10.7) are specific to mechanical systems where the actuators act directly on the generalized coordinates. For example, a robot arm typically has an actuator at each joint. A more general form of the dynamics of second-order mechanical systems is

(10.11) $T(q)f = M(q)\ddot{q} + C(q, \dot{q})\dot{q} + g(q),$

where f are the actuator forces and the $n_Q \times n_Q$ matrix $T(q)$ specifies how the actuators act on the generalized coordinates, as a function of the system configuration.

As an example, consider replacing the motors at the joints of our two-joint RP arm of example 10.1.2 with two thrusters attached to the center of mass of the second link. The location of the center of mass of the second link in the world frame is

$x = [x_1, x_2]^T$, and the thrusters provide a force $f = [f_1, f_2]^T$ expressed in the world frame. To use the dynamic equations we have already derived, we would like to express the generalized forces u at the joints as a function of f. To do this, let ϕ be the forward kinematics (see section 3.8) mapping from q to x,

$$x = \phi(q) = [q_2 \cos q_1, q_2 \sin q_1]^T.$$

The velocities are given by

$$\dot{x} = \frac{\partial \phi}{\partial q}\dot{q} = J(q)\dot{q},$$

where $J(q)$ is the manipulator Jacobian at the center of mass at the second link. Then by the analysis in section 4.7, the generalized forces u and f are related by

$$u = T(q)f = J^T(q)f$$

(10.12)
$$\begin{bmatrix} u_1 \\ u_2 \end{bmatrix} = \begin{bmatrix} -q_2 \sin q_1 & q_2 \cos q_1 \\ \cos q_1 & \sin q_1 \end{bmatrix} \begin{bmatrix} f_1 \\ f_2 \end{bmatrix}.$$

In other words, $T(q)$ is simply the transpose of the manipulator Jacobian.

If $T(q)$ is rank n_Q, dynamics of the form of equation (10.11) can be put in the form of equation (10.7) by defining "virtual" actuators $u = T(q)f$, and transforming any actuator limits on f to limits on u. This is sometimes called a *feedback transformation* since the transformation from f to u depends on q.

Finally, mechanical systems are often subject to dissipative forces such as dry Coulomb friction or viscous damping. These can be treated as external forces to be added after deriving the equations of motion using Lagrange's equations. There are many possible models of friction and damping, but in most cases these forces are a function of $\dot{q}$ and possibly q, so we write

$$u = M(q)\ddot{q} + C(q, \dot{q})\dot{q} + g(q) + b(q, \dot{q}).$$

Inertia Matrix

As we have seen, the inertia matrix $M(q)$ determines the equations of motion, except for gravitational and dissipative forces. Another way to see this is by observing that the kinetic energy of a mechanical system is determined by its inertia matrix, and can be written

(10.13)
$$K(q, \dot{q}) = \frac{1}{2}\dot{q}^T M(q)\dot{q}.$$

The fact that $M(q)$ is positive definite implies that the kinetic energy is positive for any nonzero $\dot{q}$.

Equation (10.13) shows how the kinetic energy depends on the inertia matrix. We can also derive the inertia matrix from the kinetic energy,

$$(10.14) \qquad M_{ij} = \frac{\partial^2 K(q, \dot{q})}{\partial \dot{q}_i \partial \dot{q}_j}.$$

In some cases, such as the planar body of example 10.1.1, the inertia matrix can be written independent of the configuration q, and the Christoffel symbols are zero. This means that the dynamics are invariant to the configuration—they "look" the same from any configuration. For the planar body, the inertia matrix is

$$M(q) = M = \begin{bmatrix} m & 0 & 0 \\ 0 & m & 0 \\ 0 & 0 & I \end{bmatrix}.$$

For some robots, such as a mobile manipulator consisting of a robot arm mounted on a cart, the dynamics are invariant to some configuration variables (such as the cart's position and orientation on the floor) but not others (such as the arm's configuration).

10.3   Velocity Constraints

Suppose that the mechanical system is subject to a set of k linearly independent constraints linear in velocity, i.e., of the form

$$(10.15) \qquad A(q)\dot{q} = 0,$$

where $A(q)$ is a $k \times n_Q$ matrix, and the k row vectors of $A(q)$ are written $a_j(q)$, $j = 1 \dots k$. Such constraints are called *Pfaffian* constraints. One source of Pfaffian constraints is rolling without slipping, such as in a wheeled mobile robot; a sliding constraint of this form is given in Example 10.3.1.

Since the constraints of equation (10.15) are satisfied throughout motion, we can differentiate the left-hand side and set it equal to zero:

$$A(q)\ddot{q} + \dot{A}(q)\dot{q} = 0.$$

The constrained Lagrange's equations can then be written

$$(10.16) \qquad \frac{d}{dt}\frac{\partial L}{\partial \dot{q}} - \frac{\partial L}{\partial q} = M(q)\ddot{q} + C(q, \dot{q})\dot{q} + g(q) = u + A^T(q)\lambda$$

$$(10.17) \qquad\qquad A(q)\dot{q} = \dot{A}(q)\dot{q} + A(q)\ddot{q} = 0,$$

where $\lambda = [\lambda_1, \dots, \lambda_k]^T$ are the *Lagrange multipliers*. The generalized force $\lambda_j a_j(q)$ is applied by constraint j to maintain the constraint. The constrained Lagrange's equations yield $n_Q + k$ equations to be solved for the $n_Q + k$ variables $\ddot{q}$ and λ.

If we are not interested in calculating the k constraint forces, we can use equation (10.17) to eliminate λ from equation (10.16). Solving equation (10.16) for $\ddot{q}$ and plugging into equation (10.17), dropping the dependence of M, A, and g on q and C on q, $\dot{q}$, we get

$$\dot{A}\dot{q} + AM^{-1}(u + A^T\lambda - C\dot{q} - g) = 0.$$

Now solving for λ, we get

$$\lambda = (AM^{-1}A^T)^{-1}(-\dot{A}\dot{q} + AM^{-1}(C\dot{q} + g - u)).$$

Recognizing that $-\dot{A}\dot{q} = A\ddot{q}$, plugging back into equation (10.16), and manipulating, we get

$$(\mathcal{I} - A^T(AM^{-1}A^T)^{-1}AM^{-1})(M\ddot{q} + C\dot{q} + g - u) = 0,$$

where $\mathcal{I}$ is the identity matrix. If we define

$$P_u = \mathcal{I} - A^T(AM^{-1}A^T)^{-1}AM^{-1},$$

then we get the form

(10.18) $P_u(M\ddot{q} + C\dot{q} + g) = P_u u.$

The $n_Q \times n_Q$ matrix P_u is only rank $n_Q - k$, so we cannot premultiply both sides of equation (10.18) by P_u^{-1}; if we could, we would be left with the unconstrained dynamics. The projection matrix P_u projects generalized forces to the components that do work on the system. The remaining forces, defined by the projection $(\mathcal{I} - P_u)$, are the components resisted by the constraints. These two sets of forces are orthogonal to each other with respect to the inertia matrix. In other words,

$$(P_u u)^T M^{-1}(\mathcal{I} - P_u)u = 0$$

for any u.

Defining the matrix

(10.19) $P = M^{-1}P_u M = \mathcal{I} - M^{-1}A^T(AM^{-1}A^T)^{-1}A$

and rearranging equation (10.18), we get the equivalent form

(10.20) $P\ddot{q} = PM^{-1}(u - C\dot{q} - g).$

Here the rank $n_Q - k$ matrix P projects general motions to motions satisfying the constraints of equation (10.17). The remaining motions, defined by the projection $(\mathcal{I} - P)$, are the components in the constrained directions. The projections P and $(\mathcal{I} - P)$ are orthogonal by the inertia matrix, i.e.,

$$(P\dot{q})^T M(\mathcal{I} - P)\dot{q} = 0$$

for any $\dot{q}$.

In the discussion above, we used the notion of orthogonality *by the inertia matrix*. This notion of orthogonality is the appropriate one when discussing dynamics; the inertia matrix captures the metric properties of the coordinates used in describing the system, which may mix linear and angular coordinates. This is in contrast to our usual notion of orthogonality, which says two vectors v_1 and v_2 are orthogonal if $v_1^T \mathcal{I} v_2 = v_1^T v_2 = 0$. The identity matrix $\mathcal{I}$ indicates that space looks the same in every direction. Further discussion of the geometry of mechanical systems is deferred to chapter 12, and further discussion of the projections P and P_u can be found in [60, 293, 308].

From equation (10.19) we know that $MP = P_u M$, so the matrices P and P_u satisfy the identity

$$P = P_u^T.$$

For this reason, we will only refer to the matrices P and P^T.

EXAMPLE 10.3.1   *A knife-edge can slide on a horizontal plane in the direction it is pointing or spin about an axis through the contact point and orthogonal to the plane, but it cannot slide perpendicular to its heading direction. Let $q = [q_1, q_2, q_3]^T$ represent the configuration of the knife-edge, where (q_1, q_2) denotes the contact point on the plane and q_3 denotes the heading direction (figure 10.3). If the mass of the knife is m and its inertia is I about the axis of rotation, the Lagrangian is the kinetic energy*

$$L = \frac{1}{2} m \left(\dot{q}_1^2 + \dot{q}_2^2 \right) + \frac{1}{2} I \dot{q}_3^2.$$

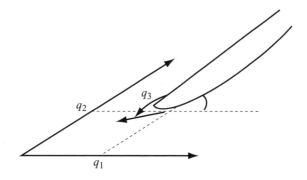

Figure 10.3   A knife-edge on a plane.

The single constraint can be written

$$\dot{q}_1 \sin q_3 - \dot{q}_2 \cos q_3 = 0,$$

or $A(q) = a_1(q) = [\sin q_3, -\cos q_3, 0]$. Differentiating this constraint, we get

(10.21) $\ddot{q}_1 \sin q_3 + \dot{q}_1 \dot{q}_3 \cos q_3 - \ddot{q}_2 \cos q_3 + \dot{q}_2 \dot{q}_3 \sin q_3 = 0.$

Applying Lagrange's equations, we get

(10.22) $m\ddot{q}_1 = u_1 + \lambda_1 \sin q_3$

(10.23) $m\ddot{q}_2 = u_2 - \lambda_1 \cos q_3$

(10.24) $I\ddot{q}_3 = u_3.$

Solving equations (10.21) through (10.24), we get

(10.25) $\ddot{q}_1 = \dfrac{1}{m}\left(u_1 \cos^2 q_3 + (u_2 - m\dot{q}_1 \dot{q}_3)\cos q_3 \sin q_3 - m\dot{q}_2 \dot{q}_3 \sin^2 q_3\right)$

(10.26) $\ddot{q}_2 = \dfrac{1}{m}\left(u_2 \sin^2 q_3 + (u_1 + m\dot{q}_2 \dot{q}_3)\cos q_3 \sin q_3 + m\dot{q}_1 \dot{q}_3 \cos^2 q_3\right)$

(10.27) $\ddot{q}_3 = \dfrac{u_3}{I}$

(10.28) $\lambda_1 = (u_2 - m\dot{q}_1 \dot{q}_3)\cos q_3 - (u_1 + m\dot{q}_2 \dot{q}_3)\sin q_3,$

where (u_1, u_2) are the forces along the (q_1, q_2)-directions, and u_3 is the torque about an axis through the contact point and orthogonal to the plane.

Alternatively, we could study the projected dynamics. The inertia matrix $M(q)$ is

$$M(q) = \begin{bmatrix} m & 0 & 0 \\ 0 & m & 0 \\ 0 & 0 & I \end{bmatrix}$$

and we calculate the projection matrix using (10.19):

$$P = \begin{bmatrix} \cos^2 q_3 & \sin q_3 \cos q_3 & 0 \\ \sin q_3 \cos q_3 & \sin^2 q_3 & 0 \\ 0 & 0 & 1 \end{bmatrix}.$$

Note that the first two rows (and columns) of P are linearly dependent, so $\mathrm{rank}(P) = n_Q - k = 2$. In this example, $C(q, \dot{q})\dot{q}$ and $g(q)$ are zero, so the projected dynamics (10.20) become

$$P\ddot{q} = PM^{-1}u,$$

which yields the two independent equations of motion in the unconstrained directions

$$\ddot{q}_1 \cos q_3 + \ddot{q}_2 \sin q_3 = \frac{1}{m}(u_1 \cos q_3 + u_2 \sin q_3)$$

$$\ddot{q}_3 = \frac{u_3}{I}.$$

The constraint equation (10.21) completes the system of three equations, and we solve to get equations (10.25), (10.26), and (10.27) above.

As the previous example shows, even for very simple systems with constraints, the expressions can quickly become unwieldy. When possible, it may be preferable to choose a reduced set of generalized coordinates to eliminate the constraints. For example, if $a_j(q) = \frac{\partial c(q)}{\partial q}$ for some function $c(q) = 0$, then the velocity constraint is actually the time derivative of a configuration constraint $c(q) = 0$, which reduces the dimension of the configuration space by one. Therefore, it is possible to reduce the number of generalized coordinates by one and eliminate the constraint. As an example, the planar motion of a point on a circle centered at the origin can be represented using coordinates (x, y) and the velocity constraint $x\dot{x} + y\dot{y} = 0$. This velocity constraint can be integrated to the configuration constraint $x^2 + y^2 = R$, however, where the radius R is defined by the initial position of the point. This configuration constraint allows an unconstrained description of the configuration using a single angle coordinate θ. When a velocity constraint can be integrated to a configuration constraint, as in this case, the constraint is called *holonomic*. *Nonholonomic* constraints are velocity constraints that cannot be integrated to a configuration constraint. Motion planning with nonholonomic constraints is left to chapter 12.

10.4 Dynamics of a Rigid Body

Until now, we have been using freely the concepts of "center of mass" and "inertia about an axis" to get to the use of Lagrange's equations as quickly as possible. We now provide formal definitions of these quantities and apply them to the dynamics of a translating and rotating rigid body.

Let $\mathcal{B}$ be a rigid body occupying a volume $V \subset \mathbb{R}^3$, r be a vector from the origin to a point in $\mathcal{B}$, and $\rho(r)$ be the mass density of $\mathcal{B}$ as a function of the location r. Then the mass of $\mathcal{B}$ is the volume integral of the mass density

$$m = \int_V \rho(r)dV,$$

and the *center of mass* is the weighted average of the mass density

$$r_{\text{cm}} = \frac{1}{m} \int_V r\rho(r)dV.$$

When a body moves freely in space, it is convenient to describe the translational position of the body by the Cartesian coordinates q of the center of mass relative to a stationary inertial frame. The translational kinetic energy of the body can be written $K = \frac{1}{2}\dot{q}^T m\dot{q} = \frac{1}{2}m\|\dot{q}\|^2$. Applying Lagrange's equations yields the familiar equation

(10.29) $f = m\ddot{q},$

where f is the linear force applied to the body expressed in the inertial frame.

10.4.1   Planar Rotation

When a body moves in a plane, a single configuration variable q can be used to describe its orientation. Such motion occurs, for example, when the body rotates about a fixed axis, or when the body slides freely on a frictionless plane. In the former case, it is convenient to define a stationary x-y-z inertial frame with the z-axis along the axis of rotation. In the latter case, it will be convenient to define an x-y-z inertial frame at the center of mass of the body, with the z-axis orthogonal to the plane of motion. (Since we are focusing on rotational motion only, the center of mass can be assumed stationary.)

The kinetic energy of a body rotating in the plane is the integral over the body of the differential kinetic energy at each point $r = (x, y, z)^T$:

$$K = \int_V \frac{1}{2}\rho(r)v^2(r)dV,$$

where q is the angle of the body, $\dot{q}$ is the angular velocity, and $v(r) = \dot{q}\sqrt{x^2 + y^2}$ is the linear velocity at r. Therefore we can write the kinetic energy in the form of equation (10.13),

(10.30) $K = \dfrac{1}{2}\dot{q}^2 \displaystyle\int_V \rho(r)(x^2 + y^2)dV = \dfrac{1}{2}\dot{q}^T I_{zz}\dot{q},$

where

(10.31) $I_{zz} = \displaystyle\int_V \rho(r)(x^2 + y^2)dV$

is the inertia of the body about the z-axis. If the body is uniform density, equation (10.31) simplifies to

$$I_{zz} = m \int_V (x^2 + y^2) dV,$$

where m is the mass of the body. Applying Lagrange's equations to equation (10.30), we get

(10.32) $\tau_z = I_{zz} \ddot{q},$

where τ_z is the torque about the z-axis.

If we choose a z-axis through the center of mass of the body and a parallel z'-axis a distance d away, then the scalar inertias I_{zz} and $I_{z'z'}$ are related by the *parallel-axis theorem* for planar rotation:

(10.33) $I_{z'z'} = I_{zz} + md^2$

The proof of this theorem is straightforward and in fact is implicit in our derivation of the equations of motion of the RP arm.

10.4.2   Spatial Rotation

This section requires extra mathematical machinery, and can be safely skipped if the reader is not interested in the dynamics of a rotating spatial body.

In our Lagrangian formulation, we first choose a set of coordinates q, express the Lagrangian in terms of q and $\dot{q}$, and derive the equations of motion. To do this for a rotating spatial body, we can choose q to be three angles describing the orientation of the body in a world frame. Then we can express the kinetic energy of the body as a function of q and $\dot{q}$ and proceed as before. If we do this, however, the inertia matrix $M(q)$ will be extremely complex for any choice of q, providing little insight into the nature of the motion. The equations of motion are rarely written this way. Another problem is that no choice of three orientation variables can provide a smooth, global coordinatization of the space of orientations. In the same way that latitude and longitude coordinates for the Earth "go bad" at the poles, where the longitude changes discontinuously, any choice of three coordinates to represent orientations will have singularities. (For motions away from these bad orientations, however, three coordinates work just fine, so this is not the most serious problem. In fact, we have a similar problem representing a single angle by a real number, which requires the use of mod2π arithmetic.)

So we will not begin by choosing angular coordinates, and instead of defining the angular velocity as the time-derivative of coordinates, we define $\omega_s = [\omega_{x_s}, \omega_{y_s}, \omega_{z_s}]^T$ to be the angular velocity of the body about the x_s-y_s-z_s axes of a stationary inertial frame at the center of mass of the body. The linear velocity at a point $r_s = (x_s, y_s, z_s)^T$

on the body is $\omega_s \times r_s$. The total kinetic energy of the body can be written

(10.34) $$K = \frac{1}{2} \int_V \rho(r_s)(\omega_s \times r_s)^T (\omega_s \times r_s) dV,$$

which can be simplified to

(10.35) $$K = \frac{1}{2}\omega_s^T \left(\int_V \rho(r_s) \begin{bmatrix} y_s^2 + z_s^2 & -x_s y_s & -x_s z_s \\ -x_s y_s & x_s^2 + z_s^2 & -y_s z_s \\ -x_s z_s & -y_s z_s & x_s^2 + y_s^2 \end{bmatrix} dV \right) \omega_s,$$

or

(10.36) $$K = \frac{1}{2}\omega_s^T \mathbb{I}_s \omega_s.$$

The matrix $\mathbb{I}_s$ is the symmetric positive definite inertia matrix for the body written in the inertial frame. Because $\mathbb{I}_s$ is defined in the stationary world frame, it changes as the body rotates.

The *angular momentum* of the body is $P = \mathbb{I}_s \omega_s$, and the torque $\tau_s = [\tau_{xs}, \tau_{ys}, \tau_{zs}]^T$ acting on the body, expressed in the inertial frame, is the rate of change of P:

$$\tau_s = \frac{dP}{dt}$$
$$= \frac{d\mathbb{I}_s}{dt}\omega_s + \mathbb{I}_s \frac{d\omega_s}{dt}.$$

The density of the body is not changing as it rotates, so the change of $\mathbb{I}_s$, $d\mathbb{I}_s/dt$, is due only to the motion of the body in the world frame, giving $d\mathbb{I}_s/dt = \omega_s \times \mathbb{I}_s$. Plugging in, we get

(10.37) $$\tau_s = \omega_s \times \mathbb{I}_s \omega_s + \mathbb{I}_s \dot{\omega}_s.$$

This is known as *Euler's equation* in the inertial frame.

To turn equation (10.37) into a matrix equation, we define the *skew-symmetric matrix representation* $\widehat{\omega}_s$ of the vector $\omega_s = [\omega_{x_s}, \omega_{y_s}, \omega_{z_s}]^T$:

$$\widehat{\omega}_s = \begin{bmatrix} 0 & -\omega_{z_s} & \omega_{y_s} \\ \omega_{z_s} & 0 & -\omega_{x_s} \\ -\omega_{y_s} & \omega_{x_s} & 0 \end{bmatrix}.$$

We can now express equation (10.37) as the matrix equation

(10.38) $$\tau_s = \widehat{\omega}_s \mathbb{I}_s \omega_s + \mathbb{I}_s \dot{\omega}_s.$$

We still do not have a representation of the orientation of the body in the world frame, however. We need an equation for the evolution of the body's orientation (the

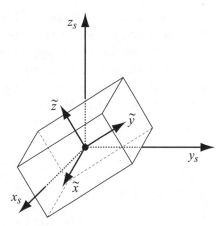

Figure 10.4 The rotation matrix for a body is obtained by expressing the unit vectors $\tilde{x}$, $\tilde{y}$, and $\tilde{z}$ of the body x-y-z frame in the inertial frame x_s-y_s-z_s.

kinematic equation) to go with equation (10.38) for the evolution of the velocity (the dynamic equation).

To do this, define a frame x-y-z attached to the body at its center of mass. As described in Chapter 3, our representation of the orientation of the body will be as a 3×3 rotation matrix

$$R = \begin{bmatrix} \tilde{x}_1 & \tilde{y}_1 & \tilde{z}_1 \\ \tilde{x}_2 & \tilde{y}_2 & \tilde{z}_2 \\ \tilde{x}_3 & \tilde{y}_3 & \tilde{z}_3 \end{bmatrix} \in SO(3),$$

where $\tilde{x} = [\tilde{x}_1, \tilde{x}_2, \tilde{x}_3]^T$ is the unit vector in the body x-direction expressed in the inertial coordinate frame. The vectors $\tilde{y}$ and $\tilde{z}$ are defined similarly (figure 10.4).

Each column vector of R moves according to the angular velocity ω_s, so the kinematics of the rotating rigid body can be written

(10.39) $\dot{R} = \omega_s \times R = \widehat{\omega}_s R.$

Together, equations (10.39) and (10.38) describe the motion of a rotating rigid body in a spatial frame. The use of the rotation matrix representation of the orientation allows us to write the kinematics in a simple and globally correct fashion, which is not possible with any choice of three coordinates.[1]

---

1. A globally correct representation of orientation can be achieved using only four numbers with quaternions (see appendix E). We use the matrix representation because it allows the convenient use of matrix multiplications.

One difficulty with the equations of motion in an inertial frame is that $\mathbb{I}_s$ changes as the body rotates. It would be more convenient and intuitive to define the equations of motion in a frame fixed to the body, where a *body inertia matrix* $\mathbb{I}$ is unchanging. To do so, we use the coordinate frame x-y-z attached to the body at its center of mass. The angular velocity in this frame is written $\omega = [\omega_x, \omega_y, \omega_z]^T$ and the external torque is written $\tau = [\tau_x, \tau_y, \tau_z]^T$. These are related to ω_s and τ_s by the following equations:

$$\omega_s = R\omega$$

$$\tau_s = R\tau$$

The inertial frame coordinates $r_s = [x_s, y_s, z_s]^T$ of a point are related to its coordinates in the body frame $r = [x, y, z]^T$ by

$$r_s = Rr.$$

The kinematic equations in the two frames are related by

$$\dot{R} = \widehat{\omega}_s R = R\widehat{\omega}.$$

Plugging these relations into equation (10.34) and simplifying, we find

(10.40) $$K = \frac{1}{2}\omega^T \mathbb{I}\omega$$

(10.41) $$\mathbb{I} = R^T \mathbb{I}_s R,$$

where the symmetric positive definite body-fixed inertia matrix is given by

(10.42) $$\mathbb{I} = \begin{bmatrix} I_{xx} & I_{xy} & I_{xz} \\ I_{yx} & I_{yy} & I_{yz} \\ I_{zx} & I_{zy} & I_{zz} \end{bmatrix} = \int_V \rho(r) \begin{bmatrix} y^2 + z^2 & -xy & -xz \\ -xy & x^2 + z^2 & -yz \\ -xz & -yz & x^2 + y^2 \end{bmatrix} dV.$$

The (possibly non-unique) eigenvectors of $\mathbb{I}$ define orthogonal *principal axes of inertia* of the body. If the body x-y-z frame is chosen so that the axes are aligned with principal axes of inertia, then all off-diagonal terms of $\mathbb{I}$ are zero, and I_{xx}, I_{yy}, I_{zz} are the principal moments of inertia. In the general case, one principal axis is the axis of maximum inertia, one principal axis is the axis of minimum inertia, and the third principal axis (the intermediate axis of inertia) is orthogonal. Because of symmetries, however, the inertia about two or three of the principal axes might be identical. Often the principal axes of inertia of a body are evident from symmetries (figure 10.5).

To derive the dynamics in the body frame, we cannot simply take the time-derivative of $\mathbb{I}\omega$, since this is not defined in an inertial frame. (The time-derivative of the momentum $\mathbb{I}_s\omega_s$ is a generalized force, while the time-derivative of $\mathbb{I}\omega$ is not.) Instead,

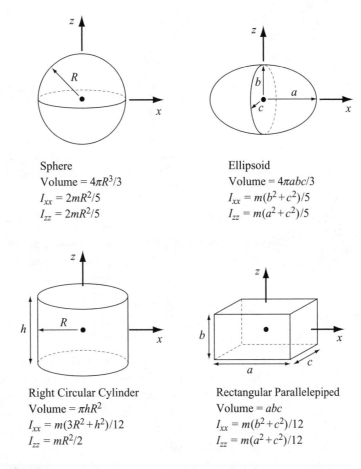

Sphere
Volume $= 4\pi R^3/3$
$I_{xx} = 2mR^2/5$
$I_{zz} = 2mR^2/5$

Ellipsoid
Volume $= 4\pi abc/3$
$I_{xx} = m(b^2 + c^2)/5$
$I_{zz} = m(a^2 + c^2)/5$

Right Circular Cylinder
Volume $= \pi h R^2$
$I_{xx} = m(3R^2 + h^2)/12$
$I_{zz} = mR^2/2$

Rectangular Parallelepiped
Volume $= abc$
$I_{xx} = m(b^2 + c^2)/12$
$I_{zz} = m(a^2 + c^2)/12$

Figure 10.5 Inertias about principal axes of inertia for four different uniform density volumes. Note that the principal axes of inertia are not unique for the sphere and cylinder.

we begin with equation (10.38),

$$\tau_s = \widehat{\omega}_s \mathbb{I}_s \omega_s + \mathbb{I}_s \dot{\omega}_s,$$

and plug in $\tau_s = R\tau$, $\omega_s = R\omega$, $\mathbb{I}_s = R\mathbb{I}R^T$, and $\widehat{\omega}_s = \dot{R}R^T$ to get

$$R\tau = \dot{R}R^T R\mathbb{I}R^T R\omega + R\mathbb{I}R^T \left(\frac{d}{dt}(R\omega) \right)$$

$$= \dot{R}R^T R\mathbb{I}R^T R\omega + R\mathbb{I}R^T (\dot{R}\omega + R\dot{\omega}).$$

Recognizing from our identities that $\dot{R}\omega = R\widehat{\omega}\omega = 0$, and premultiplying both sides by $R^{-1} = R^T$, we get

$$\tau = R^T \dot{R} R^T R \mathbb{I} R^T R\omega + R^T R \mathbb{I} R^T R\dot{\omega}.$$

Plugging in $\widehat{\omega} = R^T \dot{R}$ and noticing that $R^T R$ is the identity matrix, this simplifies to

$$\tau = \widehat{\omega}\mathbb{I}\omega + \mathbb{I}\dot{\omega}.$$

This is Euler's equation in the body frame. Note that it has the same form as Euler's equation in the spatial frame. Collecting together the kinematics and dynamics in one place, the equations of motion in the body frame are written

(10.43) $\dot{R} = R\widehat{\omega}$

(10.44) $\tau = \widehat{\omega}\mathbb{I}\omega + \mathbb{I}\dot{\omega}.$

The big advantage of this form over the spatial equations is that $\mathbb{I}$ is constant.

If the body x-y-z frame is aligned with the principal axes, making all off-diagonal terms of $\mathbb{I}$ zero, equation (10.44) simplifies to

(10.45) $$\begin{bmatrix} \tau_x \\ \tau_y \\ \tau_z \end{bmatrix} = \begin{bmatrix} I_{xx}\dot{\omega}_x + (I_{zz} - I_{yy})\omega_y\omega_z \\ I_{yy}\dot{\omega}_y + (I_{xx} - I_{zz})\omega_x\omega_z \\ I_{zz}\dot{\omega}_z + (I_{yy} - I_{xx})\omega_x\omega_y \end{bmatrix}.$$

One key implication of equations (10.44) and (10.45) is that $\dot{\omega}$ may not be zero even if τ is zero. Although the angular momentum and kinetic energy of a rotating body are constant when no external torques are applied, the angular velocity of the body may not be constant. For further interpretation of equation (10.45), see the mechanics textbooks by Symon [407] and Marsden and Ratiu [308] or the robotic manipulation textbook by Mason [312].

When it is difficult to solve the integrals of equation (10.42) directly to find $\mathbb{I}$, it may be possible to split the body into simpler components and solve for (or look up in a table) the inertia matrix at the center of mass of each component separately. If we then transform these inertia matrices to a common frame, we can simply add them to get the inertia matrix for the composite body in that common frame. This transformation can be accomplished by a translation followed by a rotation, as outlined below (see figure 10.6).

Let $\mathbb{I}_i$ be the inertia matrix of the ith component, expressed in its own local coordinate frame x_i-y_i-z_i at its center of mass. Let r_i be the vector from the origin of the local frame to the origin of the common frame x-y-z, expressed in the local frame. The inertia $\mathbb{I}_i$ can be expressed in a frame aligned with x_i-y_i-z_i, but located at the

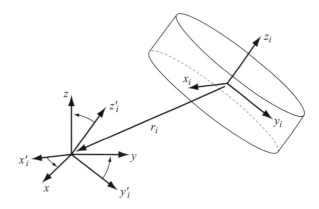

Figure 10.6   The inertia matrix of body $\mathcal{B}_i$, expressed in a frame x_i-y_i-z_i at the center of mass of the body, can be expressed in another frame x-y-z by a translation and rotation.

origin of the common frame, using the *parallel-axis theorem*

(10.46) $\quad \mathbb{I}_i' = \mathbb{I}_i + m_i \left(\|r_i\|^2 \mathcal{I} - r_i r_i^T \right),$

where m_i is the mass of the ith component and $\mathcal{I}$ is the 3×3 identity matrix. Now let R_i denote the rotation matrix describing the orientation of this translated local frame relative to the common frame. Rotating the inertia matrix into the common frame, we get

$$\mathbb{I}_i'' = R_i \mathbb{I}_i' R_i^T.$$

The matrix $\mathbb{I}_i''$ is the inertia of the ith component expressed in the common frame. Performing this translation and rotation for all k components of the body, the total inertia of the body in the common frame is $\mathbb{I} = \mathbb{I}_1'' + \cdots + \mathbb{I}_k''$.

Lagrange's Equations Revisited

We have gone to great lengths to avoid choosing three generalized coordinates and using Lagrange's equations! Now that we have done this work and developed some understanding of the body inertia matrix $\mathbb{I}$, it will be easier to see how Lagrange's equations could be applied.

Figure 10.7 shows a choice of coordinates $q = [q_1, q_2, q_3]^T$ due to Euler. The body x-y plane intersects the spatial x_s-y_s plane along a line called the *line of nodes*. The coordinate q_1 is the angle from the x_s-axis to the line of nodes, q_2 is the angle from the line of nodes to the body x-axis, and q_3 is the angle from the z_s-axis to the body z-axis. With some work, the angular velocity $\omega = [\omega_x, \omega_y, \omega_z]^T$ can be expressed in

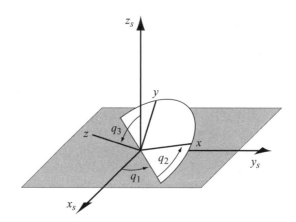

Figure 10.7   Euler angles for a rotating body.

terms of these coordinates:

(10.47) $\omega_x = \dot{q}_3 \cos q_2 + \dot{q}_1 \sin q_2 \sin q_3$

(10.48) $\omega_y = -\dot{q}_3 \sin q_2 + \dot{q}_1 \cos q_2 \sin q_3$

(10.49) $\omega_z = \dot{q}_2 + \dot{q}_1 \cos q_3.$

Plugging these into the kinetic energy $K = \frac{1}{2}\omega^T \mathbb{I}\omega$, we get the kinetic energy in the form $K = \frac{1}{2}\dot{q}^T M(q)\dot{q}$, as we are used to. From there we can apply Lagrange's equations as before to get the dynamic equations of motion for generalized torques acting along the coordinates.

A good thing about this formulation is that we have used the fewest possible numbers to represent the orientation, and the kinematics are trivial. Significant drawbacks are the complexity of the equations, as well as the singularities in the coordinate representation.

Problems

1. Represent the configuration of a point mass in a plane by polar coordinates $q = [r, \theta]^T$ and use Lagrange's equations to find the equations of motion. Then write the inertia matrix $M(q)$, derive the Christoffel symbols, and show that the dynamics of equation (10.10) are equivalent to the equations you derived using the Lagrangian method.

2. Use Lagrange's equations to derive the equations of motion of a 2R (two revolute joints) robot arm operating in a vertical plane. The first link has length L_1, mass m_1, and inertia I_1 about the center of mass, and the center of mass is a distance r_1 from the first joint.

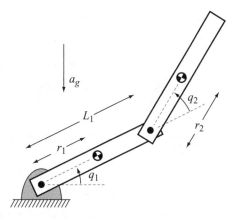

Figure 10.8 The 2R robot arm.

For the second link, m_2, I_2, and r_2 are defined similarly (figure 10.8). Put the equations of motion in the standard form of equation (10.7).

3. Find the eight Christoffel symbols for the mass matrix of problem 2.

4. Use Lagrange's equations to find the equations of motion of a PR robot arm. Provide your own drawing and parameters and solve with these parameters. Put the equations in the standard form of equation (10.7).

5. Find the inertia matrix of a round tube of length L, inner diameter d_1, outer diameter d_2, and density ρ. Choose a frame aligned with the principal axes of inertia. Remember that inertia matrices in a common frame can be added and subtracted.

6. The inertia matrix of a body in a coordinate frame x_1-y_1-z_1 at the center of mass of the body is $\mathbb{I}_1$. The orientation of this coordinate frame is R_1 relative to a frame x-y-z. The origin of x-y-z is at r_1 in the frame x_1-y_1-z_1. Transform the inertia matrix $\mathbb{I}_1$ to an inertia matrix $\mathbb{I}$ expressed in the x-y-z frame, where

$$\mathbb{I}_1 = \begin{bmatrix} 1 & 0 & 0 \\ 0 & 2 & 0 \\ 0 & 0 & 3 \end{bmatrix}, \quad r_1 = \begin{bmatrix} 3 \\ 0 \\ 2 \end{bmatrix}, \quad R_1 = \begin{bmatrix} 0 & -1 & 0 \\ 1 & 0 & 0 \\ 0 & 0 & 1 \end{bmatrix}.$$

Also provide a drawing of the two frames showing their position and orientation relative to each other.

7. Consider a barbell constructed of two spheres of radius 10 cm welded to the ends of a right circular cylinder bar of length 20 cm and radius 2 cm. Each body is a solid volume constructed of steel, with a mass density of 7850 kg/m^3. Find the approximate inertia matrix in a principal-axis frame at the center of mass.

8. Prove the parallel-axis theorem [equation (10.46)].

9. Derive equations (10.47), (10.48), and (10.49).

10. Write a program to simulate the tumbling motion of a rigid body in space.

11. A point of mass m moves in three-dimensional space $\mathbb{R}^3$, actuated by three orthogonal thrusters, with equations of motion $u = m\ddot{q}$ (no gravity). Now imagine that the mass (still with three thrusters) is constrained to move on a sphere of radius 1 centered at the origin of the inertial frame. Write the Pfaffian constraint and solve for $\ddot{q}$ and the Lagrange multiplier λ.

12. Find the projection matrix P for problem 11 and write the constrained equations of motion in the form of equation (10.20).

13. In problem 11, it is possible to reduce the number of generalized coordinates from three to two and eliminate the Lagrange multiplier. Choose latitude (q_1) and longitude (q_2) coordinates to describe the position of the point on the sphere, and use Lagrange's equations to solve for the dynamics in these coordinates. Explain what the generalized forces are. Give the Christoffel symbols in these coordinates.

11 *Trajectory Planning*

IN CHAPTER 10 we described dynamic models for robot systems. Equipped with such a dynamic model, the trajectory planning problem is to find control (force) inputs $u(t)$ yielding a trajectory $q(t)$ that avoids obstacles, takes the system to the desired goal state, and perhaps optimizes some objective function while doing so. This can be considered a complete "motion-planning" problem, as opposed to a "path-planning" problem that only asks for a feasible curve $q(s)$ in the configuration space, without reference to the speed of execution.

In this chapter we study two approaches to trajectory planning for a dynamic system: the decoupled approach, which involves first searching for a path in the configuration space and then finding a time-optimal time scaling for the path subject to the actuator limits; and the direct approach, where the search takes place in the system's state space. Examples of the latter approach include optimal control and numerical optimization, grid-based searches, and randomized probabilistic methods. In this chapter we focus on fully actuated systems—systems where there is an actuator for each degree of freedom.

In section 11.1 we provide some definitions used throughout the chapter. In section 11.2 we describe an algorithm for finding the time-optimal execution of a path subject to actuator limits, and describe how this can be used in a decoupled trajectory planner. In section 11.3 we outline several approaches to trajectory planning directly in the system state space, including optimal control, gradient-based numerical methods, and dynamic programming. Other methods, such as rapidly-exploring random trees (RRTs), are described in chapter 7.

11.1   Preliminaries

In this chapter a path $q(s)$ is assumed to be a twice-differentiable curve on the configuration space $\mathcal{Q}$, $q : [0, 1] \rightarrow \mathcal{Q}$. A *time scaling* $s(t)$ is a function $s : [0, t_f] \rightarrow [0, 1]$ assigning an s value to each time $t \in [0, t_f]$. Together, a path and a time scaling specify a trajectory $q(s(t))$, or $q(t)$ for short. The time scaling $s(t)$ should be twice-differentiable and monotonic ($\dot{s}(t) > 0$ for all $t \in (0, t_f)$). The twice-differentiability of $s(t)$ ensures that the acceleration $\ddot{q}(t)$ is well defined and bounded. Note that *uniform* time scalings $s(t) = kt$ are a subset of the more general time-scaling functions considered here.

Configurations q and forces u are both $n_{\mathcal{Q}}$-dimensional vectors.

11.2   Decoupled Trajectory Planning

Given a collision-free path $q(s)$ for a robot system, what is the fastest feasible trajectory that follows this path? In other words, what is the time-optimal time scaling $s(t)$ subject to actuator constraints? This question is of considerable importance for maximizing the productivity of robot systems when a path has been given by task specifications or found by a path planner. This problem has been solved elegantly by Shin and McKay [385] and Bobrow, Dubowsky, and Gibson [51], with subsequent enhancements by Pfeiffer and Johanni [348], Slotine and Yang [388], and Shiller and Lu [384].

Let us assume that the equations of motion of our system are in the standard form of equation (10.7) or equation (10.10) from chapter 10. The robot is subject to the actuator limits

(11.1) $u_i^{\min}(q, \dot{q}) \leq u_i \leq u_i^{\max}(q, \dot{q})$.

In general, the actuator limits may be functions of the system configuration and velocity. For example, the torque available to accelerate a DC motor decreases as its angular velocity increases. The simplest example of actuator limits are the symmetric, state-independent bounds

$|u_i| \leq u_i^{\max}$.

For a given path $q(s)$, we can substitute

(11.2) $$\frac{dq}{ds}\dot{s} = \dot{q}$$

(11.3) $$\frac{d^2q}{ds^2}\dot{s}^2 + \frac{dq}{ds}\ddot{s} = \ddot{q}$$

into equation (10.10) to get

(11.4) $$M(q(s)) \left(\frac{d^2 q}{ds^2} \dot{s}^2 + \frac{dq}{ds} \ddot{s} \right) + \left(\frac{dq}{ds} \dot{s} \right)^T \Gamma(q(s)) \left(\frac{dq}{ds} \dot{s} \right) + g(q(s)) = u$$

or

(11.5) $$\left(M(q(s)) \frac{dq}{ds} \right) \ddot{s} + \left(M(q(s)) \frac{d^2 q}{ds^2} + \left(\frac{dq}{ds} \right)^T \Gamma(q(s)) \frac{dq}{ds} \right) \dot{s}^2 + g(q(s)) = u.$$

These equations can be expressed compactly as the vector equation

(11.6) $$a(s)\ddot{s} + b(s)\dot{s}^2 + c(s) = u$$

defining the dynamics constrained to the path $q(s)$. The vector functions $a(s)$, $b(s)$, and $c(s)$ are inertial, velocity product, and gravitational terms in terms of s, respectively.

As the robot travels along the path $q(s)$, its state at any time is identified by $(s, \dot{s})$. Actuator limits can be expressed as a function of the path state by substituting equation (11.2) into equation (11.1), yielding $u^{\min}(s, \dot{s})$ and $u^{\max}(s, \dot{s})$. Therefore, at all times the system must satisfy the constraints

(11.7) $$u^{\min}(s, \dot{s}) \le a(s)\ddot{s} + b(s)\dot{s}^2 + c(s) \le u^{\max}(s, \dot{s}).$$

Let $L_i(s, \dot{s})$ and $U_i(s, \dot{s})$ be the minimum and maximum accelerations $\ddot{s}$ satisfying the ith component of equation (11.7), and define

(11.8) $$\alpha_i(s, \dot{s}) = \frac{u_i^{\max}(s, \dot{s}) - b_i(s)\dot{s}^2 - c_i(s)}{a_i(s)}, \quad \beta_i(s, \dot{s}) = \frac{u_i^{\min}(s, \dot{s}) - b_i(s)\dot{s}^2 - c_i(s)}{a_i(s)}.$$

Then $U_i(s, \dot{s}) = \alpha_i(s, \dot{s})$, $L_i(s, \dot{s}) = \beta_i(s, \dot{s})$ if $a_i(s) > 0$, and $U_i(s, \dot{s}) = \beta_i(s, \dot{s})$, $L_i(s, \dot{s}) = \alpha_i(s, \dot{s})$ if $a_i(s) < 0$. (If $a_i(s) = 0$ for any i, the system is at a *zero inertia point*, and we will set aside this possibility until subsection 11.2.1.) We define

$$L(s, \dot{s}) = \max_{i \in 1 \dots n_Q} L_i(s, \dot{s}), \quad U(s, \dot{s}) = \min_{i \in 1 \dots n_Q} U_i(s, \dot{s}).$$

The actuator limits (11.7) can then be expressed as

(11.9) $$L(s, \dot{s}) \le \ddot{s} \le U(s, \dot{s}).$$

The problem can now be stated:

Given a path $q : [0, 1] \to Q$, an initial state $(0, \dot{s}_0)$, and a final state $(1, \dot{s}_f)$, $\dot{s}_0, \dot{s}_f \ge 0$, find a monotonically increasing twice-differentiable time scaling $s : [0, t_f] \to [0, 1]$ that (1) satisfies $s(0) = 0$, $\dot{s}(0) = \dot{s}_0$, $s(t_f) = 1$, $\dot{s}(t_f) = \dot{s}_f$, and (2) minimizes the total travel time t_f along the path while respecting the actuator constraints (11.9) for all time $t \in [0, t_f]$.

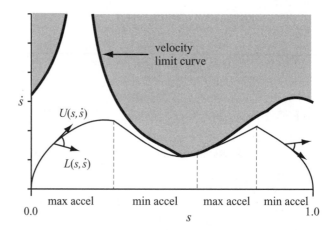

Figure 11.1   At each point $(s, \dot{s})$ in the phase plane we can draw a motion cone defined by the maximum and minimum accelerations $\ddot{s}$ satisfying the actuator limits. The time-optimal trajectory from $(0, \dot{s}_0)$ to $(1, \dot{s}_f)$ is the curve that maximizes the area underneath it while remaining on the boundary of the motion cones. In this example, the trajectory switches between maximum and minimum acceleration three times. The velocity limit curve indicates the states where the cone collapses to a single tangent vector, and the gray region represents inadmissible states where the cone disappears—no feasible actuation will keep the system on the path.

We can conveniently visualize this problem in the $(s, \dot{s})$ state space. At any state $(s, \dot{s})$, the constraints (11.9) specify a range of feasible accelerations along the path, $L(s, \dot{s}) \leq \ddot{s} \leq U(s, \dot{s})$. This range can be interpreted as a cone of tangent vectors in the state space, as illustrated in figure 11.1. The problem is to find a curve from $(0, \dot{s}_0)$ to $(1, \dot{s}_f)$ such that $\dot{s} \geq 0$ everywhere and the tangent at each state is inside the cone at that state. Further, the curve should maximize the speed $\dot{s}$ at each s to minimize the time of motion. A consequence of this is that the curve always follows the upper or lower bound of the cone (maximum or minimum acceleration) at each state.[1] This kind of trajectory is called a "bang-bang" trajectory, and at least one of the actuators is always saturated. The heart of the time-scaling problem is to find the switching points between maximum and minimum acceleration.

At some states $(s, \dot{s})$, the actuation constraints (11.9) indicate that there is no feasible acceleration that will allow the system to continue to follow the path. Such regions of the state space are shown in gray in figure 11.1. We will call these regions *inadmissible* regions. At any inadmissible state, the robot is doomed to leave the path immediately. At admissible states, the robot may still be doomed to eventually leave

---

1. Except perhaps at zero inertia points, as described in subsection 11.2.1.

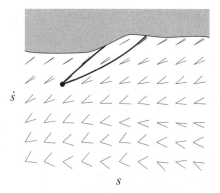

Figure 11.2 Beginning from the $(s, \dot{s})$ state represented by the dot, the system is doomed to leave the path. The set of all trajectories tangent to the motion cones is bounded by the two integral curves shown, showing that all feasible trajectories penetrate the velocity limit curve.

the path. This happens if any integral curve originating from the state, with tangents remaining inside the tangent cones, eventually must reach the inadmissible region (figure 11.2).

We will assume that, for any s, the robot is strong enough to maintain its configuration statically, so all $\dot{s} = 0$ states are admissible and the path can be executed arbitrarily slowly. We will also assume that as $\dot{s}$ increases from zero for a given s, there will be at most one switch from admissible to inadmissible. This occurs at the *velocity limit curve* $v(s)$, consisting of states $(s, \dot{s})$ satisfying

(11.10) $\quad L(s, \dot{s}) = U(s, \dot{s}).$

The velocity limit $v(s)$ is obtained by equating $L_i(s, \dot{s}) = U_j(s, \dot{s})$ for all $i, j = 1 \ldots n_Q$ and solving each equation for $\dot{s}$ (if a solution exists). Call the solution $\dot{s}_{ij}(s)$. For all i, j, keep the minimum value: $v(s) = \min_{i,j} \dot{s}_{ij}(s)$.[2]

Note that because of the $\max(\cdot)$ and $\min(\cdot)$ functions used in calculating $L(s, \dot{s})$, $U(s, \dot{s})$, and $v(s)$, these functions are generally not smooth.

As mentioned earlier, the problem is to find the switches between maximum and minimum acceleration. The following algorithm uses numerical integration to find the set of switches, expressed as the s values at which the switches occur.

---

2. In general, equation (11.10) may be satisfied for multiple values of $\dot{s}$ for a single value of s. This may occur due to friction in the system, weak actuators that cannot hold each configuration statically, or the form of the actuation limit functions. In this case, there may be inadmissible "islands" in the phase plane. This significantly complicates the problem of finding an optimal time scaling, and we ignore this possibility. See [385] for a time-scaling algorithm for this case.

Time-Scaling Algorithm

1. Initialize an empty list of switches $S = \{\}$ and a switch counter $i = 0$. Set $(s_i, \dot{s}_i) = (0, \dot{s}_0)$.

2. Integrate the equation $\ddot{s} = L(s, \dot{s})$ *backward* in time from $(1, \dot{s}_f)$ until the velocity limit curve is penetrated (reached transversally, not tangentially), $s = 0$, or $\dot{s} = 0$ at $s < 1$. There is no solution to the problem if $\dot{s} = 0$ is reached. Otherwise, call this phase plane curve F and proceed to the next step.

3. Integrate the equation $\ddot{s} = U(s, \dot{s})$ forward in time from $(s_i, \dot{s}_i)$. Call this curve A_i. Continue integrating until either A_i crosses F or A_i penetrates the velocity limit curve. (If A_i crosses $s = 1$ or $\dot{s} = 0$ before either of these two cases occurs, there is no solution to the problem.) If A_i crosses F, then increment i, let s_i be the s value at which the crossing occurs, and append s_i to the list of switches S. The problem is solved and S is the solution. If instead the velocity limit curve is penetrated, let $(s_{\text{lim}}, \dot{s}_{\text{lim}})$ be the point of penetration and proceed to the next step.

4. Search the velocity limit curve $v(s)$ forward in s from $(s_{\text{lim}}, \dot{s}_{\text{lim}})$ until finding the first point where the feasible acceleration ($L = U$ on the velocity limit curve) is tangent to the velocity limit curve. If the velocity limit is $v(s)$, then a point $(s_0, v(s_0))$ satisfies the tangency condition if $\frac{dv}{ds}\big|_{s=s_0} = U(s_0, v(s_0))/v(s_0)$. Call the first tangent point reached $(s_{\text{tan}}, \dot{s}_{\text{tan}})$.[3] From $(s_{\text{tan}}, \dot{s}_{\text{tan}})$, integrate the curve $\ddot{s} = L(s, \dot{s})$ backward in time until it intersects A_i. Increment i and call this new curve A_i. Let s_i be the s value of the intersection point. This is a switch point from maximum to minimum acceleration. Append s_i to the list S.

5. Increment i and set $(s_i, \dot{s}_i) = (s_{\text{tan}}, \dot{s}_{\text{tan}})$. This is a switch point from minimum to maximum acceleration. Append s_i to the list S. Go to step 3.

An illustration of the time-scaling algorithm is shown in figure 11.3.

11.2.1   Zero Inertia Points

Until now, we have been making the assumption that each $a_i(s)$ in (11.8) is always nonzero. In this usual case, the velocity limit occurs when $L(s, \dot{s}) = U(s, \dot{s})$. If

---

3. An alternative approach to finding $(s_{\text{tan}}, \dot{s}_{\text{tan}})$ is to choose a point $(s_{\text{lim}}, \dot{s}')$, where $\dot{s}' < \dot{s}_{\text{lim}}$, integrate L forward from there, and check if the solution penetrates the velocity limit curve. If so, choose $\dot{s}'' < \dot{s}'$; if not, choose $\dot{s}'' > \dot{s}'$. Perform the integration of L from $(s_{\text{lim}}, \dot{s}'')$. Repeat the binary search until the forward integration just touches the velocity limit curve tangentially. The tangent point is $(s_{\text{tan}}, \dot{s}_{\text{tan}})$.

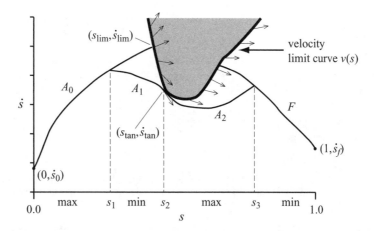

Figure 11.3   An illustration of the time-scaling algorithm. (Step 2) Beginning from $(1, \dot{s}_f)$, the minimum acceleration is integrated backward until the velocity limit curve is reached. The resulting phase plane curve is denoted F. (Step 3) Beginning from $(0, \dot{s}_0)$, the maximum acceleration is integrated forward until the velocity limit curve is reached at $(s_{\text{lim}}, \dot{s}_{\text{lim}})$. This phase plane curve is denoted A_0. (Step 4) The velocity limit curve $v(s)$ is searched forward from s_{lim} until a point is found where the feasible acceleration is tangent to the limit curve. The figure shows the feasible accelerations at points on the velocity limit curve as arrows. An arrow becomes tangent to the velocity limit curve at $(s_{\text{tan}}, \dot{s}_{\text{tan}})$. From $(s_{\text{tan}}, \dot{s}_{\text{tan}})$, the minimum acceleration is integrated backward until it reaches A_0. This curve is called A_1. The s value of the intersection is s_1 and is added to the switch list, $\mathcal{S} = \{s_1\}$. (Step 5) The point $(s_2, \dot{s}_2)$ is set equal to $(s_{\text{tan}}, \dot{s}_{\text{tan}})$, and s_2 is added to the switch list, $\mathcal{S} = \{s_1, s_2\}$. (Step 2) Maximum acceleration is integrated forward from $(s_2, \dot{s}_2)$ until it hits F. This curve is denoted A_2, and the s value of the intersection, s_3, is added to the switch list, yielding $\mathcal{S} = \{s_1, s_2, s_3\}$. The algorithm terminates.

$a_i(s) = 0$, however, the force at the ith actuator is independent of $\ddot{s}$, and therefore the ith actuator defines no acceleration constraints $L_i(s, \dot{s})$, $U_i(s, \dot{s})$. Instead, it defines directly a *velocity* constraint using (11.7):

$$(11.11) \qquad u_i^{\min}(s, \dot{s}) \le b_i(s)\dot{s}^2 + c_i(s) \le u_i^{\max}(s, \dot{s}).$$

In the case of a zero inertia point where k of the $a_i(s)$ are zero, let $\dot{s}_{\text{zip}}^{\max}(s)$ be the maximum velocity satisfying all k constraints (11.11), and let $\dot{s}^{\max}(s) = \min(v(s), \dot{s}_{\text{zip}}^{\max}(s))$. Then $\dot{s}^{\max}(s)$ is the *true* velocity limit curve, generalizing the curve $v(s)$ by allowing for the possibility of zero inertia points.

If a point on the velocity limit curve $\dot{s}^{\max}(s)$ is determined by a zero inertia point velocity constraint, then $L(s, \dot{s}) < U(s, \dot{s})$ at this point. This point is called a *critical* point. If, in addition, either $U(s, \dot{s})$ integrated forward from this point, or $L(s, \dot{s})$

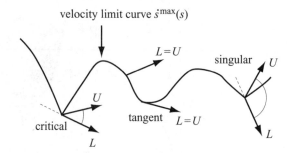

Figure 11.4   At critical and singular points, $U(s, \dot{s}) \neq L(s, \dot{s})$. At a singular point, following *U* forward or *L* backward results in penetration of the velocity limit curve.

integrated backward from this point, would result in immediate penetration of the velocity limit curve, then the point is called *singular* (figure 11.4).

At a singular point $(s_s, \dot{s}^{\max}(s_s))$, let

$$\ddot{s}_{\text{tangent}} = \dot{s}^{\max}(s_s) \frac{d\dot{s}^{\max}}{ds}\Big|_{s=s_s}$$

be the acceleration defined by the tangent to the velocity limit curve. If the velocity limit curve is not differentiable at the singular point (as is often the case), then define

$$\ddot{s}^{+}_{\text{tangent}} = \dot{s}^{\max}(s_s) \frac{d\dot{s}^{\max}}{ds}\Big|_{s=s_s^{+}}$$

$$\ddot{s}^{-}_{\text{tangent}} = \dot{s}^{\max}(s_s) \frac{d\dot{s}^{\max}}{ds}\Big|_{s=s_s^{-}}$$

to be the right and left limits, respectively. Then, to prevent penetration of the velocity limit curve, the maximum feasible acceleration at $(s_s, \dot{s}^{\max}(s_s))$ is

$$\ddot{s}^{\max} = \min(\ddot{s}^{+}_{\text{tangent}}, U).$$

Similarly, the minimum feasible acceleration is

$$\ddot{s}^{\min} = \max(\ddot{s}^{-}_{\text{tangent}}, L).$$

This is shown graphically in figure 11.5.

Critical points occur on a lower-dimensional subset of the robot's configuration space where $M(q)$ is not full rank. If the path passes through this subset transversally, the path will have isolated critical points. If the path travels along this lower-dimensional subset, however, we may have a continuous *critical arc* of critical points. Similarly, we may have *singular arcs*.

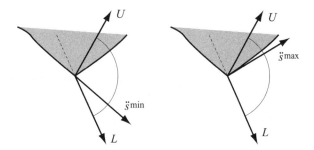

Figure 11.5 At this singular point on the velocity limit curve, integrating $L(s, \dot{s})$ backward in time would result in penetration of the velocity limit curve. The true minimum feasible acceleration at this point is $\ddot{s}^{\min}$, defined by the left tangent of the velocity limit curve at the singularity. Similarly, integrating $U(s, \dot{s})$ forward results in penetration. The true maximum feasible acceleration is $\ddot{s}^{\max}$, defined by the right tangent at the singularity.

We can now modify the time-scaling algorithm to properly account for zero inertia points. At singular points, we integrate $\ddot{s}^{\min}(s, \dot{s})$ and $\ddot{s}^{\max}(s, \dot{s})$ instead of $L(s, \dot{s})$ and $U(s, \dot{s})$, respectively. This will also allow the algorithm to "slide" along singular arcs using an acceleration between $L(s, \dot{s})$ and $U(s, \dot{s})$, instead of switching rapidly back and forth between them. In step 4, we search the velocity limit curve $\dot{s}^{\max}(s, \dot{s})$ for any critical or tangent point, not just tangent points.

EXAMPLE 11.2.1 *Consider the two-joint RP robot arm with the dynamics derived in chapter 10. We have planned a path to follow the straight line shown in figure 11.6, and we wish to find the time-optimal time scaling of the path. Let $x = [x_1, x_2]^T$ be the Cartesian coordinates of the center of mass of the second link, as shown in the figure. The path we wish to follow, parameterized by s, is expressed as*

(11.12) $x(s) = [x_1(s), x_2(s)]^T = [2s - 1, 1]^T, \quad s \in [0, 1].$

The first thing we will do is express this path in the generalized coordinates $q = [q_1, q_2]^T$ of figure 10.2 in chapter 10. To do this, we can define the forward kinematics of the robot arm to be the mapping ϕ from joint coordinates q to Cartesian coordinates x:

(11.13) $x = \phi(q)$

$[x_1, x_2]^T = [q_2 \cos q_1, q_2 \sin q_1]^T$

The inverse kinematics ϕ^{-1} gives the joint coordinates q as a function of the Cartesian coordinates x. For the RP arm, the inverse kinematics are unique within the reachable

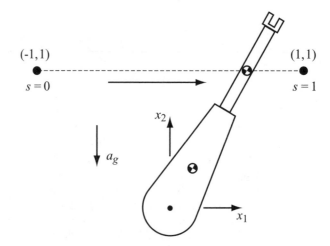

Figure 11.6   The path followed by the RP manipulator.

workspace of the robot (where $q_2 \geq 0$):

$$q = \phi^{-1}(x)$$

(11.14) $$[q_1, q_2]^T = \left[\text{atan2}(x_2, x_1), \sqrt{x_1^2 + x_2^2} \right]^T$$

where atan2(x_2, x_1) *is the two argument arctangent returning the unique angle in* $[-\pi, \pi)$ *of the Cartesian point* (x_1, x_2) *(where* $x_1^2 + x_2^2 \neq 0$*). Plugging equation (11.12) into equation (11.14), we find the parameterized path in joint coordinates*

(11.15) $$q(s) = \left[\text{atan2}(1, 2s - 1), \sqrt{4s^2 - 4s + 2} \right]^T.$$

Differentiating, we get the velocity and acceleration

(11.16) $$\dot{q} = \begin{bmatrix} \frac{-\dot{s}}{2s^2 - 2s + 1} \\ \frac{(4s-2)\dot{s}}{\sqrt{4s^2 - 4s + 2}} \end{bmatrix}$$

(11.17) $$\ddot{q} = \begin{bmatrix} \frac{(4s-2)\dot{s}^2 + (-2s^2 + 2s - 1)\ddot{s}}{(2s^2 - 2s + 1)^2} \\ \frac{\sqrt{2}(\dot{s}^2 + (4s^3 - 6s^2 + 4s - 1)\ddot{s})}{(2s^2 - 2s + 1)^{3/2}} \end{bmatrix}.$$

In chapter 10 we derived the equations of motion:

(11.18) $$\begin{aligned} u_1 = & \left(I_1 + I_2 + m_1 r_1^2 + m_2 q_2^2 \right) \ddot{q}_1 + 2 m_2 q_2 \dot{q}_1 \dot{q}_2 \\ & + a_g (m_1 r_1 + m_2 q_2) \cos q_1 \end{aligned}$$

(11.19) $$u_2 = m_2 \ddot{q}_2 - m_2 q_2 \dot{q}_1^2 + a_g m_2 \sin q_1$$

Substituting in equations (11.15), (11.16), and (11.17), we get

$$a(s)\ddot{s} + b(s)\dot{s}^2 + c(s) = u,$$

where

(11.20) $\quad a(s) = \begin{bmatrix} \dfrac{I_1 + I_2 + 2m_2 + m_1 r_1^2 - 4m_2 s + 4m_2 s^2}{-2s^2 + 2s - 1} \\[3mm] \dfrac{\sqrt{2}m_2(2s-1)}{\sqrt{2s^2 - 2s + 1}} \end{bmatrix}$

(11.21) $\quad b(s) = \begin{bmatrix} \dfrac{2(I_1 + I_2 + m_1 r_1^2)(2s-1)}{(2s^2 - 2s + 1)^2} \\[3mm] 0 \end{bmatrix}$

(11.22) $\quad c(s) = \begin{bmatrix} (m_1 r_1 + m_2\sqrt{4s^2 - 4s + 2})a_g \cos(\mathrm{atan2}(1, 2s-1)) \\[2mm] m_2 a_g \sin(\mathrm{atan2}(1, 2s-1)) \end{bmatrix}.$

Note that $s = \frac{1}{2}$ defines a zero inertia point, as $a_2(\frac{1}{2}) = 0$. (Understand this intuitively by considering the $s = \frac{1}{2}$ point in figure 11.6.) In this case there is no velocity constraint due to the second actuator, since $b_2 = 0$ (see problem 5).

We now choose the following parameters for the robot arm: $m_1 = 5$ kg, $I_1 = 0.1$ kg-m^2, $r_1 = 0.2$ m, $m_2 = 3$ kg, and $I_2 = 0.05$ kg-m^2. The actuator limits are taken to be ± 20 N-m for joint 1 and ± 40 N for joint 2, and gravity is $a_g = 9.8$ m/s^2. Figure 11.7 shows the time-optimal trajectory along the path for these choices. The minimum-time execution of the path is approximately 0.888 s. Note that one of the actuators is saturated at all times. In this example, the velocity limit curve is never reached, so there is just one switch between maximum and minimum acceleration.

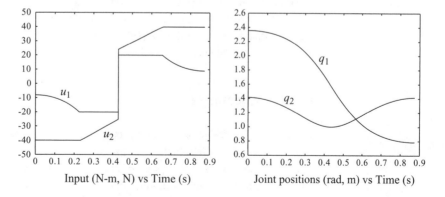

Input (N-m, N) vs Time (s)   Joint positions (rad, m) vs Time (s)

Figure 11.7 The time-optimal actuator histories and trajectory of the RP manipulator along the straight-line path of example 11.2.1.

11.2.2 Global Time-Optimal Trajectory Planning

The time-scaling algorithm finds the time-optimal trajectory along a given path. What if our real goal is to find the time-optimal trajectory between two states when we are free to choose any collision-free path? Can we use the time-scaling algorithm in conjunction with a collision-free path planner? Conceptually, imagine running the time-scaling algorithm on all possible paths between the start and goal states. Then the fastest of these is the global time-optimal trajectory.

Naturally, the problem is how to efficiently test a large number of possible paths. One approach to this problem for robot manipulators was proposed by Shiller and Dubowsky [383]. The approach is quite involved, and we only sketch it here. The first step is to define a grid on the workspace and construct all collision-free paths (without sharp turns) between the start and goal states moving along edges or diagonals of the grid. The next step is to quickly compute rough lower-bound estimates of the traveling times of these paths using a maximum velocity limit during the motion. The fastest paths are selected and smoothed by using the grid points as control points for cubic splines. The best of these paths is then submitted to the full time-scaling algorithm, generating an upper bound on the optimal travel time. All paths with lower bounds above this upper bound can be pruned. Of the remaining paths, the lower bounds are more carefully calculated, and the process continues, using increasingly accurate estimates of the lower bounds as the number of candidate paths is reduced. When the pruning process has ended, only the best path in each path-neighborhood is considered further. These best paths are submitted to a local optimization that may locally alter the paths to allow them to be executed more quickly. This process uses the travel times returned by the time-scaling algorithm as the objective function.

This approach combines collision-free path planning and time scaling in an iterative fashion to arrive at a global near-time-optimal trajectory. In the next section, we discuss methods that do not separate the path-planning and time-scaling problems, but solve directly in the state space.

11.3 Direct Trajectory Planning

In section 11.2, trajectory planning is decoupled into collision-free path planning followed by time scaling. In this section, we study methods for planning the trajectory directly in the state space. If we are interested in finding trajectories that optimize some cost function, such as motion time or expended energy, optimal control theory provides necessary conditions on the trajectories. Unfortunately, these conditions are complex for almost any robot system and cannot be solved analytically. Because of this, we consider two numerical approaches: nonlinear optimization and grid-based search.

First we describe how to transform the optimal control problem to a finite-dimensional parameter optimization problem, allowing nonlinear optimization to be used to numerically solve the optimality conditions. If the problem is well formulated (e.g., the objective and constraint functions are sufficiently smooth), nonlinear optimization may result in rapid convergence to a locally optimal trajectory. The drawbacks of this approach are that the method requires an initial guess (possibly provided by a grid-based search method), and the locally optimal solution reached generally depends heavily on this guess. Also, evaluation of constraint and objective functions, and their gradients, may be computationally costly.

We then introduce a grid-based search method that allows the user to specify how near to time-optimal the motion plan should be, while meeting "safety" requirements on obstacle avoidance. The planned motions are approximate in that the goal state may not be exactly reached, but the user has control over how large the final error can be. An advantage is that this is a global approach, unlike gradient-based nonlinear optimization. A drawback of grid-based search is that the size of the grid grows exponentially in the dimension of the state space, making the approach impractical for high-dimensional systems.

A third approach is to use RRT's, as described in chapter 7. This approach trades off optimality for planner run-time—it may be able to quickly find a feasible trajectory that is in no sense optimal.

Finally, a fourth approach is based on artificial potential fields (see chapter 4). An artificial potential field is constructed to make obstacles repulsive and the goal configuration attractive. The robot senses its current configuration, calculates the gradient of the artificial potential at this configuration, and applies the gradient forces at the actuators. (Damping forces may be included to stabilize the goal configuration.) A potential field, therefore, implicitly defines a trajectory to the goal from all initial states. This approach is fundamentally different from the previous "open-loop" approaches in that a feedback law is specified for all robot states.

11.3.1 Optimal Control

Given a dynamic system

(11.23) $$M(q)\ddot{q} + C(q, \dot{q})\dot{q} + g(q) = u,$$

we would like to find a solution $(q(t), u(t))$, $t \in [0, t_f]$ to equation (11.23) that avoids obstacles and joint limits, respects actuator limits, and takes the system from the initial state $(q_{\text{start}}, \dot{q}_{\text{start}})$ at time $t = 0$ to the final state $(q_{\text{goal}}, \dot{q}_{\text{goal}})$ at time $t = t_f$. Of all the trajectories that accomplish this, we might want to find one that minimizes some objective function J. In general, J might be a function of the controls, the trajectory, and the total motion time t_f, i.e., $J = J(u(t), q(t), t_f)$.

Before proceeding further, let's express the state of the system as $x = [q^T, \dot{q}^T]^T$ and rewrite the equations of motion in the general form

(11.24) $\dot{x}(t) = f(x(t), u(t), t),$

where f is the vector differential equation describing the kinematics and dynamics of the system. In our case, the state equations do not change with time, so $f(x, u, t) = f(x, u)$. The state equations (11.24) can be viewed as constraints defining the relationship between $x(t)$ and $u(t)$. Let the objective function J be written

(11.25) $J = \displaystyle\int_0^{t_f} \mathcal{L}(x(t), u(t), t)dt,$

where the integrand $\mathcal{L}$ is called the *Lagrangian*. (The Lagrangian $\mathcal{L}$ for optimal control is actually a generalization of the Lagrangian L for dynamics.) A typical choice of $\mathcal{L}$ is "effort," modeled as a quadratic function of the control, e.g., $\mathcal{L} = u^T W u$, where W is a positive definite weighting matrix, e.g., the identity matrix. Another common choice is to leave t_f free and choose $\mathcal{L} = 1$, implying $J = t_f$, a minimum-time problem.

For now, let us ignore the issues of actuator limits and obstacles, and assume that the final time t_f is fixed. The problem then is to find a state and control history $(x(t), u(t)), t \in [0, t_f]$ that satisfies the constraints of equation (11.24), satisfies the terminal conditions $x(t_f) = x_f$, and minimizes the cost in equation (11.25). To write a necessary condition for optimality, we define the *Hamiltonian* $\mathcal{H}$

(11.26) $\mathcal{H}(x(t), u(t), \lambda(t), t) = \mathcal{L}(x(t), u(t), t) + \lambda^T f(x(t), u(t)),$

where $\lambda(t)$ is a vector of *Lagrange multipliers*.[4] Then the *Pontryagin minimum principle*[5] says that at an optimal solution $(x^*(t), u^*(t), \lambda^*(t))$,

(11.27) $\mathcal{H}^*(t) = \mathcal{H}(x^*(t), u^*(t), \lambda^*(t), t) \leq \mathcal{H}(x^*(t), u(t), \lambda^*(t), t).$

In other words, if the control history $u^*(t)$ is optimal, then at any time t, any other feasible control $u(t)$ will give an $\mathcal{H}(t)$ greater than or equal to that of the optimal $\mathcal{H}^*(t)$. In the absence of any other constraints on the state and control, then, a necessary

4. These Lagrange multipliers play a similar role to those in dynamics with constraints—they are used to enforce constraints, here the state equation constraints.
5. Often known as the Pontryagin maximum principle. In our case, we are minimizing a cost function; we could equivalently maximize a utility function, or the negative of the cost function.

condition for optimality can be written

(11.28)
$$\frac{\partial \mathcal{H}}{\partial u} = 0.$$

This says that the linear sensitivity of the Hamiltonian to changes in u is zero, meaning that the control is *extremal,* but not necessarily optimal. A sufficient condition for local optimality of a solution is that equation (11.28) is satisfied and the Hessian of the Hamiltonian is positive definite along the trajectory of the solution:

(11.29)
$$\frac{\partial^2 \mathcal{H}}{\partial u^2} > 0$$

This is known as the convexity or Legendre-Clebsch condition. The Lagrange multipliers evolve according to the adjoint equation

(11.30)
$$\dot{\lambda} = -\frac{\partial \mathcal{H}}{\partial x}.$$

Equation (11.28) can sometimes be used to write u as a function of x and λ. In this case, optimization boils down to choosing initial conditions for equation (11.30) to ensure that the goal is reached.

EXAMPLE 11.3.1   *Consider a simple double-integrator system with one degree of freedom, $\ddot{q} = u$, such as a point mass on a line actuated by a force. Let $x = [x_1, x_2]^T = [q, \dot{q}]^T$, so the equations of motion can be written in the form*

$$\dot{x} = f(x, u), \qquad \begin{bmatrix} \dot{x}_1 \\ \dot{x}_2 \end{bmatrix} = \begin{bmatrix} x_2 \\ u \end{bmatrix}.$$

Choose the objective function

$$J = \int_0^{t_f} u^2 dt.$$

Then the Hamiltonian is

$$\mathcal{H} = u^2 + \lambda_1 x_2 + \lambda_2 u$$

and the necessary condition (11.28) is written

(11.31)
$$\frac{\partial \mathcal{H}}{\partial u} = 2u + \lambda_2 = 0.$$

The adjoint equation (11.30) is written

(11.32)
$$\dot{\lambda} = -\frac{\partial \mathcal{H}}{\partial x}, \qquad \begin{bmatrix} \dot{\lambda}_1 \\ \dot{\lambda}_2 \end{bmatrix} = -\begin{bmatrix} 0 \\ \lambda_1 \end{bmatrix}.$$

Equation (11.32) shows that λ_1 is constant and λ_2 is a linear function of time, so by equation (11.31), u is also a linear function of time, e.g., $u(t) = c_0 + c_1 t$.

Now we can specify the initial and final state for the system to solve for the control and state history. Let $x(0) = [0, 0]^T$ and $x(t_f) = [d, 0]^T$, i.e., the system starts at rest and ends at rest having moved a distance d in time t_f. Then we have the stopping conditions that the first and second integral of u(t) evaluated over $[0, t_f]$ be zero and d, respectively:

$$x_2(t_f) = \int_0^{t_f} u(t)dt = 0$$

$$x_1(t_f) = \int_0^{t_f} \int_0^t u(\eta)d\eta \, dt = d.$$

Solving, we get $c_0 = 6d/t_f^2$, $c_1 = -12d/t_f^3$ defining the extremal control history. The extremal state history is obtained by integration of the control.

To check if this extremal solution is a minimizer, we can use the convexity condition. In this case, the Hessian is the scalar $\partial^2\mathcal{H}/\partial u^2 = 2 > 0$, indicating that the solution is indeed (locally) optimal.

In the previous example, the convexity condition is satisfied. However, if an extremum is achieved ($\partial\mathcal{H}/\partial u = 0$) but convexity is not ($\partial^2\mathcal{H}/\partial u^2$ is only positive *semidefinite*), it does not mean that the control is not optimal. In this case, auxiliary conditions have to be satisfied to ensure optimality. An optimal control of this type is an example of a *singular* optimal control, as in the previous section with time-optimal control at zero inertia points.

What if there are actuator limits or obstacles? At an optimal solution, the minimum principle (11.27) will always be satisfied, but the control or state history may bump up against limits preventing equation (11.28) from being satisfied. The optimal solution may be constrained by these limits rather than the extremality condition (11.28). Consider the one-degree-of-freedom double integrator above, with the actuator limits $|u| < u^{max}$, and choose the minimum-time objective function

$$J = \int_0^{t_f} 1dt,$$

where the time t_f is left free. The Hamiltonian is $\mathcal{H} = 1 + \lambda_1 x_2 + \lambda_2 u$, and $\partial\mathcal{H}/\partial u = \lambda_2$ does not contain the control variable. Therefore, it provides no information on the choice of $u(t)$. Further, $\partial^2\mathcal{H}/\partial u^2$ is zero, so $\mathcal{H}$ is not convex. We know, however, that the optimal solution for this problem is a bang-bang trajectory, just like the bang-bang trajectories found by the time-scaling algorithm.

We can recover the bang-bang solution using the minimum principle (11.27). We write

$$\mathcal{H}^*(t) = \mathcal{H}(x^*(t), u^*(t), \lambda^*(t)) \leq \mathcal{H}(x^*(t), u(t), \lambda^*(t))$$

$$1 + \lambda_1^*(t)x_2^*(t) + \lambda_2^*(t)u^*(t) \leq 1 + \lambda_1^*(t)x_2^*(t) + \lambda_2^*(t)u(t)$$

$$\lambda_2^*(t)u^*(t) \leq \lambda_2^*(t)u(t).$$

Therefore, $u^*(t)$ is the maximum feasible value $u^{\max}$ when $\lambda_2^*(t) < 0$ and the minimum feasible value $-u^{\max}$ when $\lambda_2^*(t) > 0$. As before, λ_2 is a linear function of time, and the terminal state conditions allow us to find the complete solution.

Only for very simple systems is it possible to solve the extremality conditions analytically. In most cases it is necessary to resort to numerical methods to solve the conditions approximately. One such method is called *shooting*. The user guesses initial values for the Lagrange multipliers $\lambda(0)$, which then are numerically integrated according to the adjoint equation $\dot{\lambda} = -\partial\mathcal{H}/\partial x$, while the control vector u evolves according to $\partial\mathcal{H}/\partial u = 0$. After integrating for time t_f (for fixed-time problems), if the final state is not equal to the desired final state, the initial guess of the Lagrange multipliers is modified in some reasonable way and the process repeats. In other words, by modifying the initial conditions, we "shoot" at the goal until we hit it. Typically the initial conditions are modified using an approximation of the gradient of the map taking the initial conditions to the final state.

Other numerical methods for approximately solving for optimal controls include dynamic programming and gradient-based nonlinear optimization. In the next subsection we discuss a nonlinear optimization approach to finding the optimal parameters of a finite parameterization of the system's trajectory or controls. In subsection 11.3.3 we introduce a grid-based search method for finding near-time-optimal trajectories.

For more on optimal control, see the books by Bryson and Ho [72], Bryson [71], Kirk [236], Lewis and Syrmos [287], Stengel [396], and Pontryagin, Boltyanskii, Gamkrelidze, and Mishchenko [354].

11.3.2   Nonlinear Optimization

The general problem can be stated

(11.33)   find $t_f, q(t), u(t)$

(11.34)   minimizing $J(u(t), q(t), t_f)$

(11.35)   subject to $M(q(t))\ddot{q}(t) + C(q(t), \dot{q}(t))\dot{q} + g(q(t)) = u(t), \quad 0 \leq t \leq t_f$

(11.36) $u^{\min}(q(t), \dot{q}(t)) \leq u(t) \leq u^{\max}(q(t), \dot{q}(t)), \quad 0 \leq t \leq t_f$

(11.37) $h(q(t)) \leq 0, \quad 0 \leq t \leq t_f$

(11.38) $q(0) = q_{\text{start}}, \quad \dot{q}(0) = \dot{q}_{\text{start}}$

(11.39) $q(t_f) = q_{\text{goal}}, \quad \dot{q}(t_f) = \dot{q}_{\text{goal}},$

where $h(q) \leq 0$ are configuration inequality constraints representing obstacles and joint limits.

To approximately solve this problem by nonlinear optimization, we approximate the continuous constraints (11.36) and (11.37) by a finite number of constraints. This is typically done by ensuring that the constraints are satisfied at a fixed number of points distributed evenly over the interval $[0, t_f]$. We also choose a finite-parameter representation of the state and control histories. We have three choices of how to do this:

1. *Parameterize the trajectory $q(t)$.* In this case, we solve for the parameterized trajectory directly. The control forces u at any time are calculated using equation (11.35).

2. *Parameterize the control $u(t)$.* We solve for $u(t)$ directly, and calculating the state $(q(t), \dot{q}(t))$ requires integrating the equations of motion (11.35).

3. *Parameterize both $q(t)$ and $u(t)$.* We have a larger number of variables, since we are parameterizing both $q(t)$ and $u(t)$. Also, we have a larger number of constraints, as $q(t)$ and $u(t)$ must satisfy the dynamic equations (11.35) explicitly, typically at a fixed number of points distributed evenly over the interval $[0, t_f]$. We must be careful to choose the parameterizations of $q(t)$ and $u(t)$ to be consistent with each other, so that the dynamic equations can be satisfied at these points.

A trajectory or control history can be parameterized in any number of ways. The parameters can be the coefficients of a polynomial in time, the coefficients of a truncated Fourier series, spline coefficients, wavelet coefficients, piecewise constant acceleration or force segments, etc. For example, the control $u_i(t)$ could be represented by $p + 1$ coefficients a_j of a polynomial in time:

$$u_i(t) = \sum_{j=0}^{p} a_j t^j$$

In addition to the parameters for the state or control history, the total time t_f may be another control parameter. The choice of parameterization has implications for the efficiency of the calculation of $q(t)$ and $u(t)$ at a given time t. The choice of parameterization also determines the sensitivity of the state and control to the parameters, and whether each parameter affects the profiles at all times $[0, t_f]$ or just on a finite-time support base. These are important factors in the stability and efficiency of the numerical optimization.

Let X be the vector of the control parameters to be solved. Assuming that either $q(t)$ or $u(t)$ has been parameterized (but not both), and that the $k + 1$ constraint checks are spaced at $\Delta t = t_f / k$ intervals, the constrained nonlinear optimization can be written

(11.40)   find X

(11.41)   minimizing $J(X)$

(11.42)   subject to $u^{\min}(X, j\Delta t) \leq u(X, j\Delta t) \leq u^{\max}(X, j\Delta t), \quad j = 0 \ldots k$

(11.43) $h(X, j\Delta t) \leq 0, \quad j = 0 \ldots k$

(11.44) $q(X, 0) = q_{\text{start}}, \quad \dot{q}(X, 0) = \dot{q}_{\text{start}}$

(11.45) $q(X, t_f) = q_{\text{goal}}, \quad \dot{q}(X, t_f) = \dot{q}_{\text{goal}}.$

A variant of this formulation approximately represents the constraints (11.42)–(11.45) by penalty functions in the objective function, allowing the use of unconstrained optimization.

A nonlinear program of this type can be solved by a number of methods, including sequential quadratic programming (SQP). Any solver will require the user to provide functions to take a guess X and calculate the objective function $J(X)$ and the constraints (11.42)–(11.45). Often the objective function will have to be calculated by numerical integration. All solvers also need the gradients of the objective function and the constraints with respect to X. These can be calculated numerically by finite differences, or, if possible, analytically. Finally, most solvers make use of Hessians of the objective function and constraint functions with respect to X. Most solvers update a numerical approximation to these Hessians rather than requesting the user to provide these. Details on different methods for nonlinear optimization can be found in [164, 326, 338]. Code for nonlinear optimization includes FSQP and CFSQP [4], NPSOL [5], and routines in the IMSL [6], NAG [7], and MATLAB Optimization Toolbox libraries.

The most important point is that for any of these solvers to work, the objective and constraints must be sufficiently smooth with respect to the control parameters X. They must be at least C^1, but usually C^2 so that Hessian information can be used to speed convergence. A key part of the problem formulation is ensuring this smoothness. When possible, the gradients, and even the Hessians, should be calculated analytically. This will minimize the possibility of the solver failing to converge due to numerical problems, a very real practical concern! Even if the objective function is calculated approximately by numerical integration, it may be possible to calculate the exact gradient of this approximation analytically.

Since nonlinear optimization uses local gradient information, it will converge to a locally optimal solution. For some problems, the control parameter space will be littered with many local optima. Therefore, the solution achieved will depend heavily on the initial guess. To ensure a good solution, the process can be started from several initial guesses, keeping the best local optimum. Nonlinear optimization can also be used as a final step to locally improve a trajectory found by some other global search method. A survey of nonlinear optimization methods for trajectory generation can be found in [49].

11.3.3   Grid-Based Search

An alternative approach, specifically for time-optimal trajectory planning, uses grid search. As motivation, consider the simple double-integrator system

$$\ddot{q} = a$$
$$|a| \le a_{max},$$

where a is the acceleration control. Let q be one-dimensional, so the system can be viewed as a point mass moving on a line with a control force. The time-optimal control from an initial state $(q_{start}, \dot{q}_{start})$ to a goal state $(q_{goal}, \dot{q}_{goal})$ is bang-bang—the actuator is saturated at all times.

Things will not be so simple when we deal with multidimensional problems with obstacles and velocity limits, so let's consider a grid-based approach that we will be able to generalize to more dimensions. First, we discretize the control set to $\{-a_{max}, 0, a_{max}\}$. Next, we choose a timestep h. Now, beginning from $(q_{start}, \dot{q}_{start})$, we integrate the three controls forward in time by h to obtain three new states. Think of the initial state $(q_{start}, \dot{q}_{start})$ as the root of a tree, and the three new states as children of the root (figure 11.8). From each of these three, we integrate the controls forward to obtain a new level of the tree. We continue in a breadth-first fashion. If the trajectory to a new node in the tree passes through an obstacle or exceeds a velocity limit, this node

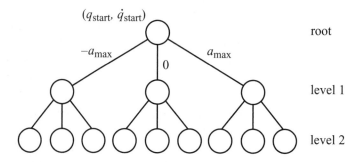

Figure 11.8   The search tree for three controls.

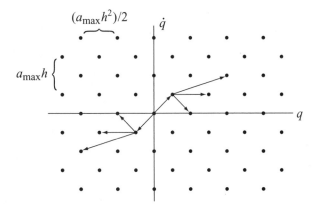

Figure 11.9   The search tree of figure 11.8 shown on the state space grid. The search begins at $(0, 0)$. The actual trajectories between vertices are not straight lines, but quadratics.

is pruned from the tree. The search continues until a trajectory reaches a state in a specified goal region. The trajectory is specified by the piecewise-constant controls to traverse the tree from the root node to this final node. Since the search is breadth-first, exploring all reachable states at time kh before moving on to time $(k + 1)h$, the trajectory is time-optimal for the chosen discretization of time and controls.

We call this a grid-based search because each of the nodes reached during the growth of the tree lies on a regular grid on the $(q, \dot{q})$ state space. From any state, the new state obtained by integrating one of the discretized accelerations for time h will involve a change in $\dot{q}$ equal to an integral multiple of $a_{\max}h$ and a change in q equal to an integral multiple of $\frac{1}{2}a_{\max}h^2$. An example of such a grid is shown in figure 11.9. The search tree shown on this grid is two levels deep, beginning at $(0, 0)$. The key point here is that, given some bounds on q and $\dot{q}$, the size of the grid is easily computed, so an upper bound on the computational complexity of the search of this grid is also easily computed.

Let us now consider a more general problem statement. The system is described by $q \in D \subset \mathbb{R}^{n_Q}$, where D is a bounded subset of $\mathbb{R}^{n_Q}$, and velocity and acceleration bounds of the form

$$|\dot{q}_i| \le v_{\max}, \quad i = 1 \dots n_Q$$
$$|\ddot{q}_i| \le a_{\max}, \quad i = 1 \dots n_Q.$$

Note that this is a very limited class of systems, as the maximum feasible $\ddot{q}_i$'s are constant and independent of the state $(q, \dot{q})$. An example of such a system might be a point in n_Q-dimensional Euclidean space with a thruster for each degree of

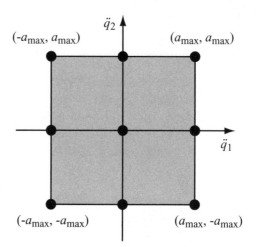

Figure 11.10   The control discretization for $n_Q = 2$.

freedom, or a Cartesian robot arm consisting of all prismatic joints and actuators with state-independent bounds. The problem is to find a collision-free, approximately time-optimal trajectory from $(q_{start}^*, \dot{q}_{start}^*)$ [at or near the desired start state $(q_{start}, \dot{q}_{start})$] to a goal state near the desired goal state $(q_{goal}, \dot{q}_{goal})$.

The algorithm uses a discretized control set A consisting of the cross products of $\{-a_{max}, 0, a_{max}\}$ for each degree of freedom, yielding 3^{n_Q} distinct controls (figure 11.10). These controls result in a regular grid on $2n_Q$-dimensional state space, similar to figure 11.9. Algorithm 21, GRID_SEARCH, is described in pseudocode below.

This algorithm is straightforward, except for one twist: the algorithm prunes a node if the trajectory passes *close* to an obstacle, not just if it passes through an obstacle. Thus the algorithm will only return trajectories that are *safe*. We define a trajectory to be $\delta_v(c_0, c_1)$-*safe* if there exists a speed-dependent ball of free configurations about each q in the trajectory, where the radius of the ball is $c_0 + c_1 \|\dot{q}\|$. The parameters c_0 and c_1 are safety parameters.

Since the algorithm uses a finite timestep h, any trajectory it finds will only be an *approximately* time-optimal safe trajectory. Instead of directly choosing h, the user could have control over a parameter ϵ, $0 < \epsilon < 1$, which defines the crudeness of the approximation. Larger values of ϵ correspond to cruder approximations, and the timestep h goes to zero as ϵ goes to zero. As we will see, ϵ may be viewed as a measure of how much we will allow $\delta_v(c_0, c_1)$-safety to be violated, giving $(1 - \epsilon)\delta_v(c_0, c_1)$-safety.

Algorithm 21 GRID_SEARCH

Input: Start node $(q^*_{start}, \dot{q}^*_{start})$, goal region $\mathcal{G}$

Output: A trajectory to $\mathcal{G}$ or FAILURE

1: Place $(q^*_{start}, \dot{q}^*_{start})$ at root of tree T (level 0)
2: *level* $\leftarrow$ 0, *solved* $\leftarrow$ FALSE, *ANS* $\leftarrow \emptyset$
3: **while** not *solved* **do**
4:   **if** no nodes in level *level* of T **then**
5:     return FAILURE
6:   **end if**
7:   **for** each *node* in level *level* of T **do**
8:     **for** each control in A **do**
9:       Integrate control for time h from *node,* getting *newnode*
10:       **if** *newnode* has not been previously reached, and trajectory does not pass close to an obstacle nor exceed v_{max} **then**
11:         add *newnode* to T as child of *node*
12:       **end if**
13:       **if** trajectory enters $\mathcal{G}$ **then**
14:         *solved* $\leftarrow$ TRUE, store *newnode* in list ANS
15:       **end if**
16:     **end for**
17:   **end for**
18:   *level* $\leftarrow$ *level* + 1
19: **end while**
20: For each node in ANS, find the trajectory that reaches $\mathcal{G}$ first

---

Although the algorithm itself is straightforward, analysis of the algorithm is quite involved. Given a desired safety margin and ϵ, we would like to know how to choose h to guarantee completeness of the algorithm, and how ϵ and the safety margin relate to the algorithm's running time. This is the problem that was studied in detail by Donald and Xavier [135], building on work by Canny, Donald, Reif, and Xavier [94, 133]. Their analysis holds for a point robot with $n_Q = 2$ or 3 moving among polygonal or polyhedral obstacles. They give us the following result.

THEOREM 11.3.2 *Let ℓ be the diameter of the robot configuration space, c_0 and c_1 be safety parameters, and v_{max} and a_{max} be the velocity and acceleration limits. Let $0 < \epsilon < 1$. Assume there exists a $\delta_v(c_0, c_1)$-safe trajectory from $(q_{start}, \dot{q}_{start})$ to $(q_{goal}, \dot{q}_{goal})$ taking time T_{opt} by some control function with $|\ddot{q}_i| \leq a_{max}$ for all i and*

$t \in [0, T_{\text{opt}}]$. *Choose the largest h so that*

$$h \leq \frac{v_{\max}}{a_{\max}}, \quad h \leq \frac{c_0 \epsilon}{2 a_{\max} c_1 (1 - \epsilon) + 5 v_{\max}},$$

and $v_{\max}$ is an integral multiple of $a_{\max} h$. Choose an approximate starting state $(q^*_{\text{start}}, \dot{q}^*_{\text{start}})$ where, for each coordinate i,

$\dot{q}^*_{\text{start},i} =$ the multiple of $a_{\max} h$ closest to $\dot{q}_{\text{start},i}$

$$q^*_{\text{start},i} = q_{\text{start},i} - \frac{h^2}{2} (\dot{q}_{\text{start},i} - \dot{q}^*_{\text{start},i}).$$

Define the goal neighborhood to be all points within $(\frac{5 a_{\max} h^2}{2}, 2 a_{\max} h)$ of $(q_{\text{goal}}, \dot{q}_{\text{goal}})$. [Note that the distance from $(q^*_{\text{start}}, \dot{q}^*_{\text{start}})$ to $(q_{\text{start}}, \dot{q}_{\text{start}})$, and the distance from any point in the goal neighborhood to $(q_{\text{goal}}, \dot{q}_{\text{goal}})$, is $O(\epsilon)$.] Then the algorithm outlined above is guaranteed to find a $(1 - \epsilon) \delta_v (c_0, c_1)$-safe trajectory taking at most time T_{opt} from $(q^*_{\text{start}}, \dot{q}^*_{\text{start}})$ to the goal neighborhood. The running time of the algorithm is

$$O \left(c^{n_Q} N \left(\frac{v_{\max}(a_{\max} c_1 + v_{\max})^3 \ell}{a_{\max}^2 c_0^3 \epsilon^3} \right)^{n_Q} \right),$$

where $n_Q = 2$ or 3, c is a constant, and N is the number of faces in the obstacles. In terms of the dimension n_Q and the approximation variable ϵ, the running time goes as $O((\frac{1}{\epsilon})^{3 n_Q})$, i.e., polynomial in ϵ and exponential in n_Q.

The proof of this theorem is beyond the scope of this chapter, and it depends on efficient goal and safety checking between grid vertices. One important property of the algorithm is that the $(1 - \epsilon) \delta_v (c_0, c_1)$-safe trajectory found by the algorithm may be quite different from the time-optimal $\delta_v (c_0, c_1)$-safe trajectory. The only guarantee is that the running time of the approximate trajectory will be no greater than T_{opt}, the time for a time-optimal $\delta_v (c_0, c_1)$-safe trajectory.

Figure 11.11 shows a cartoon example of different kinds of optimal paths for a point in the plane. The true time-optimal trajectory, with no consideration for safety, is a straight-line motion between the start and the goal. In this example, the time-optimal $\delta_v (c_0, c_1)$-safe trajectory avoids the narrow passage, as it would require unacceptably slow speeds to be safe. Finally, the algorithm outlined above finds the approximately time-optimal $(1 - \epsilon) \delta_v (c_0, c_1)$-safe trajectory from an approximate start state to an approximate goal state.

We would like to generalize the grid-search algorithm to handle more general dynamic systems of the form of equation (10.7), such as open-chain robot manipulators. Unlike the previous case, the dynamics are not decoupled generally, and the

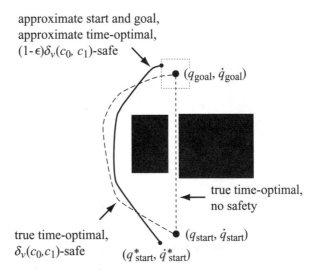

approximate start and goal,
approximate time-optimal,
$(1-\epsilon)\delta_v(c_0, c_1)$-safe

$(q_{\text{goal}}, \dot{q}_{\text{goal}})$

true time-optimal,
no safety

true time-optimal,
$\delta_v(c_0, c_1)$-safe

$(q_{\text{start}}^*, \dot{q}_{\text{start}}^*)$

$(q_{\text{start}}, \dot{q}_{\text{start}})$

Figure 11.11   A time-optimal motion, a $\delta_v(c_0, c_1)$-safe time-optimal motion, and a solution found by the algorithm. The dashed box at the goal is the the goal neighborhood projected to the configuration space.

current state of the system affects the feasible $\ddot{q}$:

(11.46) $\ddot{q} = M^{-1}(q)(u - C(q, \dot{q})\dot{q} - g(q))$

To simplify matters somewhat, we will assume constant bounds on the available controls, possibly different for each i:

(11.47) $|u_i| \leq u_i^{\max}$

At a given state $(q, \dot{q})$, equation (11.46) transforms the rectangular parallelepiped of feasible controls u implied by the constraints (11.47) to a parallelepiped in the $\ddot{q}$ space, as shown in figure 11.12. Let $\mathbf{A}(q, \dot{q})$ denote the state-dependent parallelepiped of feasible $\ddot{q}$. We assume that $\mathbf{A}(q, 0)$ contains the origin of the $\ddot{q}$ space in its interior for all $q \in \mathcal{Q}$, i.e., the actuators are strong enough to hold the robot stationary at any configuration.

To apply the algorithm from before, imagine placing a constant grid on the $\ddot{q}$ space, as shown in figure 11.12, discretizing the feasible accelerations. For a fixed time interval h, these controls will again create a regular grid of reachable points on the state space. For the current state, we use the $\ddot{q}$ grid points inside $\mathbf{A}(q, \dot{q})$ as our set of actions A. One problem is that $\mathbf{A}(q, \dot{q})$ changes during the time interval h, so that a $\ddot{q}$ that is feasible at the beginning of the timestep may no longer be feasible at the

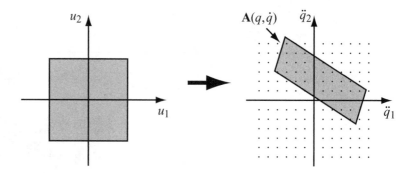

Figure 11.12 The equations of motion turn the feasible controls u into a parallelepiped $\mathbf{A}(q, \dot{q})$ of feasible accelerations $\ddot{q}$.

end. Another problem is that it might happen that there is *no* grid point that is feasible at both the beginning and end of the timestep. Worse yet, if $M^{-1}(q)$ ever loses rank, then $\mathbf{A}(q, \dot{q})$ collapses to zero volume, and no point on a fixed grid is likely to lie in it.

To prevent $\mathbf{A}(q, \dot{q})$ from collapsing, we assume an upper bound on the largest eigenvalue of $M(q)$ during any motion. This tells us how "skinny" $\mathbf{A}(q, \dot{q})$ can become, providing information on how to choose the $\ddot{q}$ grid spacing so that there are always grid points inside the region. For a given timestep h, we also have to choose a conservative approximation $\widehat{\mathbf{A}}(q, \dot{q}) \subset \mathbf{A}(q, \dot{q})$ such that any $\ddot{q}$ inside $\widehat{\mathbf{A}}(q(t), \dot{q}(t))$ stays inside $\mathbf{A}(q(t + \delta t), \dot{q}(t + \delta t))$ for all $\delta t \in [0, h]$ (figure 11.13). To construct the conservative approximation $\widehat{\mathbf{A}}(q, \dot{q})$, we need to know how quickly $M(q)$ can change, which can be bounded by global bounds on the derivatives of $M(q)$ with respect to time. In other words, properties of $M(q)$ and its derivatives must be used to avoid the problems outlined in the previous paragraph. Details can be found in [186].

The idea is to choose the $\ddot{q}$ grid spacing (which may be different for each $\ddot{q}_i$, $i = 1 \ldots n$) and the timestep h so that any feasible trajectory of the system using the full acceleration capabilities $\mathbf{A}(q, \dot{q})$ can be approximately tracked by trajectories using the discretized controls A, chosen to be the $\ddot{q}$ grid points lying inside $\widehat{\mathbf{A}}(q, \dot{q})$.[6] The allowable tracking error depends on a user-defined approximation parameter ϵ, $0 < \epsilon < 1$, where the allowable tracking error goes to zero as ϵ goes to zero.

As before, we can define a trajectory to be $\delta_v(c_0, c_1)$-safe if all real-space obstacles are avoided by a distance of at least $c_0 + c_1 \|\dot{q}\|$ at all points along the trajectory. It has been shown that if there exists a $\delta_v(c_0, c_1)$-safe trajectory from $(q_{\text{start}}, \dot{q}_{\text{start}})$

---

6. Alternatively, A could consist only of $\ddot{q}$ grid points near the boundary of $\widehat{\mathbf{A}}(q, \dot{q})$.

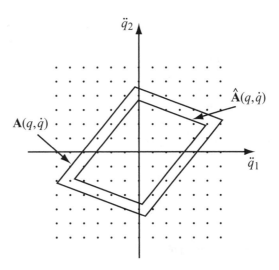

Figure 11.13   The parallelepiped $\mathbf{A}(q, \dot{q})$ represents instantaneously feasible accelerations and $\widehat{\mathbf{A}}(q, \dot{q})$ is a conservative approximation to the set of accelerations that are feasible over the entire timestep of duration h.

to $(q_{\text{goal}}, \dot{q}_{\text{goal}})$ taking time T_{opt}, then the procedure GRID_SEARCH outlined in algorithm 21 will find a $(1 - \epsilon)\delta_v(c_0, c_1)$-safe trajectory from $(q^*_{\text{start}}, \dot{q}^*_{\text{start}})$ to $(q^*_{\text{goal}}, \dot{q}^*_{\text{goal}})$ [where the errors from the desired initial and final states are $O(\epsilon)$] taking no more than time $(1 + \epsilon)T_{\text{opt}}$. The timestep h and $\ddot{q}$ grid-spacing are polynomial in ϵ, and the choice of these parameters involves lengthy calculations. The running time of the algorithm has been shown to be polynomial in ϵ and exponential in the degrees of freedom n_Q [134, 185, 205, 206].

The grid-search algorithm is attractive because it is possible to prove its completeness (using the concept of safe trajectories) and to understand how the running time depends on an approximation parameter ϵ. This allows the user to trade off computation time against the quality of the trajectory. There have been no implementations of the algorithm for more than a few degrees of freedom, however, because in practice the computation time and memory requirements grow quickly with the number of degrees of freedom.

The running time of the algorithm can be improved by using nonuniform discretization of the state space [362] or search heuristics that favor first exploring from nodes close to the goal state. The RRT and EST (see chapter 7) are methods for trajectory planning based on a heuristic that biases the search to evenly explore the state space. Like the grid-based algorithm, they discretize the controls and choose a constant, finite timestep. Unlike the grid-based algorithm, they give up on any notion of optimality and settle for probabilistic completeness, attempting to quickly find any feasible trajectory.

Problems

1. For the RP manipulator of example 11.2.1, write a program that accepts a straight-line path and draws the velocity limit curve.

2. Implement the time-scaling algorithm for the RP arm of example 11.2.1 following a straight line path specified by the user. Try to write the program in a modular fashion, so systems with different dynamics can use many of the same routines.

3. In step 2 of the time-scaling algorithm, explain why there is no solution to the time-scaling problem if $\dot{s} = 0$ is reached. Give a simple example of this case.

4. In step 3 of the time-scaling algorithm, explain why there is no solution to the time-scaling problem if the forward integration reaches $s = 1$ or $\dot{s} = 0$ before crossing the velocity limit curve or the curve F. Give a simple example of this case.

5. In example 11.2.1, explain intuitively why $b_1(s) \neq 0$ but $b_2(s) = 0$, in terms of the arm dynamics and the manipulator path.

6. At a zero inertia point s, at least one term $a_i(s)$ is equal to zero. Explain the implications if $b_i(s)$ is also zero. Should the algorithm be modified to handle this case? Remember, we are assuming the robot is strong enough to maintain any configuration statically. Does it matter if the actuator limits are state-dependent?

7. We would like to find optimal motions for the RP robot arm of example 11.2.1. The Lagrangian defining the objective function is $\mathcal{L} = w_1 u_1^2 + w_2 u_2^2$, where w_1 and w_2 are positive weights. For a fixed time of motion t_f, write the Hamiltonian, the necessary condition for optimality (11.28), and the adjoint equation (11.30).

8. For the previous problem, choose a parameterization of the joint trajectories and use a nonlinear programming package to find optimal point-to-point trajectories for the RP arm. A good initial guess for the joint trajectories would be one that exactly satisfies the start and goal state constraints. Comment on any difficulties you encounter using nonlinear programming.

9. Implement the procedure GRID_SEARCH (algorithm 21) for a point moving in a planar world with polygonal obstacles.

12 *Nonholonomic and Underactuated Systems*

EVERY ROBOT system is subject to a variety of motion constraints, but not all of these can be expressed as configuration constraints. A familiar example of such a system is a car. At low speeds, the rear wheels of the car roll freely in the direction they are pointing, but they prevent slipping motion in the perpendicular direction. This constraint implies that the car cannot translate directly to the side. We know by experience, however, that this velocity constraint does not imply a constraint on configurations; the car can reach any position and orientation in the obstacle-free plane. In fact, the prevented sideways translation can be approximated by parallel-parking maneuvers.

This no-slip constraint is a *nonholonomic constraint,* a constraint on the velocity. In addition to rolling without slipping, conservation of angular momentum is a common source of nonholonomic constraints in mechanical systems.

If, instead of viewing the car as a system subject to a motion constraint, we considered the fact that there are only two inputs (speed and steering angle) to control the car's three degrees of freedom, we might call the system *underactuated*. Underactuated systems have fewer controls than degrees of freedom. For second-order mechanical systems, such as those described in the previous chapter, underactuation implies equality constraints on the possible accelerations of the system.

In this section we study motion planning for systems that are underactuated or subject to motion constraints. Our first task is to determine if the constraints actually limit the reachable states of the robot system. This is a controllability question. The next problem is to construct algorithms that find motion plans that satisfy the motion constraints. A last problem, not addressed in this chapter, is feedback stabilization of the motion plans during execution.

We begin in section 12.1 by providing some background information on vector fields and their Lie (pronounced "lee") algebras. Section 12.2 defines the class of control systems we will consider. Section 12.3 describes different controllability notions and tests for these nonlinear systems. Section 12.4 specializes the discussion to second-order mechanical systems. Finally, section 12.5 describes a number of methods for motion planning for nonholonomic and underactuated systems.

12.1 Preliminaries

First we must decide how generally to define the state spaces of the robotic systems we will consider. For example, we could treat a very general case, allowing the state space of the system to be any smooth manifold. This would allow us to study, e.g., the motion of a spherical pendulum. The configuration space of this system is the sphere S^2. Or we could limit our treatment to systems evolving on Lie groups, particularly matrix Lie groups. This would allow us to model the orientation of a satellite as a point in $SO(3)$.

In this chapter, we restrict our attention even further to systems evolving on vector spaces $\mathcal{M} = \mathbb{R}^n$. This allows us to get to the main results as quickly as possible. Also, any n-dimensional differentiable manifold is locally "similar" (diffeomorphic) to $\mathbb{R}^n$, so, equipped with a proper set of local coordinates, any n-dimensional manifold can be treated locally as $\mathbb{R}^n$. By making this simplification, we require the use of a local coordinate system in our computations, and we may lose information about the global structure of the space. As examples, the true configuration space of a 2R robot arm is the torus $T^2 = S^1 \times S^1$, which is doughnut-shaped while $\mathbb{R}^2$ is not; and a global representation of the orientation of a satellite is $SO(3)$, which is different from a local representation using three Euler angles ($\mathbb{R}^3$). See figure 12.1 for another example.

Although we focus on vector state spaces, most of the ideas in this chapter generalize immediately to general manifolds.

In this chapter, $q \in \mathcal{Q}$ denotes the configuration of the system and $x \in \mathcal{M}$ denotes the state of the system. If the system is kinematic, then the state is simply the configuration ($\mathcal{M} = \mathcal{Q}$), and the controls are velocities. If the system is a second-order mechanical system, then x includes both configurations q and velocities $\dot{q}$, and the controls are forces (accelerations). The dimension of the configuration space $\mathcal{Q}$ is $n_{\mathcal{Q}}$, and the dimension of the state space $\mathcal{M}$ is n.

We will carry two examples throughout the chapter: a unicycle, a kinematic system; and a model of a planar spacecraft, a second-order mechanical system. We will treat

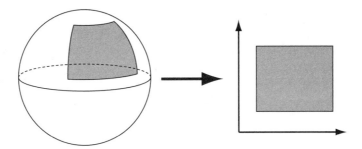

Figure 12.1 Latitude and longitude coordinates allow us to treat a patch of the sphere S^2 as a section of the plane $\mathbb{R}^2$.

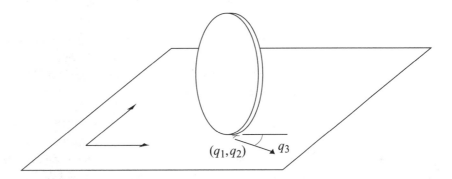

Figure 12.2 The unicycle system. The position of the point of contact is given by (q_1, q_2), and the heading direction is given by q_3.

all systems uniformly, as systems with state x on a state space $\mathcal{M}$. Only in section 12.4 and subsection 12.5.7 will we specialize our study to second-order mechanical systems such as the spacecraft model.

EXAMPLE 12.1.1   Unicycle example. *The unicycle is a wheel that rolls upright on a horizontal plane (figure 12.2). The configuration of the wheel is $q = [q_1, q_2, q_3]^T$, describing the contact point of the wheel on the plane (q_1, q_2) and the steering angle q_3 of the wheel. (We could also include the rolling angle of the wheel, i.e., the location of the air nozzle on the tire, in the description of the configuration, but we will ignore this for now.) The system is kinematic, so $x = [x_1, x_2, x_3]^T = q$, $\mathcal{M} = \mathcal{Q} = \mathbb{R}^3$, and $n_{\mathcal{Q}} = n = 3$. (Since we are dealing with local coordinates, we are ignoring the fact that the global structure of the space is $\mathbb{R}^2 \times S^1$. This will not affect the equations of motion, but requires the use of $\mathrm{mod}\, 2\pi$ arithmetic on the third coordinate.) The*

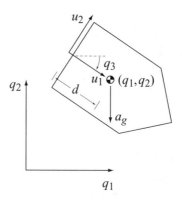

Figure 12.3   The planar body with thrusters (PBWT).

controls are the rolling speed of the wheel and the rate of change of the steering angle. Sideways translation of the wheel is prevented by the no-slip constraint imposed by the wheel. This example is sometimes known as the rolling penny, or the pizza cutter, and it is similar to a model for a car.

EXAMPLE 12.1.2   Planar body with thrusters (PBWT) example. *The body moves in a frictionless, inviscid plane by means of two thrusters fixed to the body (figure 12.3). The mass and inertia of the body (about the center of mass) are unit. The line of action of the thrust u_1 is through the center of mass, and the line of action of the thrust u_2 is perpendicular and a distance d from the center of mass. The configuration is $q = [q_1, q_2, q_3]^T$, describing the location of the center of mass (q_1, q_2) and the angle q_3 of the line of action of the first thruster relative to the world q_1-axis. The system is second-order, so $x = [x_1, x_2, x_3, x_4, x_5, x_6]^T = [q_1, q_2, q_3, \dot{q}_1, \dot{q}_2, \dot{q}_3]^T$, $\mathcal{M} = \mathbb{R}^6$, $n_Q = 3$, and $n = 2n_Q = 6$. Gravitational acceleration a_g acts in the $-q_2$-direction, and a_g may be zero.*

The rest of section 12.1 introduces concepts from differential geometry that will be useful in understanding underactuated systems. For the unicycle, e.g., we will see that its instantaneous motions can be described in terms of two "vector fields" associated with the controls to drive and steer the unicycle. Linear combinations of these two vector fields define a "distribution" describing all possible instantaneous motions of the unicycle. The "integral manifold" describes all the states the system can reach by following vector fields in the distribution. We use the "Lie bracket" to show that two vector fields in the distribution can generate a parallel-parking motion for the unicycle, effectively giving it a sideways motion, meaning that the integral

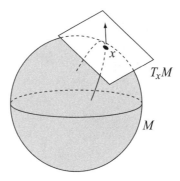

Figure 12.4 A curve on the sphere $\mathcal{M}$, a tangent vector to the curve at x, and the tangent space $T_x\mathcal{M}$ it lives in.

manifold is the entire configuration space—the velocity constraint does not reduce the reachable space.

Section 12.2 describes how a robot system can be expressed as a system of vector fields and controls, and section 12.3 uses the concepts developed in this section to study the set of states reachable by the controls.

12.1.1 Tangent Spaces and Vector Fields

Let $x : \mathbb{R} \to \mathcal{M}$ be a smooth curve on $\mathcal{M}$ parameterized by s. Then dx/ds, evaluated at $x_0 = x(s_0)$, is tangent to the curve at x_0. Call this vector V. The vector V is a *tangent vector* that is tangent to $\mathcal{M}$ at x_0. The tangent vector V lives in $T_{x_0}\mathcal{M}$, the *tangent space* of $\mathcal{M}$ at x_0. This space is an n-dimensional vector space $\mathbb{R}^n$ consisting of the tangents of all possible curves passing through x_0 (figure 12.4). The tangent spaces at different points of $\mathcal{M}$ are different spaces.

The *tangent bundle* of $\mathcal{M}$, written $T\mathcal{M}$, is the $2n$-dimensional manifold that is the union of tangent spaces at all points in $\mathcal{M}$,

$$T\mathcal{M} = \bigcup_{x \in \mathcal{M}} T_x\mathcal{M}.$$

For the systems we study, $T\mathcal{M} = \mathcal{M} \times \mathbb{R}^n = \mathbb{R}^{2n}$.[1]

A smooth *vector field* $g : \mathcal{M} \to T\mathcal{M}$ is a smooth map from points $x \in \mathcal{M}$ to tangent vectors $g(x) \in T_x\mathcal{M}$. It is possible to define C^k vector fields, but we will

---

1. We note that if $\mathcal{M}$ is a more general manifold, and it is *parallelizable* (e.g., a Lie group), then $T\mathcal{M} = \mathcal{M} \times \mathbb{R}^n$. The reader should be careful not to generalize improperly, however. For example, $TS^2 \neq S^2 \times \mathbb{R}^2$, for reasons beyond the scope of this chapter. For an intuitive discussion of this issue, see [372].

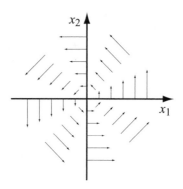

Figure 12.5   The vector field $\frac{1}{2}[-x_2, x_1]^T$.

assume that all vector fields are infinitely differentiable. (For example, the vector field $g(x) = [x_1^2, \sin x_3, x_1 x_2]^T$ is infinitely differentiable, but $[|x_1|, x_2, x_3]^T$ is only C^0.) A picture of the vector field $\frac{1}{2}[-x_2, x_1]^T$ on $\mathbb{R}^2$ is shown in figure 12.5. Tangent vectors are written as column vectors.

In the case of a kinematic system, $\mathcal{M}$ is the configuration space $\mathcal{Q}$, and $T_{x_0}\mathcal{M} = T_{q_0}\mathcal{Q}$ is the set of all possible velocities of the system at $x_0 = q_0$. In the case of a second-order system, $\mathcal{M}$ is the state space $T\mathcal{Q}$, and $T_{x_0}\mathcal{M} = T_{x_0}T\mathcal{Q}$ is the set of all possible velocities and accelerations of the system at $x_0 = [q_0^T, \dot{q}_0^T]^T$. In this case, however, the state x_0 already specifies the velocity portion $\dot{q}_0$ of the tangent vector $[\dot{q}_0^T, \ddot{q}_0^T]^T$. This implies *drift* in second-order systems, as shown in the PBWT example below.

EXAMPLE 12.1.3   Unicycle (cont.) *A tangent vector for the unicycle is given by* $\dot{x} = [\dot{x}_1, \dot{x}_2, \dot{x}_3]^T = [\dot{q}_1, \dot{q}_2, \dot{q}_3]^T$. *The unicycle is capable of rolling forward and backward and spinning in place. These two vector fields can be written* $g_1^{\text{uni}}(x) = [\cos x_3, \sin x_3, 0]^T$, *rolling forward at unit speed, and* $g_2^{\text{uni}}(x) = [0, 0, 1]^T$, *spinning counterclockwise at unit speed. The vector fields can also be written as* $g_1^{\text{uni}}(x) = (\cos x_3)\partial/\partial x_1 + (\sin x_3)\partial/\partial x_2$ *and* $g_2^{\text{uni}}(x) = \partial/\partial x_3$, *where* $\partial/\partial x_1$, $\partial/\partial x_2$, *and* $\partial/\partial x_3$ *are the canonical unit basis vectors of the tangent space, i.e., unit speed tangent vectors along the coordinates (see figure 12.6).*

EXAMPLE 12.1.4   PBWT (cont.) *A tangent vector for the PBWT is given by* $\dot{x} = [\dot{x}_1, \dot{x}_2, \dot{x}_3, \dot{x}_4, \dot{x}_5, \dot{x}_6]^T = [\dot{q}_1, \dot{q}_2, \dot{q}_3, \ddot{q}_1, \ddot{q}_2, \ddot{q}_3]^T$. *For this system, we can define three vector fields: the* drift *vector field* $g_0^{\text{pbwt}}(x)$ *corresponding to the motion of the body when no thrusters are activated, and the* control *vector fields*

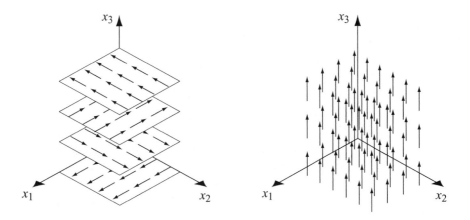

Figure 12.6 The vector fields $g_1^{\text{uni}} = [\cos x_3, \sin x_3, 0]^T$ (shown in constant x_3 layers) and $g_2^{\text{uni}} = [0, 0, 1]^T$.

$g_1^{\text{pbwt}}(x)$ and $g_2^{\text{pbwt}}(x)$ corresponding to the acceleration when thrusters 1 and 2 are fired with unit thrust, respectively. Verify that $g_0^{\text{pbwt}}(x) = [x_4, x_5, x_6, 0, a_g, 0]^T$, $g_1^{\text{pbwt}}(x) = [0, 0, 0, \cos x_3, \sin x_3, 0]^T$, and $g_2^{\text{pbwt}}(x) = [0, 0, 0, -\sin x_3, \cos x_3, -d]^T$, and write these vector fields in the canonical basis $\{\partial/\partial x_1, \partial/\partial x_2, \partial/\partial x_3, \partial/\partial x_4, \partial/\partial x_5, \partial/\partial x_6\}$. Notice if thruster 1 is fired with thrust u_1, the system follows the vector field $g_0^{\text{pbwt}}(x) + u_1 g_1^{\text{pbwt}}(x)$.

Let ϕ^g denote the *flow* of the vector field g, where $\phi_t^g(x)$ gives the system state after following the flow ϕ^g from x for a time t. The flow satisfies the equation

$$\frac{d}{dt}\phi_t^g(x) = g\big(\phi_t^g(x)\big).$$

The vector field is *complete* if its flow is defined for all x and t.

The curve $\{\phi_t^g(x) \mid t \in \mathbb{R}\}$ is the *integral curve* of g containing x. The integral curve describes the set of reachable points of $\mathcal{M}$ from x by following the vector field forward and backward in time (figure 12.7). This notion can be generalized to the *integral manifold* of a set of vector fields $\mathcal{G}$, a topic for subsection 12.1.3.

12.1.2 Distributions and Constraints

Let $\mathcal{G}$ be a set of vector fields, and let span($\mathcal{G}$) be the linear span of vector fields in $\mathcal{G}$, given by all linear combinations of vector fields in $\mathcal{G}$. At each point $x \in \mathcal{M}$, these vector fields span a linear subspace of $T_x\mathcal{M}$. The set of vector fields $\mathcal{G}$ is

Figure 12.7 An integral curve of a vector field.

said to generate a *distribution* $\mathcal{D} \subseteq T\mathcal{M}$, which is a smooth assignment of a linear subspace of $T_x\mathcal{M}$ for each $x \in \mathcal{M}$. A distribution is *regular* if the dimension of the linear subspace is the same at all x. If the dimension is m, then we say that it is an m-dimensional distribution.

Consider the two-dimensional regular distribution for the unicycle $\mathcal{D} = \text{span}(\{g_1^{\text{uni}}, g_2^{\text{uni}}\}) = u_1 g_1^{\text{uni}}(x) + u_2 g_2^{\text{uni}}(x), u_1, u_2 \in \mathbb{R}$. We might think of this as the "positive" form of the distribution—feasible motions are generated by linear combinations of the vector fields. A "negative" form of the distribution would start with all motions being feasible, then eliminate those that violate motion constraints. For instance, the unicycle distribution could be written

$$(12.1) \quad \mathcal{D}(x) = \{\dot{x} \in T_x\mathcal{M} \mid \omega(x)\dot{x} = 0\}, \quad \omega(x) = [-\sin x_3, \cos x_3, 0],$$

where $\mathcal{D}(x)$ is the linear subspace of $T_x\mathcal{M}$ defined by the distribution $\mathcal{D}$. A row vector $\omega(x)$ is called a *covector* and lives in the *cotangent space* $T_x^*\mathcal{M} = \mathbb{R}^n$, the dual of $T_x\mathcal{M}$ consisting of all linear functionals of elements of $T_x\mathcal{M}$. In other words, a *covector field* ω pairs with a vector field g to yield a real value, $\omega(x)g(x) \in \mathbb{R}$. This is sometimes called the "natural pairing" of a tangent vector and covector. The canonical basis of covector fields is $\{dx_1, \ldots, dx_n\}$, so that the constraint $\omega(x)$ in equation (12.1) can be written as $-\sin x_3 dx_1 + \cos x_3 dx_2$.

A covector field ω is sometimes known as a *one-form,* because it takes a single element of $T_x\mathcal{M}$ and produces a real number, linear in the tangent vector. A *two-form,* as we will see in section 12.4, takes two elements of $T_x\mathcal{M}$ and produces a real number, linear in each of the arguments.

The *cotangent bundle* $T^*\mathcal{M}$ is the union of cotangent spaces $T_x^*\mathcal{M}$ for all $x \in \mathcal{M}$. A set of covector fields $\{\omega_1(x), \ldots, \omega_k(x)\}$ is said to define a *codistribution* $\Omega \subseteq T^*\mathcal{M}$. If the covector fields $\omega_i(x)$, $i = 1 \ldots k$, correspond to motion constraints $\omega_i(x)\dot{x} = 0$, then Ω is called a *constraint* codistribution, and it is said to *annihilate* the distribution $\mathcal{D}$ of feasible motions, and vice versa.

Of special interest are velocity constraints of the form

$$(12.2) \quad f(q, \dot{q}) = 0$$

that cannot be integrated to yield equivalent configuration constraints. Such constraints are called *nonholonomic*. Nonholonomic constraints of the form

$$a(q)\dot{q} = 0$$

are sometimes called *Pfaffian* constraints, as discussed in chapter 10. Pfaffian constraints arise from rolling without slip [e.g., see equation (12.1)] and conservation of angular momentum. In mechanical systems, the covector field $a(q)$ can be interpreted as a generalized force, so $a(q)\dot{q}$ has units of power, and the constraint $a(q)\dot{q} = 0$ is *passive*—it does no work on the system.

In second-order underactuated systems, the underactuation implies the existence of acceleration constraints of the form

$$f(q, \dot{q}, \ddot{q}) = 0.$$

Constraints of this form that cannot be integrated to equivalent velocity constraints are sometimes referred to as "second-order nonholonomic" constraints, but this terminology is not standard.

In general, it is not easy to determine if an acceleration constraint can be integrated to yield an equivalent velocity constraint, or if a velocity constraint can be integrated to yield an equivalent configuration constraint. In the rest of this chapter, we use the "positive" form of the distribution and study the reachable set by vector fields in $\mathcal{G}$.

12.1.3 Lie Brackets

Let $\mathcal{G}$ be a set of vector fields and $\mathcal{D}$ be the distribution defined by span$(\mathcal{G})$. We would like to know the reachable set of $\mathcal{M}$ by following vector fields in $\mathcal{D}$. While this is generally difficult globally, it is possible to learn something about the reachable set *locally* by looking at the *Lie brackets* of vector fields in $\mathcal{D}$. Given two vector fields belonging to $\mathcal{D}$, the Lie bracket tells us if infinitesimal motions along these vector fields can be used to locally generate motion in a direction not contained in $\mathcal{D}$. Perhaps the best-known example is the parallel-parking maneuver for a car or, in our case, a unicycle. Direct sideways motion is prohibited by the no-slip constraint, but sideways motion can be approximated by a series of forward-backward and turning maneuvers. The implication of this is that the locally reachable set of $\mathcal{M}$ is not two-dimensional, as the two-dimensional distribution $\mathcal{D}$ might seem to indicate, but fully three-dimensional. The no-slip velocity constraint does not imply a constraint on reachable configurations.

For two vector fields $g_1, g_2 \in \mathcal{G}$, consider the state reached from $x_0 = x(0)$ by first following g_1 for a small time $\epsilon \ll 1$, then following g_2 for time ϵ, then following $-g_1$ for time ϵ, then following $-g_2$ for time ϵ. This is expressed mathematically as

$$(12.3) \qquad x(4\epsilon) = \phi_\epsilon^{-g_2}\left(\phi_\epsilon^{-g_1}\left(\phi_\epsilon^{g_2}\left(\phi_\epsilon^{g_1}(x_0)\right)\right)\right).$$

We can take a Taylor series in ϵ to solve the differential equation (12.3) approximately (see, e.g., [330] and problem 29), yielding

(12.4) $$x(4\epsilon) = x_0 + \epsilon^2 \left(\frac{\partial g_2}{\partial x} g_1(x_0) - \frac{\partial g_1}{\partial x} g_2(x_0) \right) + O(\epsilon^3),$$

where the partial derivatives are evaluated at x_0 and $O(\epsilon^3)$ indicates terms of order ϵ^3, which are dominated by the term of order ϵ^2 when ϵ is small. Note there are no $O(\epsilon)$ terms. The ϵ^2 term represents the approximate net motion of the system, and the term inside the parentheses is the Lie bracket of g_1 and g_2.

The Lie bracket of g_1 and g_2 is written $[g_1, g_2]$ and is given in local coordinates by

(12.5) $$[g_1, g_2] = \frac{\partial g_2}{\partial x} g_1 - \frac{\partial g_1}{\partial x} g_2.$$

The Lie bracket $[g_1, g_2]$ defines a new vector field, and if it is not contained in span($\mathcal{G}$), then it represents a new motion direction that can be followed approximately. Locally generating motion in this direction is "slower" than following the vector field g_1 or g_2 directly, as the net motion is only $O(\epsilon^2)$ for time $O(\epsilon)$, where $\epsilon \ll 1$. Again, parallel parking is a well-known example, as approximately generating sideways motion by forward-backward and turning motions is tedious and time-consuming. If $[g_1, g_2] = 0$, then no new motion is created, and the two vector fields are said to *commute*.

Since $[g_1, g_2]$ is a vector field, we can calculate its Lie bracket with another vector field. A *Lie product of degree k* is a bracket term where the original vector fields appear k times. For instance, $[[[g_1, g_2], g_1], g_2]$ is a Lie product of degree 4.

EXAMPLE 12.1.5   Unicycle (cont.) *The rolling and turning vector fields for the unicycle are $g_1^{\text{uni}} = [\cos x_3, \sin x_3, 0]^T$ and $g_2^{\text{uni}} = [0, 0, 1]^T$, respectively. So*

$$[g_1^{\text{uni}}, g_2^{\text{uni}}] = \frac{\partial g_2^{\text{uni}}}{\partial x} g_1^{\text{uni}} - \frac{\partial g_1^{\text{uni}}}{\partial x} g_2^{\text{uni}}$$

$$= \left[\frac{\partial g_2^{\text{uni}}}{\partial x_1} \quad \frac{\partial g_2^{\text{uni}}}{\partial x_2} \quad \frac{\partial g_2^{\text{uni}}}{\partial x_3} \right] g_1^{\text{uni}} - \left[\frac{\partial g_1^{\text{uni}}}{\partial x_1} \quad \frac{\partial g_1^{\text{uni}}}{\partial x_2} \quad \frac{\partial g_1^{\text{uni}}}{\partial x_3} \right] g_2^{\text{uni}}$$

$$= \begin{bmatrix} 0 & 0 & 0 \\ 0 & 0 & 0 \\ 0 & 0 & 0 \end{bmatrix} \begin{bmatrix} \cos x_3 \\ \sin x_3 \\ 0 \end{bmatrix} - \begin{bmatrix} 0 & 0 & -\sin x_3 \\ 0 & 0 & \cos x_3 \\ 0 & 0 & 0 \end{bmatrix} \begin{bmatrix} 0 \\ 0 \\ 1 \end{bmatrix}$$

$$= \begin{bmatrix} \sin x_3 \\ -\cos x_3 \\ 0 \end{bmatrix}.$$

Note that the Lie bracket motion is to the side, in the direction prevented by the no-slip constraint (see figure 12.8).

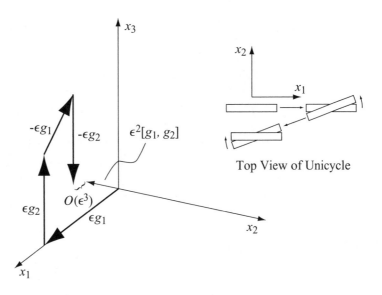

Figure 12.8 Generating a Lie bracket motion for the unicycle, starting from the origin. The net motion is approximately to the side, the Lie bracket direction. It is not exactly to the side, however, due to higher-order terms in ϵ.

The *Lie algebra* of a set of vector fields $\mathcal{G}$, written $\overline{\text{Lie}}(\mathcal{G})$, is the linear span of all Lie products, of all degrees, of vector fields in $\mathcal{G}$. To determine the Lie algebra, define $\mathcal{G}_1 = \mathcal{G}$, and the series

$$\mathcal{G}_{i+1} = \mathcal{G}_i \cup \{[g_j, g_k] \mid \forall g_j \in \mathcal{G}_1, g_k \in \mathcal{G}_i\}.$$

Then $\overline{\text{Lie}}(\mathcal{G})$ is given by the distribution span($\mathcal{G}_\infty$). For example, the series for $\mathcal{G} = \{g_1, g_2\}$ begins

$$\mathcal{G}_1 = \{g_1, g_2\}$$
$$\mathcal{G}_2 = \mathcal{G}_1 \cup \{g_3 = [g_1, g_2]\}$$
$$\mathcal{G}_3 = \mathcal{G}_2 \cup \{g_4 = [g_1, g_3], g_5 = [g_2, g_3]\}$$
$$\mathcal{G}_4 = \mathcal{G}_3 \cup \{g_6 = [g_1, g_4], g_7 = [g_1, g_5], g_8 = [g_2, g_4], g_9 = [g_2, g_5]\}$$
$$\vdots$$

The corresponding series $\mathcal{D}_1 = \text{span}(\mathcal{G}_1)$, $\mathcal{D}_2 = \text{span}(\mathcal{G}_2), \ldots$ is called the *filtration* of the distribution $\mathcal{D}_1$. The filtration is *regular* if each distribution in the filtration is regular. If the filtration is regular, then the dimension of the distribution grows at each step of the construction, or else the construction terminates. (Of course,

$\dim(\mathcal{D}_i) \le n = \dim(\mathcal{M})$ for all i.) If the filtration is regular, we are guaranteed a finite value of k such that $\mathcal{D}_k = \mathcal{D}_{k+1} = \cdots = \mathcal{D}_\infty$. This distribution is the *involutive closure* $\overline{\mathcal{D}}$ of $\mathcal{D}$, and a distribution $\mathcal{D}$ is *involutive* if $\mathcal{D} = \overline{\mathcal{D}}$.

If the filtration is not regular, then in general there is no way to know a priori a degree k at which $\mathcal{D}_k = \overline{\mathcal{D}}$. If there is a degree k at which all Lie products become zero, then the Lie algebra is called *nilpotent of order k*.

The *integral manifold* of $\mathcal{D}$ containing x_0 is the set of $\mathcal{M}$ that can be reached from x_0 by vector fields in $\mathcal{D}$, and $\mathcal{D}(x)$ is the tangent space of the integral manifold at x. By the well known *Frobenius theorem*, an m-dimensional regular distribution $\mathcal{D}$ can be integrated to yield an m-dimensional integral manifold if and only if $\mathcal{D}$ is involutive.

If a distribution $\mathcal{D}$ does not have the entire space $\mathcal{M}$ as an integral manifold, then $\mathcal{D}$ is said to generate a *foliation* of $\mathcal{M}$, and each distinct integral manifold is called a *leaf* of the foliation. Consider, e.g., the one-dimensional distribution generated by $g_2^{\text{uni}} = [0, 0, 1]^T$ (turning motion) for the unicycle (see figure 12.6). The distribution is one-dimensional, regular, and involutive, and the integral manifolds are lines in x_3 (wrapping around at 2π) with fixed position (x_1, x_2). The unicycle is confined to the same leaf of the foliation for all time if it can only follow this vector field. A more interesting example of a foliation is given in example 12.1.7 below.

The existence of integral manifolds smaller than the whole state space $\mathcal{M}$ indicates that the motion constraints actually limit the reachable state space. For example, velocity constraints on a kinematic system might be integrated to yield configuration constraints, indicating that the original constraints are actually holonomic. Similarly, acceleration constraints on a mechanical system might be integrated to yield velocity or even configuration constraints.

Lie brackets satisfy the following properties:

1. Skew-symmetry:

   $$[g_1, g_2] = -[g_2, g_1]$$

2. Jacobi identity:

   $$[g_1, [g_2, g_3]] + [g_3, [g_1, g_2]] + [g_2, [g_3, g_1]] = 0$$

Taking these properties into account, the *Philip Hall basis* gives a way to choose the smallest number of Lie products that must be considered at each degree k to generate a basis for the distribution $\mathcal{D}_k$. See the book by Serre [380] for details.

EXAMPLE 12.1.6   Unicycle (cont.) *From before, we have* $g_1^{\text{uni}} = [\cos x_3, \sin x_3, 0]^T$, $g_2^{\text{uni}} = [0, 0, 1]^T$, *and* $g_3^{\text{uni}} = [g_1^{\text{uni}}, g_2^{\text{uni}}] = [\sin x_3, -\cos x_3, 0]^T$. *The dimension of the distribution defined by* $\{g_1^{\text{uni}}, g_2^{\text{uni}}, g_3^{\text{uni}}\}$ *is three at all* $x \in \mathcal{M}$, *implying that the distribution is regular. It is also certainly involutive, since the dimension of* $\mathcal{M}$ *is* *three. To see that the three vector fields are indeed linearly independent, we define the 3×3 matrix $[g_1^{\text{uni}} \; g_2^{\text{uni}} \; g_3^{\text{uni}}]$ obtained by placing the column vectors side by side. The rank of the matrix is 3 at all $x \in \mathcal{M}$, which can be verified by the determinant*

$$\det\left[g_1^{\text{uni}} \; g_2^{\text{uni}} \; g_3^{\text{uni}}\right] = \det \begin{bmatrix} \cos x_3 & 0 & \sin x_3 \\ \sin x_3 & 0 & -\cos x_3 \\ 0 & 1 & 0 \end{bmatrix} = 1.$$

Since the distribution is regular and involutive, it has a three-dimensional integral manifold, which is the entire space $\mathcal{M}$. The distribution $\mathcal{D}_2$ is the involutive closure of $\mathcal{D}_1$. The filtration is regular.

EXAMPLE 12.1.7 *Define the vector fields $g_1(x) = [x_1 \cos x_3, x_2 \sin x_3, 0]^T$ and $g_2(x) = [0, 0, 1]^T$ on $\mathbb{R}^3$. The vector field g_2 by itself defines a regular one-dimensional involutive distribution. The vector field g_1 does not, however, as it vanishes at $x_1 = x_2 = 0$. The Lie bracket of these vector fields is $[g_1, g_2] = [x_1 \sin x_3, -x_2 \cos x_3, 0]^T$, and*

$$\det[g_1 \; g_2 \; [g_1, g_2]] = x_1 x_2.$$

This means that the distribution $\text{span}(\{g_1, g_2, [g_1, g_2]\})$ is rank 3 at points where both x_1 and x_2 are nonzero. It is not regular, as the rank is less at points where either x_1 or x_2 is zero. In fact, it is not hard to see that the integral manifold of this distribution is one-dimensional from points $[0, 0, x_3]^T$, two-dimensional from points $[x_1 \neq 0, 0, x_3]^T$ and $[0, x_2 \neq 0, x_3]^T$, and three-dimensional from all other points. The foliation is pictured in figure 12.9.

EXAMPLE 12.1.8 PBWT (cont.) *As derived previously, we have $g_0^{\text{pbwt}} = [x_4, x_5, x_6, 0, a_g, 0]^T$, $g_1^{\text{pbwt}} = [0, 0, 0, \cos x_3, \sin x_3, 0]^T$, and $g_2^{\text{pbwt}} = [0, 0, 0, -\sin x_3, \cos x_3, -d]^T$. Lie bracket computations show that*

$$g_3^{\text{pbwt}} = \left[g_0^{\text{pbwt}}, g_1^{\text{pbwt}}\right]$$
$$= [-\cos x_3, -\sin x_3, 0, -x_6 \sin x_3, x_6 \cos x_3, 0]^T$$
$$g_4^{\text{pbwt}} = \left[g_0^{\text{pbwt}}, g_2^{\text{pbwt}}\right]$$
$$= [\sin x_3, -\cos x_3, d, -x_6 \cos x_3, -x_6 \sin x_3, 0]^T$$
$$g_5^{\text{pbwt}} = \left[g_1^{\text{pbwt}}, \left[g_0^{\text{pbwt}}, g_2^{\text{pbwt}}\right]\right]$$
$$= [0, 0, 0, d \sin x_3, -d \cos x_3, 0]^T$$
$$g_6^{\text{pbwt}} = \left[g_0^{\text{pbwt}}, \left[g_1^{\text{pbwt}}, \left[g_0^{\text{pbwt}}, g_2^{\text{pbwt}}\right]\right]\right]$$
$$= [-d \sin x_3, d \cos x_3, 0, dx_6 \cos x_3, dx_6 \sin x_3, 0]^T.$$

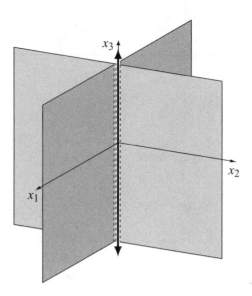

Figure 12.9   The distribution span($\{g_1, g_2, [g_1, g_2]\}$) in example 12.1.7 foliates the state space into nine separate leaves: the line defined by $x_1 = x_2 = 0$, four half-planes, and four three-dimensional quadrants.

A computation shows that

$$\det\left[g_1^{\text{pbwt}} \; g_2^{\text{pbwt}} \; g_3^{\text{pbwt}} \; g_4^{\text{pbwt}} \; g_5^{\text{pbwt}} \; g_6^{\text{pbwt}}\right] = d^4.$$

The dimension of the distribution defined by these vector fields is six at all $x \in \mathcal{M}$ (provided $d \neq 0$), so the distribution is both regular and involutive. The integral manifold is the entire space $\mathcal{M}$. The distribution $\mathcal{D}_4$ is the involutive closure of $\mathcal{D}_1$.

We now apply the ideas of this section to study controllability of underactuated systems, taking into account the fact that *controls* determine how the system vector fields are followed. It may not be possible to follow arbitrary linear combinations of system vector fields. For example, the drift vector field g_0^{pbwt} of the PBWT is fundamentally different from the control vector fields g_1^{pbwt} and g_2^{pbwt}.

12.2   Control Systems

A family of vector fields $\mathcal{G}$ on a manifold $\mathcal{M}$ is sometimes called a *dynamical polysystem*. The system is *symmetric* if for every $g \in \mathcal{G}$, $-g$ is also in $\mathcal{G}$.

The family of dynamical polysystems we will study are *control affine* nonlinear control systems, written

$$(12.6) \quad \dot{x} = g_0(x) + \sum_{i=1}^{m} g_i(x)u_i, \quad u \in \mathcal{U} \subset \mathbb{R}^m.$$

The vector field g_0 is called the *drift vector field,* defining the natural unforced motion of the system, and the g_i, $i = 1 \ldots m$, are linearly independent *control vector fields.* The control vector u belongs to the control set $\mathcal{U}$, and $u(t)$ is piecewise continuous. If $g_0 = 0$, the system is called *drift-free* or *driftless*. Kinematic systems (such as the unicycle) may be drift-free, but second-order systems (such as the PBWT) are not.

EXAMPLE 12.2.1   Unicycle (cont.) *The control system for the unicycle is written* $\dot{x} = g_1^{\text{uni}}(x)u_1 + g_2^{\text{uni}}(x)u_2$, *where u_1 is the driving speed and u_2 is the steering control.*

EXAMPLE 12.2.2   PBWT (cont.) *The control system for the PBWT is written* $\dot{x} = g_0^{\text{pbwt}}(x) + g_1^{\text{pbwt}}(x)u_1 + g_2^{\text{pbwt}}(x)u_2$, *where u_1 is the thrust force at thruster 1 and u_2 is the force at thruster 2.*

We will consider two classes of control sets:

- $\mathcal{U}_{\pm}$: This class of control sets includes any control set $\mathcal{U}$ containing the origin of $\mathbb{R}^m$ in the interior of its convex hull. In other words, the control set *positively* spans $\mathbb{R}^m$—any point in $\mathbb{R}^m$ can be generated by a positive linear combination of elements of $\mathcal{U}$. An example of such a control set is the cube centered at the origin of $\mathbb{R}^m$, $-1 \leq u_i \leq 1$, $i = 1, \ldots, m$. Another example consists of only the vertices of this cube.

- $\mathcal{U}_+$: This class of control sets includes $\mathcal{U}_{\pm}$ as a subset and includes any control set $\mathcal{U}$ that spans $\mathbb{R}^m$—any point in $\mathbb{R}^m$ can be generated by a linear combination of elements of $\mathcal{U}$. An example of such a control set is the non-negative controls $0 \leq u_i \leq 1$, $i = 1, \ldots, m$.

Examples of the control sets are shown in figure 12.10.

The system (12.6) is symmetric if it is drift-free and the control set is symmetric about the origin, e.g., a cube centered at the origin. We will abuse the term slightly and say that a drift-free system is symmetric for any positive-spanning control set $\mathcal{U} \in \mathcal{U}_{\pm}$, since the controllability properties we discuss in this chapter are the same for any $\mathcal{U} \in \mathcal{U}_{\pm}$.

If a system has drift but $g_0 \in \text{span}(\{g_1, \ldots, g_m\})$, then we may be able to choose controls $w(x) \in \mathbb{R}^m$ to always cancel the drift, thereby symmetrizing the system by

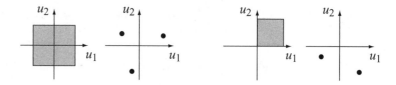

Figure 12.10   For $m = 2$ controls, the two control sets on the left belong to $\mathcal{U}_\pm$ and the two control sets on the right belong to $\mathcal{U}_+$.

the controls. In this case, the pseudocontrol $u \in \mathcal{U} \in \mathcal{U}_\pm$ can be added on top of the drift-canceling control $w(x)$, so the total control vector is $w(x) + u$, and the system is equivalent to the driftless system

$$\dot{x} = \sum_{i=1}^{m} g_i(x)u_i, \quad u \in \mathcal{U} \in \mathcal{U}_\pm.$$

As an intuitive example, imagine your motion as you walk on a conveyor. The drift vector field carries you at a constant speed in one direction. You can control your own walking speed, however, to cancel the drift and make progress in the opposite direction.

12.3   Controllability

Let V be a neighborhood of a point $x \in \mathcal{M}$ (i.e., an n-dimensional open set of $\mathcal{M}$ containing x). Let $R^V(x, T)$ indicate the set of reachable points at time T by trajectories remaining inside V and satisfying equation (12.6), and let

$$R^V(x, \leq T) = \bigcup_{0 < t \leq T} R^V(x, t).$$

We define the following four versions of nonlinear controllability (see figure 12.11):

■ The system is *controllable* from x if, for any $x_{\text{goal}} \in \mathcal{M}$, there exists a $T > 0$ such that $x_{\text{goal}} \in R^{\mathcal{M}}(x, \leq T)$. In other words, any goal state is reachable from x in finite time.

■ The system is *accessible* from x if $R^{\mathcal{M}}(x, \leq T)$ contains a full n-dimensional subset of $\mathcal{M}$ for some $T > 0$. See figure 12.11(a).

■ The system is *small-time locally accessible* (*STLA*) from x if $R^V(x, \leq T)$ contains a full n-dimensional subset of $\mathcal{M}$ for all neighborhoods V and all $T > 0$. See figure 12.11(b).

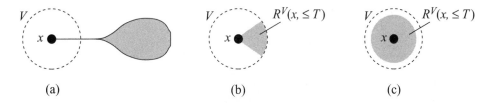

(a)   (b)   (c)

Figure 12.11 Reachable spaces for three systems on $\mathbb{R}^2$. (a) This system is accessible from x, but neither small-time locally accessible (STLA) nor small-time locally controllable (STLC). The reachable set is two-dimensional, but not while confined to the neighborhood V. (b) This system is STLA from x, but not STLC. The reachable set without leaving V does not contain a neighborhood of x. (c) This system is STLC from x.

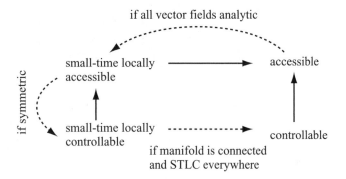

Figure 12.12 Implications among the controllability properties. Dashed arrows are conditional.

- The system is *small-time locally controllable* (STLC) from x if $R^V(x, \leq T)$ contains a neighborhood of x for all neighborhoods V and all $T > 0$. See figure 12.11(c).

The phrase "small-time" indicates that the property holds for any time $T > 0$, and "locally" indicates that the property holds for arbitrarily small (but full-dimensional) wiggle room around the initial state. For practical systems, it might take finite time to switch between controls (e.g., putting a car in reverse gear). In this case, we might say a system is locally, but not small-time, controllable. Here we ignore the switch time and retain the standard "small-time locally" terms.

If a property holds for all $x \in \mathcal{M}$, the phrase "from x" can be eliminated. Figure 12.12 shows the implications among the properties. If the vector fields are all analytic, then accessibility implies STLA.

Small-time local controllability is of special interest. STLC implies that the system can locally maneuver in any direction, and if the system is STLC at all $x \in \mathcal{M}$, then the system can follow any curve on $\mathcal{M}$ arbitrarily closely. This allows the system to maneuver through cluttered spaces, since any motion of a system with no motion constraints can be approximated by a system that is STLC everywhere. Also, if $\mathcal{M}$ is connected, then the system is controllable if it is STLC everywhere.

STLA and STLC are local concepts that can be established by looking at the behavior of the system in a neighborhood of a state. Accessibility and controllability, on the other hand, are global concepts. As a result, they may depend on things such as the topology of the space and nonlocal behavior of the system vector fields.

Some physical examples of the various properties:

- Imagine setting the minute and hour hands on a watch by turning a knob that can spin in only one direction. The configuration space of the hands is one-dimensional, since the motion of the hour hand is coupled to the motion of the minute hand. Show that this system is accessible, controllable, and STLA on the configuration space, but not STLC.

- Consider the system on $\mathbb{R}^2$ described by the drift vector field $g_0 = [x_2^2, 0]^T$ and the single control vector field $g_1 = [0, 1]^T$, where $u = u_1 \in [-1, 1]$. Show that the system is accessible and STLA from any x but neither controllable nor STLC.

- Consider the system on $\mathbb{R}^2$ described by the drift vector field $g_0 = [x_2, 0]^T$ and the single control vector field $g_1 = [0, 1]^T$, where $u = u_1 \in [-1, 1]$. This is the linear double-integrator $\ddot{q} = u$ written in the first-order form $\dot{x}_1 = x_2$, $\dot{x}_2 = u$. Convince yourself that the system is STLC only from zero-velocity states $[*, 0]^T$ (see figure 12.13).

- The unicycle satisfies all the controllability properties if $\mathcal{U} \in \mathcal{U}_{\pm}$.

- Show that the unicycle is accessible, STLA, and controllable in the obstacle-free plane, but not STLC, if $\mathcal{U}$ belongs to the class $\mathcal{U}_+$ but not $\mathcal{U}_{\pm}$.

- Any system confined to a k-dimensional integral manifold, $k < n$, satisfies none of the controllability properties.

As hinted at in the linear double-integrator example, for second-order systems with velocity variables in the state vector, STLC can only hold at zero velocity. States with nonzero velocity result in drift in the configuration variables that cannot be instantaneously compensated by finite actuation forces. *Therefore, when we talk about STLC for second-order systems, we implicitly mean STLC at zero velocity.*

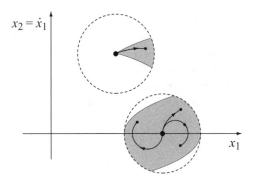

Figure 12.13 Two initial states and neighborhoods for the linear double-integrator with bounded control. The reachable sets from each initial state, by trajectories remaining in the neighborhood, are shaded, and example trajectories are shown. The system is STLC from the initial state where $\dot{x}_1 = x_2 = 0$, but not STLC from the initial state where $\dot{x}_1 = x_2 \neq 0$. Reaching a point left of this initial state (i.e., decreasing the x_1 value) requires $\dot{x}_1$ to become negative—the x_2 coordinate must leave the neighborhood.

For linear systems of the form $\dot{x} = Ax + Bu$, there is a single notion of controllability (see appendix J). For nonlinear systems, such as those we study, there are a number of notions of controllability, including the four we have defined here. A key point is that the linearizations of systems of interest to us are generally not controllable, meaning that their controllability is inherently a nonlinear phenomenon.

12.3.1 Local Accessibility and Controllability

Of the controllability properties, STLA can be checked by studying the Lie algebra of the vector fields $g_0, \ldots, g_m$.

THEOREM 12.3.1 *The system (12.6) is STLA from x if (and only if for analytic vector fields) it satisfies the* Lie algebra rank condition (LARC)—*the Lie algebra of the vector fields, evaluated at x, is the tangent space at x, or* $\overline{\mathrm{Lie}}(\{g_0, \ldots, g_m\})(x) = T_x\mathcal{M}$. *This holds for any* $\mathcal{U} \in \mathcal{U}_+$. *If the system is symmetric (drift-free and* $\mathcal{U} \in \mathcal{U}_\pm$), *then the LARC also implies small-time local controllability.*

An early version of this result is due to W.-L. Chow [112], and it is sometimes called *Chow's theorem.*

EXAMPLE 12.3.2 Unicycle (cont.) *As shown previously, the rank of the unicycle Lie algebra is three at all states, so the LARC is satisfied. Therefore, for both* $\mathcal{U} \in \mathcal{U}_+$

and $\mathcal{U} \in \mathcal{U}_{\pm}$, *the unicycle is STLA. For a control set* $\mathcal{U} \in \mathcal{U}_{\pm}$, *the system is also STLC everywhere, and therefore controllable because of the connectedness of its state manifold. It is also true that the unicycle is controllable (but not STLC) for any* $\mathcal{U} \in \mathcal{U}_{+}$, *though this cannot be shown by theorem 12.3.1. (The reader may wish to verify controllability by describing a constructive procedure to drive the unicycle to any goal location in an obstacle-free space.)*

If we eliminate one vector field from the unicycle example, allowing it only to roll forward and backward (g_1^{uni}) *or spin in place* (g_2^{uni}), *the unicycle is confined to an integral curve of the vector field, and none of the controllability properties is satisfied.*

Second-order systems with nonzero drift, such as the PBWT, are not symmetric for any control set. The system may still be STLC at zero velocity states, however, since symmetry plus the LARC is sufficient but not necessary for STLC. Sussmann [401] provided a more general sufficient condition for STLC that includes the symmetric case ($g_0 = 0$ and $\mathcal{U} \in \mathcal{U}_{\pm}$) as a special case. To understand it, we first define a Lie product term to be a *bad bracket* if the drift term g_0 appears an odd number of times in the product and each control vector field g_i, $i = 1 \ldots m$, appears an even number of times (including zero). A *good bracket* is any Lie product that is not bad. For example, $[g_1, [g_0, g_1]]$ is a bad bracket and $[g_2, [g_1, [g_0, g_1]]]$ and $[g_1, [g_2, [g_1, g_2]]]$ are good brackets. With these definitions, we can state a version of Sussmann's theorem:

THEOREM 12.3.3   *The system (12.6) is STLC at x if*

1. $g_0(x) = 0$,

2. $\mathcal{U} \in \mathcal{U}_{\pm}$,

3. the LARC is satisfied by good Lie bracket terms up to degree k, and

4. any bad bracket of degree $j \le k$ can be expressed as a linear combination of good brackets of degree less than j.

The intuition behind the theorem is the following. Bad brackets are called bad because, after generating the net motion obtained by following the Lie bracket motion prescription, we find that the controls u_i only appear in the net motion with even exponents, meaning that the vector field can only be followed in one direction. In this sense, a bad bracket is similar to a drift field, and we must be able to compensate for it. Since motions in Lie product directions of high degree are essentially "slower" than those in directions with a lower degree, we should only try to compensate for bad bracket motions by good bracket motions of lower degree. If a bad bracket of degree j can be expressed as a linear combination of good brackets of degree less than j, the good brackets are said to *neutralize* the bad bracket. For the bad bracket of degree 1 (the drift vector field g_0) there are no lower degree brackets that can be used to neutralize it, so we require $g_0(x) = 0$. Therefore, this result only holds at states x where the drift vanishes, i.e., equilibrium states.

EXAMPLE 12.3.4   PBWT (cont.) *Assume that the PBWT moves in a horizontal plane, so $a_g = 0$. As before, we define $g_3^{\text{pbwt}} = [g_0^{\text{pbwt}}, g_1^{\text{pbwt}}]$, $g_4^{\text{pbwt}} = [g_0^{\text{pbwt}}, g_2^{\text{pbwt}}]$, $g_5^{\text{pbwt}} = [g_1^{\text{pbwt}}, [g_0^{\text{pbwt}}, g_2^{\text{pbwt}}]]$, and $g_6^{\text{pbwt}} = [g_0^{\text{pbwt}}, [g_1^{\text{pbwt}}, [g_0^{\text{pbwt}}, g_2^{\text{pbwt}}]]]$. Again as before, a computation shows that*

$$\det\begin{bmatrix} g_1^{\text{pbwt}} & g_2^{\text{pbwt}} & g_3^{\text{pbwt}} & g_4^{\text{pbwt}} & g_5^{\text{pbwt}} & g_6^{\text{pbwt}} \end{bmatrix} = d^4.$$

The LARC is satisfied, so the system is STLA at all states for either control set. If $\mathcal{U} \in \mathcal{U}_\pm$, we would like to know if the system satisfies Sussmann's sufficient condition for STLC at equilibrium states $x = [q_1, q_2, q_3, 0, 0, 0]^T$, where $g_0^{\text{pbwt}}(x) = 0$. Because we use bracket terms up to degree 4 to demonstrate LARC, we must be able to neutralize all bad bracket terms of degree 4 or less. The only such bad bracket terms are the degree 3 terms

$$\left[g_1^{\text{pbwt}}, \left[g_0^{\text{pbwt}}, g_1^{\text{pbwt}} \right] \right] = [0, 0, 0, 0, 0, 0]^T$$

$$\left[g_2^{\text{pbwt}}, \left[g_0^{\text{pbwt}}, g_2^{\text{pbwt}} \right] \right] = [0, 0, 0, 2d \cos x_3, 2d \sin x_3, 0]^T = 2d g_1^{\text{pbwt}}.$$

The second term is neutralized by g_1^{pbwt}. Therefore, by Sussmann's theorem, the system is STLC at equilibrium states.

Note that in gravity, $a_g \neq 0$, so $g_0^{\text{pbwt}}(x) \neq 0$ at any state and Sussmann's theorem does not allow us to prove or disprove STLC.

Now consider the case where the PBWT is equipped with a single thruster. If the single thruster corresponds to the vector field g_1^{pbwt}, the thrust always passes through the body center of mass, and the angular velocity of the body cannot be changed. The system is not accessible. If the single thruster corresponds to the vector field g_2^{pbwt}, however, we can define the vector fields

$$\left[g_0^{\text{pbwt}}, g_2^{\text{pbwt}} \right], \quad \left[g_2^{\text{pbwt}}, \left[g_0^{\text{pbwt}}, g_2^{\text{pbwt}} \right] \right], \quad \left[g_2^{\text{pbwt}}, \left[g_0^{\text{pbwt}}, \left[g_0^{\text{pbwt}}, g_2^{\text{pbwt}} \right] \right] \right],$$

$$\left[g_2^{\text{pbwt}}, \left[g_2^{\text{pbwt}}, \left[g_0^{\text{pbwt}}, \left[g_0^{\text{pbwt}}, g_2^{\text{pbwt}} \right] \right] \right] \right],$$

$$\left[g_0^{\text{pbwt}}, \left[g_2^{\text{pbwt}}, \left[g_2^{\text{pbwt}}, \left[g_0^{\text{pbwt}}, \left[g_0^{\text{pbwt}}, g_2^{\text{pbwt}} \right] \right] \right] \right] \right]$$

and see that the determinant of the matrix formed by these columns is $-16d^8$, indicating that the system is STLA for either $\mathcal{U} \in \mathcal{U}_+$ or $\mathcal{U} \in \mathcal{U}_\pm$. Bad brackets cannot be neutralized, so theorem 12.3.3 cannot be used to show STLC. Note, however, that

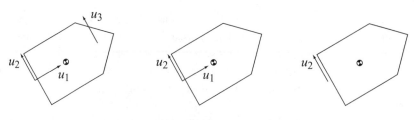

| linearly controllable | not linearly controllable, but STLC at zero velocity | not STLC, but STLA and controllable |

Figure 12.14 The PBWT in zero gravity with different numbers of thrusters. The PBWT on the left has three thrusters that can generate any force and torque combination, and it is controllable by linear control theory. Eliminating one thruster, we get the PBWT in the middle, which is no longer linearly controllable but is STLC at zero velocity. Finally, reducing the thruster count to one, we get the PBWT on the right, which is no longer STLC but remains STLA and controllable in a global sense. (Note that the PBWT with only u_1 thrust is not STLA.) All thrusters are bidirectional.

reducing to a single control vector field does not reduce the dimension of the reachable space, as it did for the kinematic unicycle case. This is because the second-order system provides a drift field with which Lie bracket terms can be generated.

Finally, the PBWT with the single control vector field g_2^{pbwt}, a control set $\mathcal{U} \in \mathcal{U}_{\pm}$, and $a_g = 0$ turns out to be (globally) controllable — any state is reachable in finite time from any other state [303]. Thus the PBWT in zero gravity provides a simple example of different controllability properties (figure 12.14). If we equip it with three independent control vector fields, e.g., a control for each coordinate, the PBWT is a linear system of three double-integrators and it is controllable by linear control theory (see appendix J). If we equip it with the two control vector fields g_1^{pbwt} and g_2^{pbwt}, it is no longer linearly controllable, but remains STLC at zero velocity. If we equip it with just the single control vector field g_2^{pbwt}, it is no longer STLC at zero velocity, but remains STLA and globally controllable.

12.3.2   Global Controllability

For kinematic systems that are STLC everywhere on a connected manifold, (global) controllability follows easily. In general, however, controllability is not easy to decide, as it may depend on nonlocal features of the control system. In the special case of a control system (12.6) with $\mathcal{U} \in \mathcal{U}_{\pm}$ and a drift vector field that repeatedly returns the system to a neighborhood of its initial state, however, demonstrating controllability is as easy as demonstrating the LARC.

First, some definitions. Consider the flow ϕ^{g_0} of the drift vector field. A point $x \in \mathcal{M}$ is called *positively Poisson stable* (PPS) for g_0 if for all $T > 0$ and any neighborhood V of x, there exists a time $t > T$ such that the flow of the vector field returns the system to V, i.e., $\phi_t^{g_0}(x) \in V$. The drift vector field g_0 is called positively Poisson stable if the set of PPS points for g_0 is dense in $\mathcal{M}$.

A point $x \in \mathcal{M}$ is called a *nonwandering point* of g_0 if for all time $T > 0$ and any neighborhood V of x there exists a time $t > T$ such that $\phi_t^{g_0}(V) \cap V \neq \emptyset$, where $\phi_t^{g_0}(V) = \{\phi_t^{g_0}(x) \mid x \in V\}$. (A positively Poisson stable point is necessarily a nonwandering point.) The *nonwandering set* of g_0 is the set of all nonwandering points of g_0. Finally, we say that the drift vector field g_0 is *weakly positively Poisson stable* (WPPS) if its nonwandering set is $\mathcal{M}$.

We now state the main theorem, taken from Lian, Wang, and Fu [289]. Related results can be found in (Jurdjevic and Sussmann [212]; Lobry [295]; Brockett [65]; Bonnard [58]; and Jurdjevic [211]).

THEOREM 12.3.5   *Assume that the drift vector field g_0 is WPPS. Then the system (12.6) with $\mathcal{U} \in \mathcal{U}_\pm$ is controllable on $\mathcal{M}$ if the LARC is satisfied.*

As an example, consider the system on $\mathbb{R}^2$ described by $\dot{x} = g_0(x) + g_1(x)u_1$, $u_1 \in [-1, 1] \in \mathcal{U}_\pm$, where $g_0(x) = \frac{1}{2}[-x_2, x_1]^T$ and $g_1(x) = [1, 0]^T$. The drift vector field (shown in figure 12.15) is WPPS, as its orbits are closed. We find that $[g_0, g_1] = [0, -\frac{1}{2}]^T$ and $\det[g_1 \ [g_0, g_1]] = -\frac{1}{2}$, so the LARC is satisfied at all x. By theorem 12.3.5, every state is reachable from every other state. Intuitively, u_1 is used to control x_1 and (waiting) time is used to "control" x_2. (In fact, in this example, it is

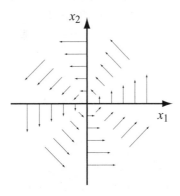

Figure 12.15   The integral curves of the WPPS drift vector field $\frac{1}{2}[-x_2, x_1]^T$ are closed (circles).

not hard to see that controllability also holds for $u_1 \in \mathcal{U} \in \mathcal{U}_+$.) This system is only STLC at the origin.

Theorem 12.3.5 is a powerful tool for establishing the global controllability of systems with drift. Systems with periodic natural unforced dynamics (such as an undamped planar pendulum or the example of figure 12.15) or energy-conserving drift on compact configuration spaces are examples of systems with WPPS drift vector fields. The latter follows from an application of Poincaré's recurrence theorem; see, e.g., the discussion by Arnold [26]. As an example, a rotating satellite moves on the compact configuration space $SO(3)$, and its natural unforced motion conserves energy. Therefore, the drift is WPPS. The LARC can be satisfied by a single body-fixed control torque, meaning that the satellite can be driven to any orientation and angular velocity with a single control vector field.

For systems with non-WPPS drift, it may be possible to construct feedback laws that always keep the system in a periodic orbit. If the system is always controllable about these periodic trajectories, i.e., if the system can reach neighborhoods of the controlled periodic trajectories, then similar reasoning can be used to demonstrate controllability of the system [87, 304].

12.4   Simple Mechanical Control Systems

This section discusses second-order mechanical control systems in greater depth, leading to simplified controllability tests and ideas that lead to reduced-complexity motion planning. We introduce the minimal set of ideas from Riemannian geometry that allows us to do this. We do not attempt to be rigorous or thorough in our treatment of mechanical systems from a geometric viewpoint. The motivated reader is instead referred to the books by Abraham and Marsden [10], do Carmo [131], Marsden and Ratiu [308], Bloch [50], Bullo and Lewis [77], and Boothby [60] for further study of differential geometry in mechanics and control. The results of this section are used in subsection 12.5.7, but otherwise this section can be skipped without affecting the reading of the rest of the chapter.

In chapter 10 we derived equations of motion of the form

(12.7) $$M(q)\ddot{q} + C(q, \dot{q})\dot{q} + g(q) = T(q)f,$$

where f is a generalized force vector, $T(q)$ defines the action of f on the coordinates, $M(q)$ is the inertia matrix, $C(q, \dot{q})\dot{q}$ are Coriolis and centrifugal terms, and $g(q)$ are potential terms. Recall that $C(q, \dot{q})\dot{q} = \dot{q}^T \Gamma(q)\dot{q}$, where $\Gamma(q)$ is the set of n_Q^3 Christoffel symbols of the inertia matrix $M(q)$, and the computation $\dot{q}^T \Gamma(q)\dot{q}$ is described in chapter 10.[2] We restrict our discussion in this section to *simple mechanical control systems* of this form with $g(q) = 0$.

Since we are considering underactuated systems, we can write $f = [u^T, 0^T]^T$, where $u \in \mathbb{R}^m$ is the control vector and 0 is an $(n_Q - m)$-vector of zeros. In this case, $T(q)$ can be written as an $n_Q \times m$ matrix, and the equations are

(12.8) $M(q)\ddot{q} + C(q, \dot{q})\dot{q} = T(q)u.$

Premultiplying both sides by $M^{-1}(q)$ (assuming full rank) and rearranging, we get

(12.9) $\ddot{q} = -M^{-1}(q)C(q, \dot{q})\dot{q} + M^{-1}(q)T(q)u.$

Writing the m columns of $M^{-1}(q)T(q)$ as $Y_i(q), i = 1 \dots m$, we get

(12.10) $\ddot{q} = -M^{-1}(q)C(q, \dot{q})\dot{q} + \sum_{i=1}^{m} Y_i(q)u_i.$

In other words, $\ddot{q}$ is the sum of a drift term due to Coriolis and centrifugal effects and a term due to the controls. According to the development so far, the control system can be expressed in the form of equation (12.6) by writing the state, drift vector field, and control vector fields as

$$x = \begin{bmatrix} q \\ \dot{q} \end{bmatrix}, \quad g_0(x) = \begin{bmatrix} \dot{q} \\ -M^{-1}(q)C(q, \dot{q})\dot{q} \end{bmatrix}, \quad g_i(x) = \begin{bmatrix} 0 \\ Y_i(q) \end{bmatrix}$$

and expressing the control system as

(12.11) $\dot{x} = g_0(x) + \sum_{i=1}^{m} g_i(x)u_i.$

We can then apply controllability tests as previously described.

12.4.1 Simplified Controllability Tests

The approach described above, while correct, ignores some structure of the equations of motion of a simple mechanical control system. Lewis and Murray [285, 286] have studied the Lie bracket structure of simple mechanical control systems to derive simplified controllability tests at equilibrium states. These tests take advantage of the Lie bracket structure to reduce the number of computations. In the rest of this section, we will assume the control set $\mathcal{U}$ belongs to the class $\mathcal{U}_{\pm}$.

---

2. Most works in the differential geometry literature define a slightly different set of Christoffel symbols Γ^* that have the inverse of the inertia matrix embedded in them, such that $\dot{q}^T \Gamma^* \dot{q} = M^{-1}\dot{q}^T \Gamma \dot{q}$. We instead use the same definition used in chapter 10.

The key simplification is that we will study vector fields *only on the configuration space Q, not the full state space* $TQ = \mathcal{M}$. This will be possible because $M(q)$, $\Gamma(q)$, and $T(q)$ all depend on q only, not $\dot{q}$. To study dynamics, however, we must be able to define derivatives (accelerations) of vector fields on Q. In particular, we need a definition of how a vector field $Y_2(q)$ on Q is changing (accelerating) along the direction of another vector field $Y_1(q)$. This is called the *covariant derivative* of $Y_2(q)$ with respect to $Y_1(q)$, and it is also a vector field on Q.

To define the covariant derivative, we need to define the acceleration of a curve $q(t)$. Often we think of $\ddot{q}(t)$ as the "coordinate" acceleration, but by this definition, whether the system is accelerating or not depends on the choice of coordinates, as we will see shortly. The vector $\ddot{q}(t)$ is not generally contained in the tangent space $T_{q(t)}Q$, and the misalignment is caused by nonzero Christoffel symbols of $M(q)$, which define how the configuration space "curves" in the coordinates q. To express the acceleration as an element of the tangent space, so that an observer living on Q can "see" it in the tangent space, we use the Christoffel symbols (i.e., the Coriolis terms) to project the coordinate acceleration $\ddot{q}$ back to the tangent space:

$$(12.12) \qquad \text{acceleration} = \ddot{q} + M^{-1}(q)C(q, \dot{q})\dot{q} = \ddot{q} + M^{-1}(q)\dot{q}^T \Gamma(q)\dot{q} \in T_{q(t)}Q$$

This is an "intrinsic" definition of acceleration independent of the coordinates chosen.

As a first example, imagine a point mass moving in a one-dimensional configuration space $Q = S^1$, visualized as a unit circle in $\mathbb{R}^2$. If the position is described by $q = [x, y]^T$ in Cartesian coordinates and θ in polar coordinates, we have

$$q = \begin{bmatrix} x \\ y \end{bmatrix} = \begin{bmatrix} \cos\theta \\ \sin\theta \end{bmatrix},$$

with second derivative

$$\ddot{q} = \begin{bmatrix} \ddot{x} \\ \ddot{y} \end{bmatrix} = \ddot{\theta}\begin{bmatrix} -\sin\theta \\ \cos\theta \end{bmatrix} + \dot{\theta}^2\begin{bmatrix} -\cos\theta \\ -\sin\theta \end{bmatrix}.$$

The $\dot{\theta}^2$ term in $\ddot{q}$ enforces the constraint that the point mass stays on the unit circle. The $\ddot{\theta}$ term expresses the change in speed tangent to the circle, and this is the only acceleration visible to an observer in Q who cannot see the $\mathbb{R}^2$ space it is embedded in. The $\ddot{\theta}$ term is the intrinsic definition of acceleration we are looking for. We subtract the $\dot{\theta}^2$ term [the centripetal acceleration, corresponding to the negative of the velocity product term in equation (12.12)] from $\ddot{q}$ to project $\ddot{q}$ to the tangent space (see figure 12.16).

For an example that does not use a manifold embedded in a higher-dimensional space, consider a point mass m moving in the plane with no forces applied to it. We can write the equations of motion in either Cartesian coordinates (x, y) or polar

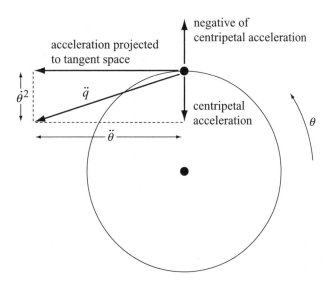

Figure 12.16 Subtracting the centripetal acceleration vector (pointing toward the center of the circle) from $\ddot{q} = [\ddot{x}, \ddot{y}]^T$ of a point mass moving around a circle gives an acceleration tangent to the circle.

coordinates (r, θ):

$$\begin{bmatrix} m & 0 \\ 0 & m \end{bmatrix} \begin{bmatrix} \ddot{x} \\ \ddot{y} \end{bmatrix} = \begin{bmatrix} 0 \\ 0 \end{bmatrix}, \quad \begin{bmatrix} m & 0 \\ 0 & m\,r^2 \end{bmatrix} \begin{bmatrix} \ddot{r} \\ \ddot{\theta} \end{bmatrix} + \begin{bmatrix} -m\,r\dot{\theta}^2 \\ 2m\,r\dot{r}\dot{\theta} \end{bmatrix} = \begin{bmatrix} 0 \\ 0 \end{bmatrix}$$

In Cartesian coordinates $q = [x, y]^T$, the mass moves such that $\ddot{x} = \ddot{y} = 0$. If we represent the configuration with polar coordinates $q = [r, \theta]^T$, however, the same motions of the mass will have $\ddot{r} \neq 0$, $\ddot{\theta} \neq 0$. So is the mass accelerating or not? The answer is that we should not think of $\ddot{q}$ as an acceleration; instead, we should think of the acceleration as being the inverse of the inertia matrix $M^{-1}(q)$ times the force. In both cases, then, since the force is zero, the acceleration is zero. The second time-derivatives $\ddot{r}$ and $\ddot{\theta}$ are not zero due to the nonzero Christoffel symbols of the inertia matrix in this choice of coordinates, as discussed in chapter 10.

Figure 12.17 shows paths of the point mass as it moves with zero force applied to it, shown in both Cartesian coordinates and polar coordinates. We say that each path in figure 12.17 is a *shortest path,* and the tangent vectors to these paths are *orthogonal* to each other at any intersection point. The notions of shortest paths (straight lines) and orthogonality are clear in our usual understanding of Euclidean geometry for the Cartesian coordinate case, but less clear in the polar coordinate case. In the polar coordinate case we need an inner product so we can define orthogonality. The inner

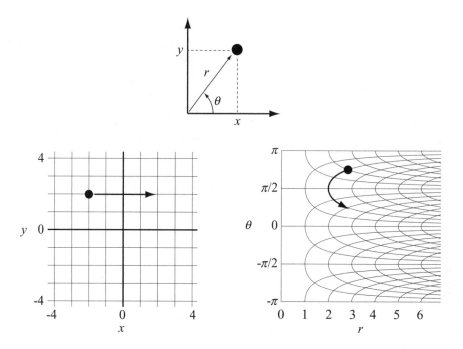

Figure 12.17   Example unforced motions of a point mass in the plane, represented in Cartesian coordinates (x, y) and polar coordinates (r, θ).

product also defines the distance metric, allowing us to talk about shortest paths between two points. For example, the length of each side of each grid "box" in the Cartesian plot of figure 12.17 is equivalent, just as the length of each side of a "box" in the polar plot is equivalent by its distance metric. Also, the "areas" of the "boxes" in the polar coordinate plot of figure 12.17 are equivalent, just as they are equivalent for each box in the Cartesian coordinate plot.

An inner product maps two tangent vectors $v_1, v_2 \in T_q \mathcal{Q}$ to $\mathbb{R}$. We write this inner product $\langle v_1, v_2 \rangle$, and v_1 and v_2 are orthogonal if $\langle v_1, v_2 \rangle = 0$. The inner product is associated with a metric $d : \mathcal{Q} \times \mathcal{Q} \to \mathbb{R}$ measuring the distance between two points. The metric and the inner product are related by the fact that the shortest path or *geodesic* between q_0 and q_f occurs for trajectories $q(t)$ minimizing the path length

$$\int_{t_0}^{t_f} \langle \dot{q}(t), \dot{q}(t) \rangle^{1/2} \, dt$$

out of all possible trajectories connecting $q(t_0) = q_0$ and $q(t_f) = q_f$. This relation allows us to call the inner product itself a metric.

As an example, the familiar Euclidean metric states that geodesics are straight lines, and the familiar inner product or "dot product" associated with the Euclidean metric is $\langle v_1, v_2 \rangle = v_1 \cdot v_2 = v_1^T v_2$. In the Euclidean metric, the rate of change of $Y_2(q)$ along the direction of $Y_1(q)$, i.e., the covariant derivative of $Y_2(q)$ with respect to $Y_1(q)$, is simply $(\partial Y_2/\partial q)Y_1$, i.e., the partial of Y_2 projected to the direction Y_1.

The Euclidean metric is natural for thinking about the motion of a point in space with standard Cartesian coordinates. Appropriate metrics for other systems are less obvious. For example, for a 2R robot arm, what does it mean for two velocities to be orthogonal? At a particular configuration of the arm, if v_1 and v_2 are Cartesian velocity vectors at the end-effector and w_1 and w_2 are the corresponding joint velocity vectors, then the condition $v_1^T v_2 = 0$ is not the same as $w_1^T w_2 = 0$. These conditions imply different and somewhat arbitrary metrics.

For the second-order mechanical systems we are interested in, the inertia matrix $M(q)$ defines a physically meaningful metric and inner product. This is sometimes called the kinetic energy metric, since it is associated with the kinetic energy $\frac{1}{2}\dot{q}^T M(q)\dot{q}$. The inner product defined by $M(q)$ is

$$\langle v_1, v_2 \rangle = v_1^T M(q)v_2, \quad v_1, v_2 \in T_q \mathcal{Q}.$$

The result has units of energy, which is physically meaningful and independent of the choice of coordinates. This metric is an example of a *Riemannian metric,* as it is bilinear (linear in each of v_1 and v_2) and symmetric and positive definite (because $M(q)$ is symmetric and positive definite). The kinetic energy metric defines a two-form, as it takes two elements of $T_q \mathcal{Q}$ and returns a real number linear in each of the two elements. When $M(q)$ is the identity matrix, the Euclidean metric is obtained.

The metric defines an *affine connection* ∇, which allows us to define the derivatives of vector fields in non-Euclidean spaces. For the affine connection ∇ associated with $M(q)$, the covariant derivative of $Y_2(q)$ with respect to $Y_1(q)$ is the vector field $\nabla_{Y_1(q)}Y_2(q)$,

$$\nabla_{Y_1(q)}Y_2(q) = \frac{\partial Y_2(q)}{\partial q}Y_1(q) + M^{-1}(q)Y_1^T(q)\Gamma(q)Y_2(q),$$

where the Christoffel symbols $\Gamma(q)$ describe how geodesics of the mechanical system "bend" in this choice of coordinates. For the Euclidean metric, the Christoffel symbols are zero.

The covariant derivative allows us to write the acceleration as an element of the tangent space to the configuration space, $\nabla_{\dot{q}(t)}\dot{q}(t) \in T_{q(t)}\mathcal{Q}$. We can write the equations of motion (12.10) in the following equivalent form:

$$\nabla_{\dot{q}}\dot{q} = \frac{\partial \dot{q}}{\partial q}\dot{q} + M^{-1}(q)\dot{q}^T\Gamma(q)\dot{q}$$

$$\nabla_{\dot{q}}\dot{q} = \ddot{q} + M^{-1}(q)C(q,\dot{q})\dot{q}$$

(12.13) $\quad\displaystyle\nabla_{\dot{q}}\dot{q} = \sum_{i=1}^{m} Y_i(q)u_i, \quad u \in \mathcal{U} \in \mathcal{U}_{\pm}$

Equation (12.13) is the familiar $a = M^{-1}f$—acceleration is equal to the inverse of the mass times the force. Here, however, the acceleration a is not just the second derivative of the coordinates $\ddot{q}$, but $\ddot{q} + M^{-1}\dot{q}^T\Gamma\dot{q}$, accounting for the noninertial coordinates. We have $a = \ddot{q}$ only if the Christoffel symbols are zero, which is the case if the inertia matrix has no dependence on the configuration q.

Unforced motions $q(t)$, i.e., motions satisfying

$$\nabla_{\dot{q}(t)}\dot{q}(t) = 0,$$

are geodesics of ∇.

The covariant derivative also allows us to define the *symmetric product* of Y_1 and Y_2, the vector field

$$\langle Y_1 : Y_2 \rangle = \nabla_{Y_1}Y_2 + \nabla_{Y_2}Y_1,$$

which is useful in controllability calculations. But why should this be so? After all, we have already seen in equation (12.11) that a $Y_i(q)$ from equation (12.10) can be turned into a control vector field $g_i(x)$ on the full state space $\mathcal{M} = T\mathcal{Q}$, and that the drift can be expressed as $g_0(x)$, allowing us to take Lie brackets as before to test controllability. If we assume the system begins from rest, however, we notice that symmetric products of the $Y_i(q)$ appear again and again in the bracket terms. In particular, we can identify the following patterns in the calculations for $i, j = 1, \ldots, m$:

$$[g_i, g_j] = \left[\begin{bmatrix} 0 \\ Y_i \end{bmatrix}, \begin{bmatrix} 0 \\ Y_j \end{bmatrix}\right] = 0,$$

$$[g_0, g_i] = \begin{bmatrix} -Y_i \\ 0 \end{bmatrix},$$

$$[g_i, [g_0, g_j]] = \begin{bmatrix} 0 \\ \langle Y_i : Y_j \rangle \end{bmatrix}$$

The first shows that no new motion directions will be created if the drift field is not included in the Lie bracket. The second shows that Lie bracketing the drift with a control vector field has the effect of taking the "acceleration" direction of the control vector field to a velocity direction (this only holds at zero velocity). The last shows that iterated Lie brackets including the drift field can be evaluated by calculating lower-degree symmetric products of the $Y_i(q)$. An added benefit is that the symmetric product operates on vector fields with half as many elements.

The key point is, for controllability computations for simple mechanical control systems beginning from rest, the symmetric product allows us to think of $Y_i(q)$ *as a control vector field on the system configuration space Q, without constructing a higher-dimensional vector field $g_i(q)$ on the full state manifold $\mathcal{M} = TQ$.* The symmetric product captures the effect of drift in the controllability computations.

For a control system (12.13) consisting of a set of control vector fields $\mathcal{Y} = \{Y_1, \ldots, Y_m\}$, let the *symmetric closure* $\overline{\mathrm{Sym}}(\mathcal{Y})$ be the distribution defined by $\mathcal{Y}$ and the iterated symmetric products of these vector fields. In this sense, the symmetric closure by the symmetric product is defined similarly to the involutive closure by the Lie bracket. Also, we can define the degree of a symmetric product to be the number of the original vector fields Y_i appearing in the expression. A symmetric product is *bad* if each of the vector fields appears an even number of times, and is *good* otherwise. With these definitions, Lewis and Murray [285, 286] proved the following theorem, building on theorem 12.3.3.

THEOREM 12.4.1 *Beginning from an equilibrium state $x = [q^T, 0^T]^T$, the system (12.13) is*

1. *STLA from x if and only if* $\overline{\mathrm{Sym}}(\mathcal{Y})(q) = T_q Q$, *and*

2. *STLC from x if* $\overline{\mathrm{Sym}}(\mathcal{Y})(q) = T_q Q$ *and every bad symmetric product can be expressed as a linear combination of good symmetric products of lower degree.*

Beginning from an equilibrium state $x = [q^T, 0^T]^T$, it is sometimes of interest to understand the locally reachable set of configurations irrespective of the velocities at those configurations. Lewis and Murray define a system to be *small-time locally configuration accessible (STLCA)* from q if the locally reachable set is full-dimensional on Q, and *small-time locally configuration controllable (STLCC)* from q if the locally reachable set on Q contains q in the interior. A stronger condition than STLCC is *small-time local equilibrium controllability (STLEC)* from q if the locally reachable set contains zero velocity states forming a neighborhood of q on Q. STLEC is stronger than STLCC, as STLEC demands that nearby configurations be reachable at zero velocity, while STLCC says nothing about the velocity. Finally, the system is *equilibrium controllable* if the system can reach any equilibrium state from any other equilibrium state.

Note that STLEC is a weaker property than STLC, as STLC requires that the locally reachable *states* contain a neighborhood of x on TQ, while STLEC only requires that the locally reachable *configurations* (at zero velocity) contain a neighborhood of q on $\mathcal{Q}$. STLEC is particularly relevant to systems subject to velocity constraints, which may be equilibrium controllable despite being confined to a reachable space of dimension less than n.

The following theorem, due to Lewis and Murray [285, 286], provides tests for STLCA, STLCC, STLEC, and equilibrium controllability.

THEOREM 12.4.2 *Beginning from an equilibrium state* $x = [q^T, 0^T]^T$, *the system (12.13) is*

1. *STLCA from* q *if and only if* $\overline{\text{Lie}}(\overline{\text{Sym}}(\mathcal{Y}))(q) = T_q\mathcal{Q}$, *and*

2. *both STLCC and STLEC from* q *if* $\overline{\text{Lie}}(\overline{\text{Sym}}(\mathcal{Y}))(q) = T_q\mathcal{Q}$ *and if every bad symmetric product can be expressed as a linear combination of good symmetric products of lower degree. If these conditions are satisfied at all* $q \in \mathcal{Q}$, *then the system is equilibrium controllable.*

Roughly speaking, $\overline{\text{Sym}}(\mathcal{Y})(q)$ corresponds to the reachable velocity directions from $[q^T, 0^T]^T$, and $\overline{\text{Lie}}(\overline{\text{Sym}}(\mathcal{Y}))(q)$ corresponds to the reachable configuration directions from $[q^T, 0^T]^T$.

EXAMPLE 12.4.3   PBWT (cont.) *The dynamics of the PBWT can be written*

$$\ddot{q} = -M^{-1}(q)C(q,\dot{q})\dot{q} + M^{-1}(q)T(q)u,$$

where $M^{-1}(q)$ *is the* 3×3 *identity matrix,* $C(q,\dot{q})\dot{q} = 0$ *(the Christoffel symbols are all zero),* $u = [u_1, u_2]^T$, *and*

$$T(q) = \begin{bmatrix} \cos q_3 & -\sin q_3 \\ \sin q_3 & \cos q_3 \\ 0 & -d \end{bmatrix}.$$

The control vector fields $\mathcal{Y} = \{Y_1(q), Y_2(q)\}$ *are the two columns of* $M^{-1}(q)T(q) = T(q)$. *Calculating the symmetric product of* $Y_1(q)$ *and* $Y_2(q)$, *we get*

$$\langle Y_1 : Y_2 \rangle = \nabla_{Y_1} Y_2 + \nabla_{Y_2} Y_1 = \begin{bmatrix} d \sin q_3 \\ -d \cos q_3 \\ 0 \end{bmatrix}.$$

We see that $\text{rank}(Y_1 \quad Y_2 \quad \langle Y_1 : Y_2 \rangle) = 3$ *for all* q, *so* $\overline{\text{Sym}}(\mathcal{Y}) = T_q\mathcal{Q}$ *for all* q, *and the system is STLA by theorem 12.4.1. The bad products are* $\langle Y_1 : Y_1 \rangle = 0$ *and* $\langle Y_2 : Y_2 \rangle = [d \cos q_3, d \sin q_3, 0]^T$, *which is neutralized by* Y_1, *so the system is also STLC by theorem 12.4.1. This confirms our previous result.*

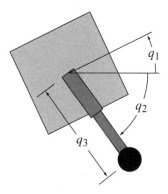

Figure 12.18   A simple single-leg hopping robot.

The reader may also wish to verify that $\mathrm{rank}(Y_2 \ \langle Y_2 : Y_2 \rangle \ \langle Y_2 : \langle Y_2 : Y_2 \rangle \rangle) = 3$, *so the PBWT is STLA with the single thruster* u_2 *by theorem 12.4.1. This also confirms our previous result.*

EXAMPLE 12.4.4   Hopper example. *Consider a simple model of a planar single-leg hopping robot in flight, ignoring the translational motion of the body. We will model the body of the hopper as a rigid body pinned to a wall by a revolute joint at its center of mass. The inertia of the body is* I *about its center of mass. An extensible massless leg is attached to the center of mass of the body, and the foot of the leg is a point mass* m. *The configuration of the system is* $q = [q_1, q_2, q_3]^T$, *where* q_1 *is the angle of the body relative to an inertial frame,* q_2 *is the angle of the leg relative to the inertial frame, and* $q_3 > 0$ *is the extension of the leg (figure 12.18). The leg has two actuators, one providing a torque to control the orientation of the leg relative to the body, and the other providing a force to control the extension of the leg. We would like to know if it is possible to control the configuration* q *using only these two actuators.*

This system is subject to a velocity constraint: the total angular momentum of the system is zero throughout the motion (problem 27). This constraint implies that the dimension of the reachable state space can be no more than five, while $n = 2n_{\mathcal{Q}} = 6$. *This rules out the possibility of STLA and STLC, so instead we focus on the reachable configurations.*

The system mass matrix and its inverse are

$$
M(q) = \begin{bmatrix} I & 0 & 0 \\ 0 & mq_3^2 & 0 \\ 0 & 0 & m \end{bmatrix}, \quad
M^{-1}(q) = \begin{bmatrix} \frac{1}{I} & 0 & 0 \\ 0 & \frac{1}{mq_3^2} & 0 \\ 0 & 0 & \frac{1}{m} \end{bmatrix},
$$

and the only nonzero Christoffel symbols are $\Gamma^2_{23} = \Gamma^2_{32} = mq_3$ and $\Gamma^3_{22} = -mq_3$. The matrix $T(q)$ describing the generalized forces from the actuators is

$$T(q) = \begin{bmatrix} 1 & 0 \\ -1 & 0 \\ 0 & 1 \end{bmatrix},$$

and $Y_1(q) = [1/I, -1/mq_3^2, 0]^T$, $Y_2(q) = [0, 0, 1/m]^T$. Calculations show that

$$\langle Y_1 : Y_1 \rangle = \begin{bmatrix} 0 \\ 0 \\ -\frac{2}{m^2 q_3^3} \end{bmatrix}, \quad \langle Y_1 : Y_2 \rangle = \langle Y_2 : Y_2 \rangle = 0, \quad [Y_1, Y_2] = \begin{bmatrix} 0 \\ -\frac{2}{m^2 q_3^3} \\ 0 \end{bmatrix}.$$

We see that Y_1, Y_2, and $[Y_1, Y_2]$ span $T_q \mathcal{Q}$ for all q with $q_3 > 0$. Also, the bad symmetric product $\langle Y_1 : Y_1 \rangle$ is neutralized by Y_2 and the bad symmetric product $\langle Y_2 : Y_2 \rangle$ is zero. Therefore, by theorem 12.4.2, the system is STLEC at all q and equilibrium controllable. We also see that the distribution $\overline{\mathrm{Sym}(\mathcal{Y})}$ is only two-dimensional, so the system is not STLA by theorem 12.4.1.

12.4.2 Kinematic Reductions for Motion Planning

In the controllability tests above, the symmetric product essentially allows us to treat the Y_i like velocity vector fields on a configuration space, thus halving the dimension of the vector fields in our controllability calculations. It is also sometimes possible to plan trajectories for underactuated mechanical systems as if they were kinematic systems. This reduction decreases the dimension of the search space by a factor of two. Since many search algorithms run in time exponential in the dimension of the search space, this reduction can greatly speed up motion planning.

Consider the original second-order mechanical system

(12.14) $$\nabla_{\dot q} \dot q = \sum_{i=1}^{m} Y_i(q) u_i, \quad u \in \mathbb{R}^m,$$

and a first-order driftless kinematic system

(12.15) $$\dot q = \sum_{i=1}^{\ell} V_i(q) w_i, \quad w \in \mathbb{R}^\ell, \quad w(t) \text{ continuous},$$

where the set of control (velocity) vector fields is written $\mathcal{V} = \{V_1, \ldots, V_\ell\}$. We make the extra stipulation that $w(t)$ be continuous because discontinuous velocities would require infinite forces. Also, in this section on kinematic reductions, the control sets are taken to be unbounded for simplicity.

A kinematic system (12.15) is a *kinematic reduction* of a mechanical system (12.14) if all feasible trajectories for the kinematic system are also feasible for the second-order system. We further say that a mechanical system (12.14) is *maximally reducible to a kinematic system* if there exists a kinematic reduction such that all feasible trajectories of the mechanical system, starting with an initial velocity in span($\mathcal{Y}$), are also trajectories of the kinematic reduction. For example, all fully actuated mechanical systems are maximally reducible to kinematic systems—we can equivalently assume the controls are either (continuous) velocities or forces.

Some underactuated mechanical systems are also maximally reducible to kinematic systems. The test is given by the following theorem due to Lewis [283].

THEOREM 12.4.5   *A second-order mechanical system (12.14) is maximally reducible to a kinematic system if and only if* $\overline{\text{Sym}}(\mathcal{Y}) = \text{span}(\mathcal{Y})$.

EXAMPLE 12.4.6   Hopper (cont.) *Our previous calculations showed that* $\langle Y_1 : Y_2 \rangle = \langle Y_2 : Y_2 \rangle = 0$ *and* $\langle Y_1 : Y_1 \rangle \in \text{span}(Y_2)$. *Therefore* $\overline{\text{Sym}}(\mathcal{Y}) = \text{span}(\mathcal{Y})$ *and the hopper is maximally reducible to a kinematic system with* $\mathcal{V} = \mathcal{Y}$. *Intuitively, this is because the underactuation constraint of the hopper can be integrated to a velocity constraint: conservation of angular momentum. Therefore, any motion possible by controlling force on the leg extension and torque on the leg rotation is also possible by driving the kinematic reduction with* Y_1 *and* Y_2 *as velocity vector fields.*

A maximal kinematic reduction that generates all trajectories of the mechanical system [with initial velocity in span($\mathcal{Y}$)] could be called a *rank m kinematic reduction*, as the controlled velocities of the reduction form an *m*-dimensional distribution $\mathcal{V}$ with span($\mathcal{V}$) = span($\mathcal{Y}$). The class of underactuated systems admitting rank *m* kinematic reductions is relatively small, however, so we would like to explore kinematic reductions for more general underactuated systems. In particular, a system that is not maximally reducible to a kinematic system may nonetheless admit a *rank 1 kinematic reduction* $\dot{q} = V_1(q) w_1(t)$. A rank 1 kinematic reduction has a single control vector field V_1, also known as a *decoupling vector field*. (The word "decoupling" stems from the fact that trajectory planning for the second-order system along an integral curve of such a vector field can be decoupled into choosing the distance traveled along the integral curve, followed by time-scaling the path according to actuator limits. This brings to mind the decoupled trajectory planning approach for fully actuated systems in the previous chapter.)

An underactuated mechanical system can follow the integral curve of a decoupling vector field $V_1(q)$ at any speed and acceleration, i.e., for any continuous $w_1(t)$. A second-order mechanical system (12.14) can have no more than *m* linearly independent decoupling vector fields at any q. For a maximally reducible mechanical system, every vector field in span($\mathcal{Y}$) is a decoupling vector field.

How do we know if a proposed decoupling vector field $V \in$ span($\mathcal{Y}$) is actually decoupling? The following theorem provides the answer (Bullo and Lynch [79]).

THEOREM 12.4.7 *A vector field V is a decoupling vector field of the second-order mechanical system (12.14) if and only if $V \in$ span($\mathcal{Y}$) and $\nabla_V V \in$ span($\mathcal{Y}$).*

Stated another way, from the definition of the kinematic reduction (12.15) for the single control vector field $V(q)$, we have

(12.16) $\dot{q} = V(q)w$

(12.17) $\ddot{q} = V(q)\dot{w} + \dfrac{\partial V}{\partial q} V(q)w^2,$

and plugging these into the equations of motion (12.8), we see that $V(q)$ is decoupling if and only if there exists a $u \in \mathbb{R}^m$ satisfying the equations of motion (12.8) for all $w, \dot{w} \in \mathbb{R}$.

EXAMPLE 12.4.8   PBWT (cont.) *Decoupling vector fields for the PBWT are $V_1 = [\cos q_3, \sin q_3, 0]^T$ and $V_2 = [-\sin q_3, \cos q_3, -d]^T$. To verify this by theorem 12.4.7, we see that $V_1 = Y_1$, $V_2 = Y_2$, $\nabla_{V_1} V_1 = 0$, and $\nabla_{V_2} V_2 = [d \cos q_3, d \sin q_3, 0]^T \in$ span(Y_1). Note, however, that $\langle Y_1 : Y_2 \rangle \notin$ span($\mathcal{Y}$), so the PBWT is not maximally reducible to a kinematic system; i.e., linear combinations of V_1 and V_2 are not decoupling vector fields.*

Consider the physical meaning of these decoupling vector fields (figure 12.19). The vector field V_1 is pure translation of the PBWT along the line of action of the thruster u_1. It is clear that translation in this direction is possible at any speed and acceleration by proper choice of the thrust. The vector field V_2 corresponds to rotation of the PBWT about a point fixed relative to the body, the center of percussion *or* center of oscillation *of the PBWT relative to the line of action of thruster u_2. The center of percussion is the instantaneously unaccelerated point for nonzero thrust u_2. This point is a distance $1/d$ from the center of mass on a line through the center of mass and perpendicular to the line of action of u_2. (In the case that the PBWT has nonunit mass m and nonunit inertia I, the center of percussion is a distance I/md from the center of mass.) Rotation about this point is possible at any speed and acceleration by using u_2 to provide the torque to rotate the PBWT and using u_1 to keep the center of percussion stationary.*

In some cases the decoupling vector fields for a second-order system are apparent by inspection. In other cases it is possible to calculate the decoupling vector fields by

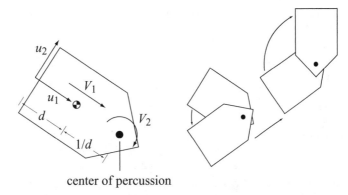

Figure 12.19 The decoupling vector fields V_1 and V_2 for the PBWT: translation along the line of action of u_1, and rotation about the center of percussion with respect to u_2. An example motion following these decoupling vector fields is shown.

solving a system of quadratic equations. To see this, recognize that any decoupling vector field $V(q)$ is contained in $\mathrm{span}(\mathcal{Y})$, so can be written in the form

$$V(q) = h_1(q)Y_1(q) + \cdots + h_m(q)Y_m(q), \quad h_i(q) \in \mathbb{R}.$$

The problem is to solve for functions $h_i(q)$ such that $\langle X_c, \nabla_V V \rangle = 0$ at all q, where the $X_c(q), c = 1 \ldots n_Q - m$, are linearly independent basis vectors of $\mathbb{R}^{n_Q}$ that are orthogonal [according to $M(q)$] to $\mathcal{Y}(q)$ at all q.

Kinematic reductions allow us to plan a path using a kinematic system, then "lift" this path to a trajectory for the full second-order system. One important issue is to understand the locally reachable configurations of the kinematic reductions. Beginning from an equilibrium state $[q^T, 0^T]^T$, we say that a second-order mechanical system (12.14) is *small-time locally kinematically controllable (STLKC)* from q if there exists a set of decoupling vector fields $V_1, \ldots, V_p$ such that $\overline{\mathrm{Lie}}(\{V_1, \ldots, V_p\})(q) = T_q Q$. This means that a kinematic system that can only move along these decoupling vector fields is STLC at q. If this holds for all $q \in Q$, the second-order system is *kinematically controllable*, meaning that it is equilibrium controllable by using only decoupling vector fields. These decoupling vector fields can therefore be used as primitives in a *kinematic* motion planner, reducing the complexity of the search for a feasible trajectory. A planned path would then consist of the concatenation of integral curves of the decoupling vector fields. Switches between decoupling vector fields must occur at zero velocity. Note that both STLC and STLKC imply STLEC, but no other implications hold generally.

The reader may easily verify that both the PBWT and the hopper are STLKC by two decoupling vector fields. For the PBWT, e.g., any configuration is reachable by concatenating integral curves of V_2, V_1, and V_2 (rotation, translation, and rotation).

12.4.3   Simple Mechanical Systems with Nonholonomic Constraints

The ideas discussed here also apply to second-order mechanical systems subject to nonholonomic constraints, such as rolling without slipping [78]. Such systems can be expressed as

$$(12.18) \qquad M(q)\ddot{q} + \dot{q}^T \Gamma(q)\dot{q} = T(q)u + A^T(q)\lambda$$

$$(12.19) \qquad\qquad A(q)\dot{q} = 0,$$

where $A(q)$ is a $k \times n_Q$ matrix describing the k Pfaffian constraints $A(q)\dot{q} = 0$ and therefore the distribution of feasible velocities $\mathcal{D}$, and $A^T(q)\lambda \in \mathbb{R}^{n_Q}$ is a set of constraint forces, where $\lambda \in \mathbb{R}^k$ is a vector of Lagrange multipliers. As described in chapter 10, we can eliminate λ from equation (12.18) using the matrix

$$P = \mathcal{I} - M^{-1}A^T(AM^{-1}A^T)^{-1}A$$

to project general motions to motions satisfying the nonholonomic constraints. We replace equation (12.18) with the $n_Q - k$ independent equations

$$(12.20) \qquad P(\ddot{q} + M^{-1}\dot{q}^T\Gamma\dot{q}) = PM^{-1}Tu.$$

In the notation of equation (12.13), we equivalently write

$$(12.21) \qquad P(q)\nabla_{\dot{q}}\dot{q} = \sum_{i=1}^{m} P(q)Y_i(q)u_i, \quad u \in \mathbb{R}^m.$$

For the constrained system of equations (12.19) and (12.20), we can define the *constrained affine connection* $\widetilde{\nabla}$, and rewrite (12.21) as

$$(12.22) \qquad \widetilde{\nabla}_{\dot{q}}\dot{q} = \sum_{i=1}^{m} \widetilde{Y}_i(q)u_i, \quad u \in \mathbb{R}^m,$$

where $\widetilde{Y}_i = PY_i$. For vector fields $\widetilde{Y}_1, \widetilde{Y}_2 \in \mathcal{D}$, the constrained affine connection $\widetilde{\nabla}$ is defined

$$\widetilde{\nabla}_{\widetilde{Y}_1}\widetilde{Y}_2 = P\nabla_{\widetilde{Y}_1}\widetilde{Y}_2.$$

Using the constrained affine connection $\widetilde{\nabla}$ and the constrained vector fields $\widetilde{\mathcal{Y}} = \{\widetilde{Y}_1, \ldots, \widetilde{Y}_m\}$ instead of ∇ and $\mathcal{Y}$, we can use the same simplified controllability tests and conditions for kinematic reductions that we used for unconstrained systems. Keep in mind that reachable velocities are confined to an $(n_Q - k)$-dimensional distribution $\mathcal{D}$ due to the k velocity constraints. Therefore, the most we can hope for is STLEC and $\overline{\mathrm{Sym}}(\widetilde{\mathcal{Y}}) = \mathcal{D}$.

Although we will not use these here, a formula for calculating the Christoffel symbols $\widetilde{\Gamma}$ of $\widetilde{\nabla}$ is given by Lewis [284], which also holds for vector fields not restricted to $\mathcal{D}$. Bullo and Zefran [80] give a computationally simpler formulation for vector fields restricted to $\mathcal{D}$. Their formulation uses an orthogonal basis of vector fields of the distribution $\mathcal{D}$ to construct modified Christoffel symbols. Choosing a basis of vector fields for the free motions of a nonholonomically constrained system is analogous to choosing a coordinate basis for the reachable configuration space of a holonomically constrained system, as discussed in chapter 10.

EXAMPLE 12.4.9   *In chapter 10 we considered the example of a knife-edge sliding on a plane. For this system, the inertia matrix $M(q)$ and the constraint matrix $A(q)$ are given by*

$$M(q) = \begin{bmatrix} m & 0 & 0 \\ 0 & m & 0 \\ 0 & 0 & I \end{bmatrix}, \quad A(q) = [\sin q_3, -\cos q_3, 0].$$

Assume that the available controls are u_1, a force in the q_1 direction, and u_3, a torque along q_3. Then

$$T(q) = \begin{bmatrix} 1 & 0 \\ 0 & 0 \\ 0 & 1 \end{bmatrix}, \quad M^{-1}(q)T(q) = \begin{bmatrix} \frac{1}{m} & 0 \\ 0 & 0 \\ 0 & \frac{1}{I} \end{bmatrix},$$

and $Y_1 = [1/m, 0, 0]^T$, $Y_2 = [0, 0, 1/I]^T$. Using the projection matrix P defined by A and M,

$$P = \begin{bmatrix} \cos^2 q_3 & \sin q_3 \cos q_3 & 0 \\ \sin q_3 \cos q_3 & \sin^2 q_3 & 0 \\ 0 & 0 & 1 \end{bmatrix},$$

we get

$$\widetilde{Y}_1 = PY_1 = \left[\frac{\cos^2 q_3}{m}, \frac{\sin q_3 \cos q_3}{m}, 0 \right]^T, \quad \widetilde{Y}_2 = PY_2 = \left[0, 0, \frac{1}{I} \right]^T,$$

$$\langle \widetilde{Y}_1 : \widetilde{Y}_1 \rangle = \langle \widetilde{Y}_2 : \widetilde{Y}_2 \rangle = 0,$$

$$\langle \widetilde{Y}_1 : \widetilde{Y}_2 \rangle = \left[-\frac{\sin q_3 \cos q_3}{mI}, -\frac{\sin^2 q_3}{mI}, 0 \right]^T = -\frac{\tan q_3}{I} \widetilde{Y}_1,$$

$$[\widetilde{Y}_1, \widetilde{Y}_2] = \left[\frac{\sin 2q_3}{mI}, -\frac{\cos 2q_3}{mI}, 0 \right]^T,$$

$$[\widetilde{Y}_2, [\widetilde{Y}_1, \widetilde{Y}_2]] = \left[2\frac{\cos 2q_3}{mI^2}, 2\frac{\sin 2q_3}{mI^2}, 0 \right]^T.$$

We see that $\overline{\mathrm{Sym}}(\{\widetilde{Y}_1, \widetilde{Y}_2\}) = \mathrm{span}(\{\widetilde{Y}_1, \widetilde{Y}_2\})$, *so the system is maximally reducible to a kinematic system by theorem 12.4.5. The control (velocity) vector fields of the kinematic reduction are* $\widetilde{Y}_1$, *motion along straight lines (when* $q_3 \neq \pm \pi/2$), *and* $\widetilde{Y}_2$, *spinning in place.*

The distribution and filtration defined by $\{\widetilde{Y}_1, \widetilde{Y}_2\}$ *are not regular, as* $\widetilde{Y}_1$ *vanishes at* $q_3 = \pm \pi/2$. *We also see that*

$$\det[\widetilde{Y}_1 \ \ \widetilde{Y}_2 \ \ [\widetilde{Y}_1, \widetilde{Y}_2]] = \frac{\cos^2 q_3}{m^2 I^2},$$

so we cannot conclude STLCA at all q until we construct $[\widetilde{Y}_2, [\widetilde{Y}_1, \widetilde{Y}_2]]$ *and see*

$$\det[\widetilde{Y}_2 \ \ [\widetilde{Y}_1, \widetilde{Y}_2] \ \ [\widetilde{Y}_2, [\widetilde{Y}_1, \widetilde{Y}_2]]] = \frac{2}{m^2 I^4},$$

so the system is STLCA and STLEC at all q by theorem 12.4.2.

12.5 Motion Planning

Motion planning for nonholonomic and underactuated systems has been the subject of a great deal of recent research, and the results could easily fill several books (see, e.g., the books edited by Li and Canny [288] and Laumond [266]). In this section we summarize a few useful approaches. The approaches can be classified by the type of robot to which they apply (e.g., the structure of the equations of motion, and with or without control constraints or drift) or the nature of the problem (with or without obstacles or cost function to be minimized). Motion-planning approaches with roots in control theory tend to apply to systems with particular structure and no obstacles, while approaches based on search algorithms are computationally intensive and are suited to finding collision-free trajectories among obstacles. Some approaches attempt to combine the benefits of control-theoretic and search-based methods.

The problem is to find a motion $(x(t), u(t))$, $t \in [0, t_f]$ satisfying the equations of motion (12.6) such that $x(0) = x_{\text{start}}$, $x(t_f) = x_{\text{goal}}$. In the presence of obstacles, where $\mathcal{Q}_{\text{free}}$ represents the free configuration space, we also require $q(t) \in \mathcal{Q}_{\text{free}}$, $t \in [0, t_f]$.

12.5.1 Optimal Control

For some simple underactuated systems, it is possible to solve analytically for optimal controls transferring the system from one state to another using the ideas developed in the previous chapter. Consider, e.g., a driftless system with $m = 2$ controls and

$n = 3$ states and the control vector fields $g_1 = [1, 0, x_2]^T$ and $g_2 = [0, 1, 0]^T$. Optimal control for such a system was first studied by Brockett [66]. The system can be written

(12.23)
$$\begin{bmatrix} \dot{x}_1 \\ \dot{x}_2 \\ \dot{x}_3 \end{bmatrix} = \begin{bmatrix} 1 & 0 \\ 0 & 1 \\ x_2 & 0 \end{bmatrix} \begin{bmatrix} u_1 \\ u_2 \end{bmatrix}.$$

Assume the time of motion is fixed, $t_f = 1$, and the objective is to minimize a measure of the control input energy:

(12.24)
$$J = \frac{1}{2} \int_0^{t_f} u^T u \, dt.$$

Then the Hamiltonian is written

$$\mathcal{H} = \frac{1}{2} \left(u_1^2 + u_2^2 \right) + \lambda_1 u_1 + \lambda_2 u_2 + \lambda_3 x_2 u_1.$$

Solving the necessary condition $\partial \mathcal{H} / \partial u = 0$, we get

(12.25) $\quad u_1 = -\lambda_1 - \lambda_3 x_2, \quad u_2 = -\lambda_2.$

The adjoint equation $\dot{\lambda} = -\partial \mathcal{H} / \partial x$ indicates that λ_1 and λ_3 are constant, and $\dot{\lambda}_2 = -\lambda_3 u_1$. Differentiating equation (12.25) with respect to time, we get

$$\dot{u}_1 = -\dot{\lambda}_1 - \lambda_3 x_2 - \lambda_3 \dot{x}_2 = -\lambda_3 u_2$$
$$\dot{u}_2 = -\dot{\lambda}_2 = \lambda_3 u_1.$$

These differential equations imply that optimal controls $u_1(t)$ and $u_2(t)$ are 90-degree out-of-phase sinusoids of the same amplitude and frequency, i.e.,

(12.26) $\quad u_1(t) = u_1(0) \cos(\lambda_3 t) - u_2(0) \sin(\lambda_3 t)$
(12.27) $\quad u_2(t) = u_1(0) \sin(\lambda_3 t) + u_2(0) \cos(\lambda_3 t).$

Given $x(0)$ and $x(1)$, the integrals of the equations of motion (12.23) define three equations to solve for λ_3, the frequency of the sinusoids, and the constants $u_1(0)$ and $u_2(0)$, defining the amplitude and phase.

These equations may be difficult to solve generally, but one simple case is of particular interest. We will choose controls so that at the end of the motion, x_1 and x_2 have returned to their initial values, while x_3 has changed from $x_3(0) = 0$ to $x_3(1) = x_{3,\text{goal}}$. In this case, $\lambda_3 = 2k\pi$, where k is any nonzero integer. This assures that

$$\int_0^1 u_1(t) dt = \int_0^1 u_2(t) dt = 0,$$

so x_1 and x_2 return to their initial values. Plugging $\lambda_3 = 2k\pi$ into the the controls [equations (12.26) and (12.27)], and putting the controls into the objective function (12.24), we find that

$$J = \frac{1}{2}(u_1(0)^2 + u_2(0)^2).$$

The cost of the motion is independent of the choice of k, so we choose $k = \pm 1$.

Integrating the equation of motion for x_3, we find that

$$x_3(1) = -\frac{u_1(0)^2 + u_2(0)^2}{4\pi}$$

for $k = 1$, and

$$x_3(1) = \frac{u_1(0)^2 + u_2(0)^2}{4\pi}$$

for $k = -1$. Therefore, if $x_{3,\text{goal}} > 0$, we choose $k = -1$ and any choice of $u_1(0)$ and $u_2(0)$ satisfying the condition $x_3(1) = x_{3,\text{goal}}$ (all choices have the same cost). If $x_{3,\text{goal}} < 0$, we choose $k = 1$ and proceed similarly. Notice that there is a one-dimensional set of solutions in the two-dimensional $(u_1(0), u_2(0))$ space, as we only need to satisfy the single equation $x_3(1) = x_{3,\text{goal}}$.

The motion described above suggests a strategy for motion planning for a more general class of systems. First, use the controls to drive m state variables directly to their goal values. Then perform motions that return these state variables to their goal values, but cause a desired net motion in the other state variables.

As an example, beginning from rest, a free-floating astronaut in space can control the orientation of his body by moving his arms in a cyclic pattern. At the end of a cycle, the *shape* (arm joint angles) of the astronaut is restored, but the orientation of his body has changed. (Keep in mind that the astronaut's total angular momentum is zero throughout the motion since there are no external forces.) We can decompose the astronaut's configuration into *shape* variables (also called *base* variables), describing the variables over which he has direct control (the arm joint angles), and *fiber* variables (the orientation of his body), which are coupled to the controls in a state-dependent manner.[3] In the example system (12.23), the shape variables are x_1 and x_2 and the fiber variable is x_3. In the single-leg hopper system of example 12.4.4, the shape variables are the leg angle q_2 and leg extension q_3, and the fiber variable is the body angle q_1 (see figure 12.20).

---

3. *Fiber controllability*, a weaker concept than complete controllability, concerns the controllability of the fiber variables without concern for the evolution of the shape variables. See [120, 233].

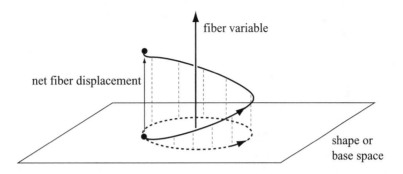

Figure 12.20 A closed loop in shape variables producing a net displacement in the fiber variable.

Many driftless systems can be transformed to a form similar to equation (12.23), allowing the control strategy of directly steering the shape variables to their goal configurations, and then performing closed loops (e.g., sinusoids) in the shape variables to achieve desired motions in the fiber variables. For example, the unicycle can be converted to this form by performing a coordinate transformation and an input feedback transformation. Define new coordinates $z = A(x)x$ and transformed controls $v = B(x)u$, where

$$A(x) = \begin{bmatrix} 0 & 0 & 1 \\ \cos x_3 & \sin x_3 & 0 \\ \sin x_3 & -\cos x_3 & 0 \end{bmatrix} \quad B(x) = \begin{bmatrix} 0 & 1 \\ 1 & x_2 \cos x_3 - x_1 \sin x_3 \end{bmatrix}.$$

Then the original unicycle system is transformed to an equivalent system of the form (12.23):

$$\begin{bmatrix} \dot{x}_1 \\ \dot{x}_2 \\ \dot{x}_3 \end{bmatrix} = \begin{bmatrix} \cos x_3 & 0 \\ \sin x_3 & 0 \\ 0 & 1 \end{bmatrix} \begin{bmatrix} u_1 \\ u_2 \end{bmatrix} \iff \begin{bmatrix} \dot{z}_1 \\ \dot{z}_2 \\ \dot{z}_3 \end{bmatrix} = \begin{bmatrix} 1 & 0 \\ 0 & 1 \\ z_2 & 0 \end{bmatrix} \begin{bmatrix} v_1 \\ v_2 \end{bmatrix}$$

A system like this is an example of a broader class of *chained-form* systems, which are the topic of the following subsection.

The motion strategy of driving the shape variables to their goal values and then performing closed loops generally results in suboptimal motions, but it is rarely possible to solve the optimality conditions analytically. In any case, the quadratic "energy-like" objective function (12.24) may not have much physical meaning for an input-transformed system such as the unicycle above. Section 12.5.3 discusses numerical methods for finding approximately optimal motion plans.

12.5.2   Steering Chained-Form Systems Using Sinusoids

Consider the following system with $m = 2$ controls and $n \geq 3$ states, generalizing the system with $n = 3$ described above:

$$(12.28) \quad \dot{x} = u_1 g_1(x) + u_2 g_2(x), \quad g_1(x) = \begin{bmatrix} 1 \\ 0 \\ x_2 \\ x_3 \\ \vdots \\ x_{n-1} \end{bmatrix}, \quad g_2(x) = \begin{bmatrix} 0 \\ 1 \\ 0 \\ 0 \\ \vdots \\ 0 \end{bmatrix}$$

Such a system is said to be in *chained form*. Considering a Lie product of the form $[g_1, [g_1, \ldots [g_1, [g_1, g_2]]]]$, where g_1 appears k times, we find that the Lie product has a value $(-1)^k$ in the $k + 2$ component of the vector field, and zeros in all other components. Therefore, the Lie algebra $\overline{\text{Lie}}(\{g_1, g_2\})$ is full rank, and the system is STLC at all x for $\mathcal{U} \in \mathcal{U}_{\pm}$.

To steer such a system, we can generalize the approach presented previously. First, drive x_1 and x_2 to their final values. Then choose controls $u_1(t)$ and $u_2(t)$ to be sinusoids at integrally related frequencies. For example, let

$$(12.29) \quad u_1(t) = a \sin 2\pi t$$

$$(12.30) \quad u_2(t) = b \cos 2k \, \pi t,$$

where k is a positive integer. Then $\dot{x}_3$ has components at frequency $2\pi(k - 1)$, $\dot{x}_4$ has components at $2\pi(k - 2)$, etc. Applying the controls for $t_f = 1$ will return the $x_1, \ldots, x_{k+1}$ variables to their initial values—the nonzero frequency means that $\dot{x}_1, \ldots, \dot{x}_{k+1}$ integrate to zero net change over the cycle. For the variables $x_{k+2}, \ldots, x_n$, however, the periodic controls will result in a nonzero DC (zero frequency) component in their time-derivatives, meaning that they will be changed over the cycle. The net change to x_{k+2} can be computed to be

$$(12.31) \quad x_{k+2}(1) - x_{k+2}(0) = \left(\frac{a}{4\pi} \right)^k \frac{b}{k!}.$$

Therefore, an algorithm to drive the system to the goal state is the following:

1. Drive the variables x_1 and x_2 directly to their goal values.

2. For each $k = 1 \ldots n - 2$, in ascending order, apply the controls (12.29) and (12.30) for time $t_f = 1$ with a and b selected according to equation (12.31) to drive x_{k+2} to its desired value, leaving all $x_1, \ldots, x_{k+1}$ unchanged.

A number of n-state 2-input systems can be transformed to the chained form (12.28), including tractor-trailer systems with multiple trailers [393]. There are also other forms similar to the chained form presented here, including forms with more than two inputs; see, e.g., [85, 330, 332, 419].

Many motion-planning algorithms based on series expansions and averaging of the equations of motion use sinusoidal inputs at appropriately chosen frequencies and phases to "tickle" the system in desired Lie bracket directions (see, e.g., [34, 50, 74–76, 175, 279, 280, 327, 328, 333, 342, 422, 423]). This idea also applies to mechanical systems with drift. For a locomotion system where motions of the shape variables induce motion of the fiber variables, these sinusoidal inputs generate "gaits" for the system.

12.5.3   Nonlinear Optimization

The previous chapter outlined a method for using gradient-based nonlinear optimization to find locally optimal trajectories for fully actuated dynamic systems. A similar approach can be applied to underactuated systems [147, 148, 343]. We typically choose a finite parameterization of the control history $u(t)$, since any a priori parameterization of the trajectory $q(t)$ will likely describe trajectories that are infeasible for the system due to the underactuation constraints. Alternatively, we could solve for $q(t)$ and enforce equality constraints at time instants throughout the trajectory to ensure that there exists a feasible $u(t)$ for the given $q(t)$.

Two features of this approach are (1) it is very general—motion planning problems for many underactuated systems, including those with drift, can be encoded as nonlinear programs; and (2) the ability to minimize an objective function may result in motions that are "efficient" in some way. Significant drawbacks, however, are as outlined in the previous chapter: (1) a good initial guess must be provided to the solver, as the solver will find only a locally optimal solution; and (2) numerical difficulties, singularities, and nonconvexity may prevent the solver from converging to a solution. The generality of the approach means that it uses little information about the particular structure of the system to ensure convergence.

For convex systems, systems with particular structure [314], or particular choices of the control parameterization, it may be possible to demonstrate favorable convergence properties for nonlinear optimization. In general, however, there are no guarantees that nonlinear optimization will be able to find any solution, let alone a good solution, to a particular problem.

12.5.4   Gradient Methods for Driftless Systems

To improve the convergence properties of gradient-based motion planning, we focus on the class of driftless systems and give up on finding optimal motions. Let $u^p = [u_1^p, \ldots, u_r^p]^T \in \mathbb{R}^r$ be a finite parameterization of the control history $u(t)$, e.g., the coefficients of truncated Fourier series for the control inputs. Let the time of motion be $t_f = 1$, and define an end-state map f that maps the initial state x_{start} and the control u^p to a final state of the system x_f:

$$x_f = f(x_{\text{start}}, u^p)$$

The end-state map f is typically obtained numerically.

Now the problem is to find a u^p so that the desired goal state x_{goal} is reached. Define the end-state error vector $e = [e_1, \ldots, e_n]^T$ to be

$$e = f(x_{\text{start}}, u^p) - x_{\text{goal}}.$$

We would like to know the direction to change u^p to move x_f in the direction $-e$, to reduce the error. This direction in the u^p space is $v \in \mathbb{R}^r$, where

$$v^T = -e^T \left[\frac{\partial f(x_{\text{start}}, u^p)}{\partial u^p} \right] = -[e_1, \ldots, e_n] \begin{bmatrix} \frac{\partial f_1}{\partial u_1^p} & \cdots & \frac{\partial f_1}{\partial u_r^p} \\ \vdots & & \vdots \\ \frac{\partial f_n}{\partial u_1^p} & \cdots & \frac{\partial f_n}{\partial u_r^p} \end{bmatrix},$$

where the partial derivatives are evaluated at the current guess for u^p. Given a current guess $u^p(i)$, we can update it as follows:

$$u^p(i+1) = u^p(i) + \alpha v(i),$$

where α is a small positive constant, perhaps chosen by a line search to maximally decrease the error. We then calculate the new vector $v(i+1)$ for $u^p(i+1)$ and iterate until we reach an iteration k such that $\| f(x_{\text{start}}, u^p(k)) - x_{\text{goal}} \| < \epsilon$ for a small constant ϵ.

This algorithm is guaranteed to converge to a solution for a sufficiently small α if there are no state or control constraints and if $\partial f / \partial u^p$ is rank n everywhere. The rank condition means that any point in a sufficiently small neighborhood of $x_f = f(x_{\text{start}}, u^p)$ is reachable by a small change to u^p, indicating that it is possible to move the error vector e in any direction. Generically, if the system is STLC everywhere and we have a rich enough control parameterization u^p, the matrix $\partial f / \partial u^p$ will have rank n (figure 12.21).

If $\partial f / \partial u^p$ loses rank at $u^p(i)$, then there are one or more directions in which we cannot move the error vector e. Such a $u^p(i)$ is a singular control and could cause the algorithm to get stuck. In this case, we add a control to the end of the motion, where

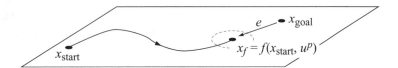

Figure 12.21   The initial state x_{start}, the state x_f reached after applying the control u^p, and the end-state error vector e. The full-rank condition on $\partial f / \partial u^p$ ensures that a neighborhood of x_f can be reached by a small variation in the control vector u^p.

the control is chosen randomly from the set of controls that result in no net motion of the system. Such a control is easy to generate because the system is driftless. Because such a control is generically nonsingular, it is said to generate a *generic loop* for the system. If we append the random control to the control u^p and treat the entire thing as the new control, the control vector is no longer singular, and the algorithm can continue.

There are many possible variants of this basic approach [129, 130, 391, 392, 400, 425]. For example, we could define a path on the error space from $e(0)$ to the origin, then choose the iterates $u^p(i)$ to force the error to track this path. We could use a metric on the error coordinates other than the standard metric implicit in the approach above. We could also perform a more sophisticated search, perhaps using the Hessian $\partial^2 f / (\partial u^p)^2$, to achieve faster convergence. Certain obstacle constraints can also be incorporated. This motion-planning method has been applied to find paths for a truck pulling trailers [129, 130, 425].

12.5.5   Differentially Flat Systems

Differentially flat systems [152, 153, 309] have a structure that makes motion planning (in the absence of control and configuration constraints such as obstacles) particularly simple. For a differentially flat system with $x \in \mathbb{R}^n$ and $u \in \mathbb{R}^m$, there exists a set of m functions y_i, $i = 1 \ldots m$, of the state, the control, and its derivatives,

$$y_i\left(x, u, \dot{u}, \ldots, u^{(r)}\right), \quad i = 1 \ldots m,$$

such that the states and control inputs can be expressed as functions of y and its time-derivatives:

$$x = \phi\left(y, \dot{y}, \ldots, y^{(p)}\right)$$
$$u = \psi\left(y, \dot{y}, \ldots, y^{(p)}\right)$$

The functions y_i are known as the *flat outputs*. Armed with a set of flat outputs, the problem of finding a feasible trajectory $(x(t), u(t))$, $x(0) = x_{start}$, $x(t_f) = x_{goal}$, $t \in [0, t_f]$ for the underactuated system is transformed to the problem of finding a curve

$y(t)$ satisfying constraints on $y(0)$, $\dot{y}(0)$, ..., $y^{(p)}(0)$ and $y(t_f)$, $\dot{y}(t_f)$, ..., $y^{(p)}(t_f)$ specified by x_{start} and x_{goal}. In other words, the problem of finding a trajectory satisfying the underactuation constraints becomes the relatively simple algebraic problem of finding a curve to fit the start and end constraints on y. Any curve $y(t)$ maps directly to a consistent pair of state and control histories $x(t)$ and $u(t)$.

The flat outputs for mechanical systems are often a function of configuration variables only, and sometimes are just the location of particular points on the system. Unfortunately, there is no systematic way to determine if a system is differentially flat, or what the flat outputs for a system are. Many important systems have been shown to be differentially flat, however. These systems include the unicycle, the PBWT, and chained-form systems.

EXAMPLE 12.5.1   Unicycle (cont.) *The flat outputs for the unicycle are $y_1 = x_1$ and $y_2 = x_2$. The state x and controls u can be derived from the flat outputs and their derivatives as follows:*

$$(12.32) \qquad [x_1, x_2, x_3]^T = \left[y_1, \, y_2, \, \tan^{-1} \frac{\dot{y}_2}{\dot{y}_1} \right]^T$$

$$(12.33) \qquad [u_1, u_2]^T = \left[\pm\sqrt{\dot{y}_1^2 + \dot{y}_2^2}, \; \frac{\dot{y}_1 \ddot{y}_2 - \ddot{y}_1 \dot{y}_2}{\dot{y}_1^2 + \dot{y}_2^2} \right]^T .$$

The orientation x_3 and the turning control u_2 are not well defined as a function of the flat outputs when the unicycle is not translating. Also, because x_3 will be 45 degrees for both $\dot{y}_1 = \dot{y}_2 = 1$ and $\dot{y}_1 = \dot{y}_2 = -1$, the sign of the forward-backward speed u_1 should be consistent with the angle x_3.

Now we would like to find a feasible trajectory from $x_{\text{start}} = [0, 0, 0]^T$ to $x_{\text{goal}} = [1, 1, 0]^T$. Since there are six state variables in the specification of the start and goal points, there are six constraints on the flat outputs y and their derivatives at the beginning and end of motion. These constraints can be written

$$
\begin{aligned}
y_1(0) &= 0 & y_2(0) &= 0 \\
y_1(t_f) &= 1 & y_2(t_f) &= 1 \\
& & \dot{y}_2(0) &= 0 \\
& & \dot{y}_2(t_f) &= 0,
\end{aligned}
$$

where the last two constraints indicate that the initial and final motion of the unicycle must be along the x-axis, indicating that the wheel is oriented with the x-axis. The simplest polynomial functions of time that have enough free coefficients to satisfy these constraints are

$$y_1(t) = a_0 + a_1 t$$
$$y_2(t) = b_0 + b_1 t + b_2 t^2 + b_3 t^3 .$$

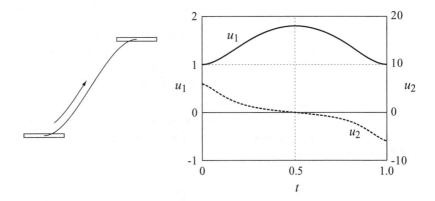

Figure 12.22   A feasible path for the unicycle from $[0, 0, 0]^T$ to $[1, 1, 0]^T$ and the controls.

Setting the time of motion $t_f = 1$ and using the constraints to solve for the polynomial coefficients, we get

$$y_1(t) = t$$
$$y_2(t) = 3t^2 - 2t^3.$$

The state $x(t)$ and control $u(t)$ can be obtained from equations (12.32) and (12.33). The unicycle motion is shown in figure 12.22.

In fitting a curve $y(t)$, we must choose a family of curves with enough degrees of freedom to satisfy the initial and terminal constraints. We may choose a family of curves with more degrees of freedom, however, and use the extra degrees of freedom to, individually or severally, (1) satisfy bounds on the control $u(t)$, (2) avoid obstacles in the configuration space, or (3) minimize a cost function. Incorporating these conditions in the calculation of $y(t)$ typically requires resorting to numerical optimization methods, and is a topic of current research. A good way to generate an initial guess for the optimization is to solve exactly for a minimal number of coefficients to satisfy the initial and terminal constraints, setting the other coefficients to zero.

For the PBWT, the flat outputs are

$$y_1 = x_1 + \frac{1}{d} \cos x_3$$
$$y_2 = x_2 + \frac{1}{d} \sin x_3.$$

The flat outputs (y_1, y_2) define a point fixed to the PBWT, at the PBWT's center of percussion with respect to the location of the thrusters.

For a car with steerable front wheels and parallel, fixed-orientation rear wheels, the flat outputs are the Cartesian coordinates of the point halfway between the rear wheels. If the car is towing a two-wheel trailer hitched midway between the rear wheels of the car, the flat outputs are the coordinates midway between the wheels of the trailer. If there are more trailers, all hitched midway between the wheels of the trailer in front, the coordinates of the midpoint of the wheels of the last trailer are flat outputs for the entire system. The orientation of the car and each trailer can be determined from sufficiently high time-derivatives of the evolution of these two coordinates.

The paper by Martin et al. [309] provides a catalog of systems known to be flat. Some notable results include the following. A system of the form (12.6) with n states and $m = n - 1$ inputs is flat if it is STLC. A driftless system (12.6) with n states and $m = n - 2$ inputs is flat if it is STLC. All chained-form systems are flat with the first and last states x_1 and x_n as the flat outputs. Other example flat systems are given in [331].

12.5.6   Cars and Cars Pulling Trailers

From a motion-planning perspective, easily the most heavily studied examples of nonholonomic systems are the kinematic car and the car pulling one or more trailers. Because of the obvious applications of automatic motion planning to systems like these, a great deal of effort has been spent in deriving efficient and complete motion planners for these systems moving in cluttered environments. The excellent book edited by Laumond [266] is focused entirely on these systems. In this subsection, we provide a brief review of important concepts and approaches to motion planning for cars and cars pulling trailers.

Cars

We focus on driftless kinematic models of cars, where the inputs are velocities and the state x is simply the configuration q. Alternatively, we could consider dynamic extensions of these models, where the inputs are accelerations.

A kinematic model of a standard car is shown in figure 12.23. The location of the point midway between the rear wheels is (x_1, x_2), the steering angle is x_3, and the orientation of the car is x_4. To ensure that the wheels do not slip, each of the front wheels must be perpendicular to the line through the wheel and the rotation center of the car. Therefore, x_3 is measured at a "virtual" wheel midway between the two front wheels. The wheelbase is L.

If the control u_2 is the rate of change of the steering angle x_3 and u_1 is the driving speed of the car, measured at the point midway between the rear wheels, the control

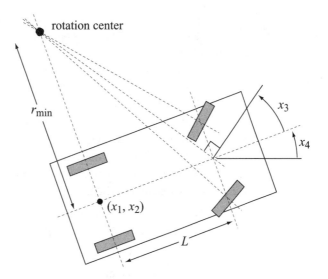

Figure 12.23 A model of a car turning at its minimum turning radius.

system can be written

$$(12.34) \quad \dot{x} = g_1(x)u_1 + g_2(x)u_2, \quad g_1(x) = \begin{bmatrix} \cos x_4 \\ \sin x_4 \\ 0 \\ \frac{1}{L}\tan x_3 \end{bmatrix}, \quad g_2(x) = \begin{bmatrix} 0 \\ 0 \\ 1 \\ 0 \end{bmatrix}.$$

Typically the steering angle is limited to $-\gamma < x_3 < \gamma$, $\gamma \in (0, \pi/2)$, giving the car a minimum turning radius $r_{\min} = L/\tan^{-1}\gamma$. The control set for this system is $\mathcal{U} \in \mathcal{U}_\pm$, and this symmetric system can be shown to be STLC by the LARC. Since the car is STLC at all configurations, the steering angle limits do not affect the reachable positions and orientations of the car. Interestingly, this means that a car that cannot turn right, for instance, can reach any position and orientation among obstacles that a fully functional car can.

Since we are primarily concerned with the position and orientation of the body of the car, we could decide to eliminate the steering angle x_3 from the representation of the configuration and treat the steering angle as part of the control. Consider the modified control inputs (v, ω), where v is the linear speed of the car and ω is the angular velocity of the body of the car. In this case, the control system becomes

$$(12.35) \quad \begin{bmatrix} \dot{x}_1 \\ \dot{x}_2 \\ \dot{x}_4 \end{bmatrix} = \begin{bmatrix} \cos x_4 \\ \sin x_4 \\ 0 \end{bmatrix} v + \begin{bmatrix} 0 \\ 0 \\ 1 \end{bmatrix} \omega,$$

which is identical to the unicycle, except that the control set satisfies the turning radius constraint $|\omega| < |v/r_{\min}|$. This does not affect the symmetry of the system, however, so it is still STLC.

As with the unicycle, if the car is limited to driving forward only, then it is globally controllable (in the absence of obstacles) but not STLC. In this section we will focus only on STLC models of cars.

Another car model of interest is the *differential-drive* car. In this case, the front wheels are replaced by casters that roll freely in any direction. The rear wheels are parallel and their speeds are controlled independently. If the speeds of the two rear wheels are u_1 and u_2, and the configuration of the car is $[x_1, x_2, x_4]^T$ as in figure 12.23, the control system can be written

$$\begin{bmatrix} \dot{x}_1 \\ \dot{x}_2 \\ \dot{x}_4 \end{bmatrix} = \begin{bmatrix} \frac{1}{2}\cos x_4 \\ \frac{1}{2}\sin x_4 \\ \frac{1}{L} \end{bmatrix} u_1 + \begin{bmatrix} \frac{1}{2}\cos x_4 \\ \frac{1}{2}\sin x_4 \\ \frac{-1}{L} \end{bmatrix} u_2.$$

With the input transformation $v = (u_1 + u_2)/2$ and $\omega = (u_1 - u_2)/L$, the system again becomes the unicycle. The major difference from the standard car model is the lack of a turning radius constraint.

Small-time local controllability for these car models implies the following important consequence: if there is a free path from q_{start} to q_{goal} for the car body moving *without any nonholonomic motion constraint,* i.e., moving as a free-flying planar body, and if there is an open set of free space about each configuration in the path, then there is a free path for the car with the motion constraint. Stated equivalently, if $\mathcal{Q}_{\text{free}}$ is connected and open so that every $q \in \mathcal{Q}_{\text{free}}$ has a neighborhood of free space, then there exists a path for the car from any $q_{\text{start}} \in \mathcal{Q}_{\text{free}}$ to any $q_{\text{goal}} \in \mathcal{Q}_{\text{free}}$. This implies that it is possible to parallel-park your car into any parking space $\epsilon > 0$ longer than your car. However, the number of direction reversals in your parking maneuver grows proportionally to $1/\epsilon^2$ [266], so you could be there a while if ϵ is small!

Let's turn our focus to the simplified standard car model [equation (12.35)] with a limited turning radius. One question that has received considerable attention is the following: Given a start and goal configuration for the car moving in an obstacle-free space, find the path that minimizes the arclength traveled by the point midway between the rear wheels. If we assume a bound on the linear velocity $|v| < v_{max}$, and no acceleration limits, then this path also corresponds to the minimum-time motion. This problem has been solved owing in large part to contributions by Reeds and Shepp [360], Sussmann and Tang [402], and Souères and Laumond [395]. See also the chapter by Souères and Boissonnat [394].

Reeds and Shepp [360] showed that the optimal path must be one of a discrete and computable set of paths. It turns out that each member of this set consists of a

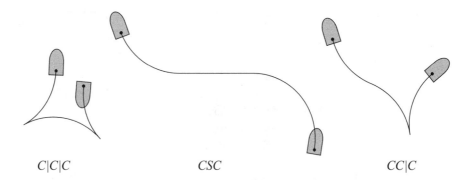

$C|C|C$ CSC $CC|C$

Figure 12.24   Three Reeds-Shepp curves.

concatenation of straight-line segments and circular arcs at the car's minimum turning radius. If C indicates a circular arc segment, S indicates a straight-line segment, and | indicates a cusp in the motion where the linear velocity v changes sign, the optimal path is guaranteed to be contained in the following set of path types:

$$\{C\,|\,C\,|\,C, \quad CC\,|\,C, \quad C\,|\,CC, \quad CC_a|C_aC, \quad C\,|\,C_aC_a\,|\,C,$$
$$C\,|\,C_{\pi/2}SC, \quad CSC_{\pi/2}\,|\,C, \quad C\,|\,C_{\pi/2}SC_{\pi/2}\,|\,C, \quad CSC\}$$

The subscript a indicates an arc of angle a. One or more of the segments may be zero length. Figure 12.24 illustrates three Reeds-Shepp curves.

In the absence of obstacles, we can simply look up the optimal path from the set above using a map indexed by the goal configuration relative to the initial configuration [394, 395]. Shortest paths may not be unique. Analogous results for time-optimal motions of a differential-drive car with wheel velocity limits were derived by Balkcom and Mason [35].

The following motion-planning methods apply to the case of carlike robots in the presence of obstacles.

Grid Search

Barraquand and Latombe [41] developed a simple planner for cars moving in a bounded (typically rectangular) subset of the plane. Define six actions for the car to be $L_{\pm}$, $R_{\pm}$, and $S_{\pm}$, for the steering wheels turned all the way to the left, turned all the way to the right, or pointed straight ahead, with the subscripts "+" and "−" indicating forward and reverse velocity, respectively. Pseudocode is given in algorithm 22, CAR_GRID_SEARCH.

Algorithm 22 CAR_GRID_SEARCH

Input: Start configuration q_{start}, goal region $\mathcal{G}(q_{goal})$

Output: A path from q_{start} to $\mathcal{G}(q_{goal})$ or FAILURE

1: Initialize search tree T and list *OPEN* with start configuration q_{start}
2: **while** *OPEN* not empty and $size(T) < MAXTREESIZE$ **do**
3:    $q \leftarrow$ first in *OPEN*, remove from *OPEN*
4:    **if** $q \in \mathcal{G}(q_{goal})$ **then**
5:       Report SUCCESS, return path
6:    **end if**
7:    **if** q is not near a previously occupied configuration **then**
8:       Mark q occupied
9:       **for all** actions in $\{L_{\pm}, R_{\pm}, S_{\pm}\}$ **do**
10:          Integrate forward a fixed time to q_{new}
11:          **if** path to q_{new} is collision-free **then**
12:             Make q_{new} a successor to q in T
13:             Compute *cost* of path to new configuration q_{new}
14:             Place q_{new} in *OPEN*, sorted by *cost*
15:          **end if**
16:       **end for**
17:    **end if**
18: **end while**
19: Report FAILURE

Conceptually, the planner keeps a tree T of configurations reached in the search and a list *OPEN* of pointers to configurations in T whose successors have not yet been generated. The pointers in the list *OPEN* are sorted by the costs of the paths to the associated configurations. The planner begins by making the car's initial configuration q_{start} the root of the tree T and initializing the list *OPEN* with a pointer to this configuration. The main loop of the planner is a simple best-first search. The planner sets the current configuration to that indicated by the minimum-cost pointer in *OPEN,* and it removes this pointer from *OPEN*. Subsequent configurations are generated by integrating each motion forward a short time, and each new collision-free configuration is added to the tree T with a record of the motion taking it there as well as a pointer to the previous configuration. Pointers to these new configurations are then inserted into the sorted list *OPEN*. This continues until one of the termination conditions is satisfied: (1) the list *OPEN* becomes empty, or the number of nodes in the tree T exceeds some user-specified value, indicating failure of the search; or (2) the planner reaches a configuration in a user-specified neighborhood $\mathcal{G}(q_{goal})$ of the goal configuration q_{goal}. Note that the planner is not exact, as it only finds a path to a goal neighborhood.

The cost of a path is a function of the number of motion steps, the number of changes in the steering direction, and the number of cusps (switches between forward and reverse linear velocity). For example, for positive weighting factors a, b, and c, the cost could be a times the number of motion steps plus b times the number of steering changes plus c times the number of reversals. For $b = c = 0$, the planner will find short paths, and for $a = b = 0$, the planner will minimize the number of cusps (see figure 12.25).

By choosing the weighting factors a, b, c to be non-negative integers, inserting pointers into the sorted list *OPEN* can be done in constant time. This is accomplished by defining a one-dimensional array with cost as the index. The configurations of cost C, then, are stored in a linked list pointed at by element C of the array. To insert a configuration of cost C into *OPEN,* we simply append it to the end of the linked list pointed at by element C of the array.

The planner discards paths that are not collision-free. For a polygonal car and obstacles, collision detection can be done exactly by recognizing that all points on

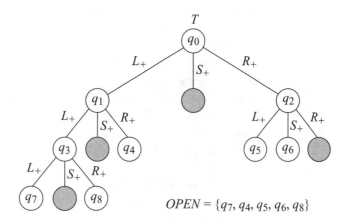

Figure 12.25   An example search tree T and list *OPEN*. (To keep the tree size reasonable, this search uses only the actions L_+, S_+, R_+.) If the path to a configuration is not free, the node is marked gray. In this example, the cost of each motion is 1, and the cost of changing steering direction is 5, to penalize excessive steering. This explains the ordering of the nodes in the sorted list *OPEN*. The cost of the path to q_7 is 3, the cost of paths to q_4, q_5, and q_6 is 7, and the cost of the path to q_8 is 8. Therefore, the planner will next remove q_7 from *OPEN* and generate its children in T. The configuration q_8 lies in the goal region, but the planner will not check this until q_8 pops off the front of the list *OPEN*. This is to ensure finding the minimum-cost path.

the car move in circular arcs or straight lines. Obstacle intersection then becomes a problem of intersecting arcs or line segments with line segments. A simpler approach is to surround the car by a disk and to only check for collision of the disk at the end of a motion step. The disk should be chosen large enough to guarantee that collision detection is *conservative*—only feasible plans will be found, but feasible paths through tight spaces could be rejected.

The planner also discards any configuration that is sufficiently close to a configuration from which the children have already been generated. Two configurations are considered sufficiently close if they occupy the same cell of a predefined grid on the configuration space. The car is assumed to be confined to a rectangular region of the plane, so $q \in (0, x_{max}) \times (0, y_{max}) \times [0, 2\pi)$, and the configuration space grid contains d^3 boxes, where d is the number of partitions of each dimension of the configuration space.

The user must specify the parameters defining the size of the goal neighborhood $\mathcal{G}(q_{\text{goal}})$, the integration time of the control steps, and the resolution of the configuration space grid used to check for prior occupancy. These parameters are interdependent. The resolution of the grid should be sufficiently fine that the application of any control step will always move the configuration to a new grid cell, and the goal neighborhood should be large enough that the car will not easily jump over it. In practice, the user should decide how much configuration error is acceptable along each dimension at the goal configuration, choose d so that each grid box is no larger (and usually somewhat smaller) than the goal region, and then choose the control step to be just long enough to guarantee that the car will exit its current grid box.

This planner is resolution complete, meaning that if the step size is sufficiently small, the planner will find a path if one exists. Because the planner uses a best-first search, the path found will be optimal for the user's cost function and the given step size. This planner actually runs faster in cluttered spaces because the obstacles prune the search tree.

One important property of the approach is that any path with p cusps (linear speed reversals) using the full motion capabilities of the car can be followed arbitrarily closely by a path with only p cusps using only motions with the steering wheel turned all the way to the left or right. This means that the discretization of the car's possible motion directions does not preclude the possibility of finding minimum-cusp motions.

A drawback to this planner is that it is not exact; paths found by the planner go to a neighborhood of the goal, not exactly to the goal. It can also be slow in open spaces due to the exponential growth of the search tree T. The maximum number of configurations that will be marked "occupied" is upper-bounded by the d^3 boxes of the occupancy grid, however, which may not be too large considering the system has only three degrees of freedom.

Examples of planner output are shown in figures 12.26 and 12.27.

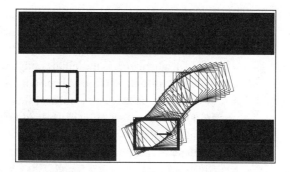

Figure 12.26 The classic parallel-parking problem.

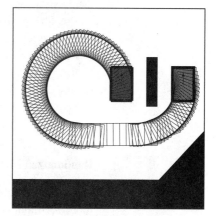

Figure 12.27 The same planning problem solved with two different cost functions. On the left, $a = 1$ and $b = c = 0$, meaning that the planner looks for the shortest path. On the right, $a = 1$, $b = 0$, and $c = 300$, meaning that the planner heavily penalizes cusps.

Omnidirectional to Nonholonomic Path Transformation

The following approach to motion planning for a car was proposed in [267]:

1. Use a path planner to find a path for a car with no motion constraints (i.e., a free-flying body).

2. Transform the path into a path satisfying the nonholonomic constraint.

3. Optimize the path.

Because the car is STLC, the path transformation in the second step is always possible if the path found in the first step does not touch any obstacles.

Step 2 proceeds as follows. Parameterize the original path returned in the first step by $s \in [0, 1]$, where $q(0)$ is the initial configuration and $q(1)$ is the final configuration. Using the lookup table of Reeds-Shepp curves, find the shortest path connecting $q(0)$ and $q(1)$. If this path is collision-free, then we have found the shortest path, and we are done. If it is not collision-free, divide the s interval $[0, 1]$ into two equal pieces, and calculate the shortest paths between $q(0)$ and $q(\frac{1}{2})$, and between $q(\frac{1}{2})$ and $q(1)$. If either of these paths is in collision, subdivide that interval again, and continue recursively until a free path is found. This procedure is guaranteed to terminate if the path found in the first step touches no obstacles. This guarantee relies on a topological property that says for any open ball $B_\delta(q)$ of radius $\delta > 0$ about a configuration q, there exists another ball $B_\epsilon(q)$ such that for any $q' \in B_\epsilon(q)$, the local path planner (Reeds-Shepp curves in this case) finds a path between q and q' that is contained in $B_\delta(q)$ (figure 12.28).

We now have a feasible path for the car, but it may be unnecessarily long. In the final step, we choose two randomly selected points along the path and replace the path in between by the shortest Reeds-Shepp path, if it is collision-free. We iterate this procedure, stopping when it fails to shorten the path a prespecified number of times in a row.

Randomized Methods

Probabilistic roadmap methods, as discussed in chapter 7, can also be applied to carlike robots. All that is required is the specification of a local planner that quickly finds a path connecting two configurations in the absence of obstacles. For example, the local planner could use Reeds-Shepp curves as described in [394, 395] to quickly find a shortest path. Two configurations are connected in the "roadmap" representation of

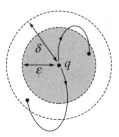

Figure 12.28   For a given open ball $B_\delta(q)$, there exists an open ball $B_\epsilon(q)$ such that paths found by the local planner from q to any point in $B_\epsilon(q)$ are completely contained in $B_\delta(q)$.

the free space if the path returned by the local planner is collision-free. This approach is probabilistically complete—the probability of finding a solution, if one exists, approaches 100% as the running time goes to infinity.

Smooth Paths

A drawback to using Reeds-Shepp curves, and in using the CAR_GRID_SEARCH planner (algorithm 22), is that the curvature of the paths is discontinuous at the transitions between straight and curved segments, even where there is no cusp. This means that either the steering angle must change instantaneously, the car must come to a stop at the transitions, or we must be willing to accept the error in execution that comes from ignoring this problem. To overcome this problem, several approaches have been proposed to planning paths with continuous steering angles. For example, a postprocessing step could be used to smooth the transitions (e.g., see [151]). A problem with this approach is that the new, smoothed path might collide with obstacles. Instead, smooth primitives can be used directly in the local planner, perhaps based on the car's differential flatness properties [159, 258, 351, 370].

Cars Pulling Trailers

Figure 12.29 shows a car pulling two trailers, with each trailer hitched between the rear wheels of the body in front. For a car pulling p trailers, the configuration of the system is written $x = [x_1, \ldots, x_{p+4}]^T$, where $x_{i+4} = \theta_i$ gives the orientation of the ith trailer relative to the body in front. The controls for this system are still u_1, the forward speed of the car, and u_2, the rate of change of the steering angle.

To derive the equations of motion for this system, let's start by looking at a single trailer (see figure 12.30). The trailer is being pulled with a linear velocity w at an

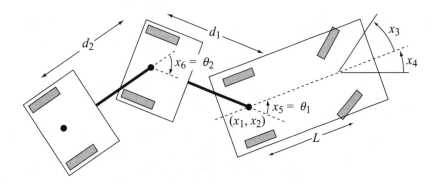

Figure 12.29 A car pulling two trailers.

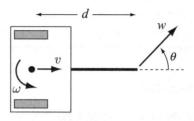

Figure 12.30   A single trailer.

angle θ at the hitch a distance d from the trailer wheels. The trailer's linear velocity is v and the angular velocity is ω. The trailer velocity (v, ω) is related to the pulling velocity w at the hitch by

(12.36) $v = w \cos \theta$

(12.37) $\omega = \dfrac{w}{d} \sin \theta.$

The linear velocity v becomes the hitch velocity w for the trailer behind this trailer, and equation (12.36) shows that the linear velocity is nonincreasing as we move back in the trailer chain.

Extending the reasoning above, the car and trailer system can be written

$$\dot{x} = g_1(x)u_1 + g_2(x)u_2,$$

where

(12.38) $$g_1(x) = \begin{bmatrix} \cos x_4 \\ \sin x_4 \\ 0 \\ \frac{1}{L}\tan x_3 \\ \frac{1}{d_1}\sin\theta_1 \\ \frac{1}{d_2}\sin\theta_2\cos\theta_1 \\ \frac{1}{d_3}\sin\theta_3\cos\theta_1\cos\theta_2 \\ \vdots \\ \frac{1}{d_p}\sin\theta_p\prod_{i=1}^{p-1}\cos\theta_i \end{bmatrix}, \quad g_2 = \begin{bmatrix} 0 \\ 0 \\ 1 \\ 0 \\ 0 \\ 0 \\ 0 \\ \vdots \\ 0 \end{bmatrix}.$$

Constructing Lie brackets of g_1 and g_2 shows that the tractor-trailer system is STLC at any x for a symmetric control set [265].

To plan motions for a tractor-trailer system among obstacles, we could extend the grid search approach of Barraquand and Latombe [41]. For example, if the car is pulling a single trailer, we can add a dimension to the configuration space grid and proceed as before with the six motion primitives $L_\pm$, $S_\pm$, and $R_\pm$. The only difference is that the equations of motion must be numerically integrated to determine the net change of the trailer orientation at the end of the motion step. Alternatively, this computation can be done offline and stored in a lookup table.

The path transformation and randomized approaches can also be extended to tractor-trailer systems by using local planners based on exact closed-form motion plans in the absence of obstacles [378, 406]. Such local planners may take advantage of the fact that all trailer systems of the form described above can be converted to chained form, allowing the use of sinusoidal controls [377, 393].

The path transformation method first finds a free path ignoring the nonholonomic constraints, and then transforms the path into one respecting the constraints. For the path transformation method to work, the local planner must satisfy the topological property discussed earlier for cars [259, 378]. A generalization of this approach turns this single transformation step into a sequence of steps, each one introducing one more nonholonomic constraint to be satisfied by the transformation [379]. This multilevel approach can lead to increased computational efficiency and shorter paths. Finally, a path can be turned into a trajectory by a time scaling $s(t)$ respecting actuator limits, as discussed in chapter 11 [259].

Figure 12.31, taken from [378], shows two paths found for a two-trailer system using a path transformation planner. The local planner used to transform the original path to a feasible path uses sinusoidal inputs and the two-trailer system's chained-form equations.

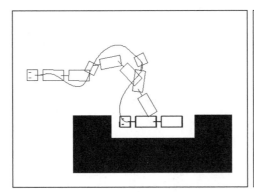

Figure 12.31 Two paths for a two-trailer system found using an omnidirectional-to-nonholonomic path transformation planner. (From Sekhavat and Laumond [378], ©1998 IEEE.)

12.5.7   Kinematic Reductions of Mechanical Systems

Subsection 12.4.2 described a class of underactuated second-order mechanical systems, called kinematically controllable systems, for which trajectory planning can be decoupled into path planning followed by time scaling according to actuator limits. The big advantage of this is that the search for a feasible motion plan can occur in the system's n_Q-dimensional configuration space instead of the $2n_Q$-dimensional state space. Since many search algorithms run in time exponential in the dimension of the search space, this reduction can greatly speed up motion planning.

We focus on systems that are not maximally reducible to kinematic systems but possess p decoupling vector fields $V_1, \ldots, V_p$ satisfying $\ldots \overline{\text{Lie}}(\{V_1, \ldots, V_p\})(q) = T_q Q$ at all $q \ldots$, meaning that the system is STLKC. The path-planning problem is to find a concatenation of integral curve segments of these vector fields to take the system from the initial configuration q_{start} to the goal configuration q_{goal}. Because the velocity must be brought to zero at switches between vector fields, in the interest of minimizing execution time, it is reasonable to design the planner to minimize the number of switches. This implicitly minimizes the use of slow Lie bracket motions. Because the second step of the procedure time-optimally time-scales the motion along the planned path, the approach produces fast trajectories in a computationally efficient manner. (Global time-optimality is precluded because of our decoupling of the problem.)

An example of a kinematically controllable system is a 3R robot arm moving in a horizontal plane with the third joint unactuated and frictionless (figure 12.32). It

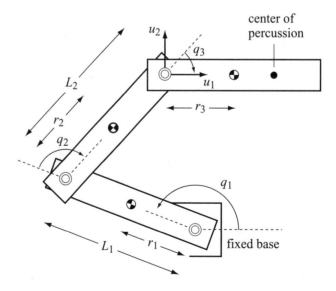

Figure 12.32   A 3R robot arm in a horizontal plane with an unactuated third joint.

is not hard to see that the third link of the arm is equivalent to the PBWT in zero gravity, except the "thruster" forces at the third joint are generated by the actuators at the first two joints. As long as the second joint is not completely extended, the arm can apply a force in any direction at the third joint. Because there is no actuator there, however, the arm cannot generate torque about this joint. The two decoupling vector fields for this system are translation along the length of the third link and rotation about the center of percussion of the third link with respect to the joint. Small-time local kinematic controllability of the PBWT implies STLKC of the 3R arm away from $q_2 = 0$ and $q_2 = \pi$.

To find motion plans minimizing the number of switches between these vector fields, we can adapt the grid search motion planner for cars, "driving" the third link around. In this problem, the four motion primitives are forward and backward translation and clockwise and counterclockwise rotation (about the center of percussion). We choose the path cost function to be the number of switches between the primitives. Inverse kinematics is used to calculate the robot's entire configuration as the third link moves, and collisions must be checked along the entire robot arm, not just at the third link. Apart from these modifications, the algorithm is the same as for carlike robots.

Once a path is found using the decoupling vector fields, the path can be time-scaled arbitrarily while respecting the underactuation constraint (zero torque at the third joint). To perform the time-optimal time scaling, we use the manipulator dynamics

$$M(q)\ddot{q} + \dot{q}^T \Gamma(q)\dot{q} = u = \begin{bmatrix} u_1 \\ u_2 \\ 0 \end{bmatrix}$$

in the time-scaling algorithm described in the previous chapter. For the 3R arm, the inertia matrix $M(q)$ is given by

$$M_{11} = I_1 + I_2 + I_3 + m_1 r_1^2 + m_2 \left(L_1^2 + r_2^2 + 2L_1 r_2 \cos q_2 \right)$$
$$+ m_3 \left(L_1^2 + L_2^2 + r_3^2 + 2L_1 L_2 \cos q_2 + 2L_2 r_3 \cos q_3 + 2L_1 r_3 \cos(q_2 + q_3) \right)$$

$$M_{12} = I_2 + I_3 + m_2 \left(r_2^2 + L_1 r_2 \cos q_2 \right)$$
$$+ m_3 \left(L_2^2 + r_3^2 + L_1 L_2 \cos q_2 + L_2 r_3 \cos q_3 + L_1 r_3 \cos(q_2 + q_3) \right)$$

$$M_{13} = I_3 + m_3 \left(r_3^2 + L_2 r_3 \cos q_3 + L_1 r_3 \cos(q_2 + q_3) \right)$$

$$M_{22} = I_2 + I_3 + m_2 r_2^2 + m_3 \left(L_2^2 + r_3^2 + 2L_2 r_3 \cos q_3 \right)$$

$$M_{23} = I_3 + m_3 \left(r_3^2 + L_2 r_3 \cos q_3 \right)$$

$$M_{33} = I_3 + m_3 r_3^2,$$

where m_i is the mass of link i, I_i is the inertia of link i about its center of mass, and r_i and L_i are defined in figure 12.32. Recall that $M_{21} = M_{12}$, $M_{31} = M_{13}$, $M_{32} = M_{23}$. The Christoffel symbols $\Gamma^i_{jk}(q)$ are derived from $M(q)$.

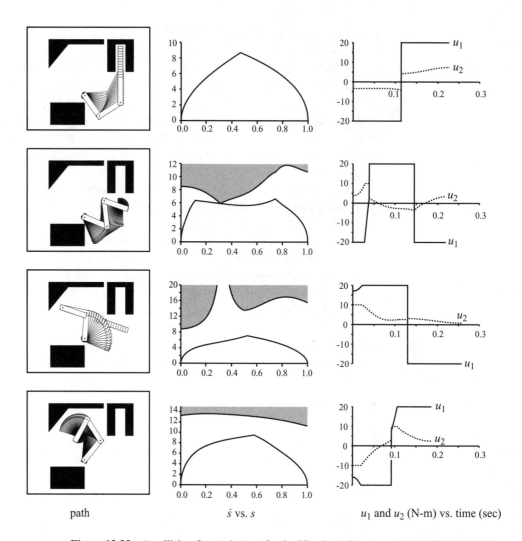

path $\dot{s}$ vs. s u_1 and u_2 (N-m) vs. time (sec)

Figure 12.33 A collision-free trajectory for the 3R robot with an unactuated third joint. The left column shows the path and the right column shows the actuator profiles. One of the actuators is always saturated. The middle column illustrates the speed along the path segment vs. a path parameter (normalized rotation angle or translation distance), as described in chapter 11.

Joint i	L_i (m)	r_i (m)	m_i (kg)	I_i (kg-m^2)	τ_i^{max} (N-m)	τ_i^{min} (N-m)
1	0.3	0.15	2.0	0.02	20	−20
2	0.3	0.15	1.0	0.01	10	−10
3		0.15	0.5	0.004125	0	0

Table 12.1   Kinematic parameters, inertial parameters, and actuator limits for the simulated robot.

Figure 12.33 shows a trajectory planned for the robot arm with the kinematic parameters, inertial parameters, and actuator limits given in table 12.1. The path consists of four separate motions along decoupling vector fields, and the motion must come to a halt at the switches. Therefore, the time-scaling problem becomes four separate problems. The complete time-optimal trajectory along the path takes 0.890 s.

This planner has been successfully implemented on an experimental underactuated 3R arm, though not using the time-optimal motions along the paths [305].

A brute-force best-first search along decoupling vector fields can be applied to any STLKC mechanical system. For systems with more degrees of freedom, however, the computational expense may be prohibitive. In this case we might give up on finding motion plans that minimize switches between decoupling vector fields. Possible approaches are multiresolution grid-search methods, probabilistic roadmap methods, or transformation of omnidirectional paths (as described for cars in subsection 12.5.6) using exact local planners based on the decoupling vector fields [200,310], and rapidly-exploring random trees (RRT's) modified to reduce the number of vector field switches [111].

12.5.8   Other Approaches

Fictitious Inputs for Drift-Free Systems

Lafferriere and Sussmann proposed a general method for steering drift-free STLC systems [253, 254]. If the original system is

$$\dot{x} = \sum_{i=1}^{m} g_i(x)u_i,$$

then an *extended* system is defined to be

$$\dot{x} = \sum_{i=1}^{r} g_i(x)v_i, \ \ r \geq n,$$

where the vector fields $g_{m+1}, \ldots, g_r$ correspond to Lie product motions of the system. These vector fields are chosen so that $\text{span}(\{g_1, \ldots, g_r\}) \, (x) = T_x \mathcal{M}$ at all x, and $v_{m+1}, \ldots, v_r$ are called *fictitious* inputs.

There are no nonholonomic constraints for the extended system, so motion planning for this system is identical to motion planning for an unconstrained system. Once we have found a path for the unconstrained system using the controls v, we transform it to a path for the original system with controls u. This transformation uses the Campbell-Baker-Hausdorff formula [330, 369] describing the motion generated by composing motions along two different vector fields, and is beyond the scope of this chapter. If the vector fields $g_1, \ldots, g_m$ are nilpotent, or *nilpotentizable* by a feedback transformation, the transformation provides an exact expression for the motion of the system with a finite number of Lie products of the two vector fields, and the transformation produces an exact motion plan. Otherwise, small errors are introduced due to higher-order terms in the Lie bracket motion prescription. These errors can be arbitrarily reduced by iterating the procedure.

This approach applies to any STLC drift-free system. The quality of the solution depends on the initial path chosen for the system, and it is in no sense optimal. For more details, see the original papers by Lafferriere and Sussmann [253, 254] or the summaries in the textbooks [330, 368].

Motion Libraries

A motion library consists of a set of canonical motions or primitives that are feasible for the underactuated system, along with a set of conditions (or transition maneuvers) for concatenating these primitives. A search for a feasible trajectory is then restricted to compositions of these primitives. The decoupling vector fields of kinematically controllable systems are examples of motion primitives that are concatenable at any configuration q. As another example, a set of primitives for an airplane might include flying level, a steady dive, and a constant climbing turn. Symmetries in the system dynamics can be exploited to minimize the number of primitives; e.g., the dynamics of flying level (in the absence of wind) are invariant to the airplane's position and orientation in a horizontal plane. The library should consist of a sufficient number of primitives to ensure controllability of the system. One formalization of these ideas is given by the Maneuver Automaton of Frazzoli, Dahleh, and Feron [160].

Rapidly Exploring Random Trees and Expansive Space Trees

RRTs and ESTs, as described in chapter 7, apply to a broad class of systems, including nonholonomic and underactuated systems. All that is required is a state equation

$\dot{x} = f(x, u)$ and a distance metric appropriate to the problem. Because no particular structure of the system is utilized, motion plans may be inefficient. The planning time may be sensitive to the chosen distance metric. (See also chapter 7, section 7.5.1.)

Problems

1. Choose a grid of points on $\mathbb{R}^2$ and sketch the tangent vectors of the vector field $[x_2, x_1^2]^T$ at those points. You may draw these with a computer if you wish. Sketch by hand or use a computer to draw a few integral curves of this vector field. Does this vector field define a regular one-dimensional distribution?

2. For the vector fields $g_1 = [x_1 + x_2, 0]^T$ and $g_2 = [0, 1 + x_2]^T$ on $\mathbb{R}^2$, sketch the integral manifolds, or foliation, defined by the distribution span($\{g_1, g_2\}$). (See figure 12.9 for a drawing of a foliation.) Is the distribution regular?

3. For the vector fields $g_1 = [1, 0, 0]^T$ and $g_2 = [0, 1, 0]^T$ on $\mathbb{R}^3$, sketch the foliation defined by the distribution span($\{g_1, g_2\}$). Is the distribution regular? Is it involutive?

4. For the vector fields $g_1 = [x_3, 0, 0]^T$ and $g_2 = [0, 1, 0]^T$ on $\mathbb{R}^3$, sketch the foliation defined by the distribution span($\{g_1, g_2\}$). Is the distribution regular? Is it involutive?

5. For the vector fields $g_1 = [1+x_3, 1-x_2, 0]^T$ and $g_2 = [0, 0, 1]^T$ on $\mathbb{R}^3$, sketch the foliation defined by the distribution span($\{g_1, g_2\}$). Is the distribution regular? Is it involutive?

6. For the vector fields $g_1 = [1+x_3, 1-x_3, 0]^T$ and $g_2 = [0, 0, 1]^T$ on $\mathbb{R}^3$, sketch the foliation defined by the distribution span($\{g_1, g_2\}$). Is the distribution regular? Is it involutive?

7. Describe physical systems with the following properties, not using the examples discussed in the chapter: accessible but not STLA; accessible but not controllable; controllable but not STLA; STLA but not STLC; STLC but not controllable.

8. For vector fields that are *linear* in the state, e.g., $g_1(x) = Ax$ and $g_2(x) = Bx$, the Lie bracket has the particularly simple form

(12.39) $[g_1, g_2](x) = (AB - BA)x,$

called the *matrix commutator* of A and B. For such vector fields, the Lie bracket can be calculated without differentiation. As an example, let $x = [x_1, x_2, x_3]^T$ and

$$A = \begin{bmatrix} 0 & 1 & 0 \\ -1 & 0 & 0 \\ 0 & 0 & 0 \end{bmatrix}, \quad B = \begin{bmatrix} 0 & 0 & 0 \\ 0 & 0 & 1 \\ 0 & -1 & 0 \end{bmatrix},$$

and let $g_1 = Ax$ and $g_2 = Bx$. Use equation (12.5) to show $[g_1, g_2](x) = (BA - AB)x$, the negative of the expression in equation (12.39). (Note that the sign of the Lie bracket is immaterial in generating the distribution.)

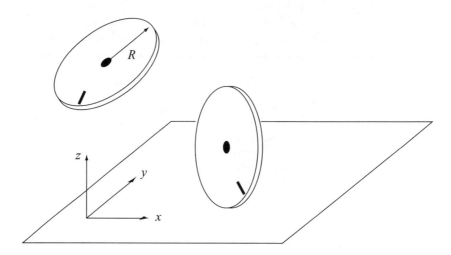

Figure 12.34 A wheel in space, then constrained to stand upright on the $z = 0$ plane. The notch on the wheel is used to keep track of the wheel's rolling angle.

9. Write Mathematica (or other symbolic software) code to take two vector fields and calculate their Lie bracket.

10. The configuration of a wheel of radius R has six degrees of freedom in three-dimensional space, described globally as $SE(3)$ or locally using x-y-z and roll-pitch-yaw coordinates. Choose six coordinates to describe the wheel's configuration in space, where the x-y-z coordinates describe the position of the center of the wheel. With these six coordinates, write two holonomic constraints that constrain the wheel to stand upright on a plane at $z = 0$ (figure 12.34). If you choose your coordinates properly, this will leave you with four coordinates to describe the configuration of the wheel on the plane. Using the time-derivatives of these coordinates, write the two nonholonomic constraints that prevent slipping at the contact between the plane and the wheel as it moves.

   This system is identical to the unicycle example in this chapter, except the configuration space is four-dimensional (the rolling angle of the wheel is included). If the two controls are the rolling angular velocity of the wheel and the turning-in-place angular velocity of the wheel, write the two control vector fields, and write the system as a control-affine nonlinear control system. Using Lie brackets of the vector fields, show that the system is STLC at any configuration if the control set belongs to $\mathcal{U}_{\pm}$.

11. For the wheel of problem 10, describe a four-step motion-planning algorithm to take the unicycle to an arbitrary configuration in its obstacle-free four-dimensional configuration space. The final step of the algorithm should drive the wheel around a circle to achieve the desired rolling angle. Your algorithm should take the final configuration as input (assuming the wheel starts from the origin configuration $[0, 0, 0, 0]^T$) and return a sequence of control values u_1 and u_2 and the times they are applied.

12. Transform the control system of problem 10 to chained form. This may be challenging!

13. Prove that the standard car model of equation (12.34) is STLC.

14. Derive the drift and control vector fields for the PBWT, assuming that it has mass m and inertia I about the center of mass. Then set $m = I = 1$ and verify that your vector fields match those given in the text.

15. Because the PBWT has three degrees of freedom but only two controls, there is a constraint on its possible accelerations. Derive this constraint, and show that it can be written in the form $\omega(x)\dot{x} = 0$ in the absence of gravity.

16. Imagine a PBWT where the control u_2 is a pure torque about the center of mass. Write the two control vector fields, put the system in the control-affine form (12.6), and use theorem 12.3.3 to show that it is STLC at zero velocity in the absence of gravity. Then put the system in the covariant derivative form of equation (12.13) and use theorem 12.4.1 to prove the same thing.

17. Flat outputs for the PBWT are $y_1 = x_1 + \frac{1}{d} \cos x_3$ and $y_2 = x_2 + \frac{1}{d} \sin x_3$. Find the maps ϕ and ψ to recover the states $x(t)$ and control inputs $u(t)$ as a function of the trajectory of the flat outputs $y(t)$.

18. Flat outputs for the car pulling trailers, described in subsection 12.5.6, are the two coordinates describing the planar location of the point midway between the two wheels of the last trailer. Find the maps ϕ and ψ to recover the states $x(t)$ and control inputs $u(t)$ as a function of the trajectory of the flat outputs.

19. Let u_1 and u_2 be the torques at the two joints of a 2R robot arm in a horizontal plane (figure 12.35). Write the dynamics of the 2R arm in the form of equation (12.6), where the masses of the first and second links are m_1 and m_2, and the inertias of the links about their centers of mass are I_1 and I_2. Because the drift vector field g_0 is energy-conserving and the arm configuration space $S^1 \times S^1 = T^2$ is compact, the drift vector field is WPPS. If possible, use theorem 12.3.5 to show that the robot arm is (globally) controllable with $u_2 = 0$; in other words, any state is reachable from any other state by using only torques at the first joint. If you cannot, explain whether you believe the arm is controllable or not, and how you might demonstrate your belief. Why is the arm not controllable if $u_2 \in \mathbb{R}$ but $u_1 = 0$?

20. Let $[u_1, u_2, u_3]^T$ be the torques at the three joints of the 3R robot arm of figure 12.32 in a horizontal plane. Write the dynamics of the 3R arm in the form of equation (12.6).

    In subsection 12.5.7, a motion planner is described for the underactuated 3R arm with $u_3 = 0$. In this case, there are two decoupling vector fields, or two rank 1 kinematic reductions. If instead only the first actuator is missing, so $u_1 = 0$, there is a rank 2 kinematic reduction—the system is maximally reducible to a kinematic system. This means that the acceleration constraint due to $u_1 = 0$ can actually be integrated to a velocity constraint: the total angular momentum about the first joint is conserved. Assuming that the 3R robot arm begins at rest, write this velocity constraint and give the rank 2 kinematic reduction in the form of equation (12.15).

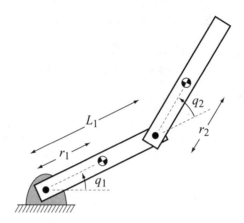

Figure 12.35   A 2R robot arm.

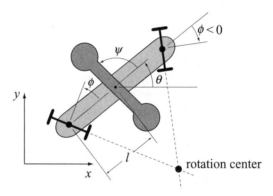

Figure 12.36   The snakeboard model.

21. The dynamics of an RP manipulator are derived at the beginning of chapter 10. Set gravity a_g to zero, and assume that the robot is missing the actuator at the prismatic joint, so $u_2 = 0$. Find the input vector field $Y_1(q)$ and use theorem 12.4.1 to show that the system is STLA when $q_2 > 0$ using only the actuator at the revolute joint. Also, provide an argument either supporting or rejecting the hypothesis that the arm is globally controllable on its state space.

22. The *snakeboard* is a commercial toy whose concept is derived from the well known skateboard. It is composed of two steerable wheeled platforms joined by a coupling bar, and the rider propels herself forward without touching the ground by steering the wheels and twisting her body back and forth. A simple model of the snakeboard is shown in figure 12.36. Here a momentum rotor simulates the rider by spinning back and forth, and by conservation of angular momentum about the rotation center chosen by the wheels, the snakeboard body

moves. The snakeboard model is an underactuated mechanical system with nonholonomic constraints, which we will write in the form of equations (12.18) and (12.19).

Let the configuration of the snakeboard be represented by $q = [x, y, \theta, \psi, \phi]^T$, where (x, y) represents the Cartesian position of the center of the snakeboard coupler, θ is its angle, and ψ and ϕ are the angle of the rotor and the steering angle of the wheels, respectively, expressed in the body frame. The inertia matrix for the snakeboard is given by

$$M = \begin{bmatrix} m & 0 & 0 & 0 & 0 \\ 0 & m & 0 & 0 & 0 \\ 0 & 0 & I + I_r + I_w & I_r & 0 \\ 0 & 0 & I_r & I_r & 0 \\ 0 & 0 & 0 & 0 & I_w \end{bmatrix},$$

where m is the total mass of the snakeboard, I is the inertia of the coupler about its center of mass, I_r is the rotor inertia, and $\frac{1}{2} I_w$ is the inertia of each set of wheels about its pivot point. (Note that because the inertia matrix is invariant to the configuration, the Christoffel symbols are zero.) The system is subject to two control inputs: a torque u_ψ that controls the rotor angle ψ, and a torque u_ϕ controlling the steering angle ϕ. Therefore $u = [u_\psi, u_\phi]^T$ and $T(q)$ can be written

$$T(q) = \begin{bmatrix} 0 & 0 \\ 0 & 0 \\ 0 & 0 \\ 1 & 0 \\ 0 & 1 \end{bmatrix}.$$

The wheels are assumed to roll without lateral slipping, and the wheel angle chooses a rotation center along a line perpendicular to the body of the snakeboard and through its center. The no-slip constraints can be manipulated into the form

$$A(q) = \begin{bmatrix} \sin\phi & 0 & -l\cos\theta\cos\phi & 0 & 0 \\ 0 & \sin\phi & -l\sin\theta\cos\phi & 0 & 0 \end{bmatrix}.$$

Write the equations of motion in the form of equations (12.18) and (12.19). Find the projection matrix $P(q)$ and the two input vector fields $\widetilde{Y}_1(q)$ and $\widetilde{Y}_2(q)$. Show that these two vector fields are decoupling vector fields and that the system is STLKC by these two decoupling vector fields. Explain what this means in terms of motion planning for this system.

23. Implement the grid search path planner CAR_GRID_SEARCH (algorithm 22).

24. For a differential-drive car, include the drive wheel angles in the description of the configuration, giving the car a five-dimensional configuration space (position and orientation of the body and two wheel orientations). Write out the control system and prove that this system is or is not STLC on this five-dimensional space.

25. Prove that all chained-form systems are differentially flat with the first and last states x_1 and x_n as flat outputs, and describe a method for finding the mappings ϕ and ψ from the flat outputs and their derivatives to the state x and the control u, respectively.

26. In example 12.4.4, perform the calculations to verify $\langle Y_1 : Y_1 \rangle$, $\langle Y_2 : Y_2 \rangle$, $\langle Y_1 : Y_2 \rangle$, and $[Y_1, Y_2]$. You may write symbolic manipulation code (e.g., in Mathematica) to do these computations for you.

27. Derive the equations of motion of the hopper in example 12.4.4 in the form of equation 12.7, and show that the underactuation implies an acceleration constraint that can be integrated to give a conservation of angular momentum constraint.

28. Write the Pfaffian constraint for the hopper in example 12.4.4 and give a driftless kinematic model of the system (the rank 2 kinematic reduction).

29. Derive the formula for the Lie bracket

$$[g_1, g_2] = \frac{\partial g_2}{\partial q} g_1 - \frac{\partial g_1}{\partial q} g_2.$$

In other words, show that the net motion obtained by following g_1 for time ϵ, g_2 for time ϵ, $-g_1$ for time ϵ, and $-g_2$ for time ϵ is

$$\epsilon^2 \left(\frac{\partial g_2}{\partial x} g_1 - \frac{\partial g_1}{\partial x} g_2 \right) + O(\epsilon^3),$$

where ϵ is small. To do this, perform a Taylor expansion to express the net motion, throwing away terms of higher order than ϵ^2. For example, after following g_1 for time ϵ, we have

$$x(\epsilon) = x(0) + \epsilon \dot{x}(0) + \frac{1}{2} \epsilon^2 \ddot{x}(0) + O(\epsilon^3).$$

Subsituting $g_1(x(0)) = \dot{x}(0)$ and $\frac{\partial g_1}{\partial x} g_1(x(0)) = \ddot{x}(0)$, where $\frac{\partial g_1}{\partial x}$ is evaluated at $x(0)$, we get

$$x(\epsilon) = x(0) + \epsilon g_1(x(0)) + \frac{1}{2} \epsilon^2 \frac{\partial g_1}{\partial x} g_1(x(0)) + O(\epsilon^3).$$

After following g_2 for time ϵ, we have

$$x(2\epsilon) = x(\epsilon) + \epsilon g_2(x(\epsilon)) + \frac{1}{2} \epsilon^2 \frac{\partial g_2}{\partial x} g_2(x(\epsilon)) + \dots$$

Leaving out terms of higher order than ϵ^2, this becomes

$$x(2\epsilon) \approx x(0) + \epsilon g_1(x(0)) + \frac{1}{2} \epsilon^2 \frac{\partial g_1}{\partial x} g_1(x(0))$$
$$+ \epsilon g_2(x(0) + \epsilon g_1(x(0))) + \frac{1}{2} \epsilon^2 \frac{\partial g_2}{\partial x} g_2(x(0)).$$

This expands to

$$x(2\epsilon) \approx x(0) + \epsilon g_1(x(0)) + \frac{1}{2} \epsilon^2 \frac{\partial g_1}{\partial x} g_1(x(0))$$
$$+ \epsilon g_2(x(0)) + \epsilon^2 \frac{\partial g_2}{\partial x} g_1(x(0)) + \frac{1}{2} \epsilon^2 \frac{\partial g_2}{\partial x} g_2(x(0)).$$

Now continue by finding $x(3\epsilon)$ and $x(4\epsilon)$ to arrive at the result.

A *Mathematical Notation*

Symbol	Meaning
$\cdot$	dot product
$\exists$	there exists
$\forall$	for all
∞	infinity
$\in$	element
$\notin$	not in
s.t.	such that
$\mathbb{R}$	real numbers
$\mathbb{R}^m$	m-dimensioned real numbers
$\cup$	union
$\cap$	intersection
$\setminus$	set difference
$\Rightarrow$	implies. $p \Rightarrow q$ is p implies q
$\Longleftrightarrow$	if and only if
S^1	a circle
∇	gradient
D	differential or distance to closest obstacle (depending on context)
d_i	distance to obstacle i in either the workspace or configuration space (depending on context)
$d(x, y)$	distance between the two points x and y
Null	null space

$O(n)$	order of n
J	Jacobian
Γ	Christoffel symbol
RM	roadmap
$\mathcal{W}$	workspace
$\mathcal{Q}$	configuration space
$\mathcal{Q}_{\text{free}}$	free space
$x(k)$	state at time k
$\|x\|$	norm of x
$\subseteq$	subset of
$\subset$	strict subset of
$\text{cl}(A)$	closure of A
T^n	n-dimensional torus
S^n	n-dimensional sphere in $\mathbb{R}^{n+1}$
$SO(n)$	special orthogonal group
$SE(n)$	special Euclidean group
$B_\epsilon(q)$	open ball of radius ϵ centered at q
Df	differential of f
∇f	gradient of f
∇	affine connection
$\nabla_{Y_1} Y_2$	covariant derivative of Y_2 with respect to Y_1
C^0	continuous
C^n	n times differentiable
$\langle x, y \rangle$	inner product of x and y
$\mathcal{I}$	identity matrix
$\text{atan2}(y, x)$	returns angle to (x, y) in the plane in range $[-\pi, \pi)$
$T_x \mathcal{M}$	tangent space of $\mathcal{M}$ at x
$T\mathcal{M}$	tangent bundle of $\mathcal{M}$
$[f, g]$	Lie bracket of vector fields f, g
$\overline{\text{Lie}}(\mathcal{G})$	the Lie algebra of a set of vector fields $\mathcal{G}$
$\overline{\mathcal{D}}$	involutive closure of the distribution $\mathcal{D}$
$\mathcal{U}_\pm$	control set positively spanning $\mathbb{R}^m$
$\mathcal{U}_+$	control set spanning $\mathbb{R}^m$
$\langle Y_1 : Y_2 \rangle$	the symmetric product of vector fields Y_1 and Y_2
$\overline{\text{Sym}}(\mathcal{Y})$	the symmetric closure of the distribution $\mathcal{Y}$
$\text{span}(\{x_1, \ldots, x_n\})$	the linear span of $\{x_1, \ldots, x_n\}$

B *Basic Set Definitions*

CONSIDER A collection of elements called a *set*. The plane is a set; the real line is a set; a point is a set; the unit interval is a set. Sets can also be listed as collections of elements, e.g., $S_1 = \{1, 4, 9\}$ and $S_2 = \{$cow, chicken, pig$\}$ are both sets. The collection of these sets is also a set, i.e., $\{\mathbb{R}^2, \mathbb{R}, [0, 1]\}$ is a set. Given two sets A and B, A is said to be a *subset* of B (denoted $A \subset B$) if every element of A is also an element of B. Of the two examples above, S_1 is a subset of the set of positive integers and S_2 is a subset of the set of animals.

Given $A \subset B$, the *complement* of A in B (denoted $B \backslash A$) is defined to be all of the elements of B that are not in A, i.e.,

$$B \backslash A = \{x \mid x \in B, x \notin A\}.$$

The *union* of A and B (denoted $A \cup B$) is to be the set of points that is in either A or B, i.e.,

$$A \cup B = \{x \mid x \in A \text{ or } x \in B\}.$$

The *intersection* of A and B (denoted $A \cap B$) is defined to be the set of all points that are in both A and B, i.e.,

$$A \cap B = \{x \mid x \in A \text{ and } x \in B\}.$$

For the remainder of this appendix we restrict the discussion to sets that are subsets of $\mathbb{R}^n$ for some n. Consider a point $x \in \mathbb{R}^n$, and define an ϵ-*neighborhood* of x to be the set

$$B_\epsilon(x) = \{y \in \mathbb{R}^n \mid d(x, y) < \epsilon\}.$$

The set $B_\epsilon(x)$ is also sometimes called an *open ball* of radius ϵ around the point x. We also sometime use the word *neighborhood* to refer to an ϵ-neighborhood with ϵ arbitrarily small.

A set $A \subset \mathbb{R}^n$ is said to be *open* if, for every point x in A, there is some ϵ so that $B_\epsilon(x)$ is also contained in A. A set A is said to be *closed* if its complement is open. Note that the concept of closure depends on the the ambient space. The set $\mathbb{R}^m$ considered by itself is open. But if $m < n$ and we consider $\mathbb{R}^m$ as a subset of the ambient space $\mathbb{R}^n$, then $\mathbb{R}^m$ is closed since its complement $\mathbb{R}^n \backslash \mathbb{R}^m$ is open. By the same token, when considered as a subset of the plane, the interval $(0, 1)$ is neither closed nor open.

The following definitions derive from open and closed sets for ACS:

DEFINITION B.0.2 (Closure/Interior/Boundary)

■ Closure *of A, denoted* cl(A), *is the intersection of all closed sets containing A.*

■ Interior *of A, denoted* int(A), *is the union of all open sets contained in A.*

■ Boundary *of A, denoted* ∂A, *is* cl$(A) \bigcap$ cl$(S \backslash A)$.

EXAMPLE B.0.3   *Consider* $[0, 1]$ *as a subset of* $\mathbb{R}^1$.

cl$([0, 1]) = [0, 1]$

int$([0, 1]) = (0, 1)$

$$\begin{aligned}
\partial[0, 1] &= [0, 1] \bigcap \text{cl}\left((-\infty, 0) \bigcup ((1, \infty))\right) \\
&= [0, 1] \bigcap ((-\infty, 0] \bigcup [1, \infty)) \\
&= \{0, 1\}
\end{aligned}$$

The following demonstrate how union and intersection operate on closures and interiors:

$A \subset B \quad \Rightarrow$ int$(A) \subset$ int(B) and cl$(A) \subset$ cl(B)

$A \subset S \quad \Rightarrow S \backslash$cl$(A) =$ int$(S \backslash A)$, $S \backslash$int$(A) =$ cl$(S \backslash A)$.

$A \subset S \quad \Rightarrow$ cl$(\emptyset) =$ int$(\emptyset) = \emptyset$, cl$(S) =$ int$(S) = S$

cl$($cl$(A)) =$ cl(A)

cl$(A \bigcup B) =$ cl$(A) \bigcup$ cl(B), int$(A) \bigcup$ int$(B) \subset$ int$(A \bigcup B)$

cl$(A \bigcap B) \subset$ cl$(A) \bigcap$ cl(B), int$(A \bigcap B) =$ int$(A) \bigcap$ int(B).

A subset A of B is *dense* if $\text{cl}(A) = B$. So $(0, 1)$ is dense in $[0, 1]$ because $\text{cl}(0, 1) = [0, 1]$. Intuitively, a subset A of B is dense if A is "almost as big" as B. The open interval and closed interval both have length 1. The set $[0, 1]\backslash\{.5\}$ is dense in $[0, 1]$. Intuitively, this means that taking away one point from an interval does not affect the size of the interval. The set of rational numbers, i.e., the set of real numbers that can be written as a fraction of two integers, is dense in the real line. A line is *not* dense in the plane. The plane, with a line removed from it, is dense in the plane.

We can also define a notion of subtraction and addition of sets. The *Minkowski sum* of A and B is

$$A \oplus B = \{x \mid x = a + b, a \in A, b \in B\}.$$

The *Minkowski difference* is

$$A \ominus B = \{x \mid x = a - b, a \in A, b \in B\}.$$

Two points in a set A are said to be within *line of sight* of each other if the straight line segment connecting them is completely contained in A. Line of sight is related to convexity. A set A is *convex* if for every $x, y \in A$, the line segment

$$\{tx + (1 - t)y \mid t \in [0, 1]\}$$

is contained in A. The *convex hull* of a set A is denoted as $\text{Co}(A)$ and is defined to be the smallest convex set that contains A. If $A \subset \mathbb{R}^n$ is a finite set with m elements $\{x_1, x_2, \ldots, x_m\}$, we can express

$$\text{Co}(A) = \left\{ y = \sum_{i=1}^{m} a_i x_i \,\middle|\, a_i \geq 0 \text{ for all } i; \; \sum_{i=1}^{m} a_i = 1 \right\}.$$

A set A is said to be *star-shaped* if there exists an $x \in A$ such that for every $y \in A$ the line segment $\{tx + (1 - t)y \mid t \in [0, 1]\}$ is contained in A. In other words, all points in A are within line of sight of at least one common point. All convex sets are star-shaped, but the converse is not true.

C *Topology and Metric Spaces*

C.1 Topology

OPERATORS act on elements of sets. In appendix B, the set complement operator was defined with respect to a superset S. Furthermore, the definitions of open and closed sets were predicated on one definition: the open neighborhood. Now we are going to reverse things. An open neighborhood will be defined in terms of open sets, and a topological space will be defined in terms of its set of elements and its open sets. This appendix is meant to be introductory. See, e.g., [9] for a complete discussion of these topics.

DEFINITION C.1.1 (Topology) *A topological space is a set S together with a collection O of subsets called* open sets *such that*

- $\emptyset \in O$ *and* $S \in O$,

- *if* $U_1, U_2 \in O$, *then* $U_1 \bigcap U_2 \in O$,

- *the union of any collection of open sets is an open set.*

Open sets can be arbitrarily designed as long as they satisfy the above three properties. The *standard topology* on $\mathbb{R}^m$ has $S = \mathbb{R}^m$ with O containing $\mathbb{R}^m$, the empty set $\emptyset$, all open rectangles, and their unions. An example is the real line with open intervals, i.e., $S = \mathbb{R}$, with O consisting of any open interval, the union of open intervals, $\mathbb{R}$, and $\emptyset$. To show this we look to the three conditions in definition C.1.1:

■ $\mathbb{R}, \emptyset \in O$ by definition,

■ $(a, b) \in O$ and $(c, d) \in O$, so

$$(a, b) \bigcap (c, d) = \begin{matrix} (c, b) \in O & \text{or,} \\ (a, d) \in O & \text{or,} \\ \emptyset \in O, \end{matrix}$$

■ any finite or infinite union of open intervals is an open interval.

The *trivial topology* on a set S consists of $O = \{\emptyset, S\}$. The *discrete topology* of a set S is defined by $O = \{A \mid A \subset S\}$. That is, the open sets are everything.

Now the definition of the open neighborhood stems from the definition of open sets. The definitions of closed sets, closure, boundary, interior, and denseness remain the same.

DEFINITION C.1.2 *A* neighborhood *of a point x, denoted* nbhd(x), *is an open set that contains x.*

C.2 Metric Spaces

The open sets of a topological space can be constructed using a distance function. In $\mathbb{R}^m$, the standard Euclidean distance function

$$d(x, y) = \left(\sum_{i=1}^{m} (x_i - y_i)^2 \right)^{\frac{1}{2}}$$

defines open sets that are open balls. More generally,

DEFINITION C.2.1 (Metric Space) *Let M be a set. A* metric *on M is a function* $d : M \times M \rightarrow \mathbb{R}^{\geq 0}$ *such that for all* $m_1, m_2, m_3 \in M$,

1. (Definiteness) $d(m_1, m_2) = 0$ *if and only if* $m_1 = m_2$

2. (Symmetry) $d(m_1, m_2) = d(m_2, m_1)$, *and*

3. (Triangle inequality) $d(m_1, m_3) \leq d(m_1, m_2) + d(m_2, m_3)$.

A metric space *is the pair* (M, d).

Note that the intuitive notion that distance must be non-negative follows directly from the three conditions above. Specifically, condition 3 allows us to write $d(m_1, m_1) \leq d(m_1, m_2) + d(m_2, m_1)$. The left-hand side of this expression is zero by condition 1 and the right-hand side is $2d(m_1, m_2)$ by condition 2, yielding $d(m_1, m_2) \geq 0$.

For $\epsilon > 0$ and $m \in M$, the *open ball* centered at m is defined to be

$$B_\epsilon(m) = \{n \in M \mid d(m, n) < \epsilon\}.$$

The set of all open balls and the union of open balls forms the *metric topology* on the metric space (M, d).

There are many distance functions other than the standard Euclidean metric. For example, the *Manhattan distance metric* is defined to be

$$d(x, y) = \sum_{i=1}^{m} |x_i - y_i|.$$

This metric is so named because it measures how far a taxicab must drive in a city grid to get from one location to another. Different metrics can be used to induce the same topology. The Manhattan and standard Euclidean metrics induce the same topology. Two metrics induce the same topology if, for any open ball at x by the first metric, there is an open ball by the second metric contained completely in the first ball, and vice versa.

C.3   Normed and Inner Product Spaces

A metric space is a special case of a topological space. Next we introduce a normed space, which is a special case of a metric space. We also introduce an inner product space, which is a special case of a normed space.

DEFINITION C.3.1   *A normed space E is a subset of a metric space M that has an operator $\| \cdot \| : E \rightarrow \mathbb{R}$ such that*

- $\|e\| \geq 0$ *for all $e \in E$, and $\|e\| = 0$ if and only if e is the zero vector (positive definiteness),*

- $\|\lambda e\| = |\lambda| \|e\|$ *for all $e \in E$ and $\lambda \in \mathbb{R}$ (homogeneity),*

- $\|e_1 + e_2\| \leq \|e_1\| + \|e_2\|$ *for all $e_1, e_2 \in E$ (triangle inequality).*

The norm can be used to define the open sets and induce a metric. A sequence $\{x_1, x_2, x_3, \ldots\}$ is said to be a *Cauchy sequence* if for any $\epsilon > 0$ there exists an integer k such that $\|x_i - x_j\| < \epsilon$ for all $i, j > k$. When a normed space has a corresponding metric for which every Cauchy sequence converges to a point in the space, we term this space a *Banach space*.

DEFINITION C.3.2   *An* inner product *on a real vector space E is a mapping* $\langle \cdot, \cdot \rangle$: $E \times E \to \mathbb{R}$ *such that*

- $\langle e, e_1 + e_2 \rangle = \langle e, e_1 \rangle + \langle e, e_2 \rangle$,

- $\langle e, \alpha e_1 \rangle = \alpha \langle e, e_1 \rangle$,

- $\langle e_2, \alpha e_1 \rangle = \langle e_1, \alpha e_2 \rangle$,

- $\langle e, e \rangle \geq 0$ *and* $\langle e, e \rangle = 0$ *if and only if e is zero.*

An inner product induces the norm $\|e\| = \langle e, e \rangle$, and a norm in turn induces a metric. When an inner product space has a corresponding metric for which every Cauchy sequence converges, we call this space a *Hilbert space*.

C.4   Continuous Functions

Paths are defined in terms of a continuous function. Let $f : S \to T$ be a mapping from the *domain S* to the *range T*. The points $f(s)$ are the *values* of f, where $s \in S$. If $U \subset S$, then the *image* of U under f is denoted $f(U) = \{f(x) \in T \mid x \in U\}$. If $V \subset T$, then the *preimage* of V under f is denoted $f^{-1}(V) = \{x \in S \mid f(x) \in V\}$. First, we introduce an abstract notion of a continuous function and then specialize it for metric spaces.

DEFINITION C.4.1   *Let S and T be topological spaces and $f : S \to T$ be a mapping. f is* continuous *at $u \in S$ if for every $V = \text{nbhd}(f(u))$ there is a $U = \text{nbhd}(u)$ such that $f(U) \subset V$. The mapping f is* continuous *if for every open subset $V \subset T$, $f^{-1}(V) = \{u \in S \mid f(u) \in V\}$ is open in S.*

Essentially, a continuous function is a function where the preimage of an open set is an open set. Now we introduce the standard "delta-epsilon" method for defining continuous functions on metric spaces: The function f is continuous at s if for every $\epsilon > 0$ there exists a $\delta > 0$ where

(C.1) $d(x, s) < \delta$ implies $d(f(x), f(s)) < \epsilon$.

EXAMPLE C.4.2 (Continuous Function)   *Let $f : \mathbb{R} \to \mathbb{R}$ be defined as $f(x) = x^2$. In order to show that f is continuous at a point s, we must find a $\delta > 0$ such that $d(f(x), f(s)) < \epsilon$ for arbitrarily small $\epsilon > 0$. Note that in $\mathbb{R}$ the distance function is $d(x, s) = |x - s|$. First, we study the quantity $|f(x) - f(s)|$:*

$$|f(x) - f(s)| = |x^2 - s^2|$$
$$= |x - s||x + s|$$
$$= |x - s||x - s + 2s|$$

Using the triangle inequality, we get

$$|f(x) - f(s)| \leq |x - s|(|x - s| + 2|s|).$$

Now we can subsitute $|x - s| < \delta$ to see that $|f(x) - f(s)|$ will be less than ϵ if

$$\delta(\delta + 2|s|) < \epsilon.$$

Using the quadratic formula, we see that this inequality can be satisfied for

$$\delta < -|s| + \sqrt{s^2 + \epsilon}.$$

The term on the right-hand side of this inequality is positive, so we can find a suitable δ. This proves that the function $f(x) = x^2$ is continuous at any point $s \in \mathbb{R}$. Note the that choice of δ depends on both s and ϵ.

The set of continuous functions is denoted C^0. If the derivative of a continuous function f is continuous, then f is said to be differentiable and belongs to a set denoted C^1. If c is k-wise differentiable, then it belongs to a set denoted C^k. If all derivatives of f exist, then f belongs to C^∞ and f is said to be *smooth*. While a path is only required to be of class C^0, a trajectory must belong to C^k, $k > 0$, to allow the definition of velocity, acceleration, etc., at all points where the system is moving.

The following are equivalent statements:

$$f : S \to T \text{ is continuous.} \iff f(\text{cl}(A)) \subset \text{cl}(f(A)) \text{ for } A \subset S$$
$$\iff f^{-1}(\text{int}(B)) \subset \text{int}(f^{-1}(B)) \text{ for } B \subset T$$

Finally, another useful property of continuous functions is that things "change" continuously. Specifically, if the scalar functions f and g are continuous at x and $f(x) < g(x)$, then there exists a nbhd(x) such that for all $y \in$ nbhd(x), $f(y) < g(y)$.

C.5   Jacobians and Gradients

Consider a vector-valued function $f : \mathbb{R}^m \rightarrow \mathbb{R}^n$ where f can be written

$$f(x) = \begin{bmatrix} f_1(x) \\ f_2(x) \\ \vdots \\ f_n(x) \end{bmatrix},$$

where $f_i : \mathbb{R}^m \rightarrow \mathbb{R}$ for $i \in \{1, 2, \ldots, n\}$.

We define the *differential* of f to be the matrix[1]

$$Df = \begin{bmatrix} \frac{\partial f_1}{\partial x_1} & \frac{\partial f_1}{\partial x_2} & \cdots & \frac{\partial f_1}{\partial x_m} \\ \frac{\partial f_2}{\partial x_1} & \frac{\partial f_2}{\partial x_2} & \cdots & \frac{\partial f_2}{\partial x_m} \\ \vdots & \vdots & \ddots & \vdots \\ \frac{\partial f_n}{\partial x_1} & \frac{\partial f_n}{\partial x_2} & \cdots & \frac{\partial f_n}{\partial x_m} \end{bmatrix}.$$

The matrix Df is denoted in a number of different ways. It is sometimes called the *Jacobian* of f and denoted J (see chapter 3). It is sometimes called the *tangent map* of f and denoted Tf. Sometimes it is necessary to specify which variables are used in the differentiation. Hence the differential can also be denoted $\frac{\partial f}{\partial x}$. Putting the variable name in the subscript serves a similar purpose. The symbols $D_x f$, J_x, and $T_x f$ all denote the differential of f with respect to the variable x.

Given a function $g : \mathbb{R}^n \rightarrow \mathbb{R}$, the *gradient* of g is defined to be

$$\nabla g = \begin{bmatrix} \frac{\partial g}{\partial x_1} \\ \frac{\partial g}{\partial x_2} \\ \vdots \\ \frac{\partial g}{\partial x_n} \end{bmatrix}.$$

As in the case of the differential, the notation $\nabla_x g$ is sometimes used to make explicit the fact that g is differentiated with respect to x. The vector $\nabla g(x)$ points in the direction that maximally increases the function at the point x. Note that by this defintition $\nabla f(x) = Df^T$. The decision as to whether the gradient should be a row vector or a column vector is somewhat arbitrary. In this book we define it as a column vector because that is the convention commonly used in the robotics community when discussing planning algorithms based on artificial potential fields.

---

1. To be technically accurate, the differential is actually a map from the tangent space of $\mathbb{R}^m$ (which happens to also be $\mathbb{R}^m$) to the tangent space of $\mathbb{R}^n$ (which is $\mathbb{R}^n$). For the purposes of this appendix, we simply represent Df as an $n \times m$ matrix.

Let $c(t)$ be a *smooth curve* in $\mathbb{R}^n$, i.e., c is a C^∞, vector-valued map $c : \mathbb{R} \to \mathbb{R}^n$. If t is time, the derivative

$$\dot{c}(t) = \frac{dc}{dt}(t) = \begin{bmatrix} \frac{dc_1}{dt}(t) \\ \frac{dc_2}{dt}(t) \\ \vdots \\ \frac{dc_n}{dt}(t) \end{bmatrix}$$

can be thought of as the velocity of a point moving along $c(t)$.

For a real-valued function $g : \mathbb{R}^n \to \mathbb{R}$, one is often interested in how the value of the function g changes as the state follows the trajectory $c(t)$. This is the same as finding the derivative of $g(c(t))$ with respect to t, $\dot{g}(t) = \frac{d}{dt}(g \circ c)(t)$, where $(g \circ c)(t) = g(c(t))$ is called the *composition* of g and c. To calculate $\dot{g}$ we can use the *chain rule,* which can be stated in a number of different ways:

$$\frac{d}{dt}(g \circ c)(t) = \sum_{i=1}^{n} \frac{\partial g}{\partial c_i} \frac{dc}{dt}(t)$$

$$= \frac{\partial g}{\partial c}\dot{c}(t)$$

$$= D_c g\dot{c}(t)$$

Note here that $\frac{\partial g}{\partial c_i}$ denotes the partial derivative of g with respect to x_i evaluated at $x_i = c_i(t)$. Likewise, $\frac{\partial g}{\partial c}$ and $D_c g$ denote the differential of g with respect to x evaluated at $x = c(t)$. When it is necessary to be explicit, these quantities are sometimes denoted as $\frac{\partial g}{\partial x}|_{x=c(t)}$ or $D_x g|_{x=c(t)}$.

Sometimes it is useful to be able to see how g changes when the curve c is represented in different coordinates than those for which g is defined. Let $h : \mathbb{R}^n \to \mathbb{R}^n$ be the smooth change of coordinates that maps from the coordinates in which c is defined (y-coordinates) into the x-coordinates required by g, i.e., $x = g(y)$. So we are interested in seeing how $g(h(c(t))) = (g \circ h \circ c)(t)$ evolves with t. Again, the chain rule allows us to express this quantity in a number of different ways:

$$\frac{d}{dt}(g \circ h \circ c)(t) = \frac{\partial g}{\partial x}\bigg|_{x=(h \circ c)(t)} \frac{\partial h}{\partial y}\bigg|_{y=c(t)} \dot{c}(t)$$

$$= \underbrace{\frac{\partial g}{\partial h}}_{(1 \times n)} \underbrace{\frac{\partial h}{\partial c}}_{(n \times n)} \underbrace{\dot{c}(t)}_{(n \times 1)}$$

$$= D_h g D_c h\dot{c}(t)$$

$$= D_c(f \circ g)\dot{c}(t)$$

In the second of these expressions the dimension of each of the three objects on the right-hand side is written below the underbrace. This is to make it clear that the dimensions are suitable for matrix multiplication and that the resulting product is indeed a scalar.

EXAMPLE C.5.1 *Consider the curve $c : \mathbb{R} \to \mathbb{R}^2$ defined in polar coordinates*

$$c(t) = \begin{bmatrix} r(t) \\ \theta(t) \end{bmatrix} = \begin{bmatrix} 2 \\ 2\pi t \end{bmatrix}.$$

Note that this curve corresponds to a point moving around a circle of constant radius 2 at a velocity of 4π, i.e., the point travels around the circle once every second. Now consider a function $g : \mathbb{R}^2 \to \mathbb{R}$ defined in Cartesian coordinates

$$g(x) = x_1.$$

In order to compute $\dot{g}$, we introduce a coordinate change $h : \mathbb{R}^2 \to \mathbb{R}^2$ that maps the vector $y = [y_1, \ y_2]^T = [r, \ \theta]^T$ in polar coordinates into Cartesian coordinates:

$$h(y) = \begin{bmatrix} y_1 \cos(y_2) \\ y_1 \sin(y_2) \end{bmatrix}$$

Using the chain rule we get

$$
\begin{aligned}
\frac{d}{dt}(g \circ h \circ c)(t) &= \frac{\partial g}{\partial h} \frac{\partial h}{\partial c} \dot{c}(t) \\[2mm]
&= \begin{bmatrix} 1 & 0 \end{bmatrix} \begin{bmatrix} \cos(y_2) & -y_1 \sin(y_2) \\ \sin(y_2) & y_1 \cos(y_2) \end{bmatrix}_{y=[2, 2\pi t]^T} \begin{bmatrix} 0 \\ 2\pi \end{bmatrix} \\[2mm]
&= \begin{bmatrix} 1 & 0 \end{bmatrix} \begin{bmatrix} \cos(2\pi t) & -2 \sin(2\pi t) \\ \sin(2\pi t) & 2 \cos(2\pi t) \end{bmatrix} \begin{bmatrix} 0 \\ 2\pi \end{bmatrix} \\[2mm]
&= -4\pi \sin(2\pi t).
\end{aligned}
$$

(C.2)

This can be checked by differentiating $(g \circ h \circ c)(t) = 2 \cos(2\pi t)$ directly to get the same answer.

The chain rule can be used in a similar manner to differentiate compositions of functions of any compatible dimension. For example, if we redefine the functions h and g so that $h : \mathbb{R}^n \to \mathbb{R}^m$ and $g : \mathbb{R}^m \to \mathbb{R}^p$, then the chain rule gives the derivative of the composition

$$\underbrace{\frac{d}{dt}(g \circ h \circ c)(t)}_{(p \times 1)} = \underbrace{\frac{\partial g}{\partial h}}_{(p \times m)} \underbrace{\frac{\partial h}{\partial c}}_{(m \times n)} \underbrace{\dot{c}(t)}_{(n \times 1)}.$$

A remark about rows and columns: In mechanics, a force vector F is usually represented by a row vector as it is a member of the cotangent space (see chapter 12). Velocity vectors belong to the tangent space and are usually represented as column vectors, e.g., v. This allows us to easily take the product Fv to get power, which is a scalar value. Many mechanics texts use up-down indicial notation to facilitate this, but such notation is not required for this book.

D   *Curve Tracing*

MANY sensor-based techniques, such as those in sections 2.3.3 and 5.2.5, are essentially curve-tracing algorithms. In both cases, the robot is, in a sense, "determining" the curve as it is being traced. Such techniques relied on two fundamental principles: the curve being traced exists and under the "right conditions," the curve can be traced with simple predictor-corrector techniques. These principles rested on two theorems, the implicit function theorem, and the Newton-Raphson convergence theorem, described below.

D.1   Implicit Function Theorem

Consider a smooth function of multiple variables, $f(x, y)$, and consider the surface that is defined by the equation $f(x, y) = z_0$ for some fixed z_0. Under certain conditions, this surface can be used to write a new function that defines the y variables in terms of the x variables, i.e., $y = g(x, z_0)$. The theorem that states these conditions is called the *implicit function theorem*.

THEOREM D.1.1 (Implicit Function Theorem)   *Let $f : \mathbb{R}^m \times \mathbb{R}^n \to \mathbb{R}^n$ be a smooth vector-valued function, $f(x, y)$. Assume that $D_y f(x_0, y_0)$ is invertible for some $x_0 \in \mathbb{R}^m$, $y_0 \in \mathbb{R}^n$. Then there exist neighborhoods X_0 of x_0 and Z_0 of $f(x_0, y_0)$ and a unique, smooth map $g : X_0 \times Z_0 \to \mathbb{R}^n$ such that*

$$f(x, g(x, z)) = z$$

for all $x \in X_0$, $z \in Z_0$.

D.2   Newton-Raphson Convergence Theorem

By numerically following the set of points where $f(x, y) = 0$, we can locally construct a curve. While there are a number of curve tracing techniques [232], consider an adaptation of a common predictor-corrector scheme. Moving in the tangent direction can serve as a prediction. However, if there is curvature, then the tangent prediction is not correct. Therefore, a correction method is used. The correction procedure occurs on a plane orthogonal to the tangent; this plane is called a correcting plane. The correction step finds the location where the curve being traced intersects the correcting plane and is an application of the Newton Convergence Theorem [232].

THEOREM D.2.1 (Newton-Raphson Convergence Theorem)   *Let $f : \mathbb{R}^n \to \mathbb{R}^n$ and $f(y^*) = 0$. For some $\rho > 0$, let f satisfy*

- $Df(y^*)$ *is nonsingular with bounded inverse, i.e.,* $\|(Df(y^*))^{-1}\| \leq \beta$

- $\|Df(x) - Df(y)\| \leq \gamma \|x - y\|$ *for all $x, y \in B_\rho(y^*)$, where $\gamma \leq \frac{2}{\rho\beta}$*

Now consider the sequence $\{y^h\}$ defined by

$$y^{h+1} = y^h - (Df(y^h))^{-1} f(y^h),$$

for any $y^0 \in B_\rho(y^*)$. Then $y^h \in B_\rho(y^*)$ for all $h > 0$, and the sequence $\{y^h\}$ quadratically converges onto y^*, i.e.,

$$\|y^{h+1} - y^*\| \leq a\|y^h - y^*\|^2$$

where $a = \frac{\beta\gamma}{2(1-\rho\beta\gamma)} < \frac{1}{\rho}$.

E

Representations of Orientation

IN CHAPTER 3, we represent orientation by matrices in $SO(3)$, which can be parameterized using three parameters. In this appendix, we describe some of the most popular methods of doing so, including Euler angles and angles with respect to a fixed frame. We also describe how orientation can be described as rotation about an arbitrary axis and by quaternions.

E.1 Euler Angles

Recall that the Euler angles ϕ, θ, ψ in chapter 3 correspond to successive rotations about body Z-Y-Z axes, and that the corresponding rotation matrix is obtained as

$$(E.1) \quad R = \begin{bmatrix} c_\phi c_\theta c_\psi - s_\phi s_\psi & -c_\phi c_\theta s_\psi - s_\phi c_\psi & c_\phi s_\theta \\ s_\phi c_\theta c_\psi + c_\phi s_\psi & -s_\phi c_\theta s_\psi + c_\phi c_\psi & s_\phi s_\theta \\ -s_\theta c_\psi & s_\theta s_\psi & c_\theta \end{bmatrix}$$

in which s_θ and c_θ denote $\sin\theta$ and $\cos\theta$ respectively.

Consider now the problem of using Euler angles to define a chart on some open set $U \subset SO(3)$. It is easy to see that a single chart cannot cover all of $SO(3)$. For example, if $R_{33} = 1$, it must be the case that $\theta = 0$, and the rotation matrix is given by

$$(E.2) \quad \begin{bmatrix} R_{11} & R_{12} & 0 \\ R_{21} & R_{22} & 0 \\ 0 & 0 & 1 \end{bmatrix} = \begin{bmatrix} c_{\phi+\psi} & -s_{\phi+\psi} & 0 \\ s_{\phi+\psi} & c_{\phi+\psi} & 0 \\ 0 & 0 & 1 \end{bmatrix}.$$

In this case, it is not possible to uniquely define ϕ and ψ, since only their sum is represented in R. A similar case occurs when $R_{33} = -1$.

To define a chart using Euler angles, we begin by defining the open set

$$U = \{R \in SO(3) \mid R_{33} \notin \{-1, 1\}\},$$

and defining the chart Φ such that

$$\Phi(R) \mapsto [\phi(R), \theta(R), \psi(R)]^T \in \mathbb{R}^3.$$

For any $R \in U$, not both of R_{13}, R_{23} are zero. Then the above equations show that $s_\theta \neq 0$. Since not both R_{13} and R_{23} are zero, then $R_{33} \neq \pm 1$, and we have $c_\theta = R_{33}$, $s_\theta = \pm\sqrt{1 - R_{33}^2}$ so

(E.3) $$\theta = \text{atan2}\left(\sqrt{1 - R_{33}^2}, R_{33}\right)$$

or

(E.4) $$\theta = \text{atan2}\left(-\sqrt{1 - R_{33}^2}, R_{33}\right).$$

The function $\theta = \text{atan2}(y, x)$ computes the arc tangent function, where x and y are the cosine and sine, respectively, of the angle θ. This function uses the signs of x and y to select the appropriate quadrant for the angle θ. Note that if both x and y are zero, atan2 is undefined.

If we choose the value for θ given by (E.3), then $s_\theta > 0$, and

(E.5) $$\phi = \text{atan2}(R_{23}, R_{13})$$
(E.6) $$\psi = \text{atan2}(R_{32}, -R_{31}).$$

If we choose the value for θ given by (E.4), then $s_\theta < 0$, and

(E.7) $$\phi = \text{atan2}(-R_{23}, -R_{13})$$
(E.8) $$\psi = \text{atan2}(-R_{32}, R_{31}).$$

Thus there are two solutions depending on the sign chosen for θ.

As described above, when $R_{33} = \pm 1$, only the sum $\phi \pm \psi$ can be determined. For $R_{33} = 1$

(E.9) $$\phi + \psi = \text{atan2}(R_{21}, R_{11})$$
$$= \text{atan2}(-R_{12}, R_{11}).$$

In this case there are infinitely many solutions. We may take $\phi = 0$ by convention. If $R_{33} = -1$, then $c_\theta = -1$ and $s_\theta = 0$, so that $\theta = \pi$. In this case (E.1) becomes

(E.10)
$$
\begin{bmatrix} -c_{\phi-\psi} & -s_{\phi-\psi} & 0 \\ s_{\phi-\psi} & c_{\phi-\psi} & 0 \\ 0 & 0 & 1 \end{bmatrix} = \begin{bmatrix} R_{11} & R_{12} & 0 \\ R_{21} & R_{22} & 0 \\ 0 & 0 & -1 \end{bmatrix}.
$$

The solution is thus

(E.11) $\phi - \psi = \text{atan2}(-R_{12}, -R_{11}) = \text{atan2}(-R_{22}, -R_{21})$.

As before there are infinitely many solutions.

There is nothing special about the choice of axes we used to define Euler angles. We could just as easily have used successive rotations about, say, the x, y, and z axes. In fact, it is easy to see that there are twelve possible ways to define Euler angles: any sequence of three axes, such that no two successive axes are the same, generates a set of Euler angles.

E.2   Roll, Pitch, and Yaw Angles

A rotation matrix R can also be described as a product of successive rotations about the world coordinate axes. These rotations define the *roll*, *pitch*, and *yaw* angles, and they are illustrated in figure E.1. Typically, the order of rotation is taken to be x-y-z: first a yaw about the world x-axis by an angle ψ, then pitch about the world y-axis by an angle θ, and finally a roll about the world z-axis by an angle ϕ^1. Since the successive rotations are relative to the world coordinate frame, the resulting rotation matrix is given by

$$R = R_{z,\phi} R_{y,\theta} R_{x,\psi}$$

$$
= \begin{bmatrix} c_\phi & -s_\phi & 0 \\ s_\phi & c_\phi & 0 \\ 0 & 0 & 1 \end{bmatrix} \begin{bmatrix} c_\theta & 0 & s_\theta \\ 0 & 1 & 0 \\ -s_\theta & 0 & c_\theta \end{bmatrix} \begin{bmatrix} 1 & 0 & 0 \\ 0 & c_\psi & -s_\psi \\ 0 & s_\psi & c_\psi \end{bmatrix}
$$

(E.12)
$$
= \begin{bmatrix} c_\phi c_\theta & -s_\phi c_\psi + c_\phi s_\theta s_\psi & s_\phi s_\psi + c_\phi s_\theta c_\psi \\ s_\phi c_\theta & c_\phi c_\psi + s_\phi s_\theta s_\psi & -c_\phi s_\psi + s_\phi s_\theta c_\psi \\ -s_\theta & c_\theta s_\psi & c_\theta c_\psi \end{bmatrix}.
$$

---

1. As with Euler angles, one can choose a different ordering for the rotations to obtain different *fixed axis* representations of orientation. The term *fixed axis* refers to the fact that successive rotations are taken with respect to axes of the fixed coordinate frame.

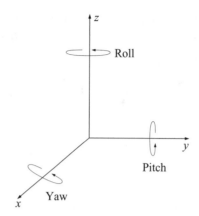

Figure E.1   Roll, pitch, and yaw angles.

The three angles, ϕ, θ, ψ, can be obtained for a given rotation matrix using a method that is similar to that used to derive the Euler angles above.

E.3   Axis-Angle Parameterization

Above we described a rotation matrix by decomposing a rotation into three successive rotations about the coordinate axes. An alternative to this is to specify a rotation matrix in terms of a rotation about an arbitrary axis in space. This provides both a convenient way to describe rotations, and an alternative parameterization for rotation matrices.

Let $k = [k_x, k_y, k_z]^T$ be a unit vector defining an axis expressed in the world frame. To determine the parameterization, we need to derive the rotation matrix $R_{k,\theta}$ representing a rotation of θ degrees about this axis. A simple way to derive this rotation matrix is to rotate the vector k into one of the coordinate axes, say the z-axis, then rotate about this axis by θ, and finally, rotate k back to its original position. As can be seen in figure E.2 we can rotate k into the world z-axis by first rotating about the world z-axis $-\alpha$, then rotating about the world y-axis by $-\beta$. Since all rotations are performed relative to the world frame, the matrix $R_{k,\theta}$ is obtained as

(E.13) $$R_{k,\theta} = R_{z,\alpha} R_{y,\beta} R_{z,\theta} R_{y,-\beta} R_{z,-\alpha}.$$

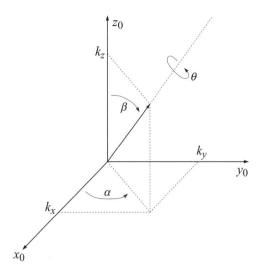

Figure E.2   Rotation about an arbitrary axis.

As can be seen in figure E.2,

(E.14) $$\sin\alpha = \frac{k_y}{\sqrt{k_x^2 + k_y^2}}$$

(E.15) $$\cos\alpha = \frac{k_x}{\sqrt{k_x^2 + k_y^2}}$$

(E.16) $$\sin\beta = \sqrt{k_x^2 + k_y^2}$$

(E.17) $$\cos\beta = k_z.$$

The final two equations follow from the fact that k is a unit vector. Substituting (E.14) through (E.17) into (E.13) we can obtain

(E.18) $$R_{k,\theta} = \begin{bmatrix} k_x^2 v_\theta + c_\theta & k_x k_y v_\theta - k_z s_\theta & k_x k_z v_\theta + k_y s_\theta \\ k_x k_y v_\theta + k_z s_\theta & k_y^2 v_\theta + c_\theta & k_y k_z v_\theta - k_x s_\theta \\ k_x k_z v_\theta - k_y s_\theta & k_y k_z v_\theta + k_x s_\theta & k_z^2 v_\theta + c_\theta \end{bmatrix},$$

in which $v_\theta = 1 - c_\theta$.

We can use this parameterization to derive a chart on $SO(3)$ as follows. Let R be an arbitrary rotation matrix with components (R_{ij}). Let $U = \{R \mid Tr(R) \neq \pm 1\}$ where

Tr denotes the trace of R. By direct calculation using (E.18) we obtain

$$\theta = \cos^{-1}\left(\frac{R_{11} + R_{22} + R_{33} - 1}{2}\right)$$

$$= \cos^{-1}\left(\frac{Tr(R) - 1}{2}\right), \text{ and}$$

$$k = \frac{1}{2\sin\theta}\begin{bmatrix} R_{32} - R_{23} \\ R_{13} - R_{31} \\ R_{21} - R_{12} \end{bmatrix}.$$

This representation is not unique since a rotation of $-\theta$ about $-k$ is the same as a rotation of θ about k, that is,

(E.19) $R_{k,\theta} = R_{-k,-\theta}.$

We can now define the mapping ϕ using k and θ. Since the axis k is a unit vector, only two of its components are independent. Therefore, only three independent quantities are required in this representation of a rotation. Thus, we can define ϕ as

(E.20) $\phi(R) = [\theta k_x, \theta k_y, \theta k_z]^T.$

Using this convention, we can recover k and θ as

(E.21) $k = \dfrac{\phi(R)}{||\phi(R)||}$ and $\theta = ||\phi(R)||.$

The angle θ is a good distance measure between two elements of $SO(3)$.

E.4   Quaternions

The axis-angle parameterization described above parameterizes a rotation matrix by three parameters (given by (E.21)). Quaternions, which are closely related to the axis-angle parameterization, can be used to define a rotation by four numbers. It is straightforward to use quaternions to define an atlas for $SO(3)$ using only four charts. Furthermore quaternion representations are very convenient for operations such as composition of rotations and coordinate transformations. For these reasons, quaternions are a popular choice for the representation of rotations in three dimensions.

Quaternions are a generalization of the complex numbers to a four-dimensional space. For this reason, we begin with a quick review of how complex numbers can be used to represent orientation in the plane. A first introduction to complex numbers often uses the example of representing orientation in the plane using unit magnitude complex numbers of the form $a + ib$, in which $i = \sqrt{-1}$. In this case, the angle θ from the real axis to the vector $(a + ib) \in \mathbb{C}$ is given by atan2(b, a), and it is easy to see that $\cos\theta = a$ and $\sin\theta = b$. Since $a, b \in \mathbb{R}$, we can consider this as an embedding of S^1 in the plane.

Using this representation, multiplication of two complex numbers corresponds to addition of the corresponding angles. This can be verified by direct calculation as

$$
\begin{aligned}
(a_1 + ib_1)(a_2 + ib_2) &= a_1a_2 + ib_1a_2 + ia_1b_2 - b_1b_2 \\
&= (a_1a_2 - b_1b_2) + i(b_1a_2 + a_1b_2) \\
&= \cos\theta_1 \cos\theta_2 - \sin\theta_1 \sin\theta_2 \\
&\quad + i(\sin\theta_1 \cos\theta_2 - \cos\theta_1 \sin\theta_2) \\
&= \cos(\theta_1 + \theta_2) + i\sin(\theta_1 + \theta_2).
\end{aligned}
$$

While a complex number $a + ib$ defines a point in the complex plane, a quaternion defines a point in a four-dimensional complex space, $q_0 + iq_1 + jq_2 + kq_3$. Here, i, j, and k represent independent square roots of negative one. They are independent in the sense that they do not combine using the normal rules of scalar multiplication. In particular, we have

(E.22) $\quad -1 = i^2 = j^2 = k^2,$

(E.23) $\quad i = jk = -kj,$

(E.24) $\quad j = ki = -ik,$

(E.25) $\quad k = ij = -ji.$

It is not a coincidence that multiplication of i, j, and k is similar to the vector cross product for the orthogonal unit basis vectors, $i = [1, 0, 0]^T$, $j = [0, 1, 0]^T$, and $k = [0, 0, 1]^T$.

Complex numbers with unit magnitude can be used to represent orientation in the plane simply by using their representation in polar coordinates. Likewise, quaternions can be used to represent rotations in 3D. In particular, for a rotation about an axis $n = [n_x, n_y, n_z]^T$ by angle θ, the corresponding quaternion, Q, is defined as

(E.26) $\quad Q = \left(\cos\frac{\theta}{2}, n_x \sin\frac{\theta}{2}, n_y \sin\frac{\theta}{2}, n_z \sin\frac{\theta}{2} \right).$

When we define the axis of rotation to be a unit vector, the corresponding quaternion has unit norm, since

$$
\begin{aligned}
\|Q\| &= \cos^2\frac{\theta}{2} + n_x^2 \sin^2\frac{\theta}{2} + n_y^2 \sin^2\frac{\theta}{2} + n_z^2 \sin^2\frac{\theta}{2}
\end{aligned}
$$

(E.27) $\quad = \cos^2\frac{\theta}{2} + \sin^2\frac{\theta}{2}\left(n_x^2 + n_y^2 + n_z^2\right)$

$\quad = 1.$

Quaternions with unit norm are sometimes referred to as rotation quaternions.

It is straightforward to apply the results from section E.3 to determine the rotation matrix $R \in SO(3)$ that corresponds to the rotation represented by a rotation quaternion. For the quaternion $Q = (q_0, q_1, q_2, q_3)$ we have

(E.28) $\quad R(Q) = \begin{bmatrix} 2(q_0^2 + q_1^2) - 1 & 2(q_1 q_2 - q_0 q_3) & 2(q_1 q_3 + q_0 q_2) \\ 2(q_1 q_2 + q_0 q_3) & 2(q_0^2 + q_2^2) - 1 & 2(q_2 q_3 - q_0 q_1) \\ 2(q_1 q_3 - q_0 q_2) & 2(q_2 q_3 + q_0 q_1) & 2(q_0^2 + q_3^2) - 1 \end{bmatrix}.$

Quaternions can be used to define an atlas for $SO(3)$ comprising four charts, (U_i, ϕ_i), with $\phi_i : U_i \to \mathbb{R}^3$. This is most easily done by using two steps. First, for a rotation matrix R, we determine the corresponding quaternion Q. Then, we use Q to determine which chart applies (i.e., we implicitly define the neighborhoods U_i in terms of Q), and use the appropriate ϕ_i to define the local coordinates.

Determining the quaternion that correponds to a rotation matrix amounts to solving the inverse of (E.28), and this can be done by a method similar to that given for the axis-angle parameterization of section E.3. In particular, for rotation matrices R such that $Tr(R) \neq \pm 1$ we have

(E.29) $\qquad q_0 = \frac{1}{2}\sqrt{1 + Tr(R)}$

(E.30) $\qquad \begin{bmatrix} q_1 \\ q_2 \\ q_3 \end{bmatrix} = \frac{1}{4q_0} \begin{bmatrix} R_{32} - R_{23} \\ R_{13} - R_{31} \\ R_{21} - R_{12} \end{bmatrix}.$

To define the four charts, we first define the four neighborhoods

$U_0 = \{Q = (q_0, q_1, q_2, q_3) \mid q_0 \geq q_1, q_2, q_3\}$

$U_1 = \{Q = (q_0, q_1, q_2, q_3) \mid q_1 \geq q_0, q_2, q_3\}$

$U_2 = \{Q = (q_0, q_1, q_2, q_3) \mid q_2 \geq q_0, q_1, q_3\}$

$U_3 = \{Q = (q_0, q_1, q_2, q_3) \mid q_3 \geq q_0, q_1, q_2\}.$

These are not actually open sets (due to the nonstrict inequality in the set definitions), but they can be used to define open sets using their interiors. Now we define the coordinate maps ϕ_i as

$\phi_0(q_0, q_1, q_2, q_3) = \left(\dfrac{q_1}{|q_0|}, \dfrac{q_2}{|q_0|}, \dfrac{q_3}{|q_0|} \right)$

$\phi_1(q_0, q_1, q_2, q_3) = \left(\dfrac{q_0}{|q_1|}, \dfrac{q_2}{|q_1|}, \dfrac{q_3}{|q_1|} \right)$

$$\phi_2(q_0, q_1, q_2, q_3) = \left(\frac{q_0}{|q_2|}, \frac{q_1}{|q_2|}, \frac{q_3}{|q_2|} \right)$$

$$\phi_3(q_0, q_1, q_2, q_3) = \left(\frac{q_0}{|q_3|}, \frac{q_1}{|q_3|}, \frac{q_2}{|q_3|} \right)$$

As we have seen above, $R_i \in SO(3)$ represents a rotation, and the composition of successive rotations, say R_1 and R_2, is represented by the rotation matrix $R = R_1 R_2$. Likewise, multiplication of quaternions corresponds to the composition of successive rotations. In particular, if Q_1 and Q_2 are two quaternions representing a rotation by θ_1 about axis n_1 and a rotation by θ_2 about axis n_2, respectively, then the result of performing these two rotations in succession is represented by the quaternion $Q = Q_1 Q_2$. Using (E.22) through (E.25) it is straightforward to determine the quaternion product. In particular, for two quaternions, X and Y, we compute their product, $Z = XY$, as

$$\begin{aligned} z_0 + iz_1 + jz_2 + kz_3 &= (x_0 + ix_1 + jx_2 + kx_3)(y_0 + iy_1 + yx_2 + yx_3) \\ &= x_0 y_0 - x_1 y_1 - x_2 y_2 - x_3 y_3 \\ &\quad + i(x_0 y_1 + x_1 y_0 + x_2 y_3 - x_3 y_2) \\ &\quad + j(x_0 y_2 + x_2 y_0 + x_3 y_1 - x_1 y_3) \\ &\quad + k(x_0 y_3 + x_3 y_0 + x_1 y_2 - x_2 y_1). \end{aligned}$$

By equating the real parts on both sides of the final equality, and by equating the coefficients of i, j, and k on both sides of the final equality, we obtain

$$\begin{aligned} z_0 &= x_0 y_0 - x_1 y_1 - x_2 y_2 - x_3 y_3 \\ z_1 &= x_0 y_1 + x_1 y_0 + x_2 y_3 - x_3 y_2 \\ z_2 &= x_0 y_2 + x_2 y_0 + x_3 y_1 - x_1 y_3 \\ z_3 &= x_0 y_3 + x_3 y_0 + x_1 y_2 - x_2 y_1. \end{aligned}$$

The quaternion $Q = (q_0, q_1, q_2, q_3)$ can be thought of as having the scalar component q_0 and the vector component $q = [q_1, q_2, q_3]^T$. Therefore, one often represents a quaternion by a pair, $Q = (q_0, q)$. Using this notation, q_0 represents the real part of Q, and q represents the imaginary part of Q. Using this notation, the quaternion product $Z = XY$ can be represented more compactly as

$$\begin{aligned} z_0 &= x_0 y_0 - x^T y \\ z &= x_0 y + y_0 x + x \times y, \end{aligned}$$

in which $\times$ denotes the vector cross product operator.

For complex numbers, the conjugate of $a + ib$ is defined by $a - ib$. Similarly, for quaternions we denote by Q^* the conjugate of the quaternion Q, and define

(E.31) $Q^* = (q_0, -q_1, -q_2, -q_3)$.

With regard to rotation, if the quaternion Q represents a rotation by θ about the axis n, then its conjugate Q^* represents a rotation by θ about the axis $-n$. It is easy to see that

(E.32) $QQ^* = (q_0^2 + ||q||^2, 0, 0, 0)$

and that

(E.33) $||QQ^*|| = ||(q_0^2 + q_1^2 + q_2^2 + q_3^2, 0, 0, 0)|| = \sum q_i^2 = ||Q||^2$.

A quaternion, Q, with its conjugate, Q^*, can be used to perform coordinate transformations. Let the point p be rigidly attached to a coordinate frame $\mathcal{F}$, with local coordinates (x, y, z). If Q specifies the orientation of $\mathcal{F}$ with respect to the base frame, and T is the vector from the world frame to the origin of $\mathcal{F}$, then the coordinates of p with respect to the world frame are given by

(E.34) $Q(0, x, y, z)Q^* + T$,

in which $(0, x, y, z)$ is a quaternion with zero as its real component. Quaternions can also be used to transform vectors. For example, if $n = (n_x, n_y, n_z)$ is the normal vector to the face of a polyhedron, then if the polyhedron is rotated by Q, the new direction of the normal is given by

(E.35) $Q(0, n_x, n_y, n_z)Q^*$.

F   *Polyhedral Robots in Polyhedral Worlds*

LINEAR REPRESENTATIONS are concise. In this appendix we consider the special case in which both the robot and all obstacles in the workspace are polygons (for two-dimensional worlds) or polyhedra (for three-dimensional worlds). Since polyhedra are three-dimensional solids whose faces are polygons, we begin by developing representations and computational methods for dealing with polygons. Although the restriction to polygonal obstacle may seem to be unrealistic, nearly all modern motion planning systems use polygonal models to represent obstacles (e.g., facet models that are common in computer graphics and so-called *polygon soup* models that are used in many CAD applications).

We begin the appendix by describing the representation of polygons in two dimensions. Following this, we describe an algorithm for determining whether two polygons intersect. This is the fundamental operation used by collision detection algorithms. We then describe an efficient algorithm that constructs a boundary representation for the configuration space obstacle region for the special case of $Q = \mathbb{R}^2$ and discuss configuration space obstacles for the case of $Q = SE(2)$.

F.1   Representing Polygons in Two Dimensions

A straight line in the plane divides the plane into three disjoint regions: the line itself, and the two regions that lie on either side of the line. To make this more precise, consider the line given by

$$(F.1) \qquad h(x, y) = ax + by - c = 0.$$

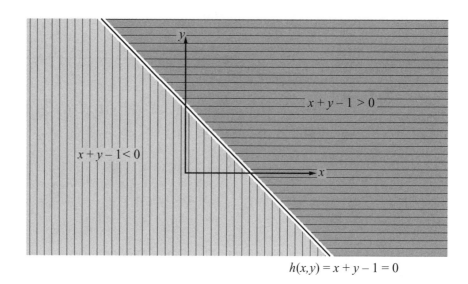

$$h(x,y) = x + y - 1 = 0$$

Figure F.1   Half-planes defined by $h(x, y) = x + y - 1$.

This equation implicitly defines a line to be the set of points whose projection onto the vector (a, b) is given by c. Thus, the vector (a, b) defines the normal to the line and c gives the signed perpendicular distance from the origin to the line. We can evaluate h for any point in the plane. Those points such that $h(x, y) \geq 0$ are said to lie in the *positive half plane,* represented by h^+. Points in h^+ are those points whose projection onto the normal is greater than the signed distance to the line. Those points such that $h(x, y) \leq 0$ are said to lie in the *negative half plane,* represented by h^-. The line itself is the intersection $h^- \cap h^+$. Figure F.1 shows an example for which the points $(0, 5)$, $(3, 5)$ lie in the negative half plane, while points $(0, 0)$, $(2, 2)$ lie in the positive half plane. Note that we can easily change the sense of the half planes by multiplying h by -1.

We can use half planes to construct polygons. In particular, we define a *convex polygonal region* in $\mathbb{R}^2$ to be the intersection of a finite number of half planes. For example, the three lines

$$h_1(x, y) = -x + y - 3$$
$$h_2(x, y) = -y$$
$$h_3(x, y) = x$$

can be used to construct a convex polygonal region by taking the intersection of the three half planes h_1^-, h_2^-, and h_3^-, as shown in figure F.2. For consistency, we will

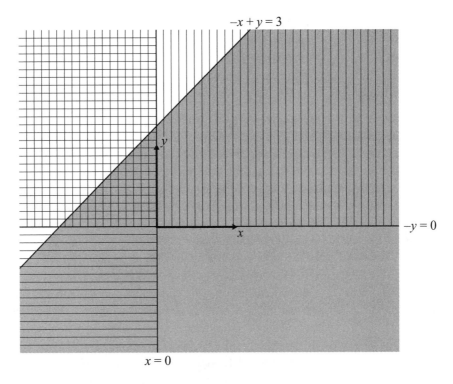

Figure F.2   A convex polygonal region constructed from the half planes h_1^-, h_2^-, and h_3^-.

always define convex polygonal regions as the intersection of negative half planes. If $h_i(x, y) \leq 0$ for each line that defines the convex polygonal region, then the point (x, y) lies inside the corresponding polygonal region. If $h_i(x, y) > 0$ for any line that defines the convex polygonal region, then the point lies outside the corresponding polygonal region. Note that a convex polygonal region need not be finite. For example, by our definition, the half space $x + y - 1 \leq 0$ is a valid convex polygonal region, even though it is unbounded (recall that a region is said to be convex if for all pairs of points in the region, the line segment connecting those points lies entirely within the region).

We define a *polygonal region* (possibly nonconvex) to be any subset of $\mathbb{R}^2$ obtained by taking the union of a finite number of convex polygonal regions. Polygonal regions need not be bounded or connected, and connected polygonal regions need not be simply connected (e.g., the union of two disjoint convex polygons is a polygonal region, but it is not connected). Finally, a *polygon* is any closed, simply connected polygonal region (alternatively, a polygonal region that is homeomorphic to a closed unit disk in the plane).

It is often convenient to represent a polygon by listing its vertices, e.g., in counter-clockwise order (it is straightforward to determine the h_i given the set of vertices). This approach is used in sections F.2 and F.3, where we discuss how to construct the configuration space obstacle and then how to determine if a robot intersects it.

F.2   Intersection Tests for Polygons

In this section, we develop an algorithm for determining whether two polygons have a nonempty intersection. Such intersection tests are the essential primitive operations for collision detection algorithms used by most all modern path planners. Furthermore, for the specific case of $Q = \mathbb{R}^2$ with polygonal obstacles, the intersection test that we develop here provides useful insight for developing an algorithm to explicitly construct the configuration space obstacle region, as described below in section F.3. We begin by considering the specific problem of testing for the intersection of a convex, polygonal robot with a specific convex, polygonal obstacle.

We will assume that the configuration of the robot is specified by $q = (x, y, \theta)$ and that the obstacle polygon is specified by a list of its vertices. It will also be convenient to explicitly represent the normal vectors for each edge of both the robot and the obstacle. We denote these normal vectors by n_i^R for the normal to edge i of the robot and n_j^W for the normal to edge j of obstacle $\mathcal{W}$. Note that the normals for the robot edges depend on the orientation (but not the x, y-coordinates) of the robot; we will often explicitly represent this dependence by the notation $n_i^R(\theta)$. We denote the vertices of the robot by r_i, and the edges by E_i^R. Similarly, we will denote the vertices of obstacle $\mathcal{W}$ by o_j and edges E_j^W. Figure F.3 illustrates the notation.

Under these conditions, the problem of determining whether the robot intersects the obstacle is equivalent to determining whether the robot configuration lies within the configuration space obstacle region. The approach that we develop here identifies the defining half spaces for the configuration space obstacle region for a fixed robot orientation, θ. If the robot configuration is contained in *each* of these half spaces, then it lies in the configuration space obstacle polygon (since this polygon is merely the intersection of the half spaces), and the robot and obstacle intersect.

The problem of identifying these defining half spaces is equivalent to determining the boundary of the configuration space obstacle polygon. Recall that for a fixed value of θ, the boundary of this polygon corresponds to the set of configurations in which the robot and obstacle touch, but do not overlap. (If the robot and obstacle overlap, then we can move the robot to any configuration in a neighborhood and remain within the obstacle polygon.) For the kth obstacle, this condition can be expressed by

(F.2) $R(q) \cap \mathcal{W} \neq \emptyset$ and $\mathrm{int}\,(R(q)) \cap \mathrm{int}\,(\mathcal{W}) = \emptyset.$

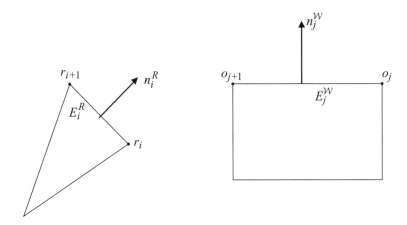

Figure F.3   Notation used to define vertices, normals and edges of the robot and obstacle polygon.

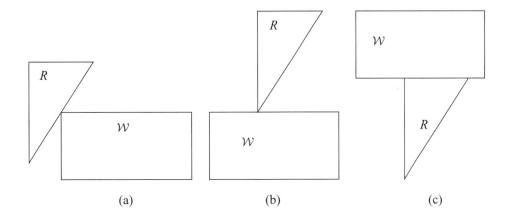

Figure F.4   (a) Type A contact, (b) Type B contact, (c) Both Type A and Type B contact.

For configurations that satisfy (F.2), there are only two possible kinds of contacts:

Type A Contact: an edge of R, say E_i^R, contains a vertex, o_j, of $\mathcal{W}$.

Type B Contact: an edge of $\mathcal{W}$, say $E_j^{\mathcal{W}}$, contains a vertex, r_i, of R.

Each possible type A or type B contact defines one half space that defines the configuration space obstacle polygon. Type A and B contacts are illustrated in figure F.4. Note that in figure F.4(c), both type A and B contacts occur simultaneously.

We begin with the case of type A contact. Type A contact between edge E_i^R and vertex o_j is possible only for certain orientations θ. In particular, such contact can occur only when θ satisfies

(F.3) $(o_{j-1} - o_j) \cdot n_i^R(\theta) \geq 0$ and $(o_{j+1} - o_j) \cdot n_i^R(\theta) \geq 0.$

This condition is sometimes referred to as an *applicability condition*. Note that the normals for the edges of R are a function of configuration, but only of θ, and not of the x, y coordinates. The condition in (F.3) can also be expressed as the condition that a negated edge normal of the robot lies between the normals of an adjacent obstacle edge. This latter formulation of the condition is used below in section F.3. Note that (F.3) is satisfied with equality when an edge of the obstacle is coincident with an edge of the robot.

Each pair, E_i^R and o_j, that satisfies (F.3), defines a half space that contains the configuration space obstacle polygon. This half space is defined by

(F.4) $f_{ij}^R(x, y, \theta) = n_i^R(q) \cdot (o_j - r_i(x, y, \theta)) \leq 0.$

This is illustrated in figure F.5.

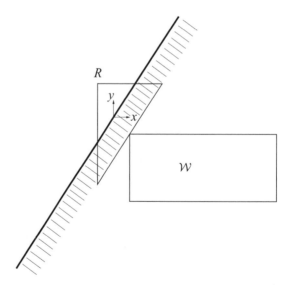

Figure F.5   The half space defined by this contact is below the thick black line that passes through the origin of the robot's coordinate frame.

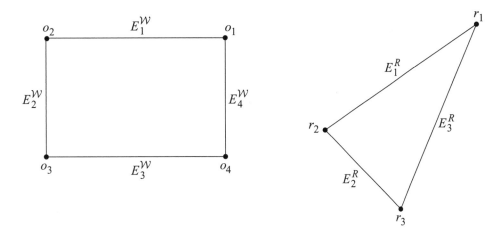

Figure F.6   The obstacle is shown on the left, and the robot on the right.

Type B contact is analogous to type A contact, but the roles of robot and obstacle are reversed. In particular, type B contact can occur between obstacle edge $E_j^{\mathcal{W}}$ and robot vertex r_i when

(F.5) $(r_{i-1}(\theta) - r_i(\theta)) \cdot n_j^{\mathcal{W}} \geq 0 \quad \text{AND} \quad (r_{i+1}(\theta) - r_i(\theta)) \cdot n_j^{\mathcal{W}} \geq 0.$

The corresponding half space is defined by

(F.6) $f_{ij}^{\mathcal{W}}(x, y, \theta) = n_j^{\mathcal{W}} \cdot (r_i(x, y, \theta) - o_j) \leq 0.$

Each type A or B contact defines one half space that contains the configuration space obstacle polygon. The configuration (x, y, θ) causes a collision only if it lies in *each* of these half spaces. Therefore, determining collision amounts to determining which i, j satisfy (F.3) or (F.5), and then verifying (F.4) or (F.6), respectively.

As an example, consider the robot and obstacle shown in figure F.6. Figure F.7(a) shows a case in which the robot and obstacle have a nonempty intersection. The following table shows the possible type A and B contacts (the first three entries of the table are the type A contacts), the definitions of the corresponding half spaces, and whether or not the half space constraints are satisfied. As can be seen, each applicable half space constraint is satisfied, and thus it is determined that the robot and obstacle are in collision.

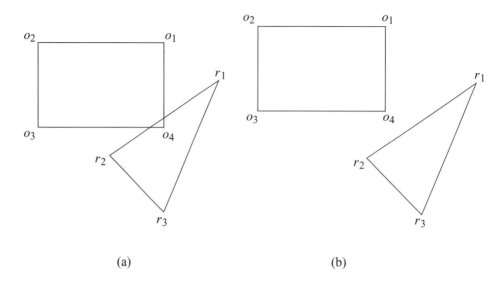

(a)   (b)

Figure F.7   The applicability conditions and half spaces for these two cases are shown in the table below.

Contact pair	half space inequality	satisfied?
E_1^R, o_4^k	$n_1^R(\theta) \cdot (o_4^k - r_1(x, y, \theta)) \leq 0$	yes
E_2^R, o_1^k	$n_2^R(\theta) \cdot (o_1^k - r_2(x, y, \theta)) \leq 0$	yes
E_3^R, o_3^k	$n_3^R(\theta) \cdot (o_3^k - r_3(x, y, \theta)) \leq 0$	yes
$E_1^{W_k}, r_3$	$n_1^{W_k} \cdot (r_3(x, y, \theta) - o_1^k) \leq 0$	yes
$E_2^{W_k}, r_3$	$n_2^{W_k} \cdot (r_3(x, y, \theta) - o_2^k) \leq 0$	yes
$E_3^{W_k}, r_1$	$n_3^{W_k} \cdot (r_1(x, y, \theta) - o_3^k) \leq 0$	yes
$E_4^{W_k}, r_2$	$n_4^{W_k} \cdot (r_2(x, y, \theta) - o_4^k) \leq 0$	yes

Figure F.7(b) shows a case in which the robot and obstacle do not intersect. The following table shows the possible type A and B contacts, the definitions of the corresponding half spaces, and whether or not the half space constraints are satisfied. As can be seen, one of the applicable half space constraints is not satisfied, and thus it is determined that the robot and obstacle are not in collision.

Contact pair	half space inequality	satisfied?
E_1^R, o_4^k	$n_1^R(\theta) \cdot \left(o_4^k - r_1(x, y, \theta)\right) \leq 0$	no
E_2^R, o_1^k	$n_2^R(\theta) \cdot \left(o_1^k - r_2(x, y, \theta)\right) \leq 0$	yes
E_3^R, o_3^k	$n_3^R(\theta) \cdot \left(o_3^k - r_3(x, y, \theta)\right) \leq 0$	yes
$E_1^{W_k}, r_3$	$n_1^{W_k} \cdot \left(r_3(x, y, \theta) - o_1^k\right) \leq 0$	yes
$E_2^{W_k}, r_3$	$n_2^{W_k} \cdot \left(r_3(x, y, \theta) - o_2^k\right) \leq 0$	yes
$E_3^{W_k}, r_1$	$n_3^{W_k} \cdot \left(r_1(x, y, \theta) - o_3^k\right) \leq 0$	yes
$E_4^{W_k}, r_2$	$n_4^{W_k} \cdot \left(r_2(x, y, \theta) - o_4^k\right) \leq 0$	yes

Suppose the robot and obstacles are not convex (note, the case of a nonconvex obstacle includes the case of multiple disconnected obstacle regions in the workspace). In this case, one can always partition the robot and obstacle into collections of convex polygons, $\{R_l\}$ and $\{W_k\}$, respectively. To determine if the robot and obstacle are in collision, we merely check to see if any pair R_l and W_k are in collision, using the method described above.

F.3   Configuration Space Obstacles in $Q = \mathbb{R}^2$: The Star Algorithm

It is sometimes convenient to explicitly represent the configuration space obstacle region in the special case of $Q = \mathbb{R}^2$ (e.g., when using the visibility graph approach described in section 5.1). For a convex robot and obstacle, it is straightforward to derive a boundary representation for the configuration space obstacle region using the ideas developed in the preceding section.

As described above, for each satisfied applicability condition, (F.3) or (F.5), one half space is defined by (F.4) or (F.6), respectively. To construct the representation of the boundary of the configuration space obstacle region, we need only find the vertices that are defined by the intersections of the lines that define these half spaces. The algorithm that we develop here, sometimes called the *star algorithm,* is a particularly efficient way to do so.

The heart of the algorithm lies in the following observations. When the applicability condition

$$(o_{j-1} - o_j) \cdot n_i^R(\theta) \geq 0 \quad \text{and} \quad (o_{j+1} - o_j) \cdot n_i^R(\theta) \geq 0$$

is satisfied and there is a contact between E_i^R and vertex o_j, of W, this contact will be maintained as the robot translates, maintaining contact with the vertex. At one extreme of this motion, the vertices o_j and r_i coincide, while at the other extreme, vertices o_j and r_{i+1} coincide. These extremes define two vertices of the configuration space obstacle region

$$o_j - r_i(0, 0, \theta), \quad \text{and} \quad o_j - r_{i+1}(0, 0, \theta).$$

Analogously, when the applicability condition

$$(r_{i-1}(\theta) - r_i(\theta)) \cdot n_j^{\mathcal{W}} \geq 0 \quad \text{and} \quad (r_{i+1}(\theta) - r_i(\theta)) \cdot n_j^{\mathcal{W}} \geq 0$$

is satisfied and there is a contact between obstacle edge $E_j^{\mathcal{W}}$ and robot vertex r_i, this contact will be maintained as the robot translates, maintaining contact with the edge. At one extreme of this motion, the vertices o_j and r_i coincide, while at the other extreme, vertices o_{j+1} and r_i coincide. These extremes define two vertices of the configuration space obstacle region

$$o_j - r_i(0, 0, \theta), \quad \text{and} \quad o_{j+1} - r_i(0, 0, \theta).$$

The enumeration of satisfied applicability conditions can be made particularly efficient by recalling that these conditions can be expressed in terms of the orientations of the robot and obstacle edge normals. We first negate the edge normals of the robot, then sort the merged list of obstacle and negated robot edge normals by orientation. We then scan this sorted list, and construct the appropriate vertices each time a negated robot edge normal lies between adjacent obstacle edge normals, or vice versa.

We note here that the algorithm described above is an implementation of the *Minkowski difference,* a useful operation in many computational geometry applications. The Minkowski difference between the robot and a convex obstacle is defined by

(F.7) $$\mathcal{WO} \ominus R(q) = \{q \in \mathcal{Q} : q = c - r \quad \text{where } r \in R(q) \quad \text{and} \quad c \in \mathcal{WO}\}$$

where $\ominus$ is the *Minkowski* difference operator [124].

F.4   Configuration Space Obstacles in $\mathcal{Q} = SE(2)$

As we have seen in chapter 3, a polygon in the plane has three degrees of freedom, two for translation and one for rotation, and its configuration space is $\mathcal{Q} = SE(2)$. Consider a polygonal robot in a workspace that contains a single obstacle. For a fixed orientation, the configuration space of the polygon is reduced to $\mathbb{R}^2$. Thus, one way to visualize this configuration space is to "stack" a set of two-dimensional configuration spaces, where each slice in the stack corresponds to the (x, y) configurations of the robot at a fixed orientation θ and the vertical axis represents the orientation of the robot. An example is shown in figure F.8.

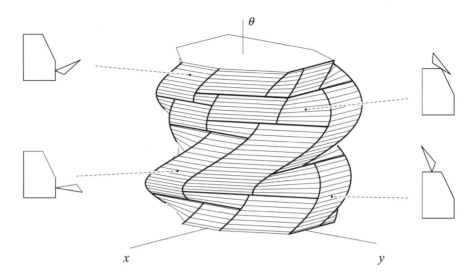

Figure F.8   The configuration space obstacle for a triangle-shaped robot in a workspace that contains a single, five-sided obstacle [69, 70].

F.5   Computing Distances between Polytopes in $\mathbb{R}^2$ and $\mathbb{R}^3$

In many applications, it is useful to know the minimum distance between two objects in addition to knowing whether or not they are in contact. We have seen in chapter 2 that knowledge of distance is essential for implementing the Bug family of algorithms. Moreover, minimum distance calculations are essential for collision detection, which is merely a special case of minimum distance calculations: if the minimum distance between two objects is zero, then they are in contact. In this section, we present an algorithm originally described by Gilbert, Johnson and Keerthi for computing the distance between convex polytopes, commonly referred to as the GJK distance computation algorithm [163].

We define the distance between polytopes A and B as

(F.8) $d(A, B) = \min\limits_{a \in A, b \in B} \|a - b\|.$

We reformulate (F.8) in terms of the Minkowski difference of two polytopes, i.e.,

(F.9) $A \ominus B = \{z \mid z = a - b, a \in A, b \in B\} = Z.$

Using (F.9) we can rewrite (F.8) as

(F.10) $d(A, B) = \min\limits_{a \in A, b \in B} \|a - b\| = \min\limits_{z \in A \ominus B} \|z\|,$

and we have reduced the problem of computing the distance between two polytopes to the problem of computing the minimum distance from one polytope to the origin.

In section F.3 we have seen an implementation of the Minkowski difference to construct the configuration space obstacle region. From this, it is easy to see that the Minkowski difference of two convex polytopes is itself a convex polytope. Since $Z = A \ominus B$ is a convex set, and since the norm, $\|z\|$, is a convex function, $z^* = \arg\min_{z \in Z} \|z\|$ is unique. Thus, there is a unique solution to (F.10). Note that the values of a and b that achieve this minimum are *not* necessarily unique.

Although finding the distance from Z to the origin may seem simpler than computing the distance between A and B, it should be noted that this is actually the case only if the necessary computations to determine $\min_{z \in Z} \|z\|$ are simpler than the computations required to compute $d(A, B)$ directly. This turns out to be the case for the GJK algorithm. Before we examine how this algorithm can be applied to the Minkowski difference of A and B, we first describe the algorithm for the case of computing the distance from Z to the origin, for Z any convex polytope.

Suppose Z is a polytope in $\mathbb{R}^n$ (i.e., $n = 2$ for polygons, and $n = 3$ for polyhedra). The GJK algorithm iteratively constructs a sequence of polytopes, each of which is the convex hull of some subset of the vertices of Z, such that at each iteration the distance from the origin to the new polytope decreases. Before describing the algorithm more formally, we define some useful terminology and notation.

The *convex hull* of a set of points in $\mathbb{R}^n$ is the smallest convex set in $\Re^n$ that contains those points. Efficient algorithms exist for computing the convex hull of general point sets, but for our purposes, we will not require such algorithms, since the GJK algorithm only deals with point sets of size three for polygons and size four for polyhedra. The convex hull of a set of three (noncollinear) points is the triangle defined by those points, and the convex hull of a set of four (noncoplanar) points is the tetrahedron defined by those points.

The GJK algorithm relies heavily on the notion of projection. In particular, for a convex set Z and a point x, the GJK algorithm computes the point $z \in Z$ with maximal projection onto x. The value of this projection operation is defined by

(F.11) $h_Z(x) = \max\{z \cdot x \mid z \in Z\}.$

and the point z^* that achieves this maximum is defined by

(F.12) $s_Z(x) = z^*   s.t.   z^* \cdot x = h_Z(x).$

The GJK algorithm for polygons is given as Algorithm 23 below. In the first step, the working vertex set V_0 is initialized to contain three arbitrarily selected vertices of the polygon, Z. At iteration k, the point x_k is determined as the point in the convex hull of the vertices in V_k that is nearest to the origin. Once x_k has been determined,

---

Algorithm 23 GJK Algorithm

Input: A polytope, $Z \subset \mathfrak{R}^2$.

Output: Minimal $\|z\|$, for $z \in Z \subset \mathfrak{R}^2$

1: $V_0 \leftarrow \{y_1, y_2, y_3\}$ with y_i vertices of Z

2: $k \leftarrow 0$

3: Compute x_k, the point in the convex hull of V_k that is nearest the origin, i.e., $x_k = \arg\min_{x \in hull(V_k)} \|x\|$.

4: Compute $h_Z(-x_k)$, and terminate if $\|x_k\| = h_Z(-x_k)$.

5: $z_k \leftarrow s_Z(-x_k)$, i.e., the projection of z_k onto x_k is nearer the origin than the projection onto x_k of any other point in Z.

6: x_k is contained in an edge of the convex hull of V_k. Let V_{k+1} contain the two vertices that bound this edge and the point z_k.

7: $k \leftarrow k + 1$

8: Go to 3.

---

in step 5 a new vertex z_k is chosen as the vertex of the original polygon, Z, whose projection onto $-x_k$ is maximal. The point z_k then replaces a vertex in the current working vertex set to obtain a new working vertex set, V_{k+1}. The algorithm terminates (step 4) when x_k is itself the closest point in Z to the origin.

It is a fairly simple matter to extend the GJK algorithm (Algorithm 23) to the case in which $Z = A \ominus B$. Note that in the GJK algorithm, we never need an explicit representation of Z. We only need to compute two functions of Z: $h_Z(x)$ and $s_Z(x)$. Each of these can be computed without explicitly constructing Z. Let $Z = A \ominus B$. We can compute $h_{A \ominus B}(x)$ as follows,

$$
\begin{aligned}
h_{A \ominus B}(x) &= \max\{z \cdot x \mid z \in Z\} \\
&= \max\{z \cdot x \mid z \in A \ominus B\} \\
&= \max\{(a - b) \cdot x \mid a \in A, b \in B\} \\
&= \max\{a \cdot x - b \cdot x \mid a \in A, b \in B\} \\
&= \max\{a \cdot x \mid a \in A\} - \min\{b \cdot x \mid b \in B\} \\
&= \max\{a \cdot x \mid a \in A\} + \max\{b \cdot (-x) \mid b \in B\} \\
&= h_A(x) + h_B(-x).
\end{aligned}
$$

(F.13)

Now, suppose that a^* achieves the value $h_A(x)$ and b^* achieves the value $h_B(-x)$. Then $z^* = a^* - b^*$, and therefore we have

(F.14) $\qquad s_{A \ominus B}(x) = s_A(x) - s_B(-x).$

Thus, we see that the GJK algorithm is easily extended to the case of the Minkowski difference of convex polygons. In steps 4 and 5, merely replace h_z and s_Z with the expressions (F.13) and (F.14). To extend the algorithm to convex polyhedra, merely replace step 1 of the algorithm by

1. $V_0 \leftarrow \{y_1, y_2, y_3, y_4\}$ with y_i vertices of Z

and step 6 of the algorithm by

6. x_k is contained in a face of the convex hull of V_k. Let V_{k+1} contain the three vertices that bound this face and the point z_k.

G   *Analysis of Algorithms and Complexity Classes*

G.1   Running Time

Yet another way to study an algorithm is to compute the running time of the algorithm purely as a function of the length of the input. *Worst-case analysis* considers the longest running time of all inputs of a particular length. *Average-time analysis* considers the average of all the running times of inputs of a particular length. The worst-case analysis is typically referred to as the *running time* of an algorithm.

Finding an expression for the exact running time is often difficult, but in most cases close estimations are possible. *Asymptotic analysis* provides the means of analyzing the running time of the algorithms for large inputs. As the lengths of the inputs become large, the high-order terms dominate the value of the expression and the low-order terms have little or no effect. Hence, a close approximation can be obtained only by considering the highest-order term in an expression. For example, the highest-order term of the function $f(n) = 2n^5 + 100n^3 + 27n + 2003$ is $2n^5$. As n becomes large, disregarding the coefficient 2, the function $f(n)$ behaves like n^5 and it is said that $f(n)$ is asymptotically at most n^5. The following are some common definitions that are useful in analyzing the asymptotic behavior of functions and hence algorithms.

DEFINITION G.1.1   *Let f and g be two functions $f, g : \mathbb{N} \to \mathbb{R}^+$.*

- *The function g is an* asymptotically upper bound *for f, denoted $f(n) \in O\,(g(n))$ and read f is big-O of g, if there* exists *a constant $c > 0$ and $n_0 \in \mathbb{N}$ such that $\forall n \geq n_0$,*

$$f(n) \leq cg(n).$$

■ *The function g is an* asymptotically strict upper bound *for f, denoted $f(n) \in o(g(n))$ and read f is small-o of g, if for* every *constant $c > 0$ and $n_0 \in \mathbb{N}$ such that $\forall n \geq n_0$,*

$$f(n) < cg(n).$$

■ *The function g is an* asymptotically lower bound *for f, denoted $f(n) \in \Omega(g(n))$ and read f is big-Omega of g, if there* exists *a constant $c > 0$ and $n_0 \in \mathbb{N}$ such that $\forall n \geq n_0$,*

$$cf(n) \geq g(n).$$

■ *The function g is an* asymptotically strict lower bound *for f, denoted $f(n) \in \omega(g(n))$ and read f is small-omega of g, if for* every *constant $c > 0$ and $n_0 \in \mathbb{N}$ such that $\forall n \geq n_0$,*

$$f(n) > cg(n).$$

■ *The function g is* asymptotically equal *to f, denoted $f(n) \in \Theta(g(n))$ and read f is theta of g, if*

$$f(n) \in O(g(n)) \quad and \quad f(n) \in \Omega(g(n)).$$

Considering only the highest terms and disregarding constant factors, the big-O notation says that the function f is no more than the function g. The big-O notation is thought of as suppressing a constant factor. For example, $f(n) = 5n^4 + 7n^3 - 4n^2 \in O(n^4)$, $f(n) = n^{\log n} + n^{100} \in O(n^{\log n})$, etc. When f is strictly less than g, the small-o notation is used. The small o-notation indicates that the function g grows much faster than the function f. For example, $f(n) = \log^{100} n \in o(n^{1/100})$, $f(n) = n^{40} \in o(2^n)$, etc. The notations Ω and ω express the opposite of O and o notations, respectively. Thus, the big-Omega notation indicates that f grows no slower than g, and the small-omega notation indicates that f grows faster than g. When the functions f and g grow at the same rate, the Θ notation is used. For example, $f(n) = 3n^5 + n^4 \in \Theta(n^5)$.

When describing the running time of different algorithms, certain terms come up frequently. The running times of common algorithms such as matrix multiplication, sorting, shortest path, etc., are $O(n^c)$, where c is some positive constant. In such cases, it is said that the running time is polynomial in the length of the input n. Other algorithms, such as satisfiability of Boolean expressions, Hamiltonian paths, decomposition of integers into prime factors, etc., are $O(2^{n^c})$, where c is some positive constant. Such algorithms are said to be running in exponential time. The following table summarizes some of the common characterizations of the running time of algorithms (c is some positive constant).

	Running time
constant	$O(1)$
logarithmic	$O(\log n)$
polylogarithmic	$O(\log^c n)$
linear	$O(n)$
polynomial	$O(n^c)$
quasipolynomial	$O\left(n^{\log^c n}\right)$
exponential	$O\left(2^{n^c}\right)$
doubly expnonential	$O\left(2^{2^{n^c}}\right)$

G.2   Complexity Theory

The goal of *complexity theory* is to characterize the amount of resources needed for the computation of specific problems. Common resources include sequential time, sequential space, number of gates in Boolean circuits, parallel time in a multiprocessor machine, etc. The exact complexity of a problem is determined by the amount of resources that is both sufficient and necessary for its solution. Sufficiency implies an upper bound on the amount of resources needed to solve the problem for every instance of the input. Necessity implies a lower bound, i.e., for some instance of the input, at least a certain amount of resources is required to solve the problem.

The amount of resources that is needed to solve a problem allows for an elegant classification of problems according to their computational complexity. Researchers have developed the notion of *complexity classes*, where a complexity class is defined by specifying (a) the type of computation model M, (b) the resource R which is measured in this model, and (c) an upper bound U on this resource. A complexity class, then, consists of all problems requiring at most an amount U of resource R for their solution in the model M. Thus, the *complexity of a problem* is determined by finding to which complexity classes it belongs (by providing upper bounds on the resource) and to which complexity classes it does not belong (by providing lower bounds). To define complexity classes more precisely, we will need to make use of definitions of alphabets, strings, and languages.

Input Representation

The amount of the resource used in a complexity class is expressed in terms of the length of the input. It is not clear, however, how to define the length of the input since it can be of different types and values, i.e., integers, names, graphs, matrices, etc. It is convenient to have a unique and clear definition of the length of the input. To this end, researchers have proposed the encoding of inputs as strings over a set of symbols and have defined the length of the input as the number of symbols of the encoding string.

DEFINITION G.2.1   *An alphabet, usually denoted by Σ, is any finite set of symbols.*

DEFINITION G.2.2   *A string s over an alphabet Σ is a sequence of symbols from Σ. The length of a string s, denoted $|s|$, is equal to the number of its symbols. The set of all strings over the alphabet Σ is denoted by Σ^*.*

The encoding of an input a is denoted by $\mathrm{enc}(a)$. To illustrate, let $\Sigma = \{0, 1\}$. Then, integers can be encoded in standard binary form, e.g., the encodings of 5 and 35 are 101 and 100001 of lengths 3 and 6, respectively. The encoding of a graph $G = (V, E)$, where $V = \{v_1, \ldots, v_n\} \subset \mathbb{N}$ and $E = \{(v'_1, v''_1), \ldots, (v'_m, v''_m)\} \subseteq V \times V$, can be obtained by concatenating the encodings of its vertex set and its edge set. The vertex set and the edge set can be encoded by concatenating the encodings of the vertices and of the edges, respectively. Special markers can be used to indicate the ending of a vertex and edge encoding. Thus, $\mathrm{enc}(G) = \mathrm{enc}(v_1) \circ \mathrm{enc}(*) \circ \cdots \circ \mathrm{enc}(v_n) \circ \mathrm{enc}(*) \circ \mathrm{enc}(+) \circ \mathrm{enc}(v'_1) \circ \mathrm{enc}(*) \circ \mathrm{enc}(v''_1) \circ \mathrm{enc}(*) \circ \cdots \circ \mathrm{enc}(v'_m) \circ \mathrm{enc}(*) \circ \mathrm{enc}(v''_m) \circ \mathrm{enc}(*)$, where $\mathrm{enc}(*)$ and $\mathrm{enc}(+)$ are the encodings of special markers used to separate vertices and indicate the start of the edge encodings, respectively, and $\circ$ denotes concatenation.

Problem Abstraction

A problem can be thought of as mapping an input instance to a solution. In many cases, we are interested in problems whose solution is either "yes" or "no." Such problems are known as *decision problems*. For example, the graph-coloring problem asks whether it is possible to color the vertices of a graph $G = (V, E)$ using k different colors such that no two vertices connected by an edge have the same color. In many other cases, we are interested in finding the best solution according to some criteria. Such problems are known as *optimization problems*. To continue our example, we may be interested in determining the minimum number of colors needed to color a graph. Generally, an optimization problem can be cast as a decision problem by imposing an upper bound. In our example, we can determine the minimum number of colors needed to color a graph $G = (V, E)$ by invoking the corresponding decision problem with $k = 1, \ldots, |V|$ until the answer to the decision problem is "yes."

In the rest of the section, we restrict our attention to decision problems since their definition is more amendable to complexity analysis and since other problems can be cast as decision problems.

Languages

Languages provide a convenient framework for expressing decision problems.

DEFINITION G.2.3 *A language L over an alphabet Σ is a set of strings over the alphabet Σ, i.e., $L \subseteq \Sigma^*$.*

The language defined by a decision problem includes all the input instances whose solution is "yes." For example, the graph-coloring problem defines the language whose elements are all the encodings of graphs that can be colored using k colors.

Acceptance of Languages

An algorithm A accepts a string $s \in \Sigma^*$ if the output of the algorithm $A(s)$ is "yes." The string s is rejected by the algorithm if its output $A(s)$ is "no." The language L accepted by an algorithm A is the set of strings accepted by the algorithm, i.e.,

$$L = \{s : s \in \Sigma^* \text{ and } A(s) = \text{"yes"}\}.$$

Note that even if L is the language accepted by the algorithm A, given some input string $s \notin L$, the algorithm will not necessarily reject s. It may never be able to determine that $s \notin L$ and thus loop forever. Language L is *decided* by an algorithm A if for every string $s \in \Sigma^*$, A accepts s if $s \in L$ and A rejects s if $s \notin L$. If L is decided by A, it guarantees that on any input string the algorithm will terminate.

DEFINITION G.2.4 *Let $t : \mathbb{N} \to \mathbb{N}$ be a function. An algorithm A decides a language L over some alphabet Σ in time $O(t(n))$ if for every string s of length n over Σ, the algorithm A in $O(t(n))$ steps accepts s if $s \in L$ or rejects s if $s \notin L$. Language L is decided in time $O(t(n))$.*

Complexity Classes

We are now ready to define some of the most important complexity classes. We start with the definition of the polynomial-time complexity class.

DEFINITION G.2.5 *A language L is in P if there exists a polynomial-time algorithm A that decides L.*

The complexity class P encompasses a wide variety of problems such as sorting, shortest path, Fourier transform, etc. Roughly speaking, P corresponds to all the problems that admit an efficient algorithm. Generally, we think of problems that are solvable by polynomial time algorithms as being tractable, or easy, and problems that require superpolynomial time as being intractable, or hard.

Indeed, for many problems there are no polynomial-time algorithms. For example, deciding whether or not a graph $G = (V, E)$ can be colored with three colors is not known to be in P. These problems can be solved by brute-force algorithms in exponential time.

DEFINITION G.2.6   *A language L is in* EXPTIME *if there exists an exponential-time algorithm A that decides L.*

Interestingly enough, many of these hard problems have a feature that is called polynomial-time verifiability. That is, although currently it is not possible to solve these problems in polynomial time, if a candidate solution to the problem, called a certificate, is given, the correctness of the solution can be verified in polynomial time. For example, a certificate for the graph-coloring problem with three colors would be a mapping that for each vertex indicates its color. The correctness can be verified in polynomial time by examining all the edges and checking for each edge that the colors of its two vertices are different. This observation is captured by the following definition.

DEFINITION G.2.7   *A language L is in NP if there exists a polynomial-time verifier algorithm A and a constant c such that for every string s there exists a certificate y of length $O(|s|^c)$ such that $A(s, y) =$ "yes" if $s \in L$ and $A(s, y) =$ "no" if $s \notin L$.*

It is clear that P $\subseteq$ NP since any language that can be decided in polynomial time can also be decided without the need of a certificate. The most fundamental question in complexity theory is whether P $\subset$ NP or P $=$ NP. After many years of extensive research the question remains unanswered. An important step was made in the 70s when Cook and Levin related the complexity of certain NP problems to the complexity of all NP problems. They were able to prove that if a polynomial-time algorithm existed for one of these problems, then a polynomial-time algorithm could be constructed for any NP problem. These special problems form an important complexity class known as NP-complete.

DEFINITION G.2.8   *A language L_1 is polynomial time reducible to a language L_2, denoted $L_1 \leq_p L_2$, if there exists a polynomial time computable function $f : \Sigma^* \to \Sigma^*$, such that for every $s \in \Sigma^*$,*

$$s \in L_1 \iff f(s) \in L_2.$$

f is called the reduction function and the algorithm F that computes f is called the reduction algorithm.

If a language L_1 is reducible to a language L_2 via some polynomial-time computable function f, and if L_2 has a polynomial-time algorithm A_2, then we can construct a polynomial-time algorithm A_1 for L_1. Given some input string s, algorithm A_1 invokes F to compute $f(s)$ and then invokes A_2 on $f(s)$ and gives the same answer as A_2. Thus, via reductions, the solution of one problem can be used to solve other problems.

DEFINITION G.2.9   *A language L is in* NP-complete *if*

1. $L \in$ NP, and

2. if $L' \in$ NP, then $L' \leq_p L$.

If L satisfies the second condition, but not necessarily the first condition, then L is NP-hard.

It is clear now that if an NP-complete problem has a polynomial-time algorithm, then via reductions it is possible to construct a polynomial-time algorithm for any problem in NP. This would imply that P = NP.

In addition to time, another common resource of interest is space. Using the same framework, complexity classes can be defined based on the amount of space the algorithms use to solve problems.

DEFINITION G.2.10   *Let $t : \mathbb{N} \rightarrow \mathbb{N}$ be a function. An algorithm A decides a language L over some alphabet Σ in space $O(t(n))$ if for every string s of length n over Σ, the algorithm A using at most $O(t(n))$ space accepts s if $s \in L$ or rejects s if $s \notin L$. The language L is decided in space $O(t(n))$.*

DEFINITION G.2.11   *A language L is in* PSPACE *if there exists a polynomial-space algorithm A that decides L.*

DEFINITION G.2.12   *A language L is in* PSPACE-complete *if*

1. $L \in$ PSPACE, and

2. if $L' \in$ PSPACE, then $L' \leq_p L$.

If L satisfies the second condition, but not necessarily the first condition, then L is PSPACE-hard.

It can be easily shown that the relationship between the different complexity classes that have been defined in this section is as follows:

$$P \subseteq NP \subseteq PSPACE \subseteq EXPTIME.$$

G.3   Completeness

When describing robotics algorithms in this book, several notions of "completeness" are used.

DEFINITION G.3.1   *An algorithm A is* complete *if in a finite amount of time, A always finds a solution if a solution exists or otherwise A determines that a solution does not exist.*

DEFINITION G.3.2   *An algorithm A is* resolution complete *if in a finite amount of time and for some small resolution step $\epsilon > 0$, A always finds a solution if a solution exists or otherwise A determines that a solution does not exist.*

DEFINITION G.3.3   *An algorithm A is* probabilistically complete *if the probability of finding a solution, if a solution exists, converges to 1, when the running time approaches infinity.*

Complete algorithms include many common algorithms such as A^*, shortest-path, scheduling problems, etc. Resolution complete algorithms have to approximate a continuous measure by discretizing it at small steps. Ray tracing from graphics algorithms and sampling-based planning algorithms that use a grid representation of the configuration space are examples of resolution complete algorithms. Probabilistic completeness guarantees that given enough time, a solution will be found (if a solution exists). If a solution does not exist, the algorithm may not be able to necessarily detect this fact and thus runs forever. In practice, probabilistic algorithms terminate and declare failure if an upper bound on the amount of time the algorithm could use has passed and a solution has not been found. The basic Probabilistic RoadMap planner (PRM) is an example of a probabilistically complete algorithm. Such an algorithm trades completeness for efficiency; in many cases, for the same problem, a probabilistically complete algorithm will find a solution faster (if a solution exists) than a complete algorithm.

H *Graph Representation and Basic Search*

H.1   Graphs

A graph is a collection of *nodes* and *edges,* i.e., $G = (V, E)$. See figure H.1. Sometimes, another term for a node is *vertex,* and this chapter uses the two terms interchangeably. We use G for graph, V for vertex (or node), and E for edge. Typically in motion planning, a node represents a salient location, and an edge connects two nodes that correspond to locations that have an important relationship. This relationship could be that the nodes are mutually accessible from each other, two nodes are within line of sight of each other, two pixels are next to each other in a grid, etc. This relationship does not have to be mutual: if the robot can traverse from nodes V_1 to V_2, but not from V_2 to V_1, we say that the edge E_{12} connecting V_1 and V_2 is directed. Such a collection of nodes and edges is called a *directed graph*. If the robot can travel from V_1 to V_2 and vice versa, then we connect V_1 and V_2 with two directed edges E_{12} and E_{21}. If for each vertex V_i that is connected to V_j, both E_{ij} and E_{ji} exist, then instead of connecting V_i and V_j with two directed edges, we connect them with a single undirected edge. Such a graph is called an *undirected graph*. Sometimes, edges are annotated with a non-negative numerical value reflective of the costs of traversing this edge. Such values are called *weights*.

A *path* or *walk* in a graph is a sequence of nodes $\{V_i\}$ such that for adjacent nodes V_i and V_{i+1}, $E_{i\ i+1}$ exists (and thus connects V_i and V_{i+1}). A graph is *connected* if for all nodes V_i and V_j in the graph, there exists a path connecting V_i and V_j. A *cycle* is a

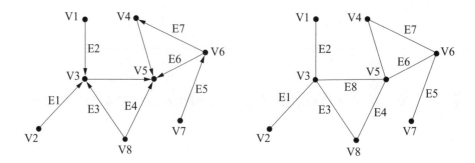

Figure H.1   A graph is a collection of nodes and edges. Edges are either directed (left) or undirected (right).

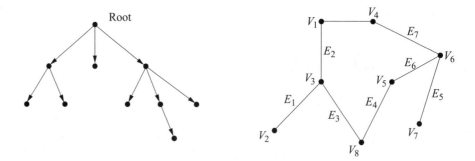

Figure H.2   A tree is a type of directed acyclic graph with a special node called the *root*. A cycle in a graph is a path through the graph that starts and ends at the same node.

path of *n* vertices such that first and last nodes are the same, i.e., $V_1 = V_n$ (figure H.2). Note that the "direction" of the cycle is ambiguous for undirected graphs, which in many situations is sufficient. For example, a graph embedded in the plane can have an undirected cycle which could be both clockwise and counterclockwise, whereas a directed cycle can have one orientation.

A *tree* is a connected directed graph without any cycles (figure H.2). The tree has a special node called the *root*, which is the only node that possesses no incoming arc. Using a parent-child analogy, a parent node has nodes below it called children; the root is a parent node but cannot be a child node. A node with no children is called a *leaf*. The removal of any nonleaf node breaks the connectivity of the tree.

Typically, one searches a tree for a node with some desired properties such as the goal location for the robot. A *depth-first search* starts at the root, chooses a child, then that node's child, and so on until finding either the desired node or a leaf. If the search encounters a leaf, the search then backs up a level and then searches through an unvisited child until finding the desired node or a leaf, repeating this process until the desired node is found or all nodes are visited in the graph (figure H.3).

Breadth-first search is the opposite; the search starts at the root and then visits all of the children of the root first. Next, the search then visits all of the grandchildren, and so forth. The belief here is that the target node is near the root, so this search would require less time (figure H.3).

A grid induces a graph where each node corresponds to a pixel and an edge connects nodes of pixels that neighbor each other. Four-point connectivity will only have edges to the north, south, east, and west, whereas eight-point connectivity will have edges to all pixels surrounding the current pixel. See figure H.4.

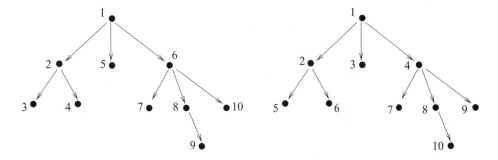

Figure H.3   Depth-first search vs. breadth-first search. The numbers on each node reflect the order in which nodes are expanded in the search.

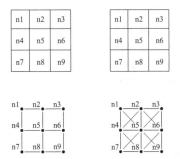

Figure H.4   Four-point connectivity assumes only four neighbors, whereas eight-point connectivity has eight.

Figure H.5   Queue vs. stack.

As can be seen, the graph that represents the grid is not a tree. However, the breadth-first and depth-first search techniques still apply. Let the *link length* be the number of edges in a path of a graph. Sometimes, this is referred to as edge depth. Link length differs from path length in that the weights of the edges are ignored; only the number of edges count. For a general graph, breadth-first search considers each of the nodes that are the same link length from the start node before going onto child nodes. In contrast, depth-first search considers a child first and then continues through the children successively considering nodes of increasing link length away from the start node until it reaches a childless or already visited node (i.e., a cycle). In other words, termination of one iteration of the depth-first search occurs when a node has no unvisited children.

The wave-front planner (chapter 4, section 4.5) is an implementation of a breadth-first search. Breadth-first search, in general, is implemented with a list where the children of the current node are placed into the list in a first-in, first-out (FIFO) manner. This construction is commonly called a *queue* and forces all nodes of the same linklength from the start to be visited first (figure H.5). The breadth-first search starts with placing the start node in the queue. This node is then *expanded* by it being popped off (i.e., removed from the front) the queue and all of its children being placed onto it. This procedure is then repeated until the goal node is found or until there are no more nodes to expand, at which time the queue is empty. Here, we expand all nodes of the same level (i.e., link length from the start) first before expanding more deeply into the graph.

Figure H.6 displays the resulting path of breadth-first search. Note that all paths produced by breadth-first search in a grid with eight-point connectivity are optimal with respect to the "eight-point connectivity metric." Figure H.7 displays the link lengths for all shortest paths between each pixel and the start pixel in the free space in Figure H.6. A path can then be determined using this information via a gradient descent of link length from the goal pixel to the start through the graph as similarly done with the wavefront algorithm.

Depth-first search contrasts breadth-first search in that nodes are placed in a list in a last-in, first-out (LIFO) manner. This construction is commonly called a *stack* and forces nodes that are of greater and greater link length from the start node to be

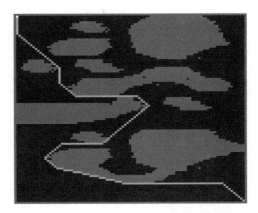

Figure H.6 White pixels denote the path that was determined with breadth-first search.

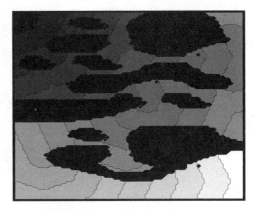

Figure H.7 A plot of linklength values from the start (upper left corner) node where colored pixels correspond to link length (where the lighter the pixel the greater the linklength in the graph) and black pixels correspond to obstacles.

visited first. Now the expansion procedure sounds the same but is a little bit different; here, we pop the stack and push all of its children onto the stack, except popping and pushing occur on the same side of the list (figure H.5). Again, this procedure is repeated until the goal node is found or there are no more nodes to expand. Here, we expand nodes in a path as deep as possible before going onto a different path.

Figure H.8 displays the resulting path of depth-first search. In this example, depth-first search did not return an optimal path but it afforded a more efficient search

Figure H.8   White pixels denote the path that was determined with depth-first search.

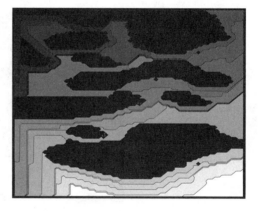

Figure H.9   A plot of linklength values from the start (upper left corner) node where colored pixels correspond to link lengths of paths defined by the depth-first search. The lighter the pixel the greater the linklengths in the graph; black pixels correspond to obstacles.

in that the goal was found more quickly than breadth-first search. Figure H.9 is similar to figure H.7, except the link lengths here do *not* correspond to the shortest path to the start; instead, the link lengths correspond to the paths derived by the depth-first search. Again, we can use a depth-first search algorithm to fill up such a map and then determine a path via gradient descent from the goal pixel to the start.

Another common search is called a *greedy search* which expands nodes that are closest to the goal. Here, the data structure is called a *priority queue* in that nodes are placed into a sorted list based on a priority value. This priority value is a heuristic that measures distance to the goal node.

H.2   *A** **Algorithm**

Breadth-first search produces the shortest path to the start node in terms of link lengths. Since the wave-front planner is a breadth-first search, a four-point connectivity wave-front algorithm produces the shortest path with respect to the Manhattan distance function. This is because it implicitly has an underlying graph where each node corresponds to a pixel and neighboring pixels have an edge length of one. However, shortest-path length is not the only metric we may want to optimize. We can tune our graph search to find optimal paths with respect to other metrics such as energy, time, traversability, safety, etc., as well as combinations of them.

When speaking of graph search, there is another opportunity for optimization: minimize the number of nodes that have to be visited to locate the goal node subject to our path-optimality criteria. To distinguish between these forms of optimality, let us reserve the term *optimality* to measure the path and *efficiency* to measure the search, i.e., the number of nodes visited to determine the path. There is no reason to expect depth-first and breadth-first search to be efficient, even though breadth-first search can produce an optimal path.

Depth-first and breadth-first search in a sense are uninformed, in that the search just moves through the graph without any preference for or influence on where the goal node is located. For example, if the coordinates of the goal node are known, then a graph search can use this information to help decide which nodes in the graph to visit (i.e., expand) to locate the goal node.

Alas, although we may have some information about the goal node, the best we can do is define a *heuristic* which hypothesizes an expected, but not necessarily actual, cost to the goal node. For example, a graph search may choose as its next node to explore one that has the shortest Euclidean distance to the goal because such a node has highest possibility, based on local information, of getting closest to the goal. However, there is no guarantee that this node will lead to the (globally) shortest path in the graph to the goal. This is just a good guess. However, these good guesses are based on the best information available to the search.

The *A** algorithm searches a graph efficiently, with respect to a chosen heuristic. If the heuristic is "good," then the search is efficient; if the heuristic is "bad," although a path will be found, its search will take more time than probably required and possibly return a suboptimal path. *A** will produce an optimal path if its heuristic is *optimistic*.

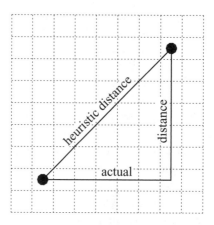

Figure H.10   The heuristic between two nodes is the Euclidean distance, which is less than the actual path length in the grid, making this heuristic optimistic.

An optimistic, or admissible, heuristic always returns a value less than or equal to the cost of the shortest path from the current node to the goal node within the graph. For example, if a graph represented a grid, an optimistic heuristic could be the Euclidean distance to the goal because the L^2 distance is always less than or equal to the L^1 distance in the plane (figure H.10).

First, we will explain the A^* search via example and then formally introduce the algorithm. See figure H.11 for a sample graph. The A^* search has a *priority queue* which contains a list of nodes sorted by priority, which is determined by the sum of the distance traveled in the graph thus far from the start node, and the heuristic.

The first node to be put into the priority queue is naturally the start node. Next, we *expand* the start node by popping the start node and putting all adjacent nodes to the start node into the priority queue sorted by their corresponding priorities. Since node B has the greatest priority, it is expanded next, i.e., it is popped from the queue and its neighbors are added (figure H.12). Note that only unvisited nodes are added to the priority queue, i.e., do not re-add the start node.

Now, we expand node H because it has the highest priority. It is popped off of the queue and all of its neighbors are added. However, H has no neighbors, so nothing is added to the queue. Since no new nodes are added, no more action or expansion will be associated with node H (figure H.12). Next, we pop off the node with greatest priority, i.e., node A, and expand it, adding all of its adjacent neighbors to the priority queue (figure H.12).

Next, node E is expanded which gives us a path to the goal of cost 5. Note that this cost is the real cost, i.e., the sum of the edge costs to the goal. At this point, there are

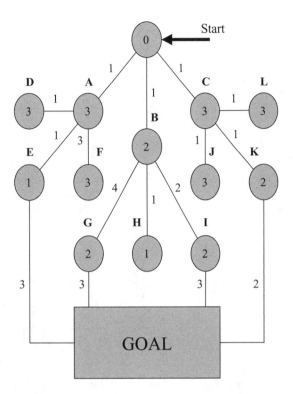

Figure H.11   Sample graph where each node is labeled by a letter and has an associated heuristic value which is contained inside the node icon. Edge costs are represented by numbers adjacent to the edges and the start and goal nodes are labeled. We label the start node with a zero to emphasize that it has the highest priority at first.

nodes in the priority queue which have a priority value greater than the cost to the goal. Since these priority values are lower bounds on path cost to the goal, all paths through these nodes will have a higher cost than the cost of the path already found. Therefore, these nodes can be discarded (figure H.12).

The explicit path through the graph is represented by a series of *back pointers*. A back pointer represents the immediate history of the expansion process. So, the back pointers from nodes A, B, and C all point to the start. Likewise, the back pointers to D, E, and F point to A. Finally, the back pointer of goal points to E. Therefore, the path defined with the back pointers is start, A, E, and goal. The arrows in figure H.12 point in the reverse direction of the back pointers.

Even though a path to the goal has been determined, *A* * is not finished because there could be a better path. *A* * knows this is possible because the priority queue

Figure H.12   (Left) Priority queue after the start is expanded. (Middle) Priority queue after the second node, B, is expanded. (Right) Three iterations of the priority queue are displayed. Each arrow points from the expanded node to the nodes that were added in each step. Since node H had no unvisited adjacent cells, its arrow points to nothing. The middle queue corresponds to two actions. Node E points to the goal which provides the first candidate path to the goal. Note that nodes D, I, F, and G are shaded out because they were discarded.

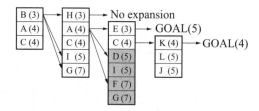

Figure H.13   Four displayed iterations of the priority queue with arrows representing the history of individual expansions. Here, the path to the goal is start, C, K, goal.

still contains nodes whose value is smaller than that of the goal state. The priority queue at this point just contains node C and is then expanded adding nodes J, K, and L to the priority queue. We can immediately remove J and L because their priority values are greater than or equal the cost of the shortest path found thus far. Node K is then expanded finding the goal with a path cost shorter than the previously found path through node E. This path becomes the current best path. Since at this point the priority queue does not possess any elements whose value is smaller than that of the goal node, this path results in the best path (figure H.13).

H.2.1   Basic Notation and Assumptions

Now, we can more formally define the A^* algorithm. The input for A^* is the graph itself. These nodes can naturally be embedded into the robot's free space and thus can have coordinates. Edges correspond to adjacent nodes and have values corresponding to the cost required to traverse between the adjacent nodes. The output of the A^*

algorithm is a back-pointer path, which is a sequence of nodes starting from the goal and going back to the start.

We will use two additional data structures, an open set O and a closed set C. The open set O is the priority queue and the closed set C contains all processed nodes. Other notation includes

- Star(n) represents the set of nodes which are adjacent to n.

- $c(n_1, n_2)$ is the length of edge connecting n_1 and n_2.

- $g(n)$ is the total length of a backpointer path from n to q_{start}.

- $h(n)$ is the heuristic cost function, which returns the estimated cost of shortest path from n to q_{goal}.

- $f(n) = g(n) + h(n)$ is the estimated cost of shortest path from q_{start} to q_{goal} via n.

The algorithm can be found in algorithm 24.

H.2.2   Discussion: Completeness, Efficiency, and Optimality

Here is an informal proof of completeness for A^*. A^* generates a search tree, which by definition, has no cycles. Furthermore, there are a finite number of acyclic paths in the tree, assuming a bounded world. Since A^* uses a tree, it only considers acyclic paths. Since the number of acyclic paths is finite, the most work that can be done,

---

Algorithm 24 A^* Algorithm

Input: A graph

Output: A path between start and goal nodes

1: **repeat**
2:     Pick n_{best} from O such that $f(n_{best}) \leq f(n), \forall n \in O$.
3:     Remove n_{best} from O and add to C.
4:     If $n_{best} = q_{goal}$, EXIT.
5:     Expand n_{best}: for all $x \in$ Star(n_{best}) that are not in C.
6:     **if** $x \notin O$ **then**
7:         add x to O.
8:     **else if** $g(n_{best}) + c(n_{best}, x) < g(x)$ **then**
9:         update x's backpointer to point to n_{best}
10:     **end if**
11: **until** O is empty

---

searching all acyclic paths, is also finite. Therefore A^* will always terminate, ensuring completeness.

This is not to say A^* will always search all acyclic paths since it can terminate as soon as it explores all paths with greater cost than the minimum goal cost found. Thanks to the priority queue, A^* explores paths likely to reach the goal quickly first. By doing so, it is efficient. If A^* does search every acyclic path and does not find the goal, the algorithm still terminates and simply returns that a path does not exist. Of course, this also makes sense if every possible path is searched.

Now, there is no guarantee that the first path to the goal found is the cheapest/best path. So, in quest for optimality (once again, with respect to the defined metric), all branches must be explored to the extent that a branch's terminating node cost (sum of edge costs) is greater than the lowest goal cost. Effectively, all paths with overall cost lower than the goal must be explored to guarantee that an even shorter one does not exist. Therefore, A^* is also optimal (with respect to the chosen metric).

H.2.3   Greedy-Search and Dijkstra's Algorithm

There are variations or special cases of A^*. When $f(n) = h(n)$, then the search becomes a *greedy* search because the search is only considering what it "believes" is the best path to the goal from the current node. When $f(n) = g(n)$, the planner is not using any heuristic information but rather growing a path that is shortest from the start until it encounters the goal. This is a classic search called *Dijkstra's algorithm*. Figure H.14 contains a graph which demonstrates Dijkstra's Algorithm. In this example, we also show backpointers being updated (which can also occur with A^*). The following lists the open and closed sets for the Dijkstra search in each step.

1. $O = \{S\}$

2. $O = \{2, 4, 1, 5\}$; $C = \{S\}$ (1, 2, 4, 5 all point back to S)

3. $O = \{4, 1, 5\}$; $C = \{S, 2\}$ (there are no adjacent nodes not in C)

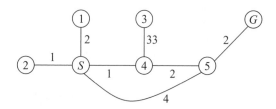

Figure H.14   Dijkstra graph search example.

4. $O = \{1,5,3\}$; $C = \{S, 2, 4\}$ (1, 2, 4 point to S; *5 points to 4*)

5. $O = \{5,3\}$; $C = \{S, 2, 4, 1\}$

6. $O = \{3, G\}$; $C = \{S, 2, 4, 1\}$ (goal points to 5 which points to 4 which points to S)

H.2.4   Example of A^* on a Grid

Figure H.15 contains an example of a grid world with a start and a goal identified accordingly. We will assume that the free space uses eight-point connectivity, and thus cell (3, 2) is adjacent to cell (4, 3), i.e., the robot can travel from (3, 2) to (4, 3). Each of the cells also has its heuristic distance to the goal where we use a modified metric which is not the Manhattan or the Euclidean distance. Instead, between free space pixels, a vertical or horizontal step has length 1 and a diagonal has length 1.4 (our approximation of $\sqrt{2}$). The cost of traveling from a free space pixel to an obstacle pixel is made to be arbitrarily high; we chose 10000. So one pixel step from a free space to an obstacle pixel along a vertical or horizontal direction costs 10000 and one pixel step along a diagonal direction costs 10000.4. Here, we are assuming that our graph connects *all* cells in the grid, not just the free space, and the prohibitively high cost of moving into an obstacle will prevent the robot from collision (figure H.16).

Note that this metric, in the free space, does not induce a true Euclidean metric because two cells sideways and one cell up is 2.4, not $\sqrt{5}$. However, this metric is quite representative of path length within the grid. This heuristic is optimistic because the actual cost to current cell to the goal will always be greater than or equal

	1	2	3	4	5	6	7
6	h=6 f= b=()	h=5 f= b=()	h=4 f= b=()	h=3 f= b=()	h=2 f= b=()	h=1 f= b=()	h=0 **Goal** b=()
5	h=6.4 f= b=()	h=5.4 f= b=()	h=4.4 f= b=()	h=3.4 f= b=()	h=2.4 f= b=()	h=1.4 f= b=()	h=1 f= b=()
4	h=6.8 f= b=()	h=5.8 f= b=()	h=4.8 f= b=()	h=3.8 f= b=()	h=2.8 f= b=()	h=2.4 f= b=()	h=2 f= b=()
3	h=7.2 f= b=()	h=6.2 f= b=()	h=5.2 f= b=()	h=4.2 f= b=()	h=3.8 f= b=()	h=3.4 f= b=()	h=3 f= b=()
2	h=7.6 f= b=()	h=6.6 f= b=()	h=5.6 f= b=()	h=5.2 f= b=()	h=4.8 f= b=()	h=4.4 f= b=()	h=4 f= b=()
1	h=8.0 f= b=()	h=7.0 f= **Start**	h=6.6 f= b=()	h=6.2 f= b=()	h=5.8 f= b=()	h=5.4 f= b=()	h=5 f= b=()
r/c	1	2	3	4	5	6	7

Figure H.15   Heuristic values are set, but backpointers and priorities have not.

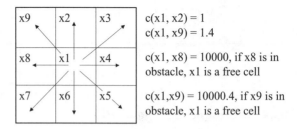

x9	x2	x3
x8	x1	x4
x7	x6	x5

c(x1, x2) = 1
c(x1, x9) = 1.4

c(x1, x8) = 10000, if x8 is in
obstacle, x1 is a free cell

c(x1,x9) = 10000.4, if x9 is in
obstacle, x1 is a free cell

Figure H.16   Eight-point connectivity and possible cost values.

h =6	h =5	h =4	h =3	h =2	h =1	h =0
f =	f =	f =	f =	f =	f =	f =
b=()	b=()	b=()	b=()	b=()	b=()	b=()
h =6.4	h =5.4	h =4.4	h =3.4	h =2.4	h =1.4	h =1
f =	f =	f =	f =	f =	f =	f =
b=()	b=()	b=()	b=()	b=()	b=()	b=()
h =6.8	h =5.8	h =4.8	h =3.8	h =2.8	h =2.4	h =2
f =	f =	f =	f =	f =	f =	f =
b=()	b=()	b=()	b=()	b=()	b=()	b=()
h =7.2	h =6.2	h =5.2	h =4.2	h =3.8	h =3.4	h =3
f =	f =	f =	f =	f =	f =	f =
b=()	b=()	b=()	b=()	b=()	b=()	b=()
h =7.6	h =6.6	h =5.6	h =5.2	h =4.8	h =4.4	h =4
f =	f =	f =	f =	f =	f =	f =
b=()	b=()	b=()	b=()	b=()	b=()	b=()
h =8.0	h =7.0	h =6.6	h=6.2	h =5.8	h =5.4	h =5
f =	f = 7.0	f=	f=	f=	f=	f=
b=()	b=()	b=()	b=()	b=()	b=()	b=()

(2,1)	7.0
State	f

Figure H.17   Start node is put on priority queue, displayed in upper right.

to the heuristic. Thus far, in figure H.15 the back pointers and priorities have not
been set.

The start pixel is put on the priority queue with a priority equal to its heuristic.
See figure H.17. Next, the start node is expanded and the priority values for each of
the start's neighbors are determined. They are all put on the priority queue sorted in
ascending order by priority. See figure H.18(left). Cell (3, 2) is expanded next, as
depicted in figure H.18(right). Here, cells (4, 1), (4, 2), (4, 3), (3, 3), and (2, 3) are
added onto the priority queue because our graph representation of the grid includes
both free space and obstacle pixels. However, cells (4, 2), (3, 3), and (2, 3) correspond
to obstacles and thus have a high cost. If a path exists in the free space or the longest
path in the free space has a traversal cost less than our arbitrarily high number chosen
for obstacles (figure H.16), then these pixels will never be expanded. Therefore, in
the figures below, we did not display them on the priority queue.

Eventually, the goal cell is reached (figure H.19 (left)). Since the priority value of
the goal is less than the priorities of all other cells in the priority queue, the resulting

Figure H.18 (Left) The start node is expanded, the priority queue is updated, and the backpointers are set, which are represented by the right bottom icon. $b = (i, j)$ points to cell (i, j). (Right) Cell (3, 2) was expanded. Note that pixels (3, 3), (2, 3), and (4, 2) are not displayed in the priority queue because they correspond to obstacles.

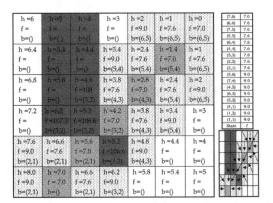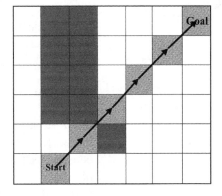

Figure H.19 (Left) The goal state is expanded. (Right) Resulting path.

path is optimal and A^* terminates. A^* traces the backpointers to find the optimal path from start to goal (figure H.19 (right)).

H.2.5 Nonoptimistic Example

Figure H.20 contains an example of a graph whose heuristic values are nonoptimistic and thus force A^* to produce a nonoptimal path. A^* puts node S on the priority queue and then expands it. Next, A^* expands node A because its priority value is 7. The goal

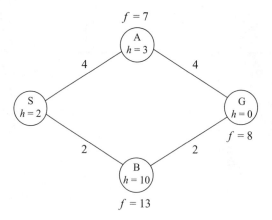

Figure H.20   A nonoptimistic heuristic leads to a nonoptimal path with A^*.

node is then reached with priority value 8, which is still less than node B's priority value of 13. At this point, node B will be eliminated from the priority queue because its value is greater than the goal's priority value. However, the optimal path passes through B, not A. Here, the heuristic is not optimistic because from B to G, $h = 10$ when the actual edge length was 2.

H.3 D^* Algorithm

So far we have only considered *static* environments where only the robot experiences motion. However, we can see that many worlds have moving obstacles, which could be other robots themselves. We term such environments *dynamic*. There are three types of dynamic obstacles: ones that move significantly slower than the robot, those that move at the same speed, and finally obstacles that move much faster than the robot. The superfast obstacle case is easy to ignore because the obstacles will be moving so fast that there probably is no need to plan for them because they will either move too fast for the planner to have time to account for them or they will be in and out of the robot's path so quickly that it does not require any consideration. In this section, we consider dynamic environments where the world changes at a speed much slower than the robot. An example can be a door opening and closing.

Consider the grid environment in figure H.21(left) which is identical to the one in figure H.15, except pixel (4, 3) is a gate which can either be a free-space pixel or an obstacle pixel. Let's assume it starts as a free-space pixel. We can run the A^* or

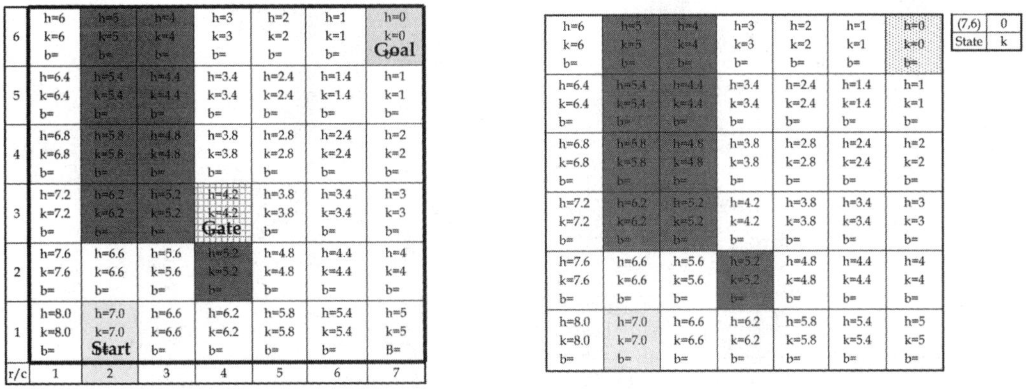

Figure H.21 (Left)

r/c	1	2	3	4	5	6	7
6	h=6 k=6 b=	h=5 k=5 b=	h=4 k=4 b=	h=3 k=3 b=	h=2 k=2 b=	h=1 k=1 b=	h=0 k=0 **Goal**
5	h=6.4 k=6.4 b=	h=5.4 k=5.4 b=	h=4.4 k=4.4 b=	h=3.4 k=3.4 b=	h=2.4 k=2.4 b=	h=1.4 k=1.4 b=	h=1 k=1 b=
4	h=6.8 k=6.8 b=	h=5.8 k=5.8 b=	h=4.8 k=4.8 b=	h=3.8 k=3.8 b=	h=2.8 k=2.8 b=	h=2.4 k=2.4 b=	h=2 k=2 b=
3	h=7.2 k=7.2 b=	h=6.2 k=6.2 b=	h=5.2 k=5.2 b=	h=4.2 k=4.2 **Gate**	h=3.8 k=3.8 b=	h=3.4 k=3.4 b=	h=3 k=3 b=
2	h=7.6 k=7.6 b=	h=6.6 k=6.6 b=	h=5.6 k=5.6 b=	h=5.2 k=5.2 b=	h=4.8 k=4.8 b=	h=4.4 k=4.4 b=	h=4 k=4 b=
1	h=8.0 k=8.0 b=	h=7.0 k=7.0 **Start**	h=6.6 k=6.6 b=	h=6.2 k=6.2 b=	h=5.8 k=5.8 b=	h=5.4 k=5.4 b=	h=5 k=5 B=

(Right)

6	h=6 k=6 b=	h=5 k=5 b=	h=4 k=4 b=	h=3 k=3 b=	h=2 k=2 b=	h=1 k=1 b=	h=0 k=0 b=
5	h=6.4 k=6.4 b=	h=5.4 k=5.4 b=	h=4.4 k=4.4 b=	h=3.4 k=3.4 b=	h=2.4 k=2.4 b=	h=1.4 k=1.4 b=	h=1 k=1 b=
4	h=6.8 k=6.8 b=	h=5.8 k=5.8 b=	h=4.8 k=4.8 b=	h=3.8 k=3.8 b=	h=2.8 k=2.8 b=	h=2.4 k=2.4 b=	h=2 k=2 b=
3	h=7.2 k=7.2 b=	h=6.2 k=6.2 b=	h=5.2 k=5.2 b=	h=4.2 k=4.2 b=	h=3.8 k=3.8 b=	h=3.4 k=3.4 b=	h=3 k=3 b=
2	h=7.6 k=7.6 b=	h=6.6 k=6.6 b=	h=5.6 k=5.6 b=	h=5.2 k=5.2 b=	h=4.8 k=4.8 b=	h=4.4 k=4.4 b=	h=4 k=4 b=
1	h=8.0 k=8.0 b=	h=7.0 k=7.0 b=	h=6.6 k=6.6 b=	h=6.2 k=6.2 b=	h=5.8 k=5.8 b=	h=5.4 k=5.4 b=	h=5 k=5 b=

(7,6)	0
State	k

Figure H.21 (Left) A pixel world similar to figure H.15, except it has a gate, heuristic, and minimum heuristic values. (Right) Goal node is expanded.

Dijkstra's algorithm to determine a path from start to goal, and then follow that path until an unexpected change occurs, which in figure H.21(left) happens at (4, 3). When the robot encounters pixel (4, 3) and determines that it changed from a free-space to an obstacle pixel, it can simply reinvoke the A^* algorithm to determine a new path. This, however, can become quite inefficient if many pixels are changing from obstacle to free space and back. The D^* algorithm was devised to "locally repair" the graph allowing for an efficient updated searching in dynamic environments, hence the term D^* [397].

D^* initially determines a path starting with the goal and working back to the start using a slightly modified Dijkstra's search. The modification involves updating a heuristic and a minimum heuristic function. Each cell in figure H.21(left) contains a heuristic cost (h) which for D^* is an *estimate* of path length from the particular cell to the goal, not necessarily the shortest path length to the goal as it was for A^*. In this example, the h values do not respect the presence of obstacles when reflecting distance to the goal node; in other words, computation of h assumes that the robot can pass through obstacles. For example, cell (1, 6) has an h value of 6. These h values will be updated during the initial Dijkstra search to reflect the existence of obstacles. The minimum heuristic values (k) are the *estimate* of the *shortest* path length to the goal. Both the h and the k values will vary as the D^* search runs, but they are equal upon initialization, and were derived from the metric described in figure H.16.

Initially, the goal node is placed on the queue with $h = 0$ and then is expanded (figure H.21, right), adding (6, 6), (6, 5), and (7, 5) onto the queue. Next, pixel (6, 6) is expanded adding cells (5, 6), (5, 5) onto the queue. Note that the k values are used

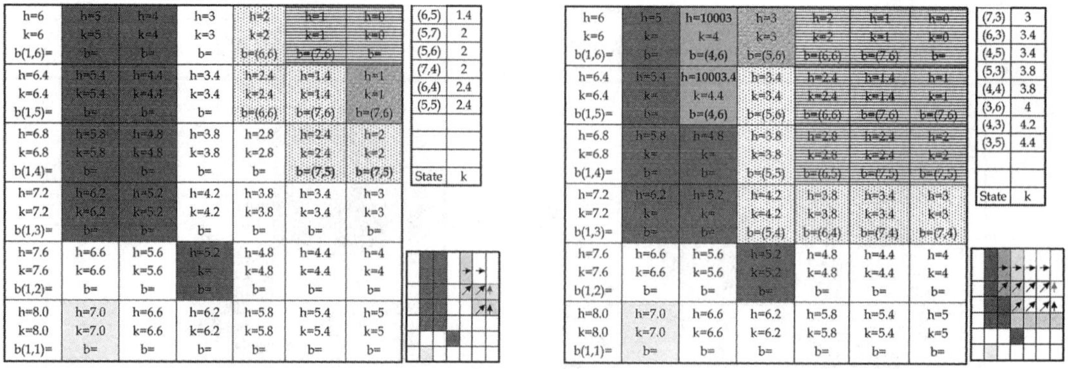

Figure H.22   (Left) First (6, 6) and then (7, 5) is expanded. (Right) The *h* values in obstacle cells that are put on priority queue are updated.

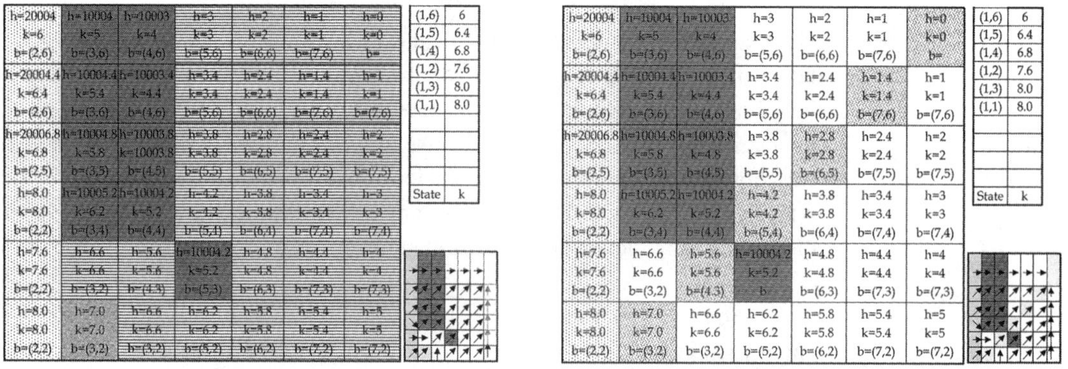

Figure H.23   (Left) Termination of Dijkstra's search phase: start cell is expanded. (Right) Tracing backpointers yields the optimal path.

to determine the priority for the Dijkstra's search (and later on for the D^* search) and that they are equal to the *h* values for the initial Dijkstra's search.

Next, pixel (7, 5) is expanded adding cells (6, 4) and (7, 4) onto the queue (figure H.22, left). More pixels are expanded until we arrive at pixel (4, 6) (figure H.22, right). When (4, 6) is expanded, pixels (3, 6) and (3, 5), which are obstacle pixels, are placed onto the priority queue. Unlike our A^* example, we display these obstacle pixels in the priority queue in figure H.22(right). Note that the *h* values of the expanded obstacle pixels are all updated to prohibitively high values which reflects the fact that they lie on obstacles.

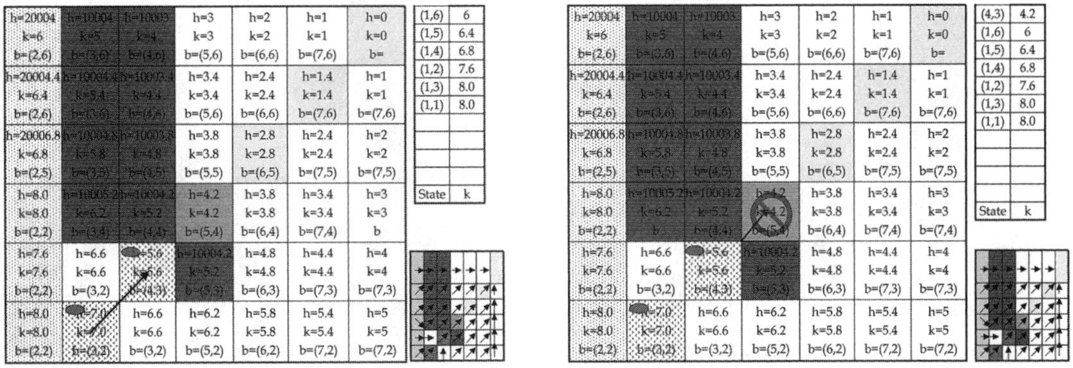

Figure H.24 (Left) The robot physically starts tracing the optimal path. (Right) The robot cannot trace the assumed optimal path: gate (4, 3) is closed.

The Dijkstra's search is complete when the start node (2, 1) is expanded (figure H.23, left). The optimal path from start to goal (assuming that the gate pixel (4, 3) is open) is found by traversing the backpointers starting from the start node to the goal node (figure H.23(right)). The optimal path is (2, 1) $\longrightarrow$ (3, 2) $\longrightarrow$ (4, 3) $\longrightarrow$ (5, 4) $\longrightarrow$ (6, 5) $\longrightarrow$ (7, 6). Note that pixels (1, 1), (1, 2), (1, 3), (1, 4), and (1, 6) are still on the priority queue.

The robot then starts tracing the optimal path from the start pixel to the goal pixel. In figure H.24(left), the robot moves from pixel (2, 1) to (3, 2). When the robot tries to move from pixel (3, 2) to (4, 3), it finds that the gate pixel (4, 3) is closed (figure H.24, left). In the initial search for an optimal path, we had assumed that the gate pixel was open, and hence the current path may not be feasible. At this stage, instead of replanning for an optimal path from the current pixel (3, 2) to goal pixel using *A**, *D** tries to make local changes to the optimal path.

D* puts the pixel (4, 3) on the priority queue because it corresponds to a discrepancy between the map and the actual environment. Note that this pixel must have the lowest minimum heuristic, i.e., *k* value, because all other pixels on the priority queue have a *k* value greater than or equal to the start and all pixels along the previously determined optimal path have a *k* value less than the start. The idea here is to put the changed pixel onto the priority queue and then expand it again, thereby propagating the possible changes in the heuristic, i.e., the *h* values, to pixels for which an optimal path to the goal passes through the changed pixel.

In the current example, pixel (4, 3) is expanded, i.e., it is popped off the priority queue and pixels whose optimal paths pass through (4, 3) are placed onto the priority

Left grid:

h=20004 k=6 b=(2,6)	h=10004 k=5 b	h=10003 k=4 b	h=3 k=3 b=(5,6)	h=2 k=2 b=(6,6)	h=1 k=1 b=(7,6)	h=0 k=0 b=		(3,2) 5.6
h=20004.4 k=6.4 b=(2,6)	h=10004.4 k=5.4 b=(3,6)	h=10003.4 k=4.4 b=(4,6)	h=3.4 k=3.4 b=(5,6)	h=2.4 k=2.4 b=(6,6)	h=1.4 k=1.4 b=(7,6)	h=1 k=1 b=(7,6)		(1,6) 6
h=20006.8 k=6.8 b=(2,5)	h=10004.8 k=5.8 b=(3,5)	h=10003.8 k=4.8 b=(4,5)	h=3.8 k=3.8 b=(5,5)	h=2.8 k=2.8 b=(6,5)	h=2.4 k=2.4 b=(7,5)	h=2 k=2 b=(7,5)		(1,5) 6.4
h=8.0 k=8.0 b=(2,2)	h=10005.2 k=6.2 b=(3,4)	h=10004.2 k=4.2 b=(4,4)	h=4.2 k=4.2 b=(5,4)	h=3.8 k=3.8 b=(6,4)	h=3.4 k=3.4 b=(7,4)	h=3 k=3 b=(7,4)		(1,4) 6.8
h=7.6 k=7.6 b=(2,2)	h=6.6 k=6.6 b=(3,2)	h=20004.6 k=5.6 b=(5,3)	h=10004.2 k=5.2 b=(5,3)	h=4.8 k=4.8 b=(6,3)	h=4.4 k=4.4 b=(7,3)	h=4 k=4 b=(7,3)		(1,2) 7.6
h=8.0 k=8.0 b=(2,2)	h=7.0 k=7.0 b=(3,2)	h=6.6 k=6.6 b=(3,2)	h=6.2 k=6.2 b=(5,2)	h=5.8 k=5.8 b=(6,2)	h=5.4 k=5.4 b=(7,2)	h=5 k=5 b=(7,2)		(1,3) 8.0
								(1,1) 8.0

State | k

Right grid:

h=20004 k=6 b=(2,6)	h=10004 k=5 b	h=10003 k=4 b=(4,6)	h=3 k=3 b=(5,6)	h=2 k=2 b=(6,6)	h=1 k=1 b=(7,6)	h=0 k=0 b=		(1,6) 6
h=20004.4 k=6.4 b=(2,6)	h=10004.4 k=5.4 b=(3,6)	h=10003.4 k=4.4 b=(4,6)	h=3.4 k=3.4 b=(5,6)	h=2.4 k=2.4 b=(6,6)	h=1.4 k=1.4 b=(7,6)	h=1 k=1 b=(7,6)		(4,1) 6.2
h=20006.8 k=6.8 b=(2,5)	h=10004.8 k=4.8 b=(3,5)	h=10003.8 k=3.8 b=(4,5)	h=3.8 k=3.8 b=(5,5)	h=2.8 k=2.8 b=(6,5)	h=2.4 k=2.4 b=(7,5)	h=2 k=2 b=(7,5)		(1,5) 6.4
h=8.0 k=8.0 b=(2,2)	h=10005.2 k=6.2 b=(3,4)	h=10004.2 k=4.2 b=(4,4)	h=4.2 k=4.2 b=(5,4)	h=3.8 k=3.8 b=(6,4)	h=3.4 k=3.4 b=(7,4)	h=3 k=3 b=(7,4)		(3,1) 6.6
h=7.6 k=7.6 b=(2,2)	h=10005.6 k=6.6 b=(3,2)	h=20004.6 k=5.6 b=(4,3)	h=10004.2 k=5.2 b=(5,3)	h=4.8 k=4.8 b=(6,3)	h=4.4 k=4.4 b=(7,3)	h=4 k=4 b=(7,3)		(2,2) 6.6
h=8.0 k=8.0 b=(2,2)	h=20006.0 k=7.0 b=(3,2)	h=10005.6 k=6.6 b=(3,2)	h=6.2 k=6.2 b=(5,2)	h=5.8 k=5.8 b=(6,2)	h=5.4 k=5.4 b=(7,2)	h=5 k=5 b=(7,2)		(1,4) 6.8
								(2,1) 7.0
								(1,2) 7.6
								(1,3) 8.0
								(1,1) 8.0

State | k

Figure H.25   (Left) The gate pixel (4, 3) is put on priority queue and expanded. The assumed optimal path from (3, 2) to goal passed through (4, 3), so the h value is increased to a high value to reflect that the assumed optimal path may not in fact be optimal. (Right) Pixel (3, 2) is expanded; the h values of (2, 2),(2, 1) and (3, 1) are updated because the assumed optimal path from these cells passed through the expanded cell. (4, 1)) remains unaffected.

queue with updated heuristic values (h values). The new h values are the h values of the changed pixel plus the path cost from the changed pixel to the given pixel. This path cost is a high number which we set to 10000.4 if it passes diagonally through an obstacle pixel. Therefore, pixel (3, 2) has an h value equal to 10004.6 (figure H.25, left).

Next, pixel (3, 2) is expanded because its k value is the smallest. However, its k value is less than its h value and we term such pixels as having a *raised state*. When a pixel is in a raised state, its back pointer may no longer point to an optimal path. Now, pixels (2, 2), (2, 1), (3, 1), and (4, 1) are on the priority queue. The h values of cells (2, 2), (2, 1), and (3, 1) are updated to high values to reflect that the estimated optimal path from these cells to the goal passed through the gate cell, and may not be optimal anymore. However, the optimal path for pixel (4, 1) did not pass through the gate, and hence its h value stays the same (figure H.25, right).

Pixel (1, 6) is expanded next, but it does not affect the h values of its neighbors. Next pixel (4, 1) is expanded and pixels (3, 2), (3, 1), (5, 1), and (5, 2) are put onto the priority queue (figure H.26(left). Now the h values of (5, 1) and (5, 2) remain unaffected, however, for cell (3, 2), the goal can now possibly be reached in $h(4, 1) + 1.4 = 6.2 + 1.4 = 7.6$ because cell (4, 1) is in a *lowered state*, i.e., it is a cell whose h values did not have to be updated because of the gate. Therefore, cell (3, 2) receives an h value of 7.6. The backpointer of (3, 2) is now set pointing toward (4, 1). The initially determined optimal path from (4, 1) to the goal did not pass through the

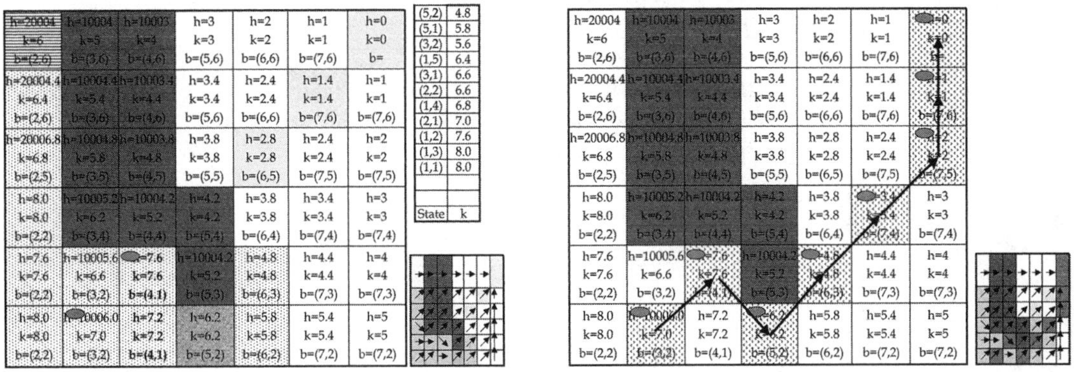

Figure H.26   (Left) (4, 1) is expanded; (5, 1) and (5, 2) remain unaffected, the *h* values of (3, 2) and (3, 1) are lowered to reflect the lowered heuristic value because of detour, while the minimum-heuristic values of these two cells are increased, and the backpointers of these cells are set pointing to (4, 1). (Right) The robot traces the new locally modified optimal path.

gate pixel, and hence it indeed is optimal even after the change of state of the gate pixel. Then, the path obtained by concatenating the (3, 2) → (4, 1) transition and the optimal path from (4, 1) will be optimal for (3, 2). Thus, the estimate for the best path from (3, 2) toward the goal, i.e., $k(3, 2)$, is now 7.6, and the process terminates. The robot then physically traces the new optimal path from (3, 2) to reach the goal (figure H.26(right)).

See algorithms 25–31 for a description of the D^* algorithm. This algorithm uses the following notation.

- *X* represents a state.

- *O* is the priority queue.

- *L* is the list of all states.

- *G* is the goal state.

- *S* is the start state.

- $t(X)$ is value of state with regards to the priority queue.

    - $t(X) = NEW$, if *X* has never been in *O*,
    - $t(X) = OPEN$, if *X* is currently in *O*, and
    - $t(X) = CLOSED$, if *X* was in *O* but currently is not.

Algorithm 25 D^* Algorithm

Input: List of all states L

Output: The goal state, if it is reachable, and the list of states L are updated so that the backpointer list describes a path from the start to the goal. If the goal state is not reachable, return NULL.

1: **for each** $X \in L$ **do**
2:    $t(X) = $ NEW
3: **end for**
4: $h(G) = 0$
5: $O = \{G\}$
6: $X_c = S$
   {The following loop is Dijkstra's search for an initial path}
7: **repeat**
8:    $k_{\min} = PROCESS - STATE(O, L)$
9: **until** $(k_{\min} = -1)$ or $(t(X_c) = CLOSED)$
10: $P = GET - BACKPOINTER - LIST(L, X_c, G)$ (algorithm 26)
11: **if** P $=$ NULL **then**
12:    Return (NULL)
13: **end if**
14: **repeat**
15:    **for each** neighbor $Y \in L$ of X_c **do**
16:       **if** $r(X_c, Y) \neq c(X_c, Y)$ **then**
17:          $MODIFY - COST(O, X_c, Y, r(X_c, Y))$
18:          **repeat**
19:             $k_{\min} = PROCESS - STATE(O, L)$
20:          **until** $(k_{\min} \geq h(X_c))$ or $(k_{\min} = -1)$
21:          $P = GET - BACKPOINTER - LIST(L, X_c, G)$
22:          **if** P $=$ NULL **then**
23:             Return (NULL)
24:          **end if**
25:       **end if**
26:    **end for**
27:    $X_c = $ the second element of P {Move to the next state in P}.
28:    $P = GET - BACKPOINTER - LIST(L, X_c, G)$
29: **until** $X_c = G$
30: Return (X_c)

Algorithm 26 *GET − BACKPOINTER − LIST(L, S, G)*

Input: A list of states *L* and two states (start and goal)

Output: A list of states from start to goal as described by the backpointers in the list of states *L*

1: **if** path exists **then**
2:     Return (The list of states)
3: **else**
4:     Return (*NULL*)
5: **end if**

Algorithm 27 *INSERT(O, X, h_{new})*

Input: Open list, a state, and an *h*-value

Output: Open list is modified

1: **if** $t(X) = NEW$ **then**
2:     $k(X) = h_{new}$
3: **else if** $t(X) = OPEN$ **then**
4:     $k(X) = \min(k(X), h_{new})$
5: **else if** $t(X) = CLOSED$ **then**
6:     $k(X) = \min(h(X), h_{new})$
7: **end if**
8: $h(X) = h_{new}$
9: $t(X) = OPEN$
10: Sort *O* based on increasing *k* values

- $c(X, Y)$ is the estimated path length between adjacent states *X* and *Y*.

- $h(X)$ is the estimated cost of a path from *X* to Goal (heuristic).

- $k(X)$ is the estimated cost of a shortest path from *X* to Goal (minimum-heuristic $= \min h(X)$ before *X* is put on *O*, values $h(X)$ takes after *X* is put on *O*).

- $b(X) = Y$ implies that *Y* is a parent state of *X*, i.e. the path is like $X \longrightarrow Y \longrightarrow G$.

- $r(X, Y)$ is the measured distance adjacent states *X* and *Y*.

Algorithm 28 $MODIFY - COST(O, X, Y, cval)$

Input: The open list, two states and a value

Output: A k-value and the open list gets updated

1: $c(X, Y) = cval$

2: **if** $t(X) = CLOSED$ **then**

3:     $INSERT(O, X, h(X))$

4: **end if**

5: Return $GET - KMIN(O)$ (algorithm 30)

Algorithm 29 $MIN - STATE(O)$

Input: The open list O

Output: The state with minimum k value in the list related values

---

1: **if** $O = \emptyset$ **then**

2:     Return (-1)

3: **else**

4:     Return $(\operatorname{argmin}_{Y \in O} k(Y))$

5: **end if**

Algorithm 30 $GET - KMIN(O)$

Input: The open list O

Output: Lowest k-value of all states in the open list

1: **if** $O = \emptyset$ **then**

2:     Return (-1)

3: **else**

4:     Return $(\min_{Y \in O} k(Y))$

5: **end if**

Algorithm 31 PROCESS-STATE

Input: List of all states L and the list of all states that are open O

Output: A $k_{\min}$, an updated list of all states, and an updated open list

---

1: $X = MIN - STATE(O)$ (algorithm 29)

2: **if** $X = NULL$ **then**

3:    Return (-1)

4: **end if**

5: $k_{\text{old}} = GET - KMIN(O)$ (algorithm 30)

6: $DELETE(X)$

7: **if** $k_{\text{old}} < h(X)$ **then**

8:    **for each** neighbor $Y \in L$ of X **do**

9:       **if** $h(Y) \le k_{\text{old}}$ and $h(X) > h(Y) + c(Y, X)$ **then**

10:          $b(X) = Y$

11:          $h(X) = h(Y) + c(Y, X);$

12:       **end if**

13:    **end for**

14: **else if** $k_{\text{old}} = h(X)$ **then**

15:    **for each** neighbor $Y \in L$ of X **do**

16:       **if** $(t(Y) = NEW)$ or $(b(Y) = X$ and $h(Y) \ne h(X) + c(X, Y))$ or $(b(Y) \ne X$ and $h(Y) > h(X) + c(X, Y))$ **then**

17:          $b(Y) = X$

18:          $INSERT(O, Y, h(X) + c(X, Y))$ (algorithm 27)

19:       **end if**

20:    **end for**

21: **else**

22:    **for each** neighbor $Y \in L$ of X **do**

23:       **if** $(t(Y) = NEW)$ or $(b(Y) = X$ and $h(Y) \ne h(X) + c(X, Y))$ **then**

24:          $b(Y) = X$

25:          $INSERT(O, Y, h(X) + c(X, Y))$

26:       **else if** $b(Y) \ne X$ and $h(Y) > h(X) + c(X, Y)$ **then**

27:          $INSERT(O, X, h(X))$

28:       **else if** $(b(Y) \ne X$ and $h(X) > h(Y) + c(X, Y))$ and $(t(Y) = CLOSED)$ and $(h(Y) > k_{\text{old}})$ **then**

29:          $INSERT(O, Y, h(Y))$

30:       **end if**

31:    **end for**

32: **end if**

33: Return $GET - KMIN(O)$ (algorithm 30)

H.4   Optimal Plans

There exists a huge number of search algorithms in the literature, with the ones discussed here being just the most basic ones. All of the techniques discussed here result in a path. A path is sufficient if the robot is able to follow it. Sometimes, randomness may push the robot off its path. One possibility is to replan, as we did in D^* (albeit for different reasons: above the environment changed and thereby mandated replanning). Another is to determine the best action for *all* nodes in the graph, not just the ones along the shortest path. A mapping from nodes to actions is called a *universal plan,* or *policy*. Techniques for finding optimal policies are known as universal planners and can be computationally more involved than the shortest path techniques surveyed here. One simple way to attain a universal plan to a goal is to run Dijkstra's algorithm backward (as in D^*): After completion, we know for each node in the graph the length of an optimal path to the goal, along with the appropriate action. Generalizations of this approach are commonly used in stochastic domains, where the outcome of actions is modeled by a probability distribution over nodes in the graph.

Statistics Primer

ELEGANT AND powerful techniques are at the fingertips of statisticians. Although difficult at first, speaking their language can be quite powerful. Probability theory provides a set of tools that can be used to quantify uncertain events. In the context of robotics, probability theory allows us make decisions in the presence of uncertainty caused by phenomena such as noisy sensor data or interaction with unpredictable humans. This section introduces a few fundamental concepts including probability, random variables, distributions, and Gaussian random vectors.

When we talk about probability, we generally talk in terms of *experiments* and *outcomes*. When an experiment is conducted, a single outcome from the set of possible outcomes for that experiment results. For example, an experiment could be flipping a coin and the set of possible outcomes is {heads, tails}. If the experiment were to take a measurement in degrees Kelvin, then the set of possible outcomes would be the interval $[0, \infty)$. An *event* is defined to be a subset of the possible outcomes.

Let S denote the set of all possible outcomes for a given experiment, and let E be an event, i.e., $E \subset S$. The *probability* of the event E occurring when the experiment is conducted is denoted $\Pr(E)$. $\Pr$ maps S to the interval $[0, 1]$. In the example of flipping a fair coin, $\Pr(\text{heads}) = 0.5$, $\Pr(\text{tails}) = 0.5$, and $\Pr(\text{heads} \cup \text{tails}) = 1$. In general, the probability must obey certain properties:

1. $0 \leq \Pr(E) \leq 1$ for all $E \subset S$.

2. $\Pr(S) = 1$.

3. $\sum_i \Pr(E_i) = Pr(E_1 \cup E_2 \cup \ldots)$ for any countable disjoint collection of sets $E_1, E_2, \ldots$. This property is known as *sigma additivity*. In particular, we have $\sum_{i=1}^n \Pr(E_i) = Pr(E_1 \cup E_2 \cup \ldots \cup E_n)$.

4. $\Pr(\emptyset) = 0$.

5. $\Pr(E^c) = 1 - \Pr(E)$, where E^c denotes the complement of E in $\mathcal{S}$.

6. $\Pr(E_1 \cup E_2) = \Pr(E_1) + \Pr(E_2) - \Pr(E_1 \cap E_2)$.

Technically, the first three axioms imply the last three.

Events may or may not depend upon each other. If the occurance of E_1 has no effect on E_2, then E_1 and E_2 are *independent;* otherwise they are dependent. We say E_1 and E_2 are independent if $\Pr(E_1 \cap E_2) = \Pr(E_1) \cdot \Pr(E_2)$. One way to express the dependence of two events is through *conditional probability*. For events E_1 and E_2, $\Pr(E_1 \mid E_2)$ is the *conditional probability* that E_1 occurs given that E_2 occurs. If E_1 and E_2 are independent and $\Pr(E_2) > 0$, then $\Pr(E_1 \mid E_2) = \Pr(E_1)$. For dependent events, Bayes' rule expresses the relationship between the conditional probabilities for two events, again assuming $\Pr(E_2) > 0$:

$$\Pr(E_1 \mid E_2) = \frac{\Pr(E_2 \mid E_1)\Pr(E_1)}{\Pr(E_2)}.$$

Bayes' rule is a useful formula; it is the foundation of the estimation methods presented in chapter 9.

I.1   Distributions and Densities

Within robotics, a somewhat simplified but nevertheless sufficient model of a *random variable* is a mapping from the set of events to the real line, usually denoted $X : \mathcal{S} \to \mathbb{R}$. A simple example of a random variable is to consider a single coin flip and define $X = 0$ when the outcome is heads and $X = 1$ when the outcome is tails. As another example, consider flipping a fair coin ten times. A random variable can be the number of heads that appeared or the number of times heads appeared sequentially, etc. Random variables are useful because they represent events as real numbers. With real numbers, we can perform calculations and analysis that are difficult or impossible to perform on the abstract events.

A *distribution* is an abstract concept that corresponds to all probability statements that can be made about a random variable. Before we discuss the various distributions used to describe random variables, we first distinguish between contiunous and discrete random variables. A random variable is said to be *discrete* if its range (the values that it maps to) is a set of discrete points. We call a random variable *continuous* if its range forms a continuum on the real line and, using a term that is defined further below, it possesses a probability density function.

Discrete random variables are commonly described using one of two types of distributions. The first is the *cumulative distribution function* (CDF) which is denoted $F_X(a) = \Pr(X \le a)$. The second is the *probability mass function* (PMF), which is defined to be $f_X(a) = \Pr(X = a)$.

Continuous random variables are described with two analogous distributions. The first is the cumulative distribution function (figure I.1), which is defined for continuous random variables exactly the same way that it is defined for discrete random variables. The second is the *probability density function* (PDF) (figure I.2), which is denoted f_X and is defined such that

$$\Pr(a \le X \le b) = \int_{x=a}^{b} f_X(x)\,dx.$$

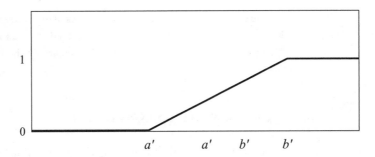

Figure I.1   Cumulative uniform distribution.

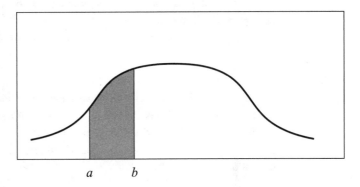

Figure I.2   Probability density function.

Note that for a continuous random variable, $Pr((X = a)) = \int_{x=a}^{a} f_X(x) \, dx = 0$. This can be disconcerting to the newcomer. Another way to view this is: consider the odds of landing exactly on a. Since the point a is a set of measure zero, it should have zero probability of occurring.

Some distributions are so common that they have their own name. For example, the uniform distribution is a family of continuous distributions over an interval. It can either be described by the CDF

$$U(x; a, b) = \begin{array}{ll} 0 & x < a \\ \frac{x-a}{b-a} & a \leq x \leq b \\ 1 & x \geq b \end{array}$$

or by the corresponding PDF

$$u(x; a, b) = \begin{array}{ll} 0 & x < a \\ \frac{1}{b-a} & a \leq x \leq b \\ 0 & x \geq b. \end{array}$$

One can calculate all sorts of probabilties using either the CDF or the PDF. Consider for $a', b' \in [a, b]$ with $a' < b'$. Using the CDF we can compute $Pr(a' \leq x \leq b') = U(b'; a, b) - U(a'; a, b)$. Alternatively we can use the PDF to compute $Pr(a' \leq x \leq b') = \int_{x=a'}^{b'} u(x; a, b) \, dx$.

I.2   Expected Values and Covariances

We previously defined a random variable to be a function that maps the event space to the real line. Similarly, we define a *random vector* to be a mapping from the event space to the space of real-valued vectors of some dimension. In other words, a random vector X is a map $X : S \rightarrow \mathbb{R}^n$. Note that a random variable is just a special case of a random vector where $n = 1$.

The *expected value* (or *mean*) for a discrete random vector is defined to be

$$E(X) = \sum_i x_i f_X(x_i),$$

where x_i is the ith value that random variable X can take and f_X is the PMF associated with X. Note that $E(X)$ is a vector in $\mathbb{R}^n$, where n is the dimension of X. It is tempting to think that the expected value is the outcome most likely to occur, but this is not generally the case. The expected value of a single fair die roll is 3.5 which, of course, cannot occur.

The expected value (or mean) of a continuous random vector is defined to be

$$E(X) = \int_{x \in \mathbb{R}^n} x f_X(x) \, dx,$$

where f_X is the PDF associated with X. As in the case of discrete random vectors, $E(X)$ is a vector in $\mathbb{R}^n$. We also denote $E(X)$ with $\bar{X}$. Expectation is a linear operator, which means that $E(aX + bY) = a E(X) + b E(Y)$.

The *variance* of a (scalar) random variable x is $E((X - \bar{X})^2)$. For a scalar random variable the variance is denoted σ^2. For a random vector we can consider the variance of each element X_i of X individually. The variance of X_i is denoted σ_i^2.

Now we want to consider the effect of one variable on another. This is termed *covariance* between two random variables X_i and X_j. Let $\sigma_{ij} = E((X_i - \bar{X}_i)(X_j - \bar{X}_j))$. By this definition σ_{ii} is the same as σ_i^2, the variance of X_i. For $i \neq j$, if $\sigma_{ij} = 0$, then X_i and X_j are independent of each other. The *covariance matrix* of a random vector X is defined to be

$$P_X = E((X - \bar{X})(X - \bar{X})^T).$$

The $n \times n$ matrix P_X contains the variances and covariances within the random vector X. Specifically, the element in the ith row, jth column of P_X will be identical to the σ_{ij} defined above.

I.3 Multivariate Gaussian Distributions

A random vector X is said to have a multivariate Gaussian distribution if it is described by the PDF

(I.1) $$f_X(x) = \frac{1}{\sqrt{(2\pi)^n |P_X|}} e^{-\frac{1}{2}(x - \bar{X})^T P_X^{-1} (x - \bar{X})},$$

where $\bar{X} \in \mathbb{R}^n$ is the mean vector and $P_X \in \mathbb{R}^{n \times n}$ is the covariance matrix. It can be verified by direct substitution that $\bar{X}$ and P_X are in fact the mean and covariance matrix of X as defined in the section above.

J Linear Systems and Control

THIS APPENDIX gives a brief review of the theory of linear time invariant (LTI) dynamical systems. Many dynamical systems that appear in science and engineering can be approximated by LTI systems, and linear systems theory provides important tools to control and observe them. We focus on the so-called *state space* formulation of LTI systems because that is the formulation used in the Kalman filter (see chapter 8). In this appendix we present some of the more fundamental concepts of LTI state-space systems, including stability, feedback control, and observability.

J.1 State Space Representation

Consider as an example the mass-spring-damper system depicted in figure J.1, where $z(t)$ denotes the position of the mass m at time t. If we assume that the spring is linear, then the force applied by the spring is given as $F_s = -kz(t)$. Likewise, if we assume that the damper is linear, then the force applied by the damper is proportional to the velocity of the mass, yielding $F_d = -\gamma \frac{dz}{dt}(t) \triangleq \gamma \dot{z}(t)$. For now we assume the externally applied force $F_{\text{ext}} = 0$. Summing these forces and applying Newton's law (force = mass × acceleration) yields

(J.1) $\qquad m\ddot{z}(t) = -\gamma \dot{z}(t) - kz(t).$

This second-order ordinary differential equation (ODE) provides a mathematical description of how the position and velocity of mass change with time. Accordingly, we call equation (J.1) a *model* of the mass-spring-damper system. If the position z and

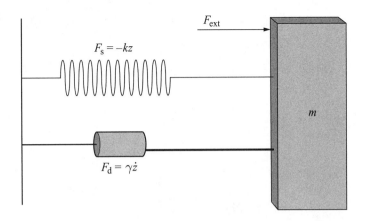

Figure J.1 Mass–spring–damper system.

velocity $\dot{z}$ are known at some instant of time t_0, then the solution to equation (J.1) subject to initial conditions $z(t_0)$ and $\dot{z}(t_0)$ will match the trajectory of the physical system.

Now define the vector

$$x(t) = \begin{bmatrix} x_1(t) \\ x_2(t) \end{bmatrix} = \begin{bmatrix} z(t) \\ \dot{z}(t) \end{bmatrix}.$$

Equation (J.1) can be rewritten in terms of x as follows:

$$\dot{x}(t) = \begin{bmatrix} \dot{z}(t) \\ \ddot{z}(t) \end{bmatrix} = \begin{bmatrix} x_2(t) \\ -\frac{1}{m}(\gamma \dot{z}(t) + kz(t)) \end{bmatrix} = \begin{bmatrix} x_2(t) \\ -\frac{1}{m}(\gamma x_2(t) + kx_1(t)) \end{bmatrix},$$

which can finally be summarized as

(J.2) $$\dot{x}(t) = \begin{bmatrix} 0 & 1 \\ -\frac{k}{m} & -\frac{\gamma}{m} \end{bmatrix} x(t).$$

Thus we have taken a second-order scalar ODE and rewritten it as a first-order vector ODE. We call this first-order vector ODE the *state-space representation* of the mass-spring-damper system, and the state vector $x(t)$ is a member of the *state space*. Since the right hand side of equation (J.2) can be written as a constant matrix multiplied by the state vector, this system is both linear and time invariant.

Generally, an LTI state-space system can be written as the vector ODE,

(J.3) $$\dot{x}(t) = Ax(t); \qquad x(t_0) = x_0,$$

where $x(t) \in \mathbb{R}^n$ and $A \in \mathbb{R}^{n \times n}$. This ODE is sometimes called a *vector field* because it assigns a vector Ax to each point x in the state space. This ODE has a unique solution, and the solution can be written in closed form,

(J.4) $x(t) = e^{A(t-t_0)} x_0,$

where the matrix exponential is defined by the Peano-Baker series

$$e^{A(t-t_0)} = \sum_{i=0}^{\infty} \frac{A^i (t-t_0)^i}{i!}$$

$$= I_{n \times n} + A(t-t_0) + \frac{A^2(t-t_0)^2}{2!} + \cdots.$$

J.2   Stability

Assuming that the matrix A has full rank, then the point $x = 0$ is the only point in the state space that satisfies the equilibrium condition $\dot{x} = 0$. The point $x = 0$ is called an equilibrium point. Note that the state will not move from an equilibrium point. If the initial condition is $x(t_0) = 0$, then $x(t)$ will remain at 0 for all time. In this section we discuss the stability of the origin of an LTI system.

We begin by defining a few notions of stability. An equilibrium point x_e (in the case of LTI systems $x_e = 0$) is said to be *stable* if for every $\epsilon > 0$ there exists a $\delta > 0$ such that whenever the initial condition satisfies $\|x_e - x(t_0)\| < \delta$ the solution $x(t)$ satisfies $\|x_e - x(t)\| < \epsilon$ for all time $t > 0$. In other words, stable means that if the initial condition starts close enough to the equilibrium, then the solution will never drift very far away. x_e is said to be *asymptotically stable* if it is stable and $\|x_e - x(t)\| \to 0$ as $t \to \infty$. Likewise, x_e is said to be *unstable* if it is neither stable nor asymptotically stable.

It is worth noting that for LTI systems, the stability properties are global. If they hold on any open subset of the state space, then they hold everywhere. Stability can be characterized in terms of the eigenvalues of the matrix A, as stated in the following theorem:

THEOREM J.2.1 (LTI stability)   *Consider the LTI system stated in equation (J.3), and let λ_i, $i \in \{1, 2, \ldots, n\}$ denote the eigenvalues of A. Let $\mathrm{re}(\lambda_i)$ denote the real part of λ_i Then the following holds:*

1. $x_e = 0$ is stable if and only if $\mathrm{re}(\lambda_i) \leq 0$ for all i.

2. $x_e = 0$ is asymptotically stable if and only if $\mathrm{re}(\lambda_i) < 0$ for all i.

3. $x_e = 0$ is unstable if and only if $\mathrm{re}(\lambda_i) > 0$ for some i.

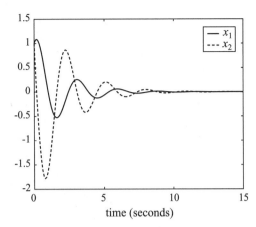 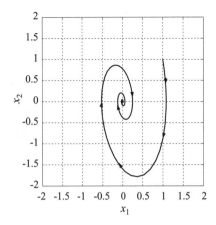

Figure J.2 Asymptotical stability. (Left) The states x_1 and x_2 (z and $\dot{z}$, respectively) plotted as time evolves. (Right) Phase plane plot of x_2 vs. x_1.

Consider the mass-spring-damper example. The eigenvalues of A are

$$\frac{-\gamma \pm \sqrt{\gamma^2 - 4km}}{2m}.$$

When the damping term is positive, the real parts of the eigenvalues are negative and the system is asymptotically stable. Figure J.2 shows two different representations of the trajectory of the mass-spring-damper system with $m = 1$, $k = 5$, and $\gamma = 1$. The figure on the left shows the values of x_1 and x_2 plotted as functions of time. As expected for an asymptotically stable system, both converge to zero. The figure on the right shows the trajectory in state space by plotting x_2 vs. x_1. This is sometimes referred to as a "phase plane" plot. The direction in which the trajectory flows is depicted by arrows. Here the trajectory starts at the initial condition and spirals into the origin. When the damping is zero, the system solution is a bounded oscillation and hence is stable but not asymptotically stable. Figure J.3 plots the time and phase plane representations of the stable trajectory that results when $m = 1$, $k = 5$, and $\gamma = 0$. Note that in the phase plane the periodic oscillation becomes a closed loop. When the damping is negative the damping term actually adds energy to the system, creating an oscillation that grows without bound. Time and phase plane plots for the case where $m = 1$, $k = 5$, and $\gamma = -0.4$ are shown in figure J.4.

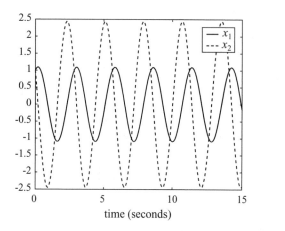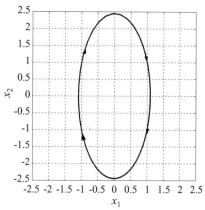

Figure J.3   Stability. (Left) The states x_1 and x_2 (z and $\dot{z}$, respectively) plotted as time evolves. (Right) Phase plane plot of x_2 vs. x_1.

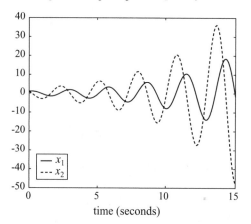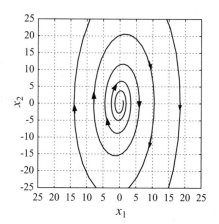

Figure J.4   Instability. (Left) The states x_1 and x_2 (z and $\dot{z}$, respectively) plotted as time evolves. (Right) Phase plane plot of x_2 vs. x_1.

J.3 LTI Control Systems

Often one has the ability to affect the behavior of a dynamical system by applying some sort of external input. For example, in the mass-spring-damper system discussed earlier we can influence the trajectory of the system by applying a time-varying external force $F(t)$ to the mass. This results in the LTI control system

(J.5) $\quad \dot{x}(t) = \begin{bmatrix} 0 & 1 \\ -\frac{k}{m} & -\frac{\gamma}{m} \end{bmatrix} x(t) + \begin{bmatrix} 0 \\ \frac{1}{m} \end{bmatrix} F(t).$

More generically, we write an LTI control system as

(J.6) $\quad \dot{x}(t) = Ax(t) + Bu(t); \qquad x(t_0) = x_0,$

where the state vector $x(t) \in \mathbb{R}^n$ and the external input vector $u(t) \in \mathbb{R}^m$. The matrix $B \in \mathbb{R}^{n \times m}$. The matrix A describes the system dynamics of the unforced system, i.e., A describes how the state would evolve if the input were zero. B describes how the inputs affect the evolution of the state.

The system described in equation (J.3) is said to be *controllable* if for any initial condition $x(t_0)$, there exists a continuous control input $u(t)$ that drives the solution $x(t)$ to the origin, $x = 0$. Note that the origin is an equilibrium point for the unforced system. This definition of controllability is equivalent to the definition of controllability for nonlinear systems presented in chapter 8, section 12.3 where the goal state is restricted to $x_{\text{goal}} = 0$.

THEOREM J.3.1 (LTI Controllability Test) *The LTI control system in equation (J.6) is controllable if and only if the matrix*

$W_c = [B \ AB \ A^2 B \ \cdots \ A^{n-1} B]$

has rank n.

Because controllability is determined solely by the matrices A and B, we can say that the pair (A, B) is controllable if the system in equation (J.6) is controllable.

One common control objective is to make the origin of a naturally unstable system stable using state feedback. Consider the control input given by the state-dependent control law

$u(t) = -Kx(t)$

for some matrix $K \in \mathbb{R}^{m \times n}$. Substituting this into equation (J.6) yields

$\dot{x}(t) = (A - BK)x(t).$

As a result, we can examine the stability of this new system in terms of the eigenvalues of the matrix $A - BK$. One of the fundamental properties of real-valued matrices is that their eigenvalues must occur in complex conjugate pairs. If $a + bi$ is an eigenvalue of a matrix, then $a - bi$ must also be an eigenvalue of that matrix. Hence we define a collection of complex numbers $\Lambda = \{\lambda_i \mid i \in \{1, 2, \ldots, n\}\}$ to be *allowable* if for each λ_i that has a nonzero imaginary part there is a corresponding conjugate λ_j. Now we are prepared to state an important result of linear control theory:

THEOREM J.3.2 (Eigenvalue Placement)   *Consider the system of equation (J.6) and assume the pair (A, B) is controllable and that the matrix B has full column rank. Let $\Lambda = \{\lambda_i \mid i \in \{1, 2, \ldots, n\}\}$ be any allowable collection of complex numbers. Then there exists a constant matrix $K \in \mathbb{R}^{m \times n}$ such that the set of eigenvalues of $(A - BK)$ is equal to Λ.*

Under the assumptions of this theorem, we can place the eigenvalues of the matrix $A - BK$ in any allowable configuration using linear feedback. The task of stabilizing an LTI system is then simply a matter of finding a K so that the corresponding eigenvalues have negative real parts. There are a number of algorithms to perform direct eigenvalue assignment (also sometimes called pole placement). Some of these are implemented in the MATLAB control systems toolbox. Similarly, the famous linear quadratic regulator (LQR) (see e.g., [396]) places the eigenvalues of $A - BK$ to optimize a user-defined cost function.

Consider as an example the mass-spring-damper system with negative damping. As was pointed out earlier, this system is unstable; solutions for initial conditions arbitrarily close to the origin will grow without bound. To use state feedback to stabilize this system, consider the matrix

$$A - BK = \begin{bmatrix} 0 & 1 \\ -\frac{k}{m} & -\frac{\gamma}{m} \end{bmatrix} - \begin{bmatrix} 0 \\ \frac{1}{m} \end{bmatrix} \begin{bmatrix} k_1 & k_2 \end{bmatrix} = \begin{bmatrix} 0 & 1 \\ -\frac{k+k_1}{m} & \frac{\gamma+k_2}{m} \end{bmatrix}.$$

The eigenvalues of $A - BK$ are

$$\frac{(-\gamma - k_2) \pm \sqrt{(-\gamma - k_2)^2 - 4(k - k_1)m}}{2m},$$

so we can ensure that the real part of both eigenvalues is negative by choosing k_2 such that $-\gamma - k_2 < 0$. This is equivalent to adding sufficient positive viscous damping to overcome the energy added by the negative damping term γ.

J.4 Observing LTI Systems

Often it is not possible to directly measure the entire state of an LTI system. Rather, the state must be observed through the use of sensors that provide some lower-dimensional measurement of the current state. If it were possible to measure only velocity in the mass-spring-damper example, then equations of motion together with the output equation for the system would be

(J.7)
$$\dot{x}(t) = \begin{bmatrix} 0 & 1 \\ -\frac{k}{m} & -\frac{\gamma}{m} \end{bmatrix} x(t) + \begin{bmatrix} 0 \\ \frac{1}{m} \end{bmatrix} F(t),$$
$$y(t) = [0 \quad 1] \, x(t),$$

where $y(t)$ represents the output signal coming from the sensor. We write a general LTI system with output equation as

(J.8)
$$\dot{x}(t) = Ax(t) + Bu(t); \qquad x(t_0) = x_0,$$
$$y(t) = Cx(t),$$

where the state vector $x(t) \in \mathbb{R}^n$, the control vector $u(t) \in \mathbb{R}^m$, and the output vector $y(t) \in \mathbb{R}^p$. The constant matrix $C \in \mathbb{R}^{p \times n}$. Note that the matrix C may not be invertible (it is usually not even square!), so the state at any instant $x(t)$ cannot be directly observed from the measurement at that instant $y(t)$. We must instead reconstruct the state by measuring the output over some interval of time and using knowledge of the system dynamics. A device that performs such a reconstruction is called an *observer*.

We say that the system of equation (J.8) is *observable* if it is possible to determine the initial state $x(t_0)$ by observing the known signals $y(t)$ and $u(t)$ over some period of time.

THEOREM J.4.1 (LTI Observability Test) *The LTI control system in equation (J.8) is observable if and only if the matrix*

$$W_o = \begin{bmatrix} C \\ CA \\ CA^2 \\ \vdots \\ CA^{n-1} \end{bmatrix}$$

has rank n.

As in the case of controllability, we say that the pair (A, C) is observable if the system in equation (J.8) is observable. Note that the pair (A, C) is observable if and only if

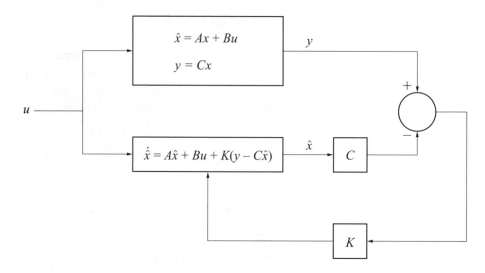

Figure J.5   Block diagram for a linear observer.

the pair (A^T, C^T) is controllable. If the pair (A, B) is controllable and the pair (A, C) is observable, then the system [and the triple (A, B, C)] is said to be *minimal*.

Now consider an observer defined by the ODE

(J.9) $\dot{\hat{x}}(t) = A\hat{x}(t) + Bu(t) + K(y(t) - C\hat{x}(t)).$

Note that this ODE requires that we know the matrices A, B, and C as well as the input $u(t)$ and output $y(t)$. The vector $\hat{x}(t)$ is called the *state estimate* produced by this observer. As shown in the block diagram in figure J.5, this observer is essentially a copy of the original dynamic system with a correcting term that is a linear function of the difference between the measured output $y(t)$ and the estimated output $C\hat{x}(t)$. The task is then to try to choose K so that the correcting term forces the state estimate to converge to the actual value.

If we define the error signal $e(t) = x(t) - \hat{x}(t)$, we can examine how the error evolves with time:

$$\begin{aligned}
\dot{e}(t) &= \dot{x}(t) - \dot{\hat{x}}(t) \\
&= Ax(t) + Bu(t) - (A\hat{x}(t) + Bu(t) + K(y(t) - C\hat{x}(t))) \\
&= A(x(t) - \hat{x}(t)) - K(Cx(t) - C\hat{x}(t)) \\
&= (A - KC)e(t)
\end{aligned}$$

If $e(t) \to 0$, then $\hat{x}(t) \to x(t)$. So the state estimate $\hat{x}(t)$ that results from the observer presented in equation (J.9) converges to the actual state $x(t)$ if K is chosen so that the unforced LTI system $\dot{e}(t) = (A - KC)e(t)$ is asymptotically stable.

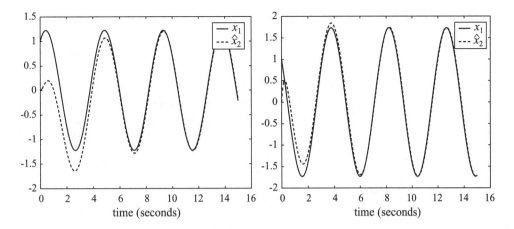

Figure J.6 Solid lines represent the actual state and the dashed line represents the state estimate determined by the observer. The left figure depicts x_1 and the right x_2.

Recall that the eigenvalues of any matrix are equal to the eigenvalues of its transpose, so the eigenvalues of $A - KC$ are identical to the eigenvalues of $A^T - C^T K^T$. According to theorem J.3.2, we can place the eigenvalues of $A^T - C^T K^T$ in any allowable configuration provided that the pair (A^T, C^T) is controllable and the matrix C^T has full column rank. This is equivalent to saying that the eigenvalues of $A - KC$ can be placed in any allowable configuration provided that the pair (A, C) is observable and C has full row rank. Under these conditions, it is possible to chose a K so that the observer estimate $\hat{x}(t)$ converges to $x(t)$.

Consider the mass-spring-damper system of equation (J.7). The matrix

$$A - KC = \begin{bmatrix} 0 & 1 - k_1 \\ -\frac{k}{m} & -\frac{\gamma}{m} - k_2 \end{bmatrix}.$$

The eigenvalues of this matrix are

$$\frac{-(\gamma + mk_2) \pm \sqrt{(-\gamma - mk_2)^2 - 4m(k - k_1)}}{2m},$$

so choosing k_2 such that $-\gamma - mk_2 < 0$ will guarantee that the observer given in equation (J.9) converges, meaning that after some initial transient, estimate $\hat{x}(t)$ will provide a good approximation of the state. For the case where $m = 1$, $k = 2$, and $\gamma = 0$, the choice of $K = [0 \ 2]^T$ will provide a convergent observer. Figure J.6 shows how the estimates $\hat{x}_1(t)$ and $\hat{x}_2(t)$ converge to $x_1(t)$ and $x_2(t)$, respectively.

J.5   Discrete Time Systems

The previous sections dealt with an LTI system whose trajectories were continuous in time. In practice, a continuous dynamical system is usually sampled at regular time intervals. The sampled or *discrete time* signal is then fed into a computer as a sequence of numbers. The computer can then use this sequence to calculate a desired control input or to estimate the state. In this section we present an overview of the theory of discrete time LTI systems and their relationship to their continuous time cousins.

Consider the continuous time signal $x(t)$. We define a sequence of vectors using the formula $x_s(k) = x(t_0 + kT)$. The sequence $x_s(k)$ is the *discrete time sampling* of the continuous signal $x(t)$. In the future, we will abuse notation and drop the s subscript on the discrete time sequence. The continuous and discrete signals can be differentiated by the letter used in their argument; $x(k)$ represents an element of the sequence and $x(t)$ denotes the continuous time signal.

Using the first-order derivative approximation

$$x(t_0 + kT) \approx \frac{x(k+1) - x(k)}{T}$$

and substituting into the continuous time LTI system of equation (J.8) yields

$$\frac{x(k+1) - x(k)}{T} \approx Ax(k) + Bu(k),$$

which leads to

$$x(k+1) \approx x(k) + TAx(k) + TBu(k).$$

Defining $F = I_{n \times n} + TA$, $G = TB$, and $H = C$, we can then write a discrete time approximation of the continuous system:

(J.10)
$$\dot{x}(k+1) = Fx(k) + Gu(k); \qquad x(0) = x_0$$
$$y(k) = Hx(k)$$

Most of the concepts from continuous LTI systems have direct analogs in discrete time LTI systems. We discuss them briefly here.

J.5.1   Stability

The discrete time notions of stability, asymptotic stability, and instability follow directly from the continuous time definitions. As in the case of continuous systems, the stability of the unforced system $x(k+1) = Fx(k)$ can be evaluated in terms of the eigenvalues of F:

THEOREM J.5.1 (Discrete Time LTI Stability) *Consider the unforced discrete time LTI system described by the equation $x(k+1) = Fx(k)$, and let λ_i, $i \in \{1, 2, \ldots, n\}$ denote the eigenvalues of F. Then the following hold:*

1. $x_e = 0$ is stable if and only if $|\lambda_i| \leq 1$ for all i.

2. $x_e = 0$ is asymptotically stable if and only if $|\lambda_i| < 1$ for all i.

3. $x_e = 0$ is unstable if and only if $|\lambda_i| > 1$ for some i.

J.5.2   Controllability and Observability

The properties of controllability and observability for the discrete time LTI system follow from the properties of the continuous time system. The controllability test is the same for both: the pair (F, G) is controllable if and only if the matrix $[G \; FG \; F^2G \; \cdots \; F^{n-1}G]$ has rank n. The pair (F, H) is observable if and only if the pair (F^T, H^T) is controllable. As in the case of continuous systems, construction of linear state feedback control laws or linear observers results in a pole placement problem which can be solved if the system is controllable or observable, respectively.

Bibliography

[1] http://www.accuray.com/ck/how9.htm and http://www.cksociety.org/.

[2] http://www.intuitivesurgical.com/about_intuitive/index.html.

[3] http://computermotion.wwwa.com/productsandsolutions/products/zeus/index.cfm.

[4] http://www.aemdesign.com.

[5] http://www.sbsi-sol-optimize.com/NPSOL.htm.

[6] http://www.vni.com.

[7] http://www.nag.com.

[8] *Webster's Ninth New Collegiate Dictionary*. Merriam-Webster, Inc., Springfield, MA, 1990.

[9] R. Abraham, J. Marsden, and T. Ratiu. *Manifolds, Tensor Analysis, and Applications*. Springer-Verlag, New York, 2 edition, 1988.

[10] R. Abraham and J. E. Marsden. *Foundations of Mechanics*. Addison-Wesley, 1985.

[11] E. U. Acar and H. Choset. Sensor-based coverage of unknown environments: Incremental construction of Morse decompositions. *International Journal of Robotics Research,* 21:345–366, April 2002.

[12] E. U. Acar, H. Choset, A. A. Rizzi, P. Atkar, and D. Hull. Morse decompositions for coverage tasks. *International Journal of Robotics Research,* 21:331–344, April 2002.

[13] S. Akella, W. Huang, K. Lynch, and M. Mason. Parts feeding on a conveyor with a one joint robot. *Algorithmica (Special Issue on Robotics),* 26(3/4):313–344, 2000.

[14] M. Akinc, K. E. Bekris, B. Chen, A. Ladd, E. Plaku, and L. E. Kavraki. Probabilistic roadmaps of trees for parallel computation of multiple query roadmaps. In *International Symposium on Robotics Research,* 2003. Book to appear.

[15] R. Alami, J. Laumond, and T. Siméon. Two manipulation planning algorithms. In K. Goldberg, D. Halperin, J. C. Latombe, and R. Wilson, editors, *Algorithmic Foundations of Robotics,* pages 109–125. A.K. Peters, 1995.

[16] R. Alami, T. Siméon, and J. P. Laumond. A geometrical approach to planning manipulation tasks. In *International Symposium on Robotics Research,* pages 113–119, 1989.

[17] P. Allen and I. Stamos. Integration of range and image sensing for photorealistic 3D modeling. In *IEEE International Conference on Robotics and Automation,* 2000.

[18] N. M. Amato, B. Bayazit, L. Dale, C. Jones, and D. Vallejo. OBPRM: An obstacle-based PRM for 3d workspaces. In P. Agarwal, L. E. Kavraki, and M. Mason, editors, *Robotics: The Algorithmic Perspective,* pages 156–168. AK Peters, 1998.

[19] N. M. Amato, O. B. Bayazit, L. K. Dale, C. Jones, and D. Vallejo. Choosing good distance metrics and local planners for probabilistic roadmap methods. In *IEEE International Conference on Robotics and Automation,* pages 630–637, 1998.

[20] N. M. Amato, K. Dill, and G. Song. Using motion planning to map protein folding landscapes and analyze folding kinetics of known native structures. In *International Conference on Research in Computational Molecular Biology,* pages 2–11, April 2002.

[21] N. M. Amato and G. Song. Using motion planning to study protein folding pathways. In *International Conference on Research in Computational Molecular Biology,* pages 287–296, 2001.

[22] E. Anshelevich, S. Owens, F. Lamiraux, and L. E. Kavraki. Deformable volumes in path planning applications. In *IEEE International Conference on Robotics and Automation,* pages 2290–2295, 2000.

[23] M. Apaydin, D. Brutlag, C. Guestrin, D. Hsu, J. C. Latombe, and C. Varm. Stochastic roadmap simulation: An efficient representation and algorithm for analyzing molecular motion. *Journal of Computational Biology,* 10:257–281, 2003.

[24] M. Apaydin, C. Guestrin, C. Varma, D. Brutlag, and J. C. Latombe. Studying protein-ligand interactions with stochastic roadmap simulation. *Bioinformatics,* 18(2):18–26, 2002.

[25] M. S. Apaydin, D. L. Brutlag, C. Guestrin, D. Hsu, and J. C. Latombe. Stochastic roadmap simulation: An efficient representation and algorithm for analyzing molecular motion. In *International Conference on Research in Computational Molecular Biology,* pages 12–21, April 2002.

[26] V. I. Arnold. *Mathematical Methods of Classical Mechanics.* Springer-Verlag, 1989.

[27] K. Arras, N. Tomatis, B. Jensen, and R. Siegwart. Multisensor on-the-fly localization: Precision and reliability for applications. *Robotics and Autonomous Systems,* 34(2-3):131–143, 2001.

[28] K. Arras and S. Vestli. Hybrid, high-precision localization for the mail distributing mobile robot system MOPS. In *IEEE International Conference on Robotics and Automation,* 1998.

[29] S. Arulampalam, S. Maskell, N. Gordon, and T. Clapp. A tutorial on particle filters for on-line non-linear/non-Gaussian Bayesian tracking. *IEEE Transactions on Signal Processing,* 50(2):174–188, 2002.

[30] S. Arya, D. M. Mount. Approximate nearest neighbor queries in fixed dimensions. In *47th Annual ACM-SIAM Symposium on Discrete Algorithms (SODA),* pages 271–280, 1993.

[31] F. Aurenhammer. Voronoi diagrams—A survey of a fundamental geometric structure. *ACM Computing Surveys,* 23:345–405, 1991.

[32] D. Avots, E. Lim, R. Thibaux, and S. Thrun. A probabilistic technique for simultaneous localization and door state estimation with mobile robots in dynamic environments. In *IEEE/RSJ International Conference on Intelligent Robots and Systems,* 2002.

[33] B. Baginski. Motion planning for manipulators with many degrees of freedom—The BB Method. Ph.D. Thesis, Technische Universität München, 1998.

[34] J. Baillieul and B. Lehman. Open-loop control using oscillatory inputs. In *CRC Control Handbook,* pages 967–980. CRC Press, Boca Raton, FL, 1996.

[35] D. J. Balkcom and M. T. Mason. Time optimal trajectories for differential drive vehicles. *International Journal of Robotics Research,* 21(3):199–217, Mar. 2002.

[36] J. Barraquand and P. Ferbach. A penalty function method for constrained motion planning. In *IEEE International Conference on Robotics and Automation,* pages 1235–1242, 1994.

[37] J. Barraquand, L. E. Kavraki, J. C. Latombe, T.-Y. Li, R. Motwani, and P. Raghavan. A random sampling scheme for robot path planning. *International Journal of Robotics Research,* 16(6):759–774, 1997.

[38] J. Barraquand, B. Langlois, and J. C. Latombe. Numerical potential field techniques for robot path planning. *IEEE Transactions on Man and Cybernetics,* 22(2):224–241, Mar/Apr 1992.

[39] J. Barraquand and J. C. Latombe. Robot motion planning: A distributed representation approach. Technical Report STAN-CS-89-1257, Stanford University, Stanford CA, 1989.

[40] J. Barraquand and J. C. Latombe. Robot motion planning: A distributed representation approach. *International Journal of Robotics Research,* 10(6):628–649, Dec. 1991.

[41] J. Barraquand and J. C. Latombe. Nonholonomic multibody mobile robots: Controllability and motion planning in the presence of obstacles. *Algorithmica,* 10:121–155, 1993.

[42] S. Basu, R. Pollack, and M.-F. Roy. *Algorithms in Real Algebraic Geometry.* Springer-Verlag, 2003.

[43] K. E. Bekris, B. Chen, A. Ladd, E. Plaku, and L. E. Kavraki. Multiple query motion planning using single query primitives. In *IEEE/RSJ International Conference on Intelligent Robots and Systems,* pages 656–661, 2003.

[44] J. Bentley. Multidimensional divide and conquer. *Communications of the ACM,* 23(4), 1980.

[45] D. Bertsekas. *Nonlinear Programming.* Athena Scientific, Belmont, MA, second edition, 1999.

[46] P. Besl and N. McKay. A method for registration of 3D shapes. *IEEE Transactions on Pattern Analysis and Machine Intelligence,* 18(14):239–256, 1992.

[47] P. Bessiere, E. Mazer, and J.-M. Ahuactzin. Planning in continuous space with forbidden regions: The Ariadne's clew algorithm. In K. Goldberg, K. Goldberg, R. Wilson, and D. Halperin, editors, *Algorithmic Foundations of Robotics (WAFR),* pages 39–47. A.K. Peters, Wellsley MA, 1995.

[48] P. Bessiere, E. Mazer, and J.-M. Ahuactzin. The ariadne's clew algorithm. *Journal of Artificial Intelligence Research (JAIR),* 9:295–316, 1998.

[49] J. T. Betts. Survey of numerical methods for trajectory optimization. *AIAA Journal of Guidance, Control, and Dynamics,* 21(2):193–207, March-April 1998.

[50] A. M. Bloch. *Nonholonomic Mechanics and Control.* Springer, New York, 2003.

[51] J. E. Bobrow, S. Dubowsky, and J. S. Gibson. Time-optimal control of robotic manipulators along specified paths. *International Journal of Robotics Research,* 4(3):3–17, Fall 1985.

[52] R. Bohlin. Path planning in practice: Lazy evaluation on a multi-resolution grid. In *IEEE/RSJ International Conference on Intelligent Robots and Systems,* 2001.

[53] R. Bohlin and L. E. Kavraki. Path planning using lazy PRM. In *IEEE International Conference on Robotics and Automation,* pages 521–528, 2000.

[54] R. Bohlin and L. E. Kavraki. A randomized algorithm for robot path planning based on lazy evaluation. In P. Pardalos, S. Rajasekaran, and J. Rolim, editors, *Handbook on Randomized Computing,* pages 221–249. Kluwer Academic Publishers, 2001.

[55] K.-F. Böhringer, B. R. Donald, L. E. Kavraki, and F. Lamiraux. Part orientation to one or two stable equilibria using programmable force fields. *IEEE Transactions on Robotics and Automation,* 16(2):731–747, 2000.

[56] K. Böhringer, B. R. Donald, and N. MacDonald. Programmable vector fields for distributed manipulation, with application to mems actuator arrays and vibratory part feeders. *International Journal of Robotics Research,* 18:168–200, Feb. 1999.

[57] J. A. Bondy and U. S. R. Murty. *Graph Theory with Applications.* John Wiley and Sons Inc., New York, NY, 2000.

[58] B. Bonnard. Contrôlabilité des systèmes nonlinéaires. *C. R. Acad. Sci. Paris,* 292:535–537, 1981.

[59] V. Boor, N. H. Overmars, and A. F. van der Stappen. The Gaussian sampling strategy for probabilistic roadmap planners. In *IEEE International Conference on Robotics and Automation,* pages 1018–1023, 1999.

[60] W. M. Boothby. *An Introduction to Differentiable Manifolds and Riemannian Geometry.* Academic Press, 1986.

[61] J. Borenstein, B. Everett, and L. Feng. *Navigating Mobile Robots: Systems and Techniques.* A.K. Peters, Ltd., Wellesley, MA, 1996.

[62] M. S. Branicky, S. M. LaValle, K. Olson, and L. Yang. Quasi-randomized path planning. In *IEEE International Conference on Robotics and Automation,* pages 1481–1487, 2001.

[63] G. E. Bredon. *Topology and Geometry.* Springer-Verlag, New York, NY, 1993.

[64] T. Bretl, J. C. Latombe, and S. Rock. Toward autonomous free climbing robots. In *International Symposium on Robotics Research,* 2003. Book to appear.

[65] R. W. Brockett. Nonlinear systems and differential geometry. *Proceedings of the IEEE,* 64(1):61–72, Jan. 1976.

[66] R. W. Brockett. Control theory and singular Riemannian geometry. In P. J. Hilton and G. S. Young, editors, *New Directions in Applied Mathematics,* pages 11–27. Springer-Verlag, 1982.

[67] R. A. Brooks and T. Lozano-Pérez. A subdivision algorithm in configuration space for findpath with rotation. *IEEE Transactions Systems, Man, and Cybernetics,* 15:224–233, 1985.

[68] R. A. Brooks. Solving the find-path problem by good representation of free space. *IEEE Transactions on Systems, Man, and Cybernetics,* 13(3):190–197, 1983.

[69] R. C. Brost. *Analysis and Planning of Planar Manipulation Tasks.* PhD thesis, Carnegie Mellon University, Jan. 1991. Available as Technical Report CMU-CS-91-149.

[70] R. C. Brost. Computing the possible rest configurations of two interacting polygons. In *IEEE International Conference on Robotics and Automation,* pages 686–693, Apr. 1991.

[71] A. E. Bryson. *Dynamic Optimization.* Addison-Wesley, 1998.

[72] A. E. Bryson and Y. C. Ho. *Applied Optimal Control.* Hemisphere Publishing, New York, 1975.

[73] J. Buhmann, W. Burgard, A. Cremers, D. Fox, T. Hofmann, F. Schneider, J. Strikos, and S. Thrun. The mobile robot RHINO. *AI Magazine,* 16(2):31–38, Summer 1995.

[74] F. Bullo. Series expansions for the evolution of mechanical control systems. *SIAM Journal on Control and Optimization,* 40(1):166–190, 2001.

[75] F. Bullo. Averaging and vibrational control of mechanical systems. *SIAM Journal on Control and Optimization,* 41:542–562, 2002.

[76] F. Bullo, N. E. Leonard, and A. D. Lewis. Controllability and motion algorithms for underactuated Lagrangian systems on Lie groups. *IEEE Transactions on Automatic Control,* 45(8):1437–1454, 2000.

[77] F. Bullo and A. D. Lewis. *Geometric Control of Mechanical Systems.* Springer, 2004.

[78] F. Bullo, A. D. Lewis, and K. M. Lynch. Controllable kinematic reductions for mechanical systems: Concepts, computational tools, and examples. In *2002 International Symposium on the Mathematical Theory of Networks and Systems,* Aug. 2002.

[79] F. Bullo and K. M. Lynch. Kinematic controllability for decoupled trajectory planning of underactuated mechanical systems. *IEEE Transactions on Robotics and Automation,* 17(4):402–412, Aug. 2001.

[80] F. Bullo and M. Žefran. On mechanical control systems with nonholonomic constraints and symmetries. *Systems and Control Letters,* 45(2):133–143, Jan. 2002.

[81] W. Burgard, A. Cremers, D. Fox, D. Hähnel, G. Lakemeyer, D. Schulz, W. Steiner, and S. Thrun. Experiences with an interactive museum tour-guide robot. *Artificial Intelligence,* 114(1-2), 2000.

[82] W. Burgard, A. Derr, D. Fox, and A. Cremers. Integrating global position estimation and position tracking for mobile robots: the dynamic Markov localization approach. In *IEEE/RSJ International Conference on Intelligent Robots and Systems,* 1998.

[83] W. Burgard, D. Fox, D. Hennig, and T. Schmidt. Estimating the absolute position of a mobile robot using position probability grids. In *Proc. of the National Conference on Artificial Intelligence (AAAI),* 1996.

[84] W. Burgard, D. Fox, H. Jans, C. Matenar, and S. Thrun. Sonar-based mapping of large-scale mobile robot environments using EM. In *Proc. of the International Conference on Machine Learning (ICML),* 1999.

[85] L. Bushnell, D. Tilbury, and S. Sastry. Steering three-input nonholonomic systems: The fire-truck example. *International Journal of Robotics Research,* 14(4):366–381, 1995.

[86] Z. J. Butler, A. A. Rizzi, and R. L. Hollis. Contact sensor-based coverage of rectilinear environments. In *Proc. of IEEE Int'l Symposium on Intelligent Control,* Sept. 1999.

[87] P. E. Caines and E. S. Lemch. On the global controllability of Hamiltonian and other nonlinear systems: Fountains and recurrence. In *IEEE International Conference on Decision and Control,* pages 3575–3580, 1998.

[88] S. Cameron. Collision detection by four-dimensional intersection testing. *IEEE Transactions on Robotics and Automation,* pages 291–302, 1990.

[89] S. Cameron. Enhancing GJK: Computing minimum distance and penetration distanses between convex polyhedra. In *IEEE International Conference on Robotics and Automation,* pages 3112–3117, 1997.

[90] J. F. Canny. *The Complexity of Robot Motion Planning.* MIT Press, Cambridge, MA, 1988.

[91] J. F. Canny. Constructing roadmaps of semi-algebraic sets I: Completeness. *Artificial Intelligence,* 37:203–222, 1988.

[92] J. F. Canny. Computing roadmaps of general semi-algebraic sets. *The Computer Journal,* 35(5):504–514, 1993.

[93] J. F. Canny and M. Lin. An opportunistic global path planner. *Algorithmica,* 10:102–120, 1993.

[94] J. F. Canny, J. Reif, B. Donald, and P. Xavier. On the complexity of kinodynamic planning. In *IEEE Symposium on the Foundations of Computer Science,* pages 306–316, White Plains, NY, 1988.

[95] J. F. Canny. Some algebraic and geometric computations in PSPACE. In *Proc. 20th ACM Symposium on the Theory of Computing,* pages 460–469, 1998.

[96] Z. L. Cao, Y. Huang, and E. Hall. Region filling operations with random obstacle avoidance for mobile robots. *Journal of Robotic systems,* pages 87–102, February 1988.

[97] J. Carpenter, P. Clifford, and P. Fernhead. An improved particle filter for nonlinear problems. *IEE Proceedings on Radar and Sonar Navigation,* 146(2-7), 1999.

[98] A. Casal. *Reconfiguration Planning for Modular Self-Reconfigurable Robots.* PhD thesis, Stanford University, Stanford, CA, 2002.

[99] J. Castellanos, J. Montiel, J. Neira, and J. Tardós. The SPmap: A probabilistic framework for simultaneous localization and map building. *IEEE Transactions on Robotics and Automation,* 15(5):948–953, 1999.

[100] J. Castellanos and J. Tardós. *Mobile Robot Localization and Map Building: A Multisensor Fusion Approach.* Kluwer Academic Publishers, Boston, MA, 2000.

[101] P. C. Chen and Y. K. Hwang. SANDROS: A motion planner with performance proportional to task difficulty. *IEEE International Conference on Robotics and Automation,* pages 2346–2353, 1992.

[102] P. C. Chen and Y. K. Hwang. SANDROS: A dynamic graph search algorithm for motion planning. *IEEE Transactions on Robotics and Automation,* 14(3):390–403, June 1998.

[103] P. Cheng and S. M. LaValle. Reducing metric sensitivity in randomized trajectory design. In *IEEE/RSJ International Conference on Intelligent Robots and Systems,* pages 43–48, 2001.

[104] H. Choset. Nonsmooth analysis, convex analysis, and their applications to motion planning. *Special Issue of the Int. Jour. of Comp. Geom. and Apps.,* 1998.

[105] H. Choset and J. Burdick. Sensor based motion planning: Incremental construction of the hierarchical generalized Voronoi graph. *International Journal of Robotics Research,* 19(2):126–148, February 2000.

[106] H. Choset and J. Burdick. Sensor based motion planning: The hierarchical generalized Voronoi graph. *International Journal of Robotics Research,* 19(2):96–125, February 2000.

[107] H. Choset and J. Y. Lee. Sensor-based construction of a retract-like structure for a planar rod robot. *IEEE Transaction of Robotics and Automation,* 17, 2001.

[108] H. Choset and K. Nagatani. Topological simultaneous localization and mapping (T-SLAM). *IEEE Transactions on Robotics Automation,* 17, April 2001.

[109] H. Choset, K. Nagatani, and A. Rizzi. Sensor based planning: Using a honing strategy and local map method to implement the generalized Voronoi graph. In *SPIE Conference on Systems and Manufacturing,* Pittsburgh, PA, 1997.

[110] H. Choset and P. Pignon. Coverage path planning: The boustrophedon decomposition. In *Proceedings of the International Conference on Field and Service Robotics,* Canberra, Australia, December 1997.

[111] P. Choudhury and K. M. Lynch. Trajectory planning for second-order underactuated mechanical systems in the presence of obstacles. In J.-D. Boissonnat, J. Burdick, K. Goldberg, and S. Hutchinson, editors, *Algorithmic Foundations of Robotics V,* pages 559–575. Springer-Verlag, 2002.

[112] W.-L. Chow. Uber systemen von linearen partiellen differentialgleichungen erster ordnung. *Math. Ann.,* 117:98–105, 1939.

[113] S. Ciarcia. An ultrasonic ranging system. *Byte Magazine,* pages 113–123, October 1984.

[114] F. H. Clarke. *Optimization and Nonsmooth Analysis.* Society of Industrial and Applied Mathematics, Philadelphia, PA, 1990.

[115] J. D. Cohen, M. C. Lin, D. Manocha, and M. K. Ponamgi. I-COLLIDE: An interactive and exact collision detection system for large-scale environments. In *Symposium on Interactive 3D Graphics,* pages 189–196, 218, 1995.

[116] J. Colegrave and A. Branch. A case study of autonomous household vacuum cleaner. In *AIAA/NASA CIRFFSS,* 1994.

[117] G. E. Collins. Quantifier elimination for real closed fields by cylindrical algebraic decomposition. In *Lecture Notes in Computer Science,* volume 33, pages 134–183. Springer-Verlag, 1975.

[118] H. Cormen, C. Leiserson, and R. Rivest. *Introduction to Algorithms.* MIT Press, Cambridge, MA, 1990.

[119] T. H. Cormen, C. E. Leiserson, R. L. Rivest, and C. Stein. *Introduction to Algorithms.* MIT Press, 2002.

[120] J. Cortes, S. Martinez, J. P. Ostrowski, and H. Zhang. Simple mechanical control systems with constraints and symmetry. *SIAM Journal on Control and Optimization,* 41(3):851–874, 2002.

[121] J. Cortés, T. Simeon, and J.-P. Laumond. A random loop generator for planning the motions of closed kinematic chains. In *IEEE International Conference on Robotics and Automation,* pages 2141–2146, 2002.

[122] J. Crowley. World modeling and position estimation for a mobile robot using ultrasound ranging. In *IEEE International Conference on Robotics and Automation,* 1989.

[123] T. Danner and L. E. Kavraki. Randomized planning for short inspection paths. In *IEEE International Conference on Robotics and Automation,* pages 971–976, San Fransisco, CA, April 2000. IEEE Press.

[124] M. de Berg, M. van Kreveld, and M. Overmars. *Computational Geometry: Algorithms and Applications.* Springer, Berlin, 1997.

[125] F. Dellaert, D. Fox, W. Burgard, and S. Thrun. Monte Carlo Localization for mobile robots. In *IEEE International Conference on Robotics and Automation,* 1999.

[126] F. Dellaert, S. Seitz, C. Thorpe, and S. Thrun. Structure from motion without correspondence. In *Proc. of the IEEE Computer Society Conference on Computer Vision and Pattern Recognition (CVPR),* 2000.

[127] A. O. Dempster, A. N. Laird, and D. B. Rubin. Maximum likelihood from incomplete data via the EM algorithm. *Journal of the Royal Statistical Society, Series B,* 39(1):1–38, 1977.

[128] G. Dissanayake, P. Newman, S. Clark, H. F. Durrant-Whyte, and M. Csorba. A solution to the simultaneous localisation and map building (SLAM) problem. *IEEE Transactions on Robotics and Automation,* 2001.

[129] A. W. Divelbiss and J. Wen. Nonholonomic path planning with inequality constraints. In *IEEE International Conference on Decision and Control,* pages 2712–2717, 1993.

[130] A. W. Divelbiss and J.-T. Wen. A path space approach to nonholonomic motion planning in the presence of obstacles. *IEEE Transactions on Robotics and Automation,* 13(3):443–451, 1997.

[131] M. P. do Carmo. *Riemannian Geometry.* Birkhäuser, Boston, MA, 1992.

[132] B. R. Donald. A search algorithm for motion planning with six degrees of freedom. *Artificial Intelligence,* 31:295–353, 1987.

[133] B. R. Donald, P. Xavier, J. Canny, and J. Reif. Kinodynamic motion planning. *Journal of the Association for Computing Machinery,* 40(5):1048–1066, Nov. 1993.

[134] B. R. Donald and P. Xavier. Provably good approximation algorithms for optimal kinodynamic planning for Cartesian robots and open chain manipulators. *Algorithmica,* 4(6):480–530, 1995.

[135] B. R. Donald and P. Xavier. Provably good approximation algorithms for optimal kinodynamic planning: robots with decoupled dynamics bounds. *Algorithmica,* 4(6):443–479, 1995.

[136] A. Doucet. On sequential simulation-based methods for Bayesian filtering. Technical report, Department of Engeneering, University of Cambridge, 1998.

[137] A. Doucet, J. de Freitas, K. Murphy, and S. Russel. Rao-Blackwellised particle filtering for dynamic Bayesian networks. In *Proc. of the Conference on Uncertainty in Artificial Intelligence (UAI),* 2000.

[138] A. Doucet, N. de Freitas, and N. Gordon. *Sequential Monte Carlo Methods in Practice.* Springer Verlag, 2001.

[139] D. Duff, M. Yim, and K. Roufas. Evolution of polybot: A modular reconfigurable robot. In *Proc. of the Harmonic Drive Intl. Symposium,* Nagano, Japan, 2001.

[140] S. Ehmann and M. C. Lin. Swift: Accelerated distance computation between convex polyhedra by multi-level Voronoi marching. In *IEEE/RSJ International Conference on Intelligent Robots and Systems,* 2000.

[141] S. A. Ehmann and M. C. Lin. Geometric algorithims: Accurate and fast proximity queries between polyhedra using convex surface decomposition. *Computer Graphics Forum—Proc. of Eurographics,* 20:500–510, 2001.

[142] A. Elfes. Sonar-based real-world mapping and navigation. *IEEE Journal of Robotics and Automation,* RA-3:249–265, June 1987.

[143] A. Elfes. *Occupancy Grids: A Probabilistic Framework for Robot Percepti on and Navigation.* PhD thesis, Department of Electrical and Computer Engineering, Carnegie Mellon University, 1989.

[144] A. Elfes. Using occupancy grids for mobile robot perception and navigation. *IEEE Computer,* pages 46–57, 1989.

[145] S. Engelson. *Passive Map Learning and Visual Place Recognition.* PhD thesis, Department of Computer Science, Yale University, 1994.

[146] M. Erdmann and M. Mason. An exploration of sensorless manipulation. *IEEE Tr. on Rob. and Autom.,* 4(4):369–379, 1988.

[147] C. Fernandes, L. Gurvits, and Z. Li. Optimal nonholonomic motion planning for a falling cat. In Z. Li and J. Canny, editors, *Nonholonomic Motion Planning.* Kluwer Academic, 1993.

[148] C. Fernandes, L. Gurvits, and Z. Li. Near-optimal nonholonomic motion planning for a system of coupled rigid bodies. *IEEE Transactions on Automatic Control,* 30(3):450–463, Mar. 1994.

[149] R. Fitch, Z. Butler, and D. Rus. Reconfiguration planning for heterogeneous self-reconfiguring robots. In *IEEE/RSJ International Conference on Intelligent Robots and Systems,* 2003.

[150] S. Fleury, P. Souères, J.-P. Laumond, and R. Chatila. Primitives for smoothing paths of mobile robots. In *IEEE International Conference on Robotics and Automation,* volume 1, pages 832–839, 1993.

[151] S. Fleury, P. Souères, J.-P. Laumond, and R. Chatila. Primitives for smoothing mobile robot trajectories. *IEEE Transactions on Robotics and Automation,* 11:441–448, 1995.

[152] M. Fliess, J. Lévine, P. Martin, and P. Rouchon. On differentially flat nonlinear systems. In *IFAC Symposium NOLCOS,* pages 408–412, 1992.

[153] M. Fliess, J. Lévine, P. Martin, and P. Rouchon. Flatness and defect of nonlinear systems: Introductory theory and examples. *International Journal of Control,* 61(6):1327–1361, 1995.

[154] A. T. Fomenko and T. L. Kunii. *Topological Modeling for Visualization.* Springer-Verlag, Tokyo, 1997.

[155] M. Foskey, M. Garber, M. Lin, and D. Manocha. A voronoi-based hybrid motion planner. In *IEEE/RSJ International Conference on Intelligent Robots and Systems,* 2001.

[156] D. Fox, W. Burgard, F. Dellaert, and S. Thrun. Monte Carlo localization: Efficient position estimation for mobile robots. In *Proc. of the National Conference on Artificial Intelligence (AAAI),* 1999.

[157] D. Fox, W. Burgard, H. Kruppa, and S. Thrun. A probabilistic approach to collaborative multi-robot localization. *Autonomous Robots,* 8(3), 2000.

[158] D. Fox, W. Burgard, and S. Thrun. Markov localization for mobile robots in dynamic environments. *Journal of Artificial Intelligence Research (JAIR),* 11:391–427, 1999.

[159] T. Fraichard and J.-M. Ahuactzin. Smooth path planning for cars. In *IEEE International Conference on Robotics and Automation,* pages 3722–3727, Seoul, Korea, 2001.

[160] E. Frazzoli, M. A. Dahleh, and E. Feron. Real-time motion planning for agile autonomous vehicles. *AIAA Journal of Guidance, Control, and Dynamics,* 25(1):116–129, 2002.

[161] C. Früh and A. Zakhor. 3D model generation for cities using aerial photographs and ground level laser scans. In *Proc. of the IEEE Computer Society Conference on Computer Vision and Pattern Recognition (CVPR),* 2001.

[162] R. Geraerts and M. Overmars. A comparative study of probabilistic roadmap planners. In J.-D. Boissonnat, J. Burdick, K. Goldberg, and S. Hutchinson, editors, *Algorithmic Foundations of Robotics V,* pages 43–58. Springer-Verlag, 2003.

[163] E. Gilbert, D. Johnson, and S. Keerthi. A fast procedure for computing distance between complex objects in three-dimensional space. *IEEE Transactions on Robotics and Automation,* 4:193–203, 1988.

[164] P. E. Gill, W. Murray, and M. H. Wright. *Practical Optimization.* Academic Press, New York, 1981.

[165] B. Glavina. Solving findpath by combination of goal-directed and randomized search. In *IEEE International Conference on Robotics and Automation,* pages 1718–1723, 1990.

[166] K. Y. Goldberg. Orienting polygonal parts without sensors. *Algorithmica,* 10:201–225, 1993.

[167] N. Gordon, D. Salmond, and A. Smith. Novel approach to nonlinear/non-Gaussian Bayesian state estimation. *IEE Procedings F,* 140(2):107–113, 1993.

[168] S. Gottschalk, M. C. Lin, and D. Manocha. OBBTree: A hierarchical structure for rapid interference detection. *Computer Graphics,* 30(Annual Conference Series):171–180, 1996.

[169] P. Grandjean and A. Robert de Saint Vincent. 3-D modeling of indoor scenes by fusion of noisy range and stereo data. In *IEEE International Conference on Robotics and Automation,* 1989.

[170] F. Gravoit, S. Cambon, and R. Alami. Asymov: a planner that deals with intricate symbolic and geometric problems. In *International Symposium on Robotics Research,* 2003. Book to appear.

[171] L. J. Guibas, C. Holleman, and L. E. Kavraki. A probabilistic roadmap planner for flexible objects with a workspace medial-axis-based sampling approach. In *IEEE/RSJ International Conference on Intelligent Robots and Systems,* pages 254–260, 1999.

[172] L. J. Guibas, J. C. Latombe, S. M. LaValle, D. Lin, and R. Motwani. A visibility-based pursuit-evasion problem. *International Journal of Computational Geometry and Applications,* 9(4/5):471–512, August/October 1999.

[173] V. Guillemin and A. Pollack, editors. *Differential Topology.* Prentice-Hall, Inc., New Jersey, 1974.

[174] K. Gupta and Z. Guo. Motion planning with many degrees of freedom: sequential search with backtracking. *IEEE Transactions on Robotics and Automation,* 6(11):897–906, 1995.

[175] L. Gurvits. Averaging approach to nonholonomic motion planning. In *IEEE International Conference on Robotics and Automation,* pages 2541–2546, 1992.

[176] J. Gutmann and D. Fox. An experimental comparison of localization methods continued. In *IEEE/RSJ International Conference on Intelligent Robots and Systems,* 2002.

[177] J.-S. Gutmann, W. Burgard, D. Fox, and K. Konolige. An experimental comparison of localization methods. In *IEEE/RSJ International Conference on Intelligent Robots and Systems,* 1998.

[178] J.-S. Gutmann and K. Konolige. Incremental mapping of large cyclic environments. In *Proc. of the IEEE Int. Symp. on Computational Intelligence in Robotics and Automation (CIRA),* 1999.

[179] J.-S. Gutmann and C. Schlegel. AMOS: Comparison of scan matching approaches for self-localization in indoor environments. In *Proc. of the 1st Euromicro Workshop on Advanced Mobile Robots.* IEEE Computer Society Press, 1996.

[180] J.-S. Gutmann, T. Weigel, and B. Nebel. A fast, accurate, and robust method for self-localization in polygonal environments using laser-range-finders. *Advanced Robotics Journal,* 14(8):651–668, 2001.

[181] D. Hähnel, W. Burgard, D. Fox, and S. Thrun. A highly efficient FastSLAM algorithm for generating cyclic maps of large-scale environments from raw laser range measurements. Submitted for publication.

[182] D. Hähnel, D. Schulz, and W. Burgard. Map building with mobile robots in populated environments. In *Proc. of the IEEE/RSJ International Conference on Intelligent Robots and Systems (IROS),* 2002.

[183] D. Halperin and M. Sharir. A near-quadratic algorithm for planning the motion of a polygon in a polygonal environment. *Discrete Computational Geometry,* 16:121–134, 1996.

[184] L. Han and N. M. Amato. A kinematics-based probabilistic roadmap for closed chain systems. In B. R. Donald, K. Lynch, and D. Rus, editors, *New Directions in Algorithmic and Computational Robotics,* pages 233–246. AK Peters, 2001.

[185] G. Heinzinger, P. Jacobs, J. Canny, and B. Paden. Time-optimal trajectories for a robot manipulator: A provably good approximation algorithm. In *IEEE International Conference on Robotics and Automation,* pages 150–156, 1989.

[186] G. Heinzinger and B. Paden. Bounds on robot dynamics. In *IEEE International Conference on Robotics and Automation,* pages 1227–1232, Scottsdale, Arizona, 1989.

[187] S. Hert, S. Tiwari, and V. Lumelsky. A Terrain-Covering Algorithm for an AUV. *Autonomous Robots,* 3:91–119, 1996.

[188] J. Hertzberg and F. Kirchner. Landmark-based autonomous navigation in sewerage pipes. In *Proc. of the First Euromicro Workshop on Advanced Mobile Robots,* 1996.

[189] H. Hirukawa, B. Mourrain, and Y. Papegay. A symbolic-numeric silhouette algorithm. In *Intelligent Robots and Systems,* pages 2358–2365, Nov 2000.

[190] C. Hofner and G. Schmidt. Path planning and guidance techniques for an autonomous mobile cleaning robot. *Robotics and Autonomous Systems,* 14:199–212, 1995.

[191] C. Holleman and L. E. Kavraki. A framework for using the workspace medial axis in PRM planners. In *IEEE International Conference on Robotics and Automation,* pages 1408–1413, 2000.

[192] D. Hsu. *Randomized Single-Query Motion Planning In Expansive Spaces.* PhD thesis, Department of Computer Science, Stanford University, 2000.

[193] D. Hsu, T. Jiang, J. Reif, and Z. Sun. The bridge test for sampling narrow passages with probabilistic roadmap planners. In *IEEE International Conference on Robotics and Automation,* 2003.

[194] D. Hsu, L. E. Kavraki, J. C. Latombe, R. Motwani, and S. Sorkin. On finding narrow passages with probabilistic roadmap planners. In e. a. P. Agarwal, editor, *Robotics: The Algorithmic Perspective,* pages 141–154. A.K. Peters, Wellesley, MA, 1998.

[195] D. Hsu, R. Kindel, J. C. Latombe, and S. Rock. Randomized kinodynamic motion planning with moving obstacles. *International Journal of Robotics Research,* 21(3):233–255, 2002.

[196] D. Hsu, J. C. Latombe, and R. Motwani. Path planning in expansive configuration spaces. In *IEEE International Conference on Robotics and Automation,* pages 2719–2726, 1997.

[197] D. Hsu, J. C. Latombe, and R. Motwani. Path planning in expansive configuration spaces. *International Journal of Computational Geometry and Applications,* 9(4/5):495–512, 1998.

[198] Y. Y. Huang, Z. L. Cao, and E. Hall. Region filling operations for mobile robot using computer graphics. In *Proceedings of the IEEE Conference on Robotics and Automation,* pages 1607–1614, 1986.

[199] T. C. Hudson, M. C. Lin, J. Cohen, S. Gottschalk, and D. Manocha. V-COLLIDE: Accelerated collision detection for VRML. In R. Carey and P. Strauss, editors, *VRML 97: Second Symposium on the Virtual Reality Modeling Language,* pages 119–125, New York City, NY, 1997. ACM Press.

[200] S. Iannitti and K. M. Lynch. Exact minimum control switch motion planning for the snakeboard. In *IEEE/RSJ International Conference on Intelligent Robots and Systems,* 2003.

[201] M. Isard and A. Blake. Condensation—conditional density propagation for visual tracking. *International Journal of Computer Vision,* 29(1), 1998.

[202] A. Isidori. *Nonlinear Control Systems: An Introduction.* Springer-Verlag, 1985.

[203] P. Isto. A two-level search algorithm for motion planning. In *IEEE International Conference on Robotics and Automation,* pages 2025–2031, 1997.

[204] P. Isto. Constructing probabilistic roadmaps with powerful local planning and path optimization. In *IEEE/RSJ International Conference on Intelligent Robots and Systems,* pages 2323–2328, 2002.

[205] P. Jacobs, G. Heinzinger, J. Canny, and B. Paden. Planning guaranteed near-time-optimal trajectories for a manipulator in a cluttered workspace. In *International Workshop on Sensorial Integration for Industrial Robots: Architectures and Applications,* Zaragoza, Spain, 1989.

[206] P. Jacobs, G. Heinzinger, J. Canny, and B. Paden. Planning guaranteed near-time-optimal trajectories for a manipulator in a cluttered workspace. Technical Report RAMP 89-15, University of California, Berkeley, Engineering Systems Research Center, Sept. 1989.

[207] K. Janich. *Topology.* Spring-Verlag, New York, NY, 1984.

[208] R. Jarvis. Collision free trajectory planning using distance transforms. *Mech Eng Trans of the IE Aust,* ME10:197–191, 1985.

[209] P. Jensfelt and S. Kristensen. Active global localisation for a mobile robot using multiple hypothesis tracking. *IEEE Transactions on Robotics and Automation,* 17(5):748–760, Oct. 2001.

[210] X. Ji and J. Xiao. Planning motion compliant to complex contact states. *International Journal of Robotics Research,* 20(6):446–465, 2001.

[211] V. Jurdjevic. *Geometric Control Theory.* Cambridge University Press, 1997.

[212] V. Jurdjevic and H. J. Sussmann. Control systems on Lie groups. *Journal of Differential Equations,* 12:313–329, 1972.

[213] L. Kaelbling, A. Cassandra, and J. Kurien. Acting under uncertainty: Discrete Bayesian models for mobile-robot navigation. In *IEEE/RSJ International Conference on Intelligent Robots and Systems,* 1996.

[214] T. Kailath. *Linear Systems.* Prentice-Hall, 1980.

[215] R. Kalman. A new approach to linear filtering and prediction problems. *Trans. of the ASME, Journal of basic engineering,* 82:35–45, March 1960.

[216] I. Kamon, E. Rimon, and E. Rivlin. Tangentbug: A range-sensor based navigation algorithm. *Int. Journal of Robotics Research,* 17(9):934–953, 1998.

[217] I. Kamon, E. Rivlin, and E. Rimon. A new range-sensor based globally convergent navigation for mobile robots. In *IEEE Int'l. Conf. on Robotics and Automation,* Minneapolis, MN, April 1996.

[218] K. Kanazawa, D. Koller, and S. Russell. Stochastic simulation algorithms for dynamic probabilistic networks. In *Proc. of the 11th Annual Conference on Uncertainty in AI (UAI),* 1995.

[219] K. Kant and S. Zucker. Toward efficient trajectory planning: Path velocity decomposition. *International Journal of Robotics Research,* 5:72–89, 1986.

[220] L. E. Kavraki. Part orientation with programmable vector fields: Two stable equilibria for most parts. In *IEEE International Conference on Robotics and Automation,* pages 20–25, Albuquerque, New Mexico, Apr. 1997.

[221] L. E. Kavraki. *Random Networks in Configuration Space for Fast Path Planning.* PhD thesis, Stanford University, 1995.

[222] L. E. Kavraki, M. Kolountzakis, and J. C. Latombe. Analysis of probabilistic roadmaps for path planning. In *IEEE International Conference on Robotics and Automation,* pages 3020–3026, 1996.

[223] L. E. Kavraki, M. N. Kolountzakis, and J. C. Latombe. Analysis of probabilistic roadmaps for path planning. *IEEE Transactions on Robotics and Automation,* 14(1):166–171, February 1998.

[224] L. E. Kavraki, F. Lamiraux, and C. Holleman. Towards planning for elastic objects. In P. Agrawal, L. E. Kavraki, and M. Mason, editors, *Robotics: The Algorithmic Perspective,* pages 313–325. A.K. Peters, 1998.

[225] L. E. Kavraki and J. C. Latombe. Randomized preprocessing of configuration space for fast path planning. Technical Report STAN-CS-93-1490, Dept. Comput. Sci., Stanford Univ., Stanford, CA, 1993.

[226] L. E. Kavraki and J. C. Latombe. Randomized preprocessing of configuration space for path planning. In *IEEE International Conference on Robotics and Automation,* pages 2138–2139, 1994.

[227] L. E. Kavraki and J. C. Latombe. Probabilistic roadmaps for robot path planning. In K. Gupta and A. P. del Pobil, editors, *Practical Motion Planning in Robotics: Current Approaches and Future Challenges,* pages 33–53. John Wiley, West Sussex, England, 1998.

[228] L. E. Kavraki, J. C. Latombe, R. Motwani, and P. Raghavan. Randomized query processing in robot motion planning. In *Proc. ACM Symp. on Theory of Computing,* pages 353–362, 1995.

[229] L. E. Kavraki, J. C. Latombe, R. Motwani, and P. Raghavan. Randomized query processing in robot path planning. *Journal of Computer and System Sciences,* 57(1):50–60, August 1998.

[230] L. E. Kavraki, J. C. Latombe, and R. Wilson. On the complexity of assembly partitioning. *Information Processing Letters,* 48:229–235, 1993.

[231] L. E. Kavraki, P. Švestka, J. C. Latombe, and M. H. Overmars. Probabilistic roadmaps for path planning in high-dimensional configuration spaces. *IEEE Transactions on Robotics and Automation,* 12(4):566–580, June 1996.

[232] H. Keller. *Lectures on Numerical Methods in Bifurcation Problems*. Tata Institute of Fundamental Research, Bombay, India, 1987.

[233] S. D. Kelly and R. M. Murray. Geometric phases and robotic locomotion. *Journal of Robotic Systems,* 12(6):417–431, 1995.

[234] O. Khatib. Real-time obstacle avoidance for manipulators and mobile robots. *International Journal of Robotics Research,* 5:90–98, 1986.

[235] R. Kindel, D. Hsu, J. C. Latombe, and S. Rock. Kinodynamic motion planning amidst moving obstacles. In *IEEE International Conference on Robotics and Automation,* pages 537–543, 2000.

[236] D. E. Kirk. *Optimal Control Theory*. Prentice-Hall Inc., 1970.

[237] H. Kitano, M. Asada, Y. Kuniyoshi, I. Noda, O. E., and H. Matsubara. RoboCup: A challenge problem for AI. *AI Magazine,* 18(1):73–85, 1997.

[238] J. T. Klosowski, M. Held, J. S. B. Mitchell, H. Sowizral, and K. Zikan. Efficient collision detection using bounding volume hierarchies of k-DOPs. *IEEE Transactions on Visualization and Computer Graphics,* 4(1):21–36, 1998.

[239] D. E. Koditschek and E. Rimon. Robot navigation functions on manifolds with boundary. *Advances in Applied Mathematics,* 11:412–442, 1990.

[240] S. Koenig and R. Simmons. A robot navigation architecture based on partially observable Markov decision process models. In D. Kortenkamp, R. Bonasso, and R. Murphy, editors, *Artificial Intelligence and Mobile Robots*. MIT/AAAI Press, Cambridge, MA, 1998.

[241] Y. Koga, K. Kondo, J. Kuffner, and J. C. Latombe. Planning motions with intentions. *Computer Graphics (SIGGRAPH'94),* pages 395–408, 1994.

[242] Y. Koga and J. C. Latombe. Experiments in dual-arm manipulation planning. In *IEEE International Conference on Robotics and Automation,* pages 2238–2245, 1992.

[243] Y. Koga and J. C. Latombe. On multi-arm manipulation planning. In *IEEE International Conference on Robotics and Automation,* pages 945–952, 1994.

[244] K. Kondo. Motion planning with six degrees of freedom by multistrategic bidirectional heuristic free-space enumeration. *IEEE Transactions on Robotics and Automation,* 7:267–277, 1991.

[245] K. Konolige. Markov localization using correlation. In *Proc. of the International Joint Conference on Artificial Intelligence (IJCAI),* 1999.

[246] J. J. Kuffner. Effective sampling and distance metrics for 3D rigid body path planning. In *IEEE International Conference on Robotics and Automation,* 2004.

[247] J. J. Kuffner, K. Nishiwaki, S. Kagami, M. Inaba, and H. Inoue. Motion planning for humanoid robots under obstacle and dynamic balance constraints. In *IEEE International Conference on Robotics and Automation,* pages 692–698, Seoul, Korea, May 2001.

[248] J. J. Kuffner, K. Nishiwaki, S. Kagami, M. Inaba, and H. Inoue. Motion planning for humanoid robots. In *International Symposium on Robotics Research,* 2003. Book to appear.

[249] J. J. Kuffner and S. M. LaValle. RRT-connect: An efficient approach to single-query path planning. In *IEEE International Conference on Robotics and Automation,* pages 995–1001, 2000.

[250] B. Kuipers and Y. Byan. A robot exploration and mapping strategy based on a semantic hierarchy of spatial representations. *Journal of Robotics and Autonomous Systems,* 8:47–63, 1991.

[251] A. M. Ladd and L. E. Kavraki. Motion planning for knot untangling. In J.-D. Boissonnat, J. Burdick, K. Goldberg, and S. Hutchinson, editors, *Algorithmic Foundations of Robotics V,* pages 7–24. Springer-Verlag, 2002.

[252] A. M. Ladd and L. E. Kavraki. Measure theoretic analysis of probabilistic path planning. *IEEE Transactions on Robotics and Automation,* 20(2):229–242, 2004.

[253] G. Lafferriere and H. Sussmann. Motion planning for controllable systems without drift. In *IEEE International Conference on Robotics and Automation,* pages 1148–1153, Sacramento, CA, 1991.

[254] G. Lafferriere and H. J. Sussmann. A differential geometric approach to motion planning. In Z. Li and J. Canny, editors, *Nonholonomic Motion Planning.* Kluwer Academic, 1993.

[255] F. Lamiraux and L. E. Kavraki. Planning paths for elastic objects under manipulation constraints. *International Journal of Robotics Research,* 20(3):188–208, 2001.

[256] F. Lamiraux and L. E. Kavraki. Positioning of symmetric and nonsymmetric parts using radial and constant fields: Computation of al equilibrium configurations. *International Journal of Robotics Research,* 20(8):635–659, 2001.

[257] F. Lamiraux and J.-P. Laumond. On the expected complexity of random path planning. In *IEEE International Conference on Robotics and Automation,* pages 3014–3019, 1996.

[258] F. Lamiraux and J.-P. Laumond. Smooth motion planning for car-like vehicles. *IEEE Transactions on Robotics and Automation,* 17(4):498–502, Aug. 2001.

[259] F. Lamiraux, S. Sekhavat, and J.-P. Laumond. Motion planning and control for Hilare pulling a trailer. *IEEE Transactions on Robotics and Automation,* 15(4):640–652, Aug. 1999.

[260] S. Land and H. Choset. Coverage path planning for landmine location. In *Third International Symposium on Technology and the Mine Problem,* Monterey, CA, April 1998.

[261] E. Larsen, S. Gottschalk, M. Lin, and D. Manocha. Fast proximity queries with swept sphere volumes. Technical Report TR99-018, Department of Computer Science, University of North Carolina at Chapel Hill, North Carolina, 1999.

[262] J. C. Latombe. *Robot Motion Planning.* Kluwer Academic Publishers, Boston, MA, 1991.

[263] J. C. Latombe. Personal communication.

[264] J.-P. Laumond and R. Alami. A geometrical approach to planning manipulation tasks: The case of a circular robot and a movable circular object amidst polygonal obstacles. Report 88314, LAAS/CNRS, Toulouse, France, 1989.

[265] J.-P. Laumond. Controllability of a multibody mobile robot. *IEEE Transactions on Robotics and Automation,* 9(6):755–763, Dec. 1993.

[266] J.-P. Laumond. *Robot motion planning and control.* Springer, 1998.

[267] J.-P. Laumond, P. E. Jacobs, M. Taïx, and R. M. Murray. A motion planner for nonholonomic mobile robots. *IEEE Transactions on Robotics and Automation,* 10(5):577–593, Oct. 1994.

[268] S. M. LaValle, J. Yakey, and L. E. Kavraki. Randomized path planning for linkages with closed kinematics chains. *IEEE Transactions on Robotics and Automation,* 17(6):951–959, 2001.

[269] S. M. LaValle and M. S. Branicky. On the relationship between classical grid search and probabilistic roadmaps. In J.-D. Boissonnat, J. Burdick, K. Goldberg, and S. Hutchinson, editors, *Algorithmic Foundations of Robotics V,* pages 59–76. Springer-Verlag, 2002.

[270] S. M. LaValle and J. J. Kuffner. Randomized kinodynamic planning. In *IEEE International Conference on Robotics and Automation,* pages 473–479, 1999.

[271] S. M. LaValle and J. J. Kuffner. Randomized kinodynamic planning. *International Journal of Robotics Research,* 20(5):378–400, May 2001.

[272] S. M. LaValle and J. J. Kuffner. Rapidly-exploring random trees: Progress and prospects. In B. R. Donald, K. Lynch, and D. Rus, editors, *New Directions in Algorithmic and Computational Robotics,* pages 293–308. AK Peters, 2001.

[273] S. M. Lavalle, D. Lin, L. J. Guibas, J. C. Latombe, and R. Motwani. Finding an unpredictable target in a workspace with obstacles. In *IEEE International Conference on Robotics and Automation,* pages 1677–1682, 1997.

[274] J. Lengyel, M. Reichert, B. R. Donald, and D. P. Greenberg. Real-time robot motion planning using rasterizing computer graphics hardware. *Computer Graphics,* 24(4):327–335, 1990.

[275] S. Lenser and M. Veloso. Sensor resetting localization for poorly modelled mobile robots. In *IEEE International Conference on Robotics and Automation,* 2000.

[276] J. J. Leonard and H. Durrant-Whyte. *Directed Sonar Sensing for Mobile Robot Navigation.* Kluwer Academic, Boston, MA, 1992.

[277] J. J. Leonard and H. Feder. A computationally efficient method for large-scale concurrent mapping and localization. In J. Hollerbach and D. Koditschek, editors, *Proceedings of the Ninth International Symposium on Robotics Research,* Salt Lake City, Utah, 1999.

[278] J. J. Leonard and H. Durrant-Whyte. Simultaneous map building and localization for an autonomous mobile robot. In *IEEE/RSJ International Workshop on Intelligent Robots and Systems,* pages 1442–1447, May 1991.

[279] N. E. Leonard. Control synthesis and adaptation for an underactuated autonomous underwater vehicle. *IEEE Journal of Oceanic Engineering,* 20(3):211–220, July 1995.

[280] N. E. Leonard and P. S. Krishnaprasad. Motion control of drift-free, left-invariant systems on Lie groups. *IEEE Transactions on Automatic Control,* 40(9):1539–1554, Sept. 1995.

[281] P. Leven and S. Hutchinson. Real-time path planning in changing environments. *International Journal of Robotics Research,* 21(12):999–1030, Dec. 2002.

[282] P. Leven and S. Hutchinson. Using manipulability to bias sampling during the construction of probabilistic roadmaps. *IEEE Transactions on Robotics and Automation,* 19(6):1020–1026, Dec. 2003.

[283] A. D. Lewis. When is a mechanical control system kinematic? In *IEEE Conference on Decision and Control,* pages 1162–1167, Dec. 1999.

[284] A. D. Lewis. Simple mechanical control systems with constraints. *IEEE Transactions on Automatic Control,* 45(8):1420–1436, 2000.

[285] A. D. Lewis and R. M. Murray. Configuration controllability of simple mechanical control systems. *SIAM Journal on Control and Optimization,* 35(3):766–790, May 1997.

[286] A. D. Lewis and R. M. Murray. Configuration controllability of simple mechanical control systems. *SIAM Review,* 41(3):555–574, 1999.

[287] F. L. Lewis and V. L. Syrmos. *Optimal Control.* John Wiley and Sons, Inc., 1995.

[288] Z. Li and J. Canny. *Nonholonomic Motion Planning.* Kluwer Academic, 1993.

[289] K. Lian, L. Wang, and L. Fu. Controllability of spacecraft systems in a central gravitational field. *IEEE Transactions on Automatic Control,* 39(12):2426–2440, Dec. 1994.

[290] M. C. Lin, D. Manocha, J. Cohen, and S. Gottschalk. Collision detection: Algorithms and applications. In J.-P. Laumond and M. Overmars, editors, *Algorithms for Robotic Motion and Manipulation,* pages 129–142. A K Peters, Wellesley, MA, 1997.

[291] S. R. Lindemann and S. M. LaValle. Incremental low-discrepancy lattice methods for motion planning. In *IEEE International Conference on Robotics and Automation,* pages 2920–2927, 2003.

[292] S. R. Lindemann and S. M. LaValle. Current issues in sampling-based motion planning. In *International Symposium on Robotics Research,* 2003. Book to appear.

[293] G. Liu and Z. Li. A unified geometric approach to modeling and control of constrained mechanical systems. *IEEE Transactions on Robotics and Automation,* 18(4):574–587, Aug. 2002.

[294] Y. Liu and S. Arimoto. Path planning using a tangent graph for mobile robots among polygonal and curved obstacles. *International Journal of Robotics Research,* 11(4):376–382, 1992.

[295] C. Lobry. Controllability of nonlinear systems on compact manifolds. *SIAM Journal on Control,* 12(1):1–4, 1974.

[296] I. Lotan, F. Schwarzer, D. Halperin, and J. C. Latombe. Efficient maintenance and self-collision testing for kinematic chains. In *Proceedings of the 18th annual Symposium on Computational geometry,* pages 43–52. ACM Press, 2002.

[297] T. Lozano-Pérez. A simple motion-planning algorithm for general robot manipulators. *IEEE Journal of Robotics and Automation,* RA-3(3):224–238, 1987.

[298] T. Lozano-Perez and M. Wesley. An algorithm for planning collision-free paths among polyhedral obstacles. *Communications of the ACM,* 22(10):560–570, 1979.

[299] F. Lu and E. Milios. Globally consistent range scan alignment for environment mapping. *Autonomous Robots,* 4:333–349, 1997.

[300] V. Lumelsky, S. Mukhopadhyay, and K. Sun. Dynamic path planning in sensor-based terrain acquisition. *IEEE Transactions on Robotics and Automation,* 6(4):462–472, August 1990.

[301] V. Lumelsky and A. Stepanov. Path planning strategies for point mobile automaton moving amidst unknown obstacles of arbitrary shape. *Algorithmica,* 2:403–430, 1987.

[302] J. E. Luntz, W. Messner, and H. Choset. Distributed manipulation using discrete actuator arrays. *International Journal of Robotics Research,* 20(7):553–582, 2001.

[303] K. M. Lynch. Controllability of a planar body with unilateral thrusters. *IEEE Transactions on Automatic Control,* 44(6):1206–1211, June 1999.

[304] K. M. Lynch and C. K. Black. Recurrence, controllability, and stabilization of juggling. *IEEE Transactions on Robotics and Automation,* 17(2):113–124, Apr. 2001.

[305] K. M. Lynch, N. Shiroma, H. Arai, and K. Tanie. Collision-free trajectory planning for a 3-DOF robot with a passive joint. *International Journal of Robotics Research,* 19(12):1171–1184, Dec. 2000.

[306] D. K. M. Ben-Or and J. Reif. The complexity of elementary algebra and geometry. *Journal of Computational Sciences,* 32:251–264, 1986.

[307] J. Marsden. *Elementary Classical Analysis.* W. H. Freeman and Company, New York, 1974.

[308] J. Marsden and T. Ratiu. *Introduction to Mechanics and Symmetry.* Springer-Verlag, New York, 1994.

[309] P. Martin, R. M. Murray, and P. Rouchon. Flat systems. In G. Bastin and M. Gevers, editors, *1997 European Control Conference Plenary Lectures and Mini-Courses.* 1997.

[310] S. Martinez, J. Cortés, and F. Bullo. A catalog of inverse-kinematics planners for underactuated systems on matrix Lie groups. In *IEEE/RSJ International Conference on Intelligent Robots and Systems,* 2003.

[311] M. T. Mason. *Manipulation by grasping and pushing operations.* PhD thesis, MIT, Artificial Intelligence Laboratory, 1982.

[312] M. T. Mason. *Mechanics of Robotic Manipulation.* MIT Press, 2001.

[313] P. Maybeck. The Kalman filter: An introduction to concepts. In *Autonomous Robot Vehicles.* Springer verlag, 1990.

[314] M. B. Milam, K. Mushambi, and R. M. Murray. A new computational approach to real-time trajectory generation for constrained mechanical systems. In *IEEE International Conference on Decision and Control,* 2000.

[315] J. Milnor. *Morse Theory.* Princeton University Press, Princeton, NJ, 1963.

[316] B. Mirtich. V-clip: Fast and robust polyhedral collision detection. *ACM Transactions on Graphics,* 17(3):177–208, 1998.

[317] M. Moll, K. Goldberg, M. A. Erdmann, and R. Fearing. Aligning parts for micro assemblies. *Assembly Automation,* 22(1):46–54, Feb. 2002.

[318] M. Moll and L. E. Kavraki. Path planning for minimal energy curves of constant length. In *IEEE International Conference on Robotics and Automation,* pages 2826–2831, 2004.

[319] M. Moll and L. E. Kavraki. Path planning for variable resolution minimal energy curves of constant length. In *IEEE International Conference on Robotics and Automation,* 2005.

[320] M. Montemerlo and S. Thrun. Simultaneous localization and mapping problem with unknown data association using FastSLAM. In *IEEE International Conference on Robotics and Automation,* 2003.

[321] M. Montemerlo, S. Thrun, D. Koller, and B. Wegbreit. FastSLAM: A factored solution to the simultaneous localization and mapping problem. In *Proc. of the National Conference on Artificial Intelligence (AAAI),* 2002.

[322] M. Montemerlo, S. Thrun, and W. Whittaker. Conditional particle filters for simultaneous mobile robot localization and people tracking. In *IEEE International Conference on Robotics and Automation,* 2002.

[323] M. Morales, S. Rodriguez, and N. M. Amato. Improving the connectivity of prm roadmaps. In *IEEE International Conference on Robotics and Automation,* pages 4427–4432, 2003.

[324] H. Moravec. Sensor fusion in certainty grids for mobile robots. *AI Magazine,* pages 61–74, Summer 1988.

[325] H. Moravec and A. Elfes. High resolution maps from wide angle sonar. In *IEEE International Conference on Robotics and Automation,* 1985.

[326] J. J. Moré and S. J. Wright. *Optimization Software Guide.* SIAM, Philadelphia, PA, 1993.

[327] K. A. Morgansen. *Temporal patterns in learning and control.* PhD thesis, Harvard University, 1999.

[328] K. A. Morgansen, P. A. Vela, and J. W. Burdick. Trajectory stabilization for a planar carangiform robot fish. In *IEEE International Conference on Robotics and Automation,* 2002.

[329] K. Murphy. Bayesian map learning in dynamic environments. In *Neural Info. Proc. Systems (NIPS),* 1999.

[330] R. M. Murray, Z. Li, and S. S. Sastry. *A Mathematical Introduction to Robotic Manipulation.* CRC Press, 1994.

[331] R. M. Murray, M. Rathinam, and W. Sluis. Differential flatness of mechanical control systems: A catalog of prototype systems. In *ASME Int Mech Eng Congress and Expo,* 1995.

[332] R. M. Murray and S. S. Sastry. Nonholonomic motion planning: Steering using sinusoids. *IEEE Transactions on Automatic Control,* 38(5):700–716, 1993.

[333] Y. Nakamura, T. Suzuki, and M. Koinuma. Nonlinear behavior and control of a nonholonomic free-joint manipulator. *IEEE Transactions on Robotics and Automation,* 13(6):853–862, 1997.

[334] P. Newman, J. Leonard, J. Neira, and J. Tardós. Explore and return: Experimental validation of real time concurrent mapping and localization. In *IEEE International Conference on Robotics and Automation,* 2002.

[335] C. Nielsen and L. E. Kavraki. A two level fuzzy PRM for manipulation planning. In *IEEE/RSJ International Conference on Intelligent Robots and Systems,* pages 1716–1722, Japan, 2000.

[336] D. Nieuwenhuisen and M. H. Overmars. Useful cycles in probabilistic roadmap graphs. In *IEEE International Conference on Robotics and Automation,* pages 446–452, 2004.

[337] C. Nissoux, T. Simeon, and J.-P. Laumond. Visibility based probabilistic roadmaps. *Advanced Robotics Journal,* 14(6), 2000.

[338] J. Nocedal and S. J. Wright. *Numerical Optimization.* Springer Verlag, 1999.

[339] I. Nourbakhsh, R. Powers, and S. Birchfield. DERVISH an office-navigating robot. *AI Magazine,* 16(2), 1995.

[340] C. Ó'Dúnlaing and C. Yap. A "retraction" method for planning the motion of a disc. *Algorithmica,* 6:104–111, 1985.

[341] M. Ollis and A. Stentz. First results in vision-based crop line tracking. In *IEEE International Conference on Robotics and Automation,* 1996.

[342] J. P. Ostrowski and J. W. Burdick. The geometric mechanics of undulatory robotic locomotion. *International Journal of Robotics Research,* 17(7):683–701, July 1998.

[343] J. P. Ostrowski, J. P. Desai, and V. Kumar. Optimal gait selection for nonholonomic locomotion systems. *International Journal of Robotics Research,* 19(3):225–237, Mar. 2000.

[344] M. Overmars. A random approach to motion planning. Technical Report RUU-CS-92-32, Dept. Comput. Sci., Utrecht Univ., Utrecht, the Netherlands, Oct. 1992.

[345] M. Overmars and P. Švestka. A probabilistic learning approach to motion planning. In K. Goldberg, D. Halperin, J. C. Latombe, and R. Wilson, editors, *Algorithmic Foundations of Robotics (WAFR),* pages 19–37. A. K. Peters, Ltd, 1995.

[346] R. Parr and A. Eliazar. DP-SLAM: Fast, robust simultaneous localization and mapping without predetermined landmarks. In *Proc. of the International Joint Conference on Artificial Intelligence (IJCAI),* 2003.

[347] J. Pearl. *Probabilistic Reasoning in Intelligent Systems: Networks of Plausible Inference*. Morgan Kaufmann Publishers, Inc., 1988.

[348] F. Pfeiffer and R. Johanni. A concept for manipulator trajectory planning. *IEEE Journal of Robotics and Automation,* RA-3(2):115–123, 1987.

[349] J. M. Phillips, N. Bedrossian, and L. E. Kavraki. Guided expansive spaces trees: A search strategy for motion- and cost-constrained state spaces. In *IEEE International Conference on Robotics and Automation,* pages 3968–3973, 2004.

[350] J. Phillips, L. Kavraki, and N. Bedrossian. Spacecraft rendezvous and docking with real-time, randomized optimization. In *AIAA Guidance, Navigation, and Control,* 2003.

[351] A. Piazza, M. Romano, and C. G. L. Bianco. G^3-splines for the path planning of wheeled mobile robots. In *European Control Conference,* 2003.

[352] C. Pisula, K. Hoff, M. Lin, and D. Manocha. Randomized path planning for a rigid body based on hardware accelarated Voronoi sampling. In B. R. Donald, K. Lynch, and D. Rus, editors, *New Directions in Algorithmic and Computational Robotics*. AK Peters, 2001.

[353] E. Plaku and L. E. Kavraki. Distributed sampling-based roadmap of trees for large-scale motion planning. In *IEEE International Conference on Robotics and Automation,* 2005.

[354] L. S. Pontryagin, V. G. Boltyanskii, R. V. Gamkrelidze, and E. F. Mishchenko. *The Mathematical Theory of Optimal Processes*. Interscience Publishers, 1962.

[355] C. Pradalier, J. Hermosillo, C. Koike, C. Braillon, P. P. Bessière, and C. Laugier. Safe and autonomous navigation for a car-like robot among pedestrian. In *IARP Int. Workshop on Service, Assistive and Personal Robots,* 2003.

[356] F. Preparata and M. I. Shamos. *Computational Geometry: An Introduction*. Springer-Verlag, 1985. p198–257.

[357] S. Quinlan. Efficient distance computation between nonconvex objects. In *IEEE International Conference on Robotics and Automation,* pages 3324–3329, 1994.

[358] A. Rao and K. Goldberg. Manipulating algebraic parts in the plane. *IEEE Tr. on Rob. and Autom.,* 11:598–602, 1995.

[359] N. Rao, N. Stolzfus, and S. Iyengar. A retraction method for learned navigation in unknown terrains for a circular robot. *IEEE Transactions on Robotics and Automation,* 7:699–707, October 1991.

[360] J. A. Reeds and L. A. Shepp. Optimal paths for a car that goes both forwards and backwards. *Pacific Journal of Mathematics,* 145(2):367–393, 1990.

[361] J. Reif. Complexity of the mover's problem and generalizations. In *Proc. 20th IEEE Symposium on Foundations of Computer Science,* pages 421–427, 1979.

[362] J. H. Reif and H. Wang. Nonuniform discretization for kinodynamic motion planning and its applications. *SIAM Journal of Computing,* 30(1):161–190, 2000.

[363] D. Reznik, E. Moshkivich, and J. F. Canny. Building a universal planar manipulator. In K.-F. Böhringer and H. Choset, editors, *Distributed Manipulation,* pages 147–171. Kluwer Academic Publishers, Boston, 2000.

[364] E. Rimon and D. E. Koditschek. Exact robot navigation using artificial potential functions. *IEEE Transactions on Robotics and Automation,* 8(5):501–518, October 1992.

[365] T. Röfer. Using histogram correlation to create consistent laser scan maps. In *IEEE/RSJ International Conference on Intelligent Robots and Systems,* 2002.

[366] H. Rohnert. Shortest path in the plane with convex polygonal obstacles. *Information Processing Letters,* 23:71–76, 1986.

[367] G. Sánchez and J. C. Latombe. On delaying collision checking in prm planning: Application to multi-robot coor dination. *International Journal of Robotics Research,* 21(1):5–26, 2002.

[368] S. S. Sastry. *Nonlinear Systems: Analysis, Stability, and Control.* Springer-Verlag, New York, 1999.

[369] D. H. Sattinger and O. L. Weaver. *Lie Groups and Algebras with Applications to Physics, Geometry, and Mechanics.* Springer-Verlag, 1986.

[370] A. Scheuer and T. Fraichard. Collision-free and continuous-curvature path planning for car-like robots. In *IEEE/RSJ International Conference on Intelligent Robots and Systems,* pages 1304–1311, Osaka, Japan, 1997.

[371] B. Schiele and J. Crowley. A comparison of position estimation techniques using occupancy grids. In *IEEE International Conference on Robotics and Automation,* 1994.

[372] B. Schutz. *Geometrical methods of mathematical physics.* Cambridge University Press, 1980.

[373] J. T. Schwartz and M. Sharir. On the piano movers' problem: II. General techniques for computing topological properties of real algebraic manifolds. *Advances in Applied Mathematics,* 4:298–351, 1983.

[374] J. T. Schwartz and M. Sharir. On the piano movers' problem: V. The case of a rod moving in three-dimensional space amidst polyhedral obstacles. *Communications on Pure and Applied Mathematics,* 37:815–848, 1984.

[375] J. T. Schwartz and M. Sharir. A survey of motion planning and related geometric algorithms. *Artificial Intelligence.*, 37:157–169, 1988.

[376] F. Schwarzer, M. Saha, and J. C. Latombe. Exact collision checking of robot paths. In J.-D. Boissonnat, J. Burdick, K. Goldberg, and S. Hutchinson, editors, *Algorithmic Foundations of Robotics V,* pages 25–42. Springer-Verlag, 2002.

[377] S. Sekhavat and J.-P. Laumond. Topological property of trajectories computed from sinusoidal inputs for nonholonomic chained form systems. In *IEEE International Conference on Robotics and Automation,* pages 3383–3388, 1996.

[378] S. Sekhavat and J.-P. Laumond. Topological property for collision-free non-holonomic motion planning: the case of sinusoidal inputs for chained form systems. *IEEE Transactions on Robotics and Automation,* 14(5):671–680, Oct. 1998.

[379] S. Sekhavat, P. Švestka, J.-P. Laumond, and M. H. Overmars. Multilevel path planning for nonholonomic robots using semiholonomic subsystems. *International Journal of Robotics Research,* 17(8):840–857, Aug. 1998.

[380] J.-P. Serre. *Lie Algebras and Lie Groups.* W. A. Benjamin, New York, 1965.

[381] J. Sethian. *Level Set Methods and Fast Marching Methods.* Cambridge University Press, Cambridge, UK, 1999.

[382] H. Shatkay and L. Kaelbling. Learning topological maps with weak local odometric information. In *Proceedings of IJCAI-97.* IJCAI, Inc., 1997. 1997.

[383] Z. Shiller and S. Dubowsky. On computing the global time-optimal motions of robotic manipulators in the presence of obstacles. *IEEE Transactions on Robotics and Automation,* 7(6): 785–797, Dec. 1991.

[384] Z. Shiller and H.-H. Lu. Computation of path constrained time optimal motions with dynamic singularities. *ASME Journal of Dynamic Systems, Measurement, and Control,* 114:34–40, Mar. 1992.

[385] K. G. Shin and N. D. McKay. Minimum-time control of robotic manipulators with geometric path constraints. *IEEE Transactions on Automatic Control,* 30(6):531–541, June 1985.

[386] R. Simmons and S. Koenig. Probabilistic robot navigation in partially observable environments. In *Proc. of the International Joint Conference on Artificial Intelligence (IJCAI),* 1995.

[387] A. Singh, J. C. Latombe, and D. Brutlag. A motion planning approach to flexible ligand binding. In *Intelligent Systems for Molecular Biology,* pages 252–261, 1999.

[388] J.-J. E. Slotine and H. S. Yang. Improving the efficiency of time-optimal path-following algorithms. *IEEE Transactions on Robotics and Automation,* 5(1):118–124, Feb. 1989.

[389] R. Smith and P. Cheeseman. On the representation and estimation of spatial uncertainty. *The International Journal of Robotics Research,* 5(4):56–68, 1986.

[390] R. Smith, M. Self, and P. Cheeseman. Estimating uncertain spatial relationships in robotics. In I. Cox and G. Wilfong, editors, *Autonomous Robot Vehicles.* Springer Verlag, 1990.

[391] E. Sontag. Gradient techniques for systems with no drift: A classical idea revisited. In *IEEE International Conference on Decision and Control,* pages 2706–2711, 1993.

[392] E. D. Sontag. Control of systems without drift via generic loops. *IEEE Transactions on Automatic Control,* 40(7):1210–1219, July 1995.

[393] O. J. Sørdalen. Conversion of a car with n trailers into a chained form. In *IEEE International Conference on Robotics and Automation,* pages 1382–1387, 1993.

[394] P. Souères and J.-D. Boissonnat. Optimal trajectories for nonholonomic mobile robots. In J.-P. Laumond, editor, *Robot Motion Planning and Control.* Springer, 1998.

[395] P. Souères and J.-P. Laumond. Shortest paths synthesis for a car-like robot. *IEEE Transactions on Automatic Control,* 41(5):672–688, May 1996.

[396] R. F. Stengel. *Optimal control and estimation.* Dover, New York, 1994.

[397] A. Stentz. Optimal and efficient path planning for unknown and dynamic environments. *International Journal of Robotics and Automation,* 10, 1995.

[398] G. Strang. *Linear Algebra and Its Applications.* Orlando: Academic Press, 1980.

[399] A. Sudsang and L. Kavraki. A geometric approach to designing a programmable force field with a unique stable equilibrium for parts in the plane. In *IEEE International Conference on Robotics and Automation (ICRA),* pages 1079–1085, Seoul, 2001.

[400] H. Sussmann. A continuation method for nonholonomic path-finding problems. In *IEEE International Conference on Decision and Control,* pages 2718–2723, 1993.

[401] H. J. Sussmann. A general theorem on local controllability. *SIAM Journal on Control and Optimization,* 25(1):158–194, Jan. 1987.

[402] H. J. Sussmann and W. Tang. Shortest paths for the Reeds-Shepp car: a worked out example of the use of geometric techniques in nonlinear optimal control. Technical Report SYCON-91-10, Rutgers University, 1991.

[403] I. Suzuki and M. Yamashita. Searching for a mobile intruder in a polygonal region. *SIAM Journal of Computing,* 21(5):863–888, October 1992.

[404] P. Švestka. A probabilistic approach to motion planning for car-like robots. Technical Report RUU-CS-93-18, Dept. Comput. Sci., Utrecht Univ., Utrecht, the Netherlands, 1993.

[405] P. Švestka and M. H. Overmars. Coordinated motion planning for multiple car-like robots using probabilistic roadmaps. In *IEEE International Conference on Robotics and Automation,* pages 1631–1636, 1995.

[406] P. Švestka and J. Vleugels. Exact motion planning for tractor-trailer robots. In *IEEE International Conference on Robotics and Automation,* pages 2445–2450, 1995.

[407] K. R. Symon. *Mechanics.* Addison-Wesley, 1971.

[408] X. Tang, B. Kirkpatrick, S. Thomas, G. Song, and N. M. Amato. Using motion planning to study rna folding kinetics. In *International Conference on Research in Computational Molecular Biology,* 2004.

[409] M. Teodoro, G. N. Phillips, and L. E. Kavraki. Molecular docking: A problem with thousands of degrees of freedom. In *IEEE International Conference on Robotics and Automation,* pages 960–966, 2001.

[410] J. Thorpe. *Elementary Topics in Differential Geometry.* Springer-Verlag, 1985.

[411] S. Thrun. Exploration and model building in mobile robot domains. In *Proc. of the IEEE International Conference on Neural Networks,* 1993.

[412] S. Thrun. A probabilistic online mapping algorithm for teams of mobile robots. *International Journal of Robotics Research,* 20(5):335–363, 2001.

[413] S. Thrun. Learning occupancy grids with forward sensor models. *Autonomous Robots,* 2002.

[414] S. Thrun, M. Bennewitz, W. Burgard, A. Cremers, F. Dellaert, D. Fox, D. Hähnel, C. Rosenberg, N. Roy, J. Schulte, and D. Schulz. MINERVA: A second generation mobile tour-guide robot. In *IEEE International Conference on Robotics and Automation,* 1999.

[415] S. Thrun, A. Bücken, W. Burgard, D. Fox, T. Fröhlinghaus, D. Hennig, T. Hofmann, M. Krell, and T. Schimdt. Map learning and high-speed navigation in RHINO. In D. Kortenkamp, R. Bonasso, and R. Murphy, editors, *AI-based Mobile Robots: Case studies of successful robot systems.* MIT Press, Cambridge, MA, to appear.

[416] S. Thrun, W. Burgard, and D. Fox. A real-time algorithm for mobile robot mapping with applications to multi-robot and 3D mapping. In *IEEE International Conference on Robotics and Automation,* 2000.

[417] S. Thrun, D. Fox, and W. Burgard. A probabilistic approach to concurrent mapping and localization for mobile robots. *Machine Learning and Autonomous Robots (joint issue),* 31(1–3):29–53, 1998.

[418] S. Thrun, J.-S. Gutmann, D. Fox, W. Burgard, and B. Kuipers. Integrating topological and metric maps for mobile robot navigation: A statistical approach. In *Proc. of the National Conference on Artificial Intelligence (AAAI),* 1998.

[419] D. Tilbury, R. Murray, and S. Sastry. Trajectory generation for the n-trailer problem using Goursat normal form. In *IEEE International Conference on Decision and Control,* 1993.

[420] G. van den Bergen. Efficient collision detection of complex deformable models using AABB trees. *Journal of Graphics Tools: JGT,* 2(4):1–14, 1997.

[421] G. van den Bergen. A fast and robust GJK implementation for collision detection of convex objects. *Journal of Graphics Tools: JGT,* 4(2):7–25, 1999.

[422] P. Vela and J. W. Burdick. Control of biomimetic locomotion via averaging theory. In *IEEE International Conference on Robotics and Automation,* 2003.

[423] P. A. Vela, K. A. Morgansen, and J. W. Burdick. Underwater locomotion from oscillatory shape deformations. In *IEEE International Conference on Decision and Control,* 2002.

[424] G. Weiß, C. Wetzler, and E. von Puttkamer. Keeping track of position and orientation of moving indoor systems by correlation of range-finder scans. In *IEEE/RSJ International Conference on Intelligent Robots and Systems,* pages 595–601, 1994.

[425] J. T. Wen. Control of nonholonomic systems. In W. S. Levine, editor, *The Control Handbook,* pages 1359–1368. CRC Press, 1996.

[426] J. Wiegley, K. Goldberg, M. Peshkin, and M. Brokowski. A complete algorithm for designing passive fences to orient parts. In *Proc. Int. Conf. on Rob. and Autom.,* pages 1133–1139, 1996.

[427] S. Wilmarth, N. M. Amato, and P. Stiller. MAPRM: A probabilistic roadmap planner with sampling on the medial axis of th e free space. In *IEEE International Conference on Robotics and Automation,* pages 1024–1031, 1999.

[428] R. Wilson and J. C. Latombe. Geometric reasoning about mechanical assembly. *Artificial Intelligence,* 71:371–396, 1995.

[429] R. H. Wilson, L. E. Kavraki, J. C. Latombe, and T. Lozano-Pérez. Two-handed assembly sequencing. *International Journal of Robotics Research,* 14:335–350, 1995.

[430] B. Yamauchi and P. Langley. Place recognition in dynamic environments. *Journal of Robotic Systems,* 14(2):107–120, 1997.

[431] M. Yim and A. Berlin. Two approaches to distributed manipulation. In K.-F. Böhringer and H. Choset, editors, *Distributed Manipulation,* pages 237–261. Kluwer Academic Publishers, Boston, 2000.

[432] T. Yoshikawa. Manipulability of robotic mechanisms. *International Journal of Robotics Research,* 4(2):3–9, Apr. 1985.

[433] M. Zhang and L. E. Kavraki. A new method for fast and accurate derivation of molecular conformations. *Journal of Chemical Information and Computer Sciences,* 42(1):64–70, 2002.

Index